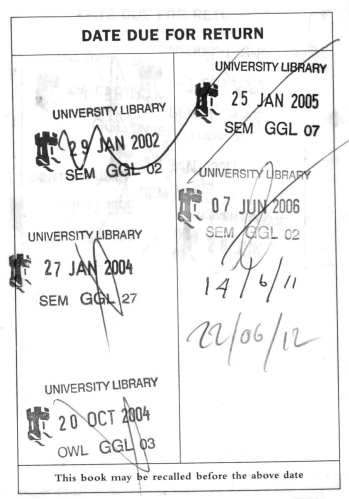

PARASITISM AND THE PLATYHELMINTHS

Frontispiece The lower surface of a common sole (*Solea solea*) infected with two adult specimens of the monogenean *Entobdella soleae*. Source: colour lithograph from a water-colour by Annie Willis in Cunningham, 1890.

PARASITISM AND THE PLATYHELMINTHS

GRAHAM C. KEARN

CHAPMAN & HALL

London · Glasgow · New York · Tokyo · Melbourne · Madras

Published by Chapman & Hall, 2–6 Boundary Row, London SE1 8HN, UK

Chapman & Hall, 2–6 Boundary Row, London SE1 8HN, UK

Chapman & Hall GmbH, Pappelallee 3, 69469 Weinheim, Germany

Chapman & Hall USA, 115 Fifth Avenue, New York, NY 10003, USA

Chapman & Hall Japan, ITP-Japan, Kyowa Building, 3F, 2–2–1 Hirakawacho, Chiyoda-ku, Tokyo 102, Japan

Chapman & Hall Australia, 102 Dodds Street, South Melbourne, Victoria 3205, Australia

Chapman & Hall India, R. Seshadri, 32 Second Main Road, CIT East, Madras 600 035, India

First edition 1998

© 1998 G. C. Kearn

Typset in India by Pure Tech India Ltd, Pondicherry
Printed in Great Britain at the University Press, Cambridge

ISBN 0 412 80460 3   1001373456

A Catalogue record for this book is available from the British Library

Library of Congress Catalog Card Number: 97–75126

This book is dedicated to my wife Margaret for her untiring support, and to my parents who encouraged my interest in natural history.

Contents

Preface

This book is not designed for any specific group of people. It is an attempt to bring together for the first time in one volume information on all the platyhelminth groups that are involved in symbiotic,* mainly parasitic, relationships. It is for me a labour of love and in writing this book it has been my intention to share with others the enthusiasm and interest that I have for these animals. They have been a source of excitement and fascination for me for much of my life, both by way of my own research and through the discoveries of others.

The influence of Jack Llewellyn has been important in the writing of this book. His publications helped me to focus my interest on the parasitic platyhelminths, in particular on the monogeneans, while an undergraduate student at London University's Imperial College in the late 1950s, and it was his suggestion that I should embark on a study of the monogenean *Entobdella soleae* for my PhD programme under his supervision. During the next 3 years at Birmingham University he shaped the way I looked at parasites and thought about them. Up to now, monogeneans have received less than their fair share of attention in text books, and the importance of monogeneans is reflected in the fact that four chapters of this book are devoted to them (with one chapter on *E. soleae*). The generous gift by Jack Llewellyn of his collection of reprints on cestodes and digeneans was also invaluable.

More than 30 years of undergraduate teaching have also had an important influence on this book. With competition for teaching time in an expanding School of Biological Sciences, my contribution on parasitology was restricted to about 20 lectures. I considered that any attempt to deal with all the major parasitic groups of animals would be doomed, with time for only a glimpse of each. I was left with two possibilities: a systems approach, in which topics such as feeding and attachment are considered in a comparative way, and an organismal–systematic approach involving an in-depth study of a single selected group of parasites.

The advantage of the organismal–systematic approach is that we are able to appreciate how all aspects of the biology of an organism are integrated and to chart the course of evolutionary progress through the chosen group. The organismal–systematic approach, including the study of 'text-book types', is regarded by many as outdated but, in my view, still offers the student of parasitology the most satisfying and effective route to an understanding of parasite–host relationships. Therefore, it was this approach that I chose to follow in my undergraduate teaching, using the parasitic platyhelminths as a vehicle and the four contrasting life styles of *Entobdella soleae, Hymenolepis diminuta, Fasciola hepatica* and *Schistosoma mansoni* as evolutionary 'milestones'. The book embraces the same approach, with the addition of a chapter on the remarkable range of symbiotic associations indulged in by the turbellarians.

* Used here in the original broad sense to include parasitism as well as mutualism and other kinds of association (see section 1.5).

Reconstructing the possible evolutionary steps that link the major groups of parasitic platyhelminths and tracing the possible evolutionary development within each group are contentious exercises. I have deliberately chosen not to enter into discussion on the merits of rival evolutionary schemes because I believe that such considerations would be disruptive and make the account unbearably tedious for the reader. The evolutionary story that I have selected to provide the backbone for most of this book leans heavily on the wisdom of parasitologists like Jack Llewellyn and John Pearson, and is one that seems to me to be logical and feasible. Undoubtedly there will be many who disagree, some with the major tenets of the scheme and others with details.

A common reaction from new undergraduate recruits to my course in parasitology is to express horror at the seemingly endless list of often unpronounceable names. I have tried to define the descriptive terms as they are encountered and I have also included an outline classification of the parasitic playthelminths in an appendix. Included in this classification scheme are the names of the genera that feature significantly in the book.

There is one other aspect of parasitology that I have always found informative and fascinating. We tend to take for granted now the complexity of many life cycles and yet no more than 150 years ago most of these were unknown. The accounts of the discovery of these life cycles, especially those that involve human hosts, are as exciting as any detective story and conceal many lessons about the development of the scientific method and indeed about human conduct and relationships. I regard the historical background as important, have always included it in my course and have tried to do likewise in the book.

During the preparation of this book many people have offered willing assistance and to these I am most grateful. Dr Peter Croghan (University of East Anglia) took on the onerous task of critically reading the first draft of the manuscript. I am indebted to him for his invaluable comments; any errors remaining in the text are entirely my responsibility. Dr Ian Whittington (University of Queensland) deserves special mention because of his untiring patience and cheerful cooperation in the face of repeated, seemingly never-ending requests for help. Ms Sheila Davies and Mr Rick Evans-Gowing of the University of East Anglia skilfully dealt with matters related to photography and electron miscroscopy. I have also received support in various ways from the following: Professors David Halton, Peter Pappas, John Pearson, Alan Wilson; Drs Tom Cribb, George Duncan, David Gibson, Audrey Glauert, Arlene Jones, Kazuo Ogawa. I have had assistance with translation from Birgit Hennig, Marion Platen and Alison Tomlinson, and my daughters Sarah and Frances have helped in various ways. Those who have kindly contributed photographs are acknowledged in the appropriate places.

Graham C. Kearn

Norwich, May 1997

CHAPTER 1

The platyhelminth panorama

1.1 OVERVIEW

Borradaile *et al.* (1961) defined the Phylum Platyhelminthes as follows: free-living and parasitic, bilaterally symmetrical, triploblastic Metazoa, usually flattened dorsoventrally, without anus, coelom or haemocoel, with a flame-cell system and with complicated, usually hermaphroditic, organs of reproduction. Hyman (1951) offered a more economical definition, describing them as acoelomate Bilateria without a definitive anus. We perhaps could add to these definitions that they have neither exoskeleton or endoskeleton, and their soft, often ciliated, bodies are highly flexible. The popular term 'flatworm' is invariably used for all platyhelminths, even for those that are cylindrical.

There are two life styles in particular to which platyhelminths show exceptional adaptability. One of these is symbiosis and especially parasitism, with three of the major groups of flatworms being exclusively parasitic, namely the monogeneans, cestodes (tapeworms) and digeneans (flukes). The other is the interstitial habit, i.e. the adaptation of small free-living platyhelminths for life in the maze of interconnecting spaces between sand grains, especially in the marine littoral zone. Free-living flatworms are also widespread in other benthic, marine and freshwater environments and a few have colonized humid terrestrial habitats. Some platyhelminths are pelagic, the smaller ones using their surface cilia to remain in suspension and larger ones propelling themselves by undulating their flat bodies. Most free-living flatworms have microphagous, predatory or scavenging habits (Fig. 1.1).

Traditionally, all the platyhelminths that did not belong to the major, exclusively parasitic groups, were placed in the taxon 'Turbellaria', but the current specialist view is that this 'group' is polyphyletic and not a valid taxon (Ehlers, 1986). Nevertheless, 'turbellarians' remains a convenient collective term to describe those groups of platyhelminths such as acoels, rhabdocoels, triclads and polyclads that are largely free-living in life style.

The acoel turbellarians (Fig. 1.2*a*) are small, ciliated and mostly one to several millimetres in length; they are exclusively marine and mostly benthic. They usually have a statocyst in the head region, but their most distinctive feature is the lack of a digestive cavity, the mouth opening directly, or sometimes by way of a pharynx, into a syncytial vacuolated mass of tissue filling the interior of the body. Rhabdocoel turbellarians (Fig. 1.2*b*) are also small ciliated organisms, but their pharynx communicates with an intestine that is usually sac-shaped and their epidermal cells contain secretory bodies called rhabdoids that are often rod-shaped. Rhabdocoels are abundant in fresh water and in the sea. The triclad turbellarians or planarians

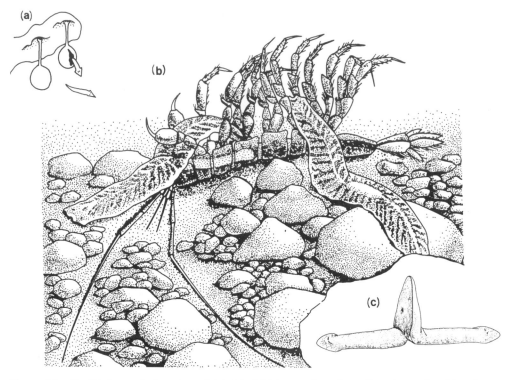

Figure 1.1 The biology of planarians (triclads). (*a*) Young planarians, *Dugesia* sp., hatching from a stalked capsule. (*b*) Freshwater planarians, *Dendrocoelum lacteum*, attacking a crustacean, *Asellus* sp. Note the pharynx of one planarian inside the prey. (*c*) Copulation in *Dugesia polychroa*. Source: (*a*) redrawn from Buchsbaum, 1957; (*b*), (*c*) reproduced from Ball and Reynoldson, 1981 (drawn by Julian Mulock), with permission from Cambridge University Press.

(Figs 1.1, 1.2*c*) are significantly larger, ranging from 2–3 mm to over 50 mm in length. The epidermis, or part of it, is ciliated and contains rhabdoids, but their most conspicuous features are the tubular, folded (plicate) pharynx, the three-branched hence triclad intestine and strongly flattened shape. They inhabit fresh water and the sea and they are also found in humid terrestrial environments. The polyclad turbellarians (Fig. 1.2*d*) are marine and most of them are flat benthic animals of relatively large size (2–3 mm to several centimetres in length). They have a ciliated, rhabdoid-containing epidermis and a highly folded pharynx opening into a central intestinal cavity from which tubular gut caeca radiate outwards through the body (hence 'polyclad').

The monogeneans, cestodes and digeneans are all parasites and all exchange their ciliated epidermis for an unciliated syncytial tegument (Fig. 1.3) at the end of larval life. Monogeneans (Fig. 1.2*e*) range from less than 0.5 mm to 1–2 cm in length. Most of them parasitize the skin or gills of fishes, to which they attach themselves by a posterior, hook-bearing attachment organ or haptor. They have no intermediate host, the ciliated larva seeking out new fish hosts. Larval cestodes too have a hook-bearing haptor, but this is lost and the adults (Fig. 1.2*f*) acquire a new anterior attachment organ or scolex to attach their often greatly elongated, ribbon-like bodies (more than 1 m long in some cases) to the lining of the vertebrate intestine. Cestodes have no gut and must absorb all their nutrients through the body surface. They use the food chain by stowing away in one or more intermediate hosts to spread to new

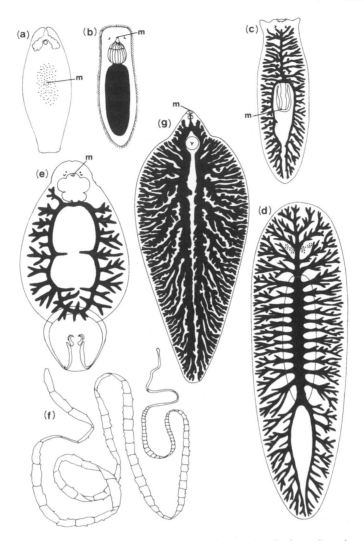

Figure 1.2 The platyhelminth panorama, showing the gut (if present) and other salient features. (*a*) An acoel, *Anaperus* sp. (*b*) A rhabdocoel, *Provortex* sp. (*c*) A triclad, *Dendrocoelum lacteum*. (*d*) A polyclad, *Notoplana* sp. (*e*) A monogenean, *Entobdella soleae*. (*f*) A cestode, *Taenia* sp. (*g*) A digenean, *Fasciola hepatica*. m = mouth. Modified from various sources.

definitive hosts. Digeneans (Fig. 1.2*g*) are similar in size range to monogeneans. They use muscular oral and ventral suckers (they have no haptor or scolex) to attach to their vertebrate hosts. Many digeneans inhabit the vertebrate intestine (others occur in organs as diverse as the lungs, liver and blood system), but they retain their own gut. The second phase of their life cycle involves a complex and curious process of multiplication spent in the haemocoel of a mollusc. Although the mollusc is often described as an intermediate host (it will be referred to as the first intermediate host in this book), there is no reason to regard it as different in status from the vertebrate host and there is no predator–prey relationship between the mollusc and the vertebrate as there is between the hosts in cestode life cycles. However, additional hosts interpolated between the mollusc and the vertebrate are akin to the intermediate hosts of cestodes, since these additional hosts are preyed upon by the vertebrate,

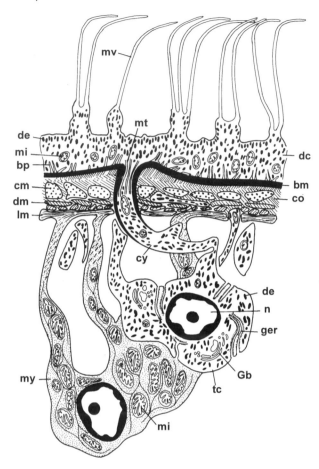

Figure 1.3 Diagrammatic longitudinal section through the dorsal body wall of the monogenean *Entobdella soleae*. bm = basement lamina; bp = basal plasma membrane; cm = circular muscle; co = interstitial connective tissue; cy = cytoplasmic connection between cell body and surface syncytium; dc = distal cytoplasm (syncytium) of tegument; de = dense granule; dm = diagonal muscle; Gb = Golgi body; ger = granular endoplasmic reticulum; lm = longitudinal muscle; mi = mitochondrion; mt = microtubule; mv = microvillus; my = myocyton; n = nucleus; tc = tegumentary cyton (cell body). Source: reproduced from Lyons, 1970, with permission from Cambridge University Press.

thereby permitting digenean stages inside them to reach the gut lumen of the vertebrate.

1.2 HISTORICAL PERSPECTIVE

The history of the study of flatworms is a tangled web which has been explored by Hyman (1951) and will be summarized briefly here. In the 18th and early 19th centuries taxonomists were more impressed by similarities in external appearance than by comparisons of internal anatomy, and a worm-like shape with the body longer than wide was a feature that dominated most classifications. Since some modern phyla are entirely made up of worm-like animals and virtually all phyla have some representatives that have arrived at this shape by convergence, it is not surprising that Linnaeus grouped all the invertebrates, with the exception of insects, in the Vermes (worms). Naturally, those modern platyhelminths that were known at the time were included in this group, but within it the tapeworms were separated from the rest (the flukes and planarians). An added complication was the creation a little later of the Infusoria, which included, as well as protozoans, rotifers and microscopic nematodes, the free-swimming stages (cercariae) of digeneans. Lamarck had a better understanding of animal relationships than many of his predecessors and contemporaries. In 1809 he separated the annelids and arthropods other than insects from the worms, creating the Annelida and later (1816) united the tapeworms and flukes under the name 'vers molasses' or soft worms.

This was followed by a period of confusion, with flatworms for a time being linked with leeches, and then in 1851 Vogt made an important step forward by recognizing that leeches are annelids and separating the nematodes, acanthocephalans and other groups from the Platyelmia, which included the flatworms and nemertines. In 1859, the name was changed to Platyelminthes by Gegenbaur, but he maintained the composition established by Vogt, namely turbellarians (including the nemertines), flukes and tapeworms. The phylum Vermes was then temporarily resurrected, but in 1876 Minot recognized the many differences between the nemertines and the flatworms and his phylum Plathelminthes excluded the nemertines and corresponds with the modern conception of the phylum. According to Hyman (1951) the etymologically correct form of the phylum name is Platyhelminthes, from the Greek *platys*, 'flat', and *helminthes*, 'worms'.

1.3 THE PLATYHELMINTHS AND INVERTEBRATE EVOLUTION

Platyhelminths occupy an important position near the base of the evolutionary 'ladder' and have figured frequently in theoretical considerations of the origins and divergence of the multicellular animals (metazoans). A popular view is that they were the first invertebrates to establish bilateral symmetry, recognized by the inclusion of the platyhelminths in the Bilateria as opposed to the Radiata, which encompasses the sponges, ctenophores and cnidarians. However, Willmer (1990) argued that the phylum Cnidaria is unlikely to have been primitively radial, identified features of bilaterality in the sponges and pointed out that, although superficially radial, ctenophores often have superimposed bilateral features and might be better described as biradial. Willmer concluded that a biradial condition is more likely to have been a primitive feature of early invertebrates and that transition between dominant radiality and dominant bilaterality has probably taken place repeatedly in most phyla.

Platyhelminths are also traditionally regarded as the first to acquire a mesoderm or middle population of cells between the outer covering layer of cells or ectoderm and the single layer of cells or endoderm lining the gut, a condition which is described as triploblastic (three-layered) as opposed to the diploblastic (two-layered) condition of the sponges, ctenophores and cnidarians. Willmer has also challenged this concept, questioning the grounds for regarding the mesoderm as being absent in most so-called diploblastic animals. However, it is true that the mesoderm in the platyhelminths is a well-developed and densely packed population of cells that obliterates the embryonic blastocoel and renders them acoelomate.

The origins of the platyhelminths have been the subject of numerous theories. A popular view, the planuloid–acoeloid theory, is that the flatworms were derived

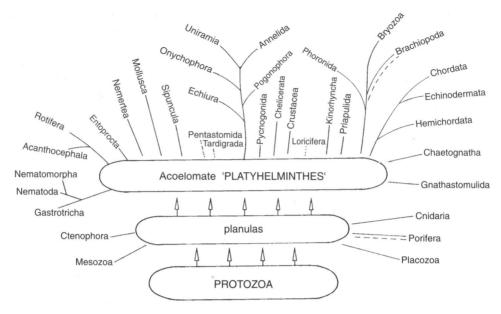

Figure 1.4   A possible phylogeny for invertebrate animals. Source: reproduced from Willmer, 1990, with permission from Cambridge University Press.

from a small, ovoid, solid (acoelomate), ciliated creature resembling the planula larva of cnidarians. An alternative hypothesis, the archecoelomate theory, proposes that fluid-filled mesodermal cavities or coeloms, which are conventionally considered as later developments in ancestors of 'higher' invertebrates such as annelids and echinoderms, put in a much earlier appearance. This scheme envisages the platyhelminths as derivatives of relatively large archecoelomates which became acoelomate as a consequence of a reduction in size related to colonization of the interstitial environment between sand grains. Willmer (1990) analysed, compared and considered these hypotheses and favoured for the platyhelminths a primary acoelomate condition, the product of a direct planuloid ancestry, and she saw the evolutionary trend being from small and unspecialized to larger and more highly organized animals.

Willmer regarded the acoelomate platyhelminths as a kind of platform from which the other metazoan groups have arisen (Fig. 1.4). She likened the metazoan evolutionary panorama not to a vine or a branched tree but to a field of grass and more specifically to an old-fashioned meadow, with some hardy perennials flourishing and branching among the shorter grasses. If her picture is correct then the proto-platyhelminths contained enormous potential, giving rise not only to a wide range of platyhelminths but more importantly to a remarkable multiplicity of other types that went on to produce the bulk of the animals on this planet.

1.4   FEATURES OF THE PLATYHELMINTHS

1.4.1   Size, shape and locomotion

The trend from small to large within the platyhelminths has important consequences. Many platyhelminths have a ciliated epidermis, and small platyhelminths are able to use these cilia for swimming freely. Larger platyhelminths are unable to swim freely by means of their surface ciliation. They are forced to take up a benthic life style and because of their flattened shapes the area of contact between the animal and the substrate is relatively large. This permits a special kind of ciliary locomotion, the ventral epidermal cilia exerting their propulsive force by beating in a layer of mucus secreted by the animal and creating the beautiful and apparently effortless gliding motion typical of planarians. In some platyhelminths, similar gliding is brought about not by cilia but by propagating waves of longitudinal muscle contraction on the ventral surface. This benthic, creeping way of life accentuates the bilateral symmetry of these animals, leading to elaboration of the sensory equipment on the leading, head end, with corresponding concentration of nervous tissue in the anterior 'brain' (see Fig. 1.9) and further differentiation between the dorsal and ventral surfaces.

There is, however, another development in locomotion in platyhelminths with wide applications throughout the Animal Kingdom; this is the employment of a hydrostatic system. Cnidarians use the fluid-filled gut (coelenteron) as a hydrostatic sac, which permits circular and longitudinal muscles to antagonize each other and to change the shape of the body, but in the platyhelminths it is the cellular mesoderm that forms the bulk of the semi-fluid hydrodynamic mass. This is associated with a different arrangement of body musculature. Antagonistic layers of longitudinal and circular muscle lie adjacent to one another beneath the epidermis (Fig. 1.3), while in cnidarians these layers are separated and are developments of the epithelia covering the animal and lining the gut. Elongation and contraction of the platyhelminth body are brought about by these muscles, over-extension or over-contraction of the animal being prevented presumably by a criss-cross arrangement of inextensible fibres in the basement lamina beneath the epidermis (see Fig. 1.3 and Barrington, 1979). The cellular contents of the body are distorted by the contraction of the body wall muscles in much the same way as a liquid-filled coelenteron, pseudo-coelom or coelom. Most texts regard the muscle-based locomotion of platyhelminths as slow and attribute this to the damping down effect of the cellular body contents on the pressure changes generated by muscle contraction (see, for example, Smith *et al.*, 1971). However, some platyhelminths, for example some monogeneans, undergo leech-like locomotion with surprising speed and agility, and there are monogeneans (section 4.12) and polyclad turbellarians (Prudhoe, 1985) that are able to swim freely by means of strong body undulations. Some monogeneans are able to maintain indefinitely relatively vigorous undulating body movements for breathing (section 4.6). Few platyhelminths burrow, and this may reflect the reduced efficiency of a cell-filled sac compared with a liquid-filled cavity for exerting the kind of concentrated force needed to penetrate into the substrate. Nevertheless, many platyhelminths are small enough to adopt an interstitial habit.

1.4.2 Size, shape and gaseous exchange

Size and shape have a profound influence on gaseous exchange. Platyhelminths have no gills or other special respiratory organs and no blood system in their acoelomate bodies. Alexander (1979) calculated that the maximum possible diameter for a cylindrical 'flatworm', using oxygen at a specified rate, is 1.5 mm. Consequently one would expect small platyhelminths to be cylindrical, as indeed many of them are, but increase in size can only be achieved in the same circumstances by becoming flat. Alexander calculated that a platyhelminth with a flat body would need to be no more than 0.5 mm thick if oxygen diffuses in only from the dorsal surface and 1.0 mm if oxygen diffuses in also from the ventral surface.

1.4.3   Size, shape and gut morphology

Size and shape also influence gut morphology. In a small cylindrical platyhelminth like a rhabdocoel, it is sufficient to have a simple sac-shaped intestine (Fig. 1.2*b*), since the diffusion distances for nutrients from the intestine to all the body cells are relatively short. With increase in size comes body flattening; the consequence of this is that body cells would become more and more remote from a central saccular intestine. The solution is to develop intestinal branching. In moderately-sized platyhelminths there are two lateral intestinal limbs (crura), but in larger species three or more limbs may be necessary and these may give off side branches that ramify throughout the flat body to supply remote peripheral as well as central regions (Fig. 1.2*c–e*, *g*). This progressive correlation between development of intestinal branches and size and shape is particularly well illustrated in the larval development of large platyhelminths (Fig. 1.5).

The large surface area of a flattened parasite and the short diffusion distances favour absorption of nutrients through the surface, and total loss of the gut may occur, as in cestodes.

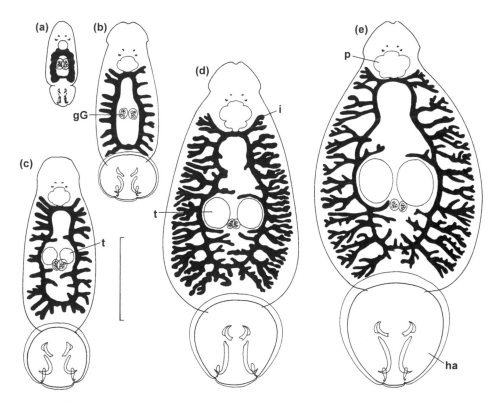

Figure 1.5(*a*)–(*e*)   Stages in early development of the monogenean *Entobdella soleae*, showing progressive elaboration of the gut diverticula. Drawn from living animals stained with neutral red. gG = glands of Goto; ha = haptor; i = intestine; p = pharynx; t = testis. Scale bar: 0.5 mm.

1.4.4 Diet and the gut

Jennings (1997) has emphasized the enormous range of organisms used by free-living flatworms for food, extending from bacteria and unicellular algae through protozoans and virtually all types of invertebrates to the eggs and young stages of fishes and amphibians. Jennings commented that the ability to cope with such a wide range of prey organisms is all the more remarkable because platyhelminths lack the segmental appendages or buccal armature used by other predatory animals to handle their prey. In addition, the parasitic platyhelminths feed on the organs or ingest the body fluids or gut contents of hosts of all kinds, and some parasites, having lost the gut, are able to obtain all their nutrients from the host by absorption of small molecules through their external surfaces.

The platyhelminth gut is essentially a blind intestinal sac with an anterior or ventral mouth. The sac is filled by a muscular pump, the pharynx, located close to the buccal cavity. Like the corresponding organ in the nematodes, the platyhelminth pharynx is well suited for ingesting fluids, but has much greater flexibility, since in many flatworms the organ can be protruded or everted and is capable of penetrating large prey or ingesting whole organisms or fragments of host tissue.

The intestinal sac is lined by an endodermal cell monolayer (gastrodermis) comprising cells often of a single type. In spite of the relative simplicity of this arrangement, some or all of the digestive process and the whole of absorption are carried out by these remarkable cells (Fig. 1.6). Their phagocytic properties permit intracellular digestion, and extracellular digestion may also occur, either by way of enzymes secreted into the gut lumen by the gastrodermal cells or by secretions from separate gland cells opening into the pharynx, oesophagus or elsewhere. Continuity between gastrodermal cells and parenchyma cells (the packing cells that fill the spaces between the organ systems) is provided by junctional complexes (Fig. 1.6).

1.4.5 Reproduction and the eggshell

In cnidarians, the reproductive system is nothing more than the gonads themselves, but in platyhelminths an elaborate ducting system has developed with specializations permitting insemination, (sperm storage) and assembly and shaping of eggshells. Most platyhelminths are hermaphrodite and both cross-insemination (Fig. 1.1c) and self-insemination occur. Parthenogenesis and asexual reproduction are also encountered.

The demands of mate finding, copulation and insemination have led to a much greater behavioural sophistication in the platyhelminths, with the corresponding development of a remarkable array of intromittent organs, including cirruses, which, when brought into action, turn inside out like the finger of a glove, and penises, which are simply protrusible and often incorporate hardened (sclerotized) tubes, hooks or pincers. Spermatozoa are brought to the intromittent organ from the testis (or testes) by a gonoduct, the vas deferens, part of which may be dilated to provide a sperm store or vesicula seminalis. There may be accessory glands associated with the male copulatory apparatus. These glands are usually called 'prostate glands', but this is an inappropriate name because there is no evidence that their secretion functions like that of mammalian prostate glands. The function of these secretions in platyhelminths is generally unknown, although in a few species that

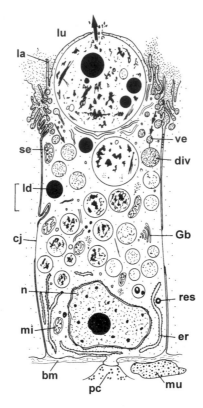

Figure 1.6   Diagrammatic drawing of a median section through a gastrodermal cell of the monogenean *Calicotyle kroyeri*. Small arrows indicate route of uptake by endocytosis of food material from the gut lumen (lu) for transfer *via* vesicles (ve) to vacuoles (div) for intracellular digestion. Large arrow indicates release by exocytosis of indigestible residues from a large apical vacuole into the lumen. bm = basement lamina; cj = close junction; er = endoplasmic reticulum; Gb = Golgi body; la = lamella; ld = lipid droplet; mi = mitochondrion; mu = muscle; n = nucleus; pc = parenchyma cell; res = residual body; se = septate desmosome. Scale bar: 2 μm. Source: redrawn from Halton and Stranock, 1976.

exchange spermatophores rather than engage in intromission they may contribute to the substance of the spermatophore.

A characteristic feature of the platyhelminths is that when the embryos are entrusted to the outside world they are enclosed in a strong case. More than one embryo may be so enclosed (as in some planarians) and the case is usually then referred to as a capsule or cocoon (Fig. 1.1*a*); the latter term will not be used here because it has been used for a different structure in the fecampiid turbellarians (Chapter 2). Single embryos are usually described as being enclosed by an eggshell.

Within the platyhelminths there has been an important shift in the source of material for the shell–capsule and in the source of provisions for the developing embryo. In the so-called 'archoophorans', which include the acoel and polyclad turbellarians, the egg cells are endolecithal, i.e. food (yolk) for the developing embryo is included in the cytoplasm of the egg cell and the egg capsule material is partly derived from a shell gland and partly from peripherally situated cortical granules in the egg cell (see review by Shinn (1993) and Ishida and Teshirogi (1986)). In the 'neoophorans', which include most of the platyhelminths that indulge in symbiotic associations, an ectolecithal condition has arisen by the creation of 'division of labour' within the ovary. The production of egg cells destined to develop into new individuals is restricted to a separate region of the ovary known as the germarium, but these egg cells no longer contain any appreciable amount of yolk. This is provided by the other region of the ovary, known as the vitellarium, which produces modified egg cells that no longer contribute genetic material to the germinal line but are abundantly supplied with yolk. Moreover, the cortical granules of the egg cell no longer make any contribution to the bulk of the shell–capsule material, although they may retain a specialized role in creation of the shell sutures that eventually permit fracturing of the shell–capsule and escape of the larva(e) (see Shinn, 1993 and section 2.3.1). It is the cortical granules of the vitelline cells that now take on full responsibility for production of shell–capsule material and the shell gland becomes redundant.

There are some platyhelminths that foreshadow the 'neoophoran' condition – in these animals the ovary retains its compact form, but is functionally a germovitellarium with some egg cells differentiating into yolk-producing follicle cells that enclose and provision a yolkless germ cell, which alone will undergo embryogenesis (Hyman, 1951).

The germarium is usually a single compact body, whereas the vitellarium is commonly a racemose follicular organ spreading throughout the animal (see, for example, Fig. 4.2). Closer scrutiny of those platyhelminths with an alimentary canal usually reveals that the vitelline follicles are closely associated with the ramifications of the gut, following any side branches as well as the main intestinal limbs. In the absence of a blood system this is a significant arrangement, since nutrients absorbed by the gut are available with minimal transport costs to the most active organ in the body, where synthesis of eggshell–capsule material and resources destined for the growing embryos is continually proceeding.

The ducts from the vitelline follicles coalesce, transporting vitelline cells from the follicles to a vitelline duct, which joins the oviduct from the germarium to form the ovovitelline duct. The vitelline duct may enlarge to form a storage reservoir for vitelline cells. Distally, the ovovitelline duct expands to form the ootype, which is highly muscular and supplied with glands (Mehlis's gland). It is this organ that receives the egg cell and numerous vitelline cells and, after release by the vitelline cells of the shell-forming droplets, ensures the coalescence of the droplets around the ootype contents. The alternative name of egg mould for the ootype reflects its role in shaping the newly formed eggshell. The amount of shell material per egg will be proportional to the surface area of the vitelline cells and hence to their number, and the more vitelline cells that are sequestered with the egg cell the more resources there will be for embryogenesis.

The eggshell (or capsule enclosing one or more embryos) is water-white when first made, but undergoes fairly rapidly a chemical reaction, the visual manifestation of which is a gradual change in colour from water-white to golden brown. This is accompanied by a change in the properties of the material from soft and pliable to hard and tough; considerable resistance develops to physical and chemical destruction and in particular to attack by proteolytic enzymes. The general view is that capsular material is proteinaceous and that the observed changes are attributable to the process of quinone-tanning or sclerotization.

Waite (1990) has summarized current knowledge of the quinone-tanning process and the following account is mostly extracted from his review. The chemistry accompanying the process is unclear but probably follows the path outlined in

Fig. 1.7, in which at least five reactants are necessary: (A) catechols, (B) catechol oxidase, (C) oxygen, (D) proteins and (E) fillers. If the presence of catechols, catechol oxidase and proteins is regarded as a sound indicator of quinone-tanning then the process is widespread in the Animal Kingdom, occurring in most of the major groups. It is prevalent in the walls of spores, cysts, egg capsules and cocoons in such a wide range of organisms as protozoans, platyhelminths, nematodes, molluscs, annelids, arthropods and elasmobranch fishes, and is important in other structures such as cnidarian skeletons, molluscan byssus threads and shells, and arthropod cuticles.

Fillers or reinforcing molecules such as collagen (byssus threads) and chitin (arthropod cuticle) are involved in quinone-tanned composites, but are not known to be present in platyhelminth shell material. It is also important to distinguish between two major types of catechols: those with side chains R having a low molecular weight (<200) and those with R having a high molecular weight (>200). Known examples of the latter involve peptide- or protein-bound 3,4-dihydroxyphe-nyl-L-alanine (DOPA), i.e. the R group has a peptide backbone. So, when the quinone-tanning process involves DOPA-containing proteins there is no need for additional proteins (D). DOPA-containing proteins have been identified in shell material from platyhelminths (the triclad *Bdelloura candida* and the digeneans

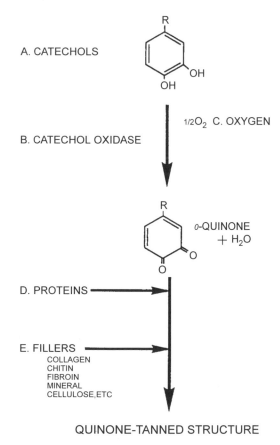

Figure 1.7 Reaction scheme summarizing the chemical essentials of quinone-tanning in living organisms. Source: reproduced from Waite, 1990.

Fasciola hepatica and *Schistosoma mansoni*), and appear to be the preferred option in strictly aquatic organisms, while the terrestrial insects have switched to low-molecular-weight catechols.

The idea that 'autotanning' might be involved in the hardening of platyhelminth eggshells was introduced by Smyth (1954). This involves the direct action of catechol oxidase on tyrosyl groups in the protein molecules to form cross-links, and is attractive because of the absence of low-molecular-weight catechols in the system. However, no confirmation of this model in any organism displaying quinone-tanning has yet been achieved.

Waite pointed out that rigorous proof of the involvement of DOPA-containing proteins in the process of quinone-tanning is not yet forthcoming and, because of the reactivity of quinones, it is hard to predict what the chemistry of the poly-merization process might be. In fact, Vincent and Hillerton (1979) have put forward an alternative. They suggested that the degree of stiffening produced during sclerotization of insect cuticle could be achieved by dehydration of the protein matrix, and they proposed that this is brought about by the injection of quinones into the matrix and H-bonding of these to the cuticular protein. This dehydration theory of Vincent and Hillerton has attracted relatively little attention, but deserves consideration by students of the sclerotization process in platyhelminth eggshells. In addition, there have been claims that disulphide bonds may play a part in the polymerization of eggshell proteins in some platyhelminths (see Smyth and Halton, 1983).

The DOPA-containing proteins and the latent catechol oxidase are present together in the peripheral droplets of the vitelline cells of platyhelminths, or in the cortical granules of archoophoran egg cells. Waite (1990) pointed out the great importance of thorough mixing of catalyst and resin in commercial glues and suggested that co-packaging of catechol oxidase and its substrate promotes interac-tion between the two, although nothing is known about how the two different proteins are stored. After release of the droplets and at an appropriate time the enzyme is activated and acts upon the DOPA residues in the protein. Waite reported that the enzyme precursors in the byssus and periostracum of molluscs and in the egg capsules of elasmobranch fishes are probably activated by chymotrypsin, and he suggested that latent catecholoxidases of flatworm eggshells may also require pro-teolytic treatment for activation. In the context of the mixing problem and limited diffusibility of the two protein components, Waite pointed out that high enzyme–substrate ratios might be expected; consistent with this is the finding that in byssus material each enzyme molecule is accompanied by two to three molecules of the DOPA-containing protein. At these concentrations, the cohesiveness of the material must depend on the participation of the enzyme for its structural integrity.

Whatever the mechanism of sclerotization is, there is no doubt that the eggshells and capsules are physically strong and chemically resistant. The sclerotin eggshells–capsules provide a sterile, sealed compartment to hold vitelline cells and egg cells together in close proximity and prevent external microorganisms from gaining access to resources destined for the embryo. Moreover, the eggs of the monogenean *Entobdella soleae* are able to pass through the gut of a predator and, provided that the shells are not punctured, the contained embryos are alive at the end of this experi-ence and are capable of continuing development and hatching (Macdonald, quoted by Kearn, 1975). This provides experimental support for the suggestion made by Llewellyn (1965) in the context of monogenean parasites, that the sclerotized shells of freely deposited eggs confer protection on platyhelminth embryos swallowed by predators.

1.4.6   Excretion and osmoregulation

There is a need for organisms to regulate the composition of their body fluids and to eliminate metabolic wastes. The protonephridial system (Fig. 1.8) is believed to deal with these functions in platyhelminths. Following Goodrich (1946), a protonephridium typically comprises a ciliated excretory canal that may branch at its inner end, with specialized cells (flame cells) at the internal terminations of the branches. The excretory canal opens to the exterior (sometimes *via* a bladder) by a nephridiopore. Flame cells (or cyrtocytes) in platyhelminths are usually about 10 μm long by 3 μm wide (Wilson and Webster, 1974), and are readily identified in living animals by the beating of the 'flame', a bundle of flagella that undulate in unison. Typically, platyhelminths have a pair of protonephridia (communicating with each other in the monogenean *Entobdella soleae*; Fig. 1.8*a*), but this simple pattern is subject to great variation in detail in the different platyhelminth groups (see Goodrich, 1946) and protonephridia are absent in some turbellarians (e.g. acoels).

The protonephridium appears to be highly successful, being widely distributed throughout the acoelomate phyla (Platyhelminthes, Rotifera, Nemertinea, Acanthocephala, Priapulida, Entoprocta, Gastrotricha, Echinodera) and in many adult or larval coelomates (Annelida, Echiuroidea, Mollusca, Phoronida, Cephalochordata) (Riegel, 1972). Indeed, it is difficult to imagine how cnidarians and platyhelminths such as acoels manage without a protonephridial system. According to Willmer (1970; in Wilson and Webster, 1974), the regulation of internal fluid composition prior to the appearance of the protonephridium was conducted by the epithelia covering the body surface and lining the gut, and cnidarians and acoels probably function in this way. This functional relationship between surface epithelia and protonephridia is consistent with the view that protonephridia arose by centripetal extension or invagination of the surface epithelium (ectoderm) into the interior of the animal (Willmer, in Wilson and Webster, 1974). Such a development would permit elimination of toxic wastes from the compact parenchyma (mesoderm) more efficiently and rapidly than by diffusion, even in flattened animals with increased surface area and shorter diffusion distances. The protonephridial system could support larger, more active animals and according to Wilson and Webster (1974) may have permitted the relative permeability of the body surface of marine, freshwater, terrestrial and parasitic flatworms.

The following account of the structure of a platyhelminth flame cell is based on a typical example from one of the parasitic groups. Each flame cell body (Fig. 1.8*b*) has a projecting collar resembling a palisade and made of cytoplasmic rods. This collar fits neatly inside the expanded open end of the first tubule cell, the region overlapping with the flame-cell collar being similarly subdivided to form a palisade of cytoplasmic rods. The tubule cell is usually a sheet of cytoplasm rolled up to form a tube, with a desmosome linking together the contiguous free edges of the tube.

In transverse section through the region where the flame-cell collar and the tubule cell interlock (Fig. 1.8*c*), it can be seen that the rods of the two palisades alternate and form two concentric circles, the rods from the flame cell lying inside those from the tubule. Adjacent inner and outer rods are joined by extracellular membranous or

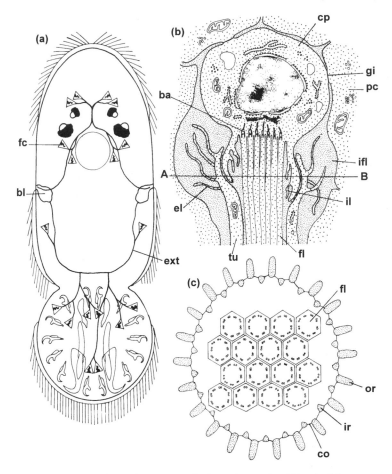

Figure 1.8   (*a*) The protonephridial system in the larva (oncomiracidium) of the monogenean *Entobdella soleae*. (*b*) Diagrammatic longitudinal section, based on electron micrographs, through a platyhelminth flame cell and associated structures (based on the digenean *Fasciola hepatica*). (*c*) Diagram of a cross-section of a flame cell through the 'flame', in the plane AB shown in (*b*). ba = basal body of flagellum; bl = bladder; co = fibrous connection; cp = cap cell; el = external leptotrich; ext = excretory tubule; fc = flame cell; fl = flagellum; gj = gap junction; ifl = interstitial (intercellular) fluid; il = internal leptotrich; ir = inner cytoplasmic rib; or = outer cytoplasmic rib; pc = parenchyma cell; tu = tubule cell. Source: (*b*), (*c*) modified from Smyth and Halton, 1983.

fibrous material resembling fine tonofibrils, and the whole of this interlocking region is sometimes called the weir. Finger-like projecting processes or leptotriches arise from the cytoplasmic rods of both the flame cell and the tubule cell. Those from the tubule cell are long, extend outwards into the intercellular space and may serve to keep neighbouring cells away. The leptotriches from the flame cell are short and project towards the flame, perhaps serving to prevent the flame from occluding the weir. The flagella comprising the flame are reported in some digeneans to be closely packed, forming a hexagonal array, with the spaces between the flagella filled by material that may serve to hold the flagella together and promote their function as a single entity (Wilson and Webster, 1974).

The most likely function of the flame cell is to filter interstitial (intercellular) fluid. There is a pathway for the entrance of this fluid into the tubule lumen between the interdigitating rods of the weir and through the extracellular membranous or filamentous material filling the gaps between the rods. It is this extra-cellular material that is regarded as the ultra-filter, permitting passage of water and small molecules but excluding macromolecules. Wilson and Webster (1974) estimated that a platyhelminth flame cell is likely to filter slightly more than its own fluid volume per second. It is envisaged that the beating flame is responsible for drawing the interstitial fluid through the filter and propelling it down the tube, but further progress towards the nephridiopore may be aided by tufts of tubule flagella that project towards the nephridiopore end of the system and by peristaltic movements of the body. The importance of body movement for the propulsion of tubule fluid in *Hymenolepis diminuta* was identified by Webster (1971), and this may also be the function of slow-moving, anteriorly directed waves of body contraction observed by Kearn and Whittington (unpublished) in the monogenean *Benedenia rohdei*.

Control of composition of body fluids has special significance for symbiotic platyhelminths that undergo drastic changes in environment during their life cycles. However, there is no definitive experimental evidence for or against a role for flatworm protonephridia in the regulation of body fluid composition, even in those that live in fresh water where they are likely to be subjected to ionic and osmotic stress (Wilson and Webster, 1974; Hertel, 1993). The living epidermis is likely to be of special importance in freshwater environments and experimental work on the freshwater-inhabiting miracidium larva of the digenean *Fasciola hepatica* indicates that the ciliated epidermis acts as an important permeability barrier, preventing uptake of water by the animal (section 13.3.2; Wilson, Pullin and Denison, 1971). Endoparasitic flatworms such as tapeworms appear to lack the ability to osmo-regulate and Wilson and Webster (1974) suggested that adaptations of the epidermis for nutrient uptake may limit its osmoregulatory capacity. Tapeworms have a well developed protonephridial system, so this must have another function.

There is better evidence for an excretory role for the flame cell system, especially in the tapeworm *Hymenolepis diminuta* (see Wilson and Webster, 1974). The presence in the protonephridial canals of lactate, succinate and urea, all known to be organic end products of metabolism, supports the suggestion that protonephridia are responsible for removing organic wastes from the deeper tissues. The presence of sodium as the major cation in the fluid, with much lower levels of potassium, points to an interstitial origin by filtration rather than by secretion for the canal fluid. Ammonia levels are low relative to the total amount of nitrogen excreted by the worm, and the bulk of excreted ammonia may leave by diffusion across the body surface. Other organic solutes such as amino acids and glucose are sparsely repre-sented in the protonephridial fluid and the resorption of glucose, lactate and urea by the tubules has been demonstrated. The increased surface area of the lining of the tubule, produced by microvilli or folds, may facilitate the resorption process. Alka-line phosphatase activity has been demonstrated in association with the tubules, flame cells and bladders of some platyhelminths. Formerly this activity was thought to be involved in the transport mechanism, but it is more probable that it is concerned with the hydrolysis of sugar phosphates prior to the separate uptake of the sugar and phosphate moieties (see Wilson and Webster, 1974; section 9.3.2(c)).

1.4.7   The nervous system and sensilla

The ladder-like nervous systems of platyhelminths are readily derived from the diffuse nerve nets of cnidarians by condensation of nervous elements to form an

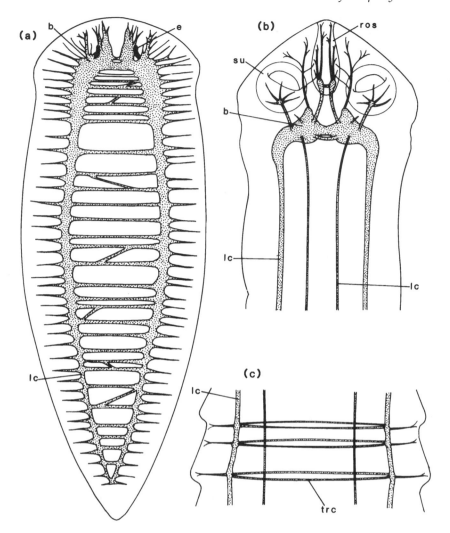

Figure 1.9 (*a*) Diagrammatic illustration of the nervous system of a planarian such as *Dugesia*. (*b*) Nervous system of the scolex and (*c*) nervous system of part of the body (strobila) of the tapeworm *Hymenolepis diminuta*; (*c*) shows the three transverse ring commissures (trc) linking the longitudinal nerve cords (lc) in a proglottis, but only four of the ten longitudinal cords are shown in (*b*) and (*c*). b = 'brain'; e = eye; ros = rostellum; su = sucker. Source: (*a*) based on Ude, 1908; (*b*), (*c*) based on Wilson and Schiller, 1969 and Lumsden and Specian, 1980.

anteriorly situated 'brain' from which longitudinal nerve cords run in a posterior direction. These cords are connected by transverse nerves to form a so-called orthogonal pattern (Gustafsson, 1992; Fig. 1.9). There may be several cords, but some are usually better developed than others, especially those near the ventral surface in active, creeping flatworms. Peripheral and central regions of the body are innervated by nerves from the brain and from the cords. At the ultrastructural level, flatworm neurons distinguish themselves by their high content of secretory vesicles (Reuter and Gustafsson, 1995). Since there is no coelom or circulatory system, neuroactive substances must be released from the neurons at synaptic and non-synaptic release sites close to their target cells or organs, or must travel to these targets *via* the intercellular matrix.

According to Reuter and Gustafsson (1995), the flatworm nervous system is characterized by diversity at all levels. Multiplicity is found in the layout of the nervous system, in the range of sensory receptors, in the types of nerve cells (uni-, bi- and multipolar) and in the variety of neuroactive secretions, which includes many peptides as well as classical neurotransmitters such as 5-hydroxytryptamine and acetylcholine (Halton *et al.*, 1994; Reuter and Gustafsson, 1995; Shaw, Maule and Halton, 1996). These different neuroactive substances may occur in their own neuronal compartments (Reuter and Gustafsson, 1995).

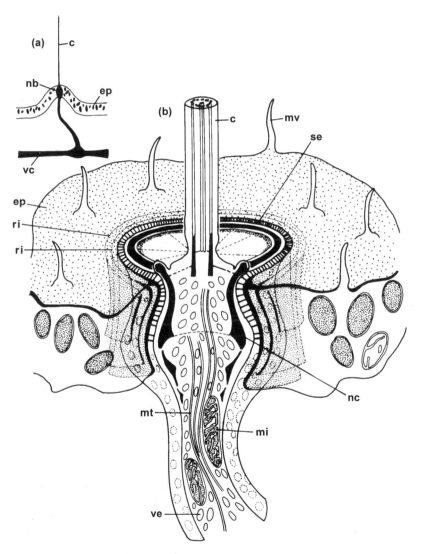

Figure 1.10   A platyhelminth sensillum, based on a contact (?) receptor from the monogenean *Gyrodactylus* sp. (*a*) Diagram showing the nervous connection to the lateral ventral nerve cord. (*b*) Three-dimensional reconstruction from electron micrographs.   c = cilium;   ep = epidermis;   mi = mitochondrion; mt = microtubules; mv = microvillus; nb = nerve bulb; nc = thickening of nerve collar; np = nerve process; ri = strengthening (?) rings around attachment site in tegument; se = septate desmosome; vc = ventral nerve cord; ve = vesicles in nerve process. Source: reproduced from Lyons, 1969a, with permission from Cambridge University Press.

Reference has already been made to the role of cilia in propulsion, especially of small platyhelminths, and to the part played by flagella in flame cells and their ducts. Cilia have another important role in platyhelminths and that is as the terminal transducer elements of sensory nerve endings. Cilia with this sensory function (sensilla) may have the typical 9 + 2 longitudinal microtubule framework of motile cilia and indeed may retain their motility, but the internal structure of most of them is modified in some way, this usually coinciding with the loss of powers of movement. These ciliated, sensory nerve endings project from the surface of the animal between the ciliated epidermal cells or, in platyhelminths with a syncytial unciliated tegument, penetrate through the tegument (Fig. 1.10). They are held in place by desmosomal junctions.

Experimental evidence pertaining to the function of these sensilla is non-existent, but there is no shortage of speculation; chemoreceptive and mechanoreceptive functions, the latter including responsiveness to contact and to water currents, have been attributed to them.

Internal organs containing modified cilia have been interpreted as cilia-based eyes (sections 4.10.4, 13.3.1), but most pigment-shielded photoreceptors ('eye spots') in platyhelminths appear to be based on microvilli (so-called rhabdomeric eyes).

1.5 ASSOCIATIONS BETWEEN PLATYHELMINTHS AND OTHER ORGANISMS

There is a striking tendency in the platyhelminths that sets them apart from most other groups of invertebrates and that is their inclination to develop symbiotic associations. As proposed by Starr (1975), the term 'symbiosis' is used here in its original sense as defined by de Bary (1879), i.e. *'des Zusammenlebens ungleichnamiger Organismen'* ('the living together of differently named organisms'). This definition clearly encompasses mutualistic relationships, conferring benefits on both partners, and one-sided relationships in which only one of the partners gains, the effects on the other partner ranging from insignificant to pathogenic. In recent times there has been a tendency to limit the scope of the term 'symbiosis' to mutually beneficial relationships, but this tendency will be resisted in this book. For the sake of simplicity of language, the term 'host' will be used for platyhelminth co-symbiotes, even in non-parasitic relationships, with the exception of situations where the co-symbiote inhabits the tissues of the flatworm.

Most invertebrate phyla have representatives that indulge in parasitic relationships. At one end of the spectrum protozoans, nematodes and arthropods have major commitments to parasitism, while at the other end the cnidarians, molluscs and echinoderms have few or no parasitic forms. Other kinds of association such as mutualism and commensalism (food sharing) are frequent and widespread throughout the Animal Kingdom. However, the platyhelminths are exceptional in having not only a major commitment to parasitism but also an involvement in a great range of associations of other kinds. Mutualistic associations occur between some turbellarians and algae, and associations with animals may be inquilinistic, with the host providing a safe haven from predators. Some platyhelminths are merely epizoic, using the animal host as a platform from which the symbiote captures free-living prey, while others supplement this diet with food discarded by the host, so introducing an ectocommensal dimension to the partnership. Further development of the commensal tendency is favoured in relationships with hosts, such as decapod crustaceans, xiphosurans and freshwater turtles, that tear their food and inevitably drop some of it, and associations with hosts with soft surface tissues, such as the epidermis of fishes, have led more than once to the ectoparasitic habit. A very early ectoparasitic association with fishes gave rise to the monogeneans, one of the three major groups of platyhelminths that are entirely parasitic. These ectoparasites of fishes have exploited, on more than one occasion, mobile, copepod fish parasites as a means of transport to new fish hosts, a phenomenon known as phoresy.

Many turbellarians have become endosymbiotic, in particular with slow-moving invertebrates such as the molluscs and the echinoderms. Some of these relationships can be described as endocommensal, with the symbiote living in the gut of the partner and ingesting host food, but there are associations in which this diet is supplemented by components from the host, introducing a parasitic tendency, which has gone all the way in some turbellarians with the loss of their own gut. Some of these parasites have colonized other habitats such as body cavities, and in some turbellarians escape from this habitat involves rupture of the body wall and death of the host. Parasites that are responsible for the death of the host as a normal event in their life cycles are known as parasitoids.

In addition to the intriguing and diverse endoparasitic relationships between turbellarians and invertebrates, the platyhelminths have given rise to two major groups of vertebrate endoparasites. These are the cestodes and the digeneans. Like the monogeneans, these groups are entirely parasitic and, with the monogeneans, represent a major proportion of modern platyhelminth species. It is likely that cestode endoparasitism has its roots in proto-monogenean ectoparasitism of the early fish-like vertebrates, and monogenean–cestode (= oncophorean) evolution spans virtually the whole of vertebrate evolution. This in itself is remarkable, but it is even more surprising to learn that digeneans have independently colonized the gut of fishes and, like the cestodes, have diversified into all the major groups of vertebrates. However, they have probably reached this position by a different route, their ancestors developing an intimate endoparasitic relationship with molluscs before the vertebrates came on the scene.

It will be clear from this brief survey that the platyhelminths probably have more to offer the student of animal associations, and parasitism in particular, than any other group of invertebrates. The range of kinds of association is wide and promises to throw light on the evolution of these relationships and possible transitions from one to the other. In addition, the monogenean–cestode lineage and that of the digeneans have their roots in the very early days of vertebrate evolution; this provides a unique opportunity to explore the way that the parasites have kept pace with the enormous diversification of the vertebrates and in particular the colonization of land. These parasites provide abundant raw material for the exploration of morphological, physiological, behavioural and ecological adaptations to parasitism, and the monogeneans and cestodes allow comparison of the ectoparasitic and endo-parasitic ways of life – the monogeneans are superbly adapted, specialist fish parasites, while their relatives the cestodes are equally highly specialized for life in one of the harshest of environments, the vertebrate intestine. Work on the tapeworms and on the schistosome digeneans has also thrown light on the way the vertebrate immune system operates against multicellular parasites, and has revealed much information on how parasites respond to and cope with front-line host defences. It is this range of material that will be explored further in the rest of this book.

CHAPTER 2

Symbiosis and the turbellarians

The parasitic life styles adopted by turbellarians are diverse, some of them with intriguing features setting them apart from other parasitic platyhelminths. First, however, it is worth exploring some of the range of non-parasitic relationships in which they are involved. The reader is referred to the Appendix for a classification of the turbellarians.

2.1 MUTUALISM – *CONVOLUTA* AND THE ALGA *TETRASELMIS*

There is no better description of the habitat of the acoel *Convoluta roscoffensis* than that of Keeble (1910), and I quote:

> An observer, walking at low tide seaward across a golden beach in Brittany, ... some yards landward of the thin line of green *Cladophora* which lies bleaching in the sun,...may see dark spinach-green glistening patches – the colonies of *C. roscoffensis*. He must tread softly lest the patches melt away at his approach. The colonies may extend for many yards as dark green, irregular strips running more or less parallel with the shore-line, or they may consist of apparently disconnected patches varying in size from one inch or so to a yard or more across.

These splashes of colour are created by dense populations of acoels that harbour in their bodies green symbiotic algal cells belonging to a single species, usually *Tetraselmis convolutae*. The melting away of the patches is a synchronized downward movement of the acoels into the sand in response to vibration. This, combined with a response to light direction and an inherent tidal rhythmicity, ensures that the animals are at the sand surface and exposed to maximum illumination during the day when the tide is out, and deeper in the sand at other times. The symbiotes are intracellular, inhabiting vacuoles in the syncytial parenchyma of the acoel (Fig. 2.1*a*), and mostly located beneath the body wall musculature with extensions into the epidermis (see Smith and Douglas, 1987).

The physiological interactions of the partners have been summarized by Smith and Douglas (1987) and are briefly as follows. There is good evidence that amino acids and glucose generated by the photosynthetic activities of the alga move from the alga to the acoel and that photosynthetically produced oxygen is available for the flatworm's respiration. There are also indications that fatty acids and sterols pass from alga to flatworm and that nitrogenous waste from the flatworm is recycled by the algal symbiote.

This mutualistic symbiosis is all the more remarkable because when the young acoels emerge from the gelatinous capsules laid by the adults they are free of algae

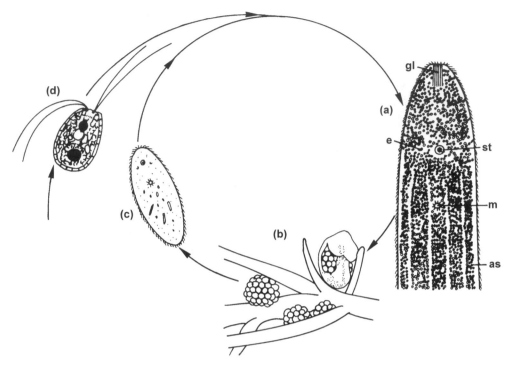

Figure 2.1 The life cycle of an acoel of the genus *Convoluta*. (*a*) Adult. (*b*) Capsules containing eggs are attached to weed or deposited in sand. (*c*) Freshly hatched juvenile, with no algal cells in the body. Juveniles ingest free-swimming algal cells, *Tetraselmis convolutae*. (*d*) as = algal symbiotes; e = eye; gl = apical glands; m = mouth; st = statocyst. Source: based on Keeble, 1910, de Beauchamp, 1961, Apelt, 1969 and Smith and Douglas, 1987.

(Fig. 2.1); if they fail to acquire symbiotes, they will die before reaching maturity. Free-living algal cells are available in the surrounding water and some of these settle and adhere to the surface of the egg capsules; other algal cells are released by the parent during egg laying. This provides a rich source of algal cells in and around the egg capsules; some of these cells are ingested, together with diatoms, by the newly-hatched juvenile worms. Significant changes occur in the algal cells after ingestion. Free-living cells of *Tetraselmis* have a pectinaceous cell wall, an eyespot and four flagella (Fig. 2.1*d*) but these features are lost when the algal cell is phagocytosed by the acoel. According to Keeble (1910), *C. roscoffensis* eventually ceases to ingest food and presumably relies entirely on its symbiotes for support, but the closely related *C. paradoxa* continues to feed throughout its life (Fig. 2.2).

An interesting feature is that the association only involves algae of the genus *Tetraselmis* and yet the available free-living algae are likely to include species of other genera. Although free-living *Tetraselmis* spp. have flagella and are motile, there is no evidence that they exhibit chemotaxis towards egg capsules of *Convoluta* and it seems that some kind of selection of potential symbiotes is made by the young worms. All *Tetraselmis* species so far tested form a viable association with *C. roscoffensis* in the laboratory, but there is some variation in the growth rates of juveniles infected with different algal species.

Selection on the part of the flatworm can also occur at a later stage in the development of the association because a well developed infection of *C. roscoffensis*

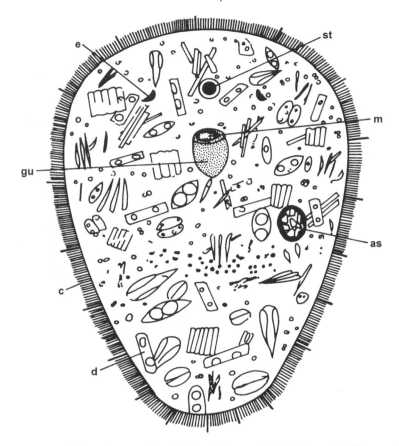

Figure 2.2 A juvenile *Convoluta paradoxa*. The body is crowded with ingested diatoms (d). as = single algal symbiote; c = cilia; e = eye; gu = gullet; m = mouth; st = statocyst. Source: from Keeble, 1910.

with *T. marinus*, an alga that supports relatively poor flatworm growth, is entirely replaced when the worm is experimentally exposed to *T. convolutae*. Thus, *Convoluta* appears to be actively involved in selecting and maintaining the species of algal symbiote that promotes the best growth.

2.2   EPIZOIC AND ECTOCOMMENSAL RELATIONSHIPS

2.2.1   Polyclads and invertebrates

Several polyclads are known to associate with invertebrates. For example, *Emprosthopharynx rasae* wraps itself around the abdomen of the hermit crab *Calcinus latens* inhabiting shells of the gastropod *Trochus sandwichensis* (see Prudhoe, 1968), *Apidioplana apluda* adheres closely to the outer surfaces of gorgonians and matches the orange–red colour of the host (Cannon, 1990), and *Notoplana comes* is consistently found upon or immediately below the brittlestar *Ophiocoma scolopendrina* (see Cannon and Grygier, 1991). In none of these associations is there any information on the nature of the relationship between the two partners, but trophic advantages and protection from predators are possibilities.

2.2.2 Temnocephalans and crustaceans

Most temnocephalans are ectosymbiotes of freshwater decapod and isopod crustaceans. Some live on the gills while others occur on the carapace and limbs (Fig. 2.3). They have important structural modifications including a posterior attachment disc and multiple prehensile tentacles at the anterior end of the body (Fig. 2.4). Temnocephalans have extensible bodies and undergo leech-like locomotion using the posterior disc and the tentacles (see below), but the tentacles also have a central role in feeding.

Temnocephalans are basically epizoic predators and use their tentacles to capture free-living and co-symbiotic invertebrates of suitable size. Jennings (1968a) has given a detailed account of feeding in *Temnocephala brenesi* and *T. novae-zealandiae*. Prey organisms, which may be protozoans, rotifers, oligochaetes, insect larvae or copepod and ostracod crustaceans, are seized by the extended tentacles, which immediately shorten, pulling the prey towards the body. The body then contracts and the body edges and the tentacles curl ventrally, creating an attitude that assists the symbiote to push the food into the mouth. Small organisms are swallowed within 14 s of capture and their bodies are partially disrupted as they pass through the pharynx into the intestine. Larger prey items such as insect larvae are held against the mouth by the tentacles, and the slightly protrusible pharynx serves to rupture their integument in some way and to ingest their body contents, a process that is completed in 2–3 min.

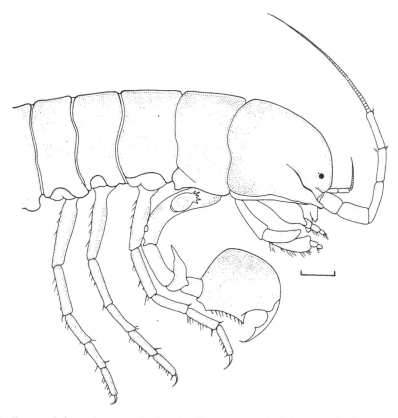

Figure 2.3 *Temnocephala geonoma* attached to the first peraeopod of the isopod *Phreatoicopsis terricola*. Note egg attached to second peraeopod. Scale bar: 2 mm. Source: reproduced from Williams, 1980, with permission from Taylor & Francis.

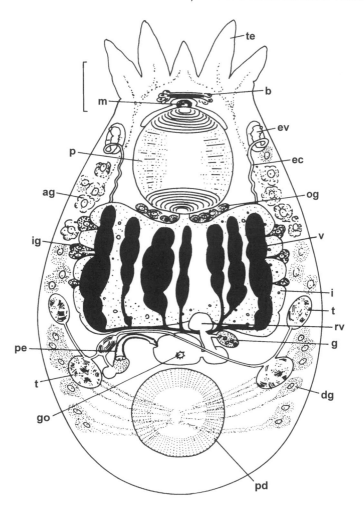

Figure 2.4 *Temnocephala geonoma,* in dorsal view. ag = anterior glands; b = 'brain'; dg = glands of the posterior disc; ec = excretory canal; ev = excretory vesicle; g = germarium; go = genital opening; i = intestine; ig = intestinal glands; m = mouth; og = oesophageal glands; p = pharynx; pd = posterior attachment disc; pe = penis; rv = resorptive vesicle; t = testis; te = tentacle; v = vitellarium. Scale bar: 200 μm. Source: modified from Williams, 1980.

Diatoms and algal filaments found in the gut of the temnocephalans were regarded by Jennings as food items from within the prey, since both species of temnocephalan consistently refused to ingest plant material in the laboratory.

Jennings (1968a) discovered a particularly interesting feature of the nutrition of young *T. brenesi*. Embryonated eggs and empty eggshells of the temnocephalan are found in abundance cemented to the gills of the shrimp host. These eggs act as a trap for suspended matter and, as soon as they are laid, detritus begins to gather around them. This detritus is a substrate for the settlement and growth of protozoans and rotifers and these organisms provide the principal food for newly hatched *T. brenesi* less than 1.5 mm in length (adults are 3 mm long).

The complex nutritional interactions of a whole community of ectosymbiotes of crustaceans were unravelled by Cannon and Jennings (1987) and were summarized diagrammatically by Jennings (1988; Fig. 2.5). Four ectoparasites on Australian

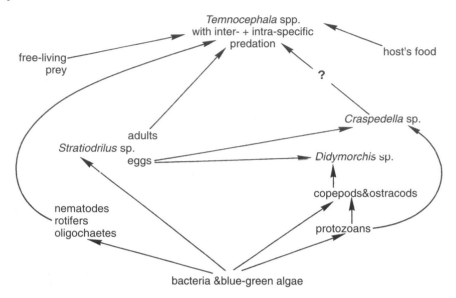

Figure 2.5 Nutrient flow among some ectosymbiotes of freshwater crayfish. Source: reproduced from Jennings, 1988, with permission from Gustav Fischer Verlag.

crayfish of the genus *Cherax* were studied. The gill chamber is occupied by the temnocephalan *Craspedella spenceri* which subsists largely on co-symbiotic protozoa, but also eats the eggs of the polychaete *Stratiodrilus novaehollandiae*. This polychaete feeds on the gill chamber microflora and its eggs are also consumed, together with ostracods and copepods, by the epizoic, rhabdocoel turbellarian *Didymorchis cherapsis*. A second temnocephalan, *Temnocephala minor*, lives on the external body surface and feeds mainly on free-living oligochaetes and small arthropods, but also displays a tendency towards commensalism because it ingests fragments of the host's food. Adult *Stratiodrilus* is probably too large for consumption by *C. spenceri* or *D. cherapsis*, but *Stratiodrilus* also participates in a similar epizoic community on the crayfish *Euastacus hystricosus* and here the larger *Temnocephala comes* does prey on the polychaete. One additional dimension to the story is the observation that *T. comes* swallows other temnocephalans and even other individuals of its own species.

Thus, apart from occasional ectocommensal tendencies, temnocephalans are predators like their free-living turbellarian relatives and their intestinal morphology and digestive physiology are also similar (Jennings, 1968a). In *T. brenesi* and *T. novaezealandiae* columnar phagocytic cells predominate in the single-layered gastrodermis, with the remaining glandular cells being the source of an endopeptidase that initiates digestion of the prey in the lumen. Supplementary gland cells occur beneath the basement membrane of the gastrodermis. After preliminary digestion and disintegration of the food in the intestinal lumen, it is phagocytosed by the columnar cells. Digestion continues within the food vacuoles, which receive from the cytoplasm more endopeptidase followed by exopeptidase, and, if the diet contains fat, lipase. Acid phosphatase activity in the cytoplasm is linked with the intracellular phase of endopeptidase digestion, but there is a change to alkaline phosphatase activity in the cytoplasm when exopeptidase appears in the food vacuoles. Fat is deposited in the vitellaria of the temnocephalans and glycogen is laid down in the

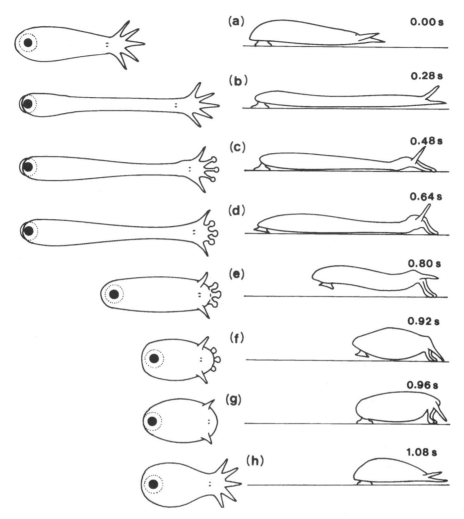

Figure 2.6(*a*)–(*h*) Successive stages in locomotion of the temnocephalan *Craspedella* sp., seen in dorsal and lateral views. The time in seconds (s) is given from zero for each stage and in the dorsal views the solid black circle represents the diameter of the peduncle. Source: reproduced from Sewell and Whittington, 1995, with permission from Taylor & Francis.

testes and musculature of the pharynx and posterior disc; in this respect temnocephalans resemble free-living predatory turbellarians (see also section 20.2).

A detailed study of attachment and 'low looping' locomotion was made by Sewell and Whittington (1995) on a species of *Craspedella* from the branchial cavity of the redclaw crayfish *Cherax quadricarinatus* (Fig. 2.6). They found that the posterior disc is able to generate suction but also has glands with ducts that convey cement to an annular zone around the periphery of the disc. The branchial chamber offers a variety of substrates and the versatility of the disc permits attachment to all of them. The sticky peripheral region allows the animal to attach itself to the gill cleaning setae, which are too small to accommodate more than part of the disc. Flat areas inside the chamber, such as the body wall, accommodate the whole disc and are suitable sites for suctorial attachment. There are five anterior tentacles and, during locomotion, temporary attachment of the anterior end is achieved by adhesive pits near the tips of the three middle ones. The pits are supplied with rhabdoids which are assumed to be adhesive.

An interesting feature of the reproductive system of temnocephalans is the presence of a resorptive vesicle (or vesicula resorbiens) which is embedded in, or adjacent to, the intestine (Fig. 2.4). This vesicle has been reported to contain yolk globules and/or degenerating spermatozoa (Jennings, 1968b; Williams, 1980) and is regarded as a recycling site for excess sexual products. The detection by Jennings (1968a) of acid phosphatase and exopeptidase in the wall of this vesicle is consistent with this suggestion, but Jennings found little or no enzyme activity in the lumen of the vesicle and suggested that initial degradation of sexual products is by autolysis, with further enzymatic breakdown after absorption of the products of autolysis by the vesicle wall. The adjacent gastrodermis may be a source of the vesicle enzymes and the intestinal lumen may provide a recycling route to the vitellarium.

Temnocephala geonoma has an unciliated epidermis (Williams, 1980) while *Diceratocephala boschmai* has an almost complete covering of cilia (Jones and Lester, 1992). In fact, juvenile *D. boschmai* are said to move about on the host by ciliary gliding. Non-motile cilia, which probably have a sensory function, are abundant where they would be expected to occur, on the tentacles and around the mouth (see, for example, Williams, 1978). Elongated motile cilia are often associated with tufts of ciliary receptors in *T. dendyi* and Williams has suggested that their activity may enhance the presumed chemoreceptive capabilities of the non-motile cilia.

There is some uncertainty about how temnocephalans spread to new hosts. The eggs are cemented to the host and there is no evidence that the hatchlings are able to swim freely or that they are able to transfer between hosts *via* the substrate. When experimentally separated from their hosts, adult temnocephalans can survive for a long time, and this is not surprising in view of the widespread occurrence of prey items similar to those available when attached to their host. Detached adult specimens of *T. brenesi* survived on a diet of small oligochaetes and crustaceans for more than 2 months (Jennings, 1968b), and *D. boschmai* survived for a similar length of time (Jones and Lester, 1992). Detached specimens of *T. brenesi* laid eggs, but *D. boschmai* only laid eggs when attached to the host.

In spite of this ability to survive prolonged separation from the host, there is no evidence that new hosts are colonized by detached individuals. The presence on the gills of their shrimp hosts of more eggs of *T. brenesi* than could be accounted for by the small number of adults present led Jennings (1968b) to suggest that the flatworm spends part of its life away from the host, but exhaustive searches of the pools favoured by the shrimps failed to reveal any free individuals, although there are some isolated, unconfirmed reports of free-living temnocephalans (see Jones and Lester, 1992). Williams (1980) commented that mature and immature specimens of *T. geonoma* never appear to relinquish contact with the host and she came to the conclusion that transfer to new isopods is probably accomplished when the hosts make contact with each other, or with cast exoskeletons from infected individuals.

2.2.3   The triclad *Bdellasimilis* and turtles

The association between the unusual triclad *Bdellasimilis barwicki* and Australian freshwater turtles has been studied by Jennings (1985) and by Sluys (1990). Ball (in Jennings, 1985) regarded the flatworm as a relative of the marine triclads, implying an invasion of fresh water from the sea, but Sluys believes it is derived from fresh-water planarians.

At the posterior end of the body the animal possesses two ventrally directed cups situated close together but each joined to the body by its own peduncle (Fig. 2.7). These are attachment organs (suckers?), which are used to fix the animals to the skin lining the limb pits of the host. Jennings (1985) identified the remains of aquatic oligochaetes and insect larvae in the gut of specimens preserved within minutes of removal from the host, and confirmed that specimens separated from the host and attached in a container by their posterior cups readily ingest living prey items of this kind. He reported that the flatworms respond to mechanical disturbance created by active prey items and to stimulation by substances liberated by potential food, by extending the body and making several rapid sweeping movements at different levels through the water. If contact is made, the prey item is trapped by curling of the forebody and within a few seconds is pushed through the open mouth into the

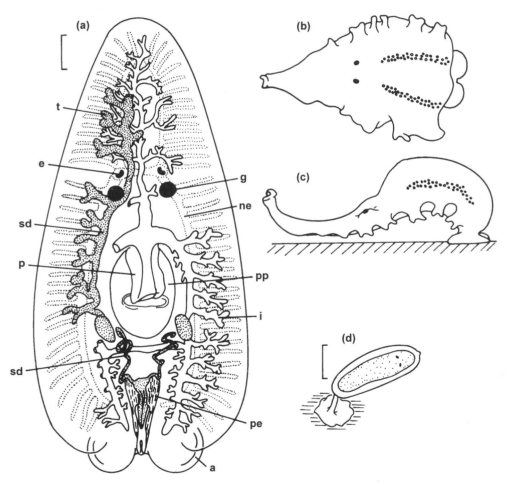

Figure 2.7 The triclad *Bdellasimilis barwicki* from the armpits of freshwater turtles. (*a*) General morphology. (*b*), (*c*) Attached living animal in feeding posture, seen from above and from the side. (*d*) Egg capsule. a = attachment cup; e = eye; g = germarium; i = intestine; ne = nerve cord; p = pharynx; pe = penis; pp = peripharyngeal cavity; sd = sperm duct; t = testis. Vitellarium not shown. Scale bars: (*a*) 1 mm; (*d*) 0.5 mm. Source: reproduced from Richardson, 1968, with permission from the Linnean Society of New South Wales.

peripharyngeal chamber that encloses the pharynx. The prey is retained in this chamber with the mouth closed; the pharynx is pushed into the body of the prey and proceeds to suck out fluids and tissue fragments. In this respect, *Bdellasimilis* differs from its free-living triclad relatives, which do not take food into the peripharyngeal chamber but protrude the pharynx through the mouth before inserting it into the prey (Fig. 1.1*b*). This change in prey handling may reflect the special problems of the ectosymbiotic life style of *Bdellasimilis*. According to Jennings, free-living triclads restrain the prey by pinning it to the substrate with the body, a procedure which might prove difficult for a symbiote living on the flexible and shifting surface of the host's limb pit skin.

Other potential sources of food were considered by Jennings. The limb pits of the host have their own epifauna and epiflora comprising bacteria, filamentous algae, protozoans, rotifers and nematodes, and patches of epidermal tissue are also shed from the pit lining, but Jennings found no evidence that *Bdellasimilis* ingests any of this material. On the other hand, *Bdellasimilis* readily takes fragments of any soft food of animal origin and this, combined with the presence in the gut of *Bdellasimilis* of fragments that seem likely to be vertebrate connective tissue, suggests that the symbiote may also feed opportunistically on shredded food material lost by the host. When feeding, the turtles extend their forelimbs, exposing the anterior limb pits and their occupants, and use the clawed forelimbs and the beak to tear their prey apart. Fragments drifting from the mouth would be readily available for capture by *Bdellasimilis*. So, *Bdellasimilis* is an ectosymbiotic predator that probably supplements its diet by opportunistic commensalism. It does not feed on co-symbiotes or host tissue.

2.3   ENDOCOMMENSALS AND ENDOPARASITES

2.3.1   Umagillid rhabdocoels and echinoderms

The elegant investigations of Shinn (references below) have provided illuminating insight into the life cycles and biology of symbiotic umagillid turbellarians with contrasting hosts and habitats. Umagillids belonging to the genera *Syndisyrinx* and *Syndesmis* inhabit the intestines of sea urchins. Shinn (1981) has demonstrated that *Syndisyrinx franciscanus* from *Strongylocentrotus* spp. ingests not only symbiotic ciliates but also host intestinal tissue, although he was unable to determine unequivocally whether the host intestinal cells were taken before or after sloughing by the host. *Syndesmis dendrastrorum* from *Dendraster excentricus* consumes intestinal tissue and material ingested by the host, but another species of *Syndesmis* from *Strongylocentrotus* spp. was found to subsist entirely on host intestinal tissue. This trend towards parasitism reaches its climax in *Fallacohospes inchoatus*, which inhabits the gut of a crinoid and has entirely lost its own gut (Kozloff, 1965; see below).

The adult *Syndisyrinx franciscanus* is hermaphrodite (Fig. 2.8*a*) and produces egg capsules one at a time at a maximum rate of one per one and a half days (Shinn, 1983a). Each capsule (Fig. 2.8*b*, *c*) from a large adult contains six to eight eggs and several hundred vitelline cells, and has a long filament fashioned from a cement gland secretion deposited at one end of the capsule. The egg capsules are white and sticky when first produced, but turn amber as tanning proceeds and lose their

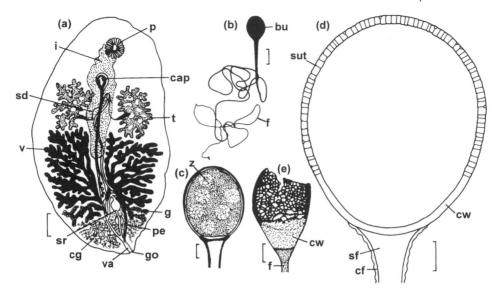

Figure 2.8 The umagillid *Syndisyrinx franciscanus* from the intestine of sea urchins. (*a*) Whole animal. (*b*) Egg capsule (bu = bulb; f = filament). (*c*) Bulb of egg capsule. (*d*) Diagram of a section through the bulb of the egg capsule and the base of the filament. (*e*) Bulb of egg capsule after hatching; the upper part of the capsule is shattered but the lower part is intact. cap = capsule inside uterus; cf = coating of filament; cg = cement gland; cw = part of capsule wall devoid of sutures; g = germarium; go = common genital opening; i = intestine; p = pharynx; pe = penis; sd = sperm duct; sf = sclerotized core of filament; sr = seminal receptacle; sut = hatching suture; t = testis; v = vitellarium; va = vagina; z = zygote. Scale bars: (*a*), (*b*) 200 μm; (*c*) 50 μm; (*d*) 25 μm; (*e*) 50 μm. Source: (*a*) redrawn from Hyman, 1951; (*b*), (*c*) drawn from photographs in Shinn and Cloney, 1986; (*d*) reproduced from Shinn and Cloney, 1986, copyright © 1986 Alan R. Liss, Inc., reprinted by permission of Wiley-Liss Inc., a subsidiary of John Wiley & Sons, Inc.; (*e*) drawn from a photograph in Shinn, 1983a.

stickiness before they are laid. Shinn showed that the egg capsules pass out of the sea urchin with the faeces and undergo most of embryogenesis in the sea.

Embryogenesis is complete after 2 months, but spontaneous hatching is rare and the embryos are able to survive for 10–12 months. Shinn offered evidence strongly suggesting that there is no intermediate host and that sea urchins become infected by ingesting egg capsules with their food. His experiments reveal that it is the intestinal fluid of the urchins and not their jaw apparatus that is involved in hatching, not, as might be supposed, by the direct action of the fluid on the shell but probably indirectly by stimulating the embryos to produce a hatching enzyme. The consequence of this is that material filling a fine network of shell sutures at the opposite end from the filament is weakened, leading to disintegration of this region of the shell (Fig. 2.8*d*, *e*) and liberation of the embryos. Capsules containing immature embryos are likely to be ingested, but Shinn found that their shells remained intact, and that the embryos survived passage through the intestine and on completing embryogenesis were capable of hatching when reingested by an urchin.

Adequate tanning is probably also important for protection of embryos in freshly laid capsules passing through the host's digestive system. The need for tanning to reach completion before laying places a constraint on the rate of capsule assembly, but this seems to have been solved in umagillids such as *Wahlia pulchella* by the provision of a separate uterine compartment ('secondary uterus'; Fig. 2.9). This chamber stores freshly made capsules while tanning proceeds, and this leaves the way clear for assembly of new capsules. Consequently, the rate of capsule release by

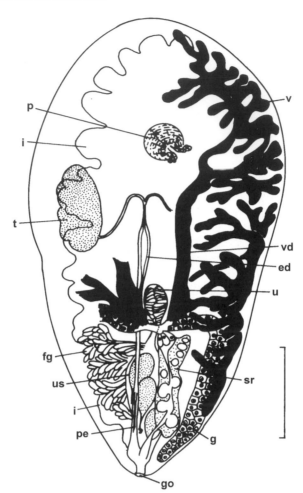

Figure 2.9 The umagillid *Wahlia pulchella*, in dorsal view, from the intestine of the sea cucumber *Stichopus californicus*. Vitellarium and germarium omitted from left side, testis omitted from right side. ed = ejaculatory duct; fg = capsule filament gland; g = germarium; go = genital opening; i = intestine; p = pharynx; pe = penis stylet; sr = seminal receptacle; t = testis; u = uterus; us = secondary uterus containing egg capsules; v = vitellarium; vd = vitelline duct. Scale bar: 0.5 mm. Source: modified from Kozloff and Shinn, 1987.

Wahlia is 10–15 times greater than that of *S. franciscanus*, which has no secondary uterus and must store eggs in the capsule-assembly part of the female tract until they are fully tanned (Shinn, 1986a). *Wahlia* inhabits the intestine of the Pacific sea cucumber *Stichopus californicus* and like *S. franciscanus* is propagated by ingestion of unhatched capsules followed by hatching in the intestine.

Some umagillids inhabit the coelom of holothuroid echinoderms, raising the question of the escape route for egg capsules. Many holothuroids eject the intestine and respiratory trees (evisceration) in the laboratory and may also do so in the field. Spontaneous autumnal evisceration has been regarded as a dependable means of escape for infective stages of parasites infecting the coelom of some holothuroids. However, Shinn (1985a) demonstrated that evisceration is not required for the escape of infective stages of *Anoplodium hymanae* (Fig. 2.10) from the coelom of *Stichopus californicus*, and there is no seasonal restriction on release. He made the interesting discovery that the perivisceral coelom communicates with the rectal lumen by way of pores in the wall of the posterior rectum. Egg capsules laid by adults in the coelom become enclosed by host coelomocytes, and the resulting so-called 'brown bodies' are ejected by way of the rectal pores and the anus. The eggs of the coelom-inhabiting monogenean *Dictyocotyle coeliaca* escape from their ray host *via* abdominal pores that provide communication between the coelom and the outside (see Table 7.1; Hunter and Kille, 1950).

Most of the stalked egg capsules of *A. hymanae* contain only one zygote and they have a single circular suture at the end opposite to the root of the stalk (Fig. 2.10c). Like the umagillids described previously, the capsules rarely hatch spontaneously, and developed embryos can survive in their capsules for 10–11 months (Shinn, 1985b). The host, *S. californicus*, is an epibenthic holothuroid that gathers detritus, probably including egg capsules of *A. hymanae*, by means of mucus-covered, oral tube feet. The food-laden mucus is then transferred by the feeding appendages to the mouth. Shinn showed that hatching is stimulated by host digestive fluids and that larvae reach the coelom by penetrating the walls of the intestine or the respiratory trees. Prominent gland cells in newly hatched larvae (Fig. 2.10d) were not found in the smallest coelomic individuals, indicating that they may have a part to play in penetration.

Shinn (1983b) found cells that he believed to be host coelomocytes in the intestine of *A. hymanae* and thought it likely that the parasite ingests coelomic fluid.

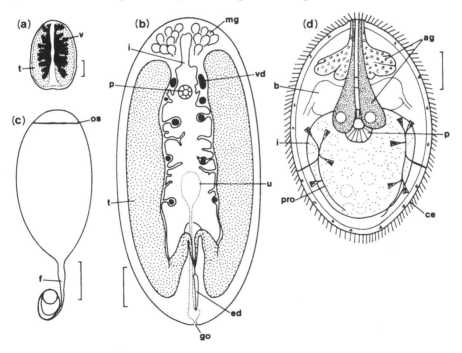

Figure 2.10 The umagillid *Anoplodium hymanae* from the coelom of the sea cucumber *Stichopus californicus*. (*a*) Whole specimen, drawn from life. (*b*) Whole animal in dorsal view, showing intestine, male reproductive system and uterus. (*c*) Egg capsule. (*d*) Newly hatched individual in dorsal view. ag = anterior glands; b = 'brain'; ce = ciliated epidermis; ed = ejaculatory duct; f = filament; go = genital opening; i = intestine; mg = male accessory glands; os = opercular suture; p = pharynx; pro = protonephridial system; t = testis; u = uterus; v = vitellarium; vd = vitelline ductule. Scale bars: (*a*) 0.5 mm; (*b*) 250 μm; (*c*) 50 μm; (*d*) 25 μm. Source: (*a*), (*b*) redrawn from Shinn, 1983b; (*c*) drawn from a photograph in Shinn, 1985a; (*d*) redrawn from Shinn, 1985b.

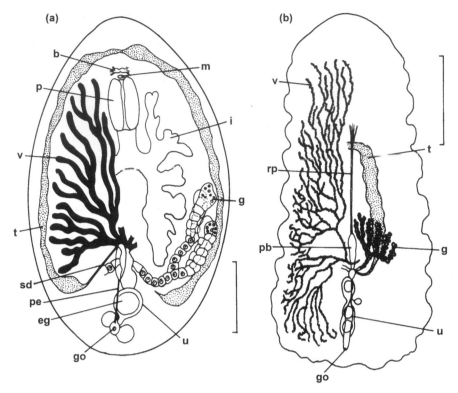

Figure 2.11 Umagillids from the intestine of crinoids. (*a*) *Desmote inops*, in ventral view. Vitellarium omitted on left side and germarium omitted on right. (*b*) *Fallacohospes inchoatus*, in dorsal view. The vitellarium conceals the gonads on the left side and has been omitted on the right side. b = 'brain'; eg = egg; g = germarium; go = genital opening; i = intestine; p = pharynx; pb = penis bulb; pe = penis; rp = penis bulb retractor muscle; sd = sperm duct; t = testis; u = uterus; v = vitellarium. Scale bars: (*a*) 250 μm; (*b*) 0.5 mm. Source: redrawn from Kozloff, 1965.

Some umagillids occur in the intestines of suspension-feeding crinoids. According to Kozloff (1965), *Desmote inops* and *Fallacohospes inchoatus* are both found in the intestine of *Florometra serratissima*; *D. inops* has a prominent pharynx and branched intestine while *F. inchoatus* lacks a gut (Fig. 2.11). The latter reaches a length of 2.5 mm and a breadth of 1.5 mm but is markedly flattened, being only 220 μm thick at the most. This flattening and concomitant increase in surface area must greatly facilitate the uptake of nutrients *via* the ciliated epidermis and reduce diffusion distances for these nutrients within the body. Moreover, a further increase in area is apparently achieved by conspicuous ruffling of the body edges, a development reminiscent of some gyrocotylid fish parasites (section 7.2.2).

In spite of its small size (length about 1 mm) and well-developed gut, *D. inops* is also flattened. This may indicate that this organism supplements the food taken in by the gut with nutrients absorbed through the surface. *F. inchoatus*, which, according to Kozloff (1965) is closely related to *D. inops*, may have taken this development one step further, absorbing all its nutrient requirements through the surface and losing its gut.

Shinn (1986b) found that *F. inchoatus* produces egg capsules typically containing two zygotes that hatch spontaneously after a development period in the sea of 40–42 days. This suggests that suspension-feeding crinoids are infected by ingesting free-swimming larvae taken in with the host's feeding current, rather than by eating capsules containing embryos, as in umagillids exploiting deposit-feeding hosts. Shinn suggested that the mode of infection employed by crinoid-inhabiting umagillids may be primitive for the family.

Jennings and Cannon (1987) identified physiologically active intracellular haemoglobins in four separate umagillid rhabdocoels that inhabit the gut of holothurians. They found differences between these species in the location of the pigment. In *Cleistogamia longicirrus* and *C. heronensis* the haemoglobin is located in the parenchymal cells around the brain and pharynx and in the pharyngeal muscles; in *C. heronensis* it also occurs around the germarium and the secondary uterus. In *Paranotothrix queenslandensis* it occurs in the subepidermal musculature, while in *Seritia stichopi* it is uniformly distributed in the body. Jennings and Cannon pointed out that diffusion can satisfy the oxygen requirements of even the most deep-seated cells in free-living and ectosymbiotic flatworms living in well aerated habitats, but may not be sufficient to maintain similarly-sized endosymbiotes living in oxygen-poor environments. They suggested that the gut of holothurians may be such an environment and that intracellular haemoglobin would give the symbiote an advantage over host gut cells in the competition for oxygen, facilitating the diffusion of the gas into their bodies. This hypothesis is consistent with the lack of pigment in small individuals of three of the species; small specimens have a larger surface area / volume ratio and a correspondingly smaller problem in meeting their oxygen needs by diffusion. The hypothesis is also consistent with the observation that the endosymbiote with the smallest haemoglobin content lives in the hind gut, where oxygen levels are likely to be higher because of the proximity of the openings of the respiratory trees and the absence of host digestive activity. Jennings and Cannon also maintained that haemoglobin in these animals might have a secondary role as an oxygen store for use during temporary hypoxia. It will be shown later (Chapter 4) that some monogenean ectoparasites whose hosts inhabit anoxic environments have found a different way of enhancing oxygen uptake by performing breathing movements.

Seritia stichopi contains more haemoglobin than its relatives and even small specimens are pigmented. This species appears to be much more dependent on its haemoglobin for survival, since it is the only species studied by Jennings and Cannon (1987) that died after treatment with carbon monoxide and conversion of its haemoglobin to carboxyhaemoglobin. Jennings and Cannon pointed out that hosts contain large numbers of *S. stichopi* and suggested that larger quantities of haemoglobin may support denser populations of this flatworm. Intracellular haemoglobins are found in other turbellarian symbiotes (see below).

2.3.2 Graffillid rhabdocoels and molluscs

Many graffillid rhabdocoels are endosymbiotes in the gut of bivalve and gastropod molluscs and they have special features that appear to be related to the peculiarities of the molluscan digestive system. Jennings and Phillips (1978) and Jennings (1981) have studied in detail the fascinating relationship between the graffillid *Paravortex scrobiculariae* and its host the intertidal bivalve *Scrobicularia plana*.

The host is found at a depth of 15–20 cm in mud near high water mark and is submerged for only 2–3 h in each 12 h tidal cycle. *S. plana* is a deposit- and filter-feeder, active while submerged at high water; specimens collected as the tide recedes have full stomachs and the organic contents of the stomachs show extensive extracellular digestion. Semi-digested soluble and finely particulate organic material enters the main and secondary ducts of the digestive gland and then the blind-ending distal tubules. The amount of material following this route increases with time until about low water, when the stomach is usually empty. Cells lining the distal digestive tubules phagocytose food, forming new food vacuoles or phagosomes, and these fuse with primary lysosomes containing enzymes, so forming heterolysosomes where intracellular digestion occurs. The heterolysosomes shrink during the digestive cycle but at the end of the process contain one or more residual bodies still with much enzymic activity. The cells then disintegrate and fragments containing residual bodies leave the tubules. The host enzymes in the residual bodies initiate extracellular digestion in the first phase of the next digestive cycle.

P. scrobiculariae lives in the host's gut and is hermaphrodite and viviparous. The related endosymbiote *Paravortex gemellipara* from the digestive gland of *Modiolus demissus* is shown in Fig. 2.12. In *P. scrobiculariae* as many as 40 capsules, each containing two embryos and numerous vitelline cells, enter the parenchyma from the female atrium; fully developed embryos leave the capsules and move freely in the parenchyma. They enter the intestine and presumably escape from the parent through the mouth.

Adult symbiotes in the host's gut have a pronounced tidal migration pattern. On the ebb tide they swim forward, using their external ciliation, from the anterior intestine of the host through the stomach to the secondary ducts of the digestive gland. The lack of attachment organs in these symbiotes undoubtedly reflects the fact that transport of gut contents in the host is slow, being brought about by cilia not by muscle contraction. They remain in the digestive gland during low water and migrate back to the intestine during the flood tide. *P. scrobiculariae* has an anterior subterminal mouth, opening ventrally through a small ciliated buccal cavity into a muscular pharynx. A short ciliated oesophagus links the pharynx with the simple saccate intestine. *P. scrobiculariae* feeds partly on semi-digested components of the host food, which it ingests as it travels from the stomach into the digestive gland, and partly on residual bodies and cellular debris released from the host's blind-ending digestive tubules at the end of the digestive cycle. Thus, the feeding habits of *P. scrobiculariae* are partly commensal and partly parasitic, but the work of Jennings and Phillips (1978) indicated that there is an extra dimension to the parasitic aspect of its life cycle.

P. scrobiculariae kept *in vitro* in yeast and bacterial suspensions ingested little and did not survive for long. Jennings and Phillips noted the absence of extrinsic and intrinsic glands associated with the pharynx and the lack of gland cells among the gastrodermal cells (cf. temnocephalans, section 2.2.2). Moreover, they were unable to demonstrate enzymic activity in the pharynx and oesophagus, apart from acid phosphatase in the intrinsic and extrinsic pharyngeal muscles, and noted that specimens kept in filtered sea water for 4 days had only small amounts of endogenous enzymes in the gastrodermal cells. In marked contrast, individuals collected from the host digestive gland had greatly enhanced enzymic activity. It seems that *P. scrobiculariae* ingests the still-active host enzymes released by cell disintegration, or by separation of cytoplasm by constriction, at the end of the intracellular phase of the host's digestive cycle. These host enzymes are able to continue the digestion of host food in the seclusion of the symbiote's own gut, releasing nutrients, which are

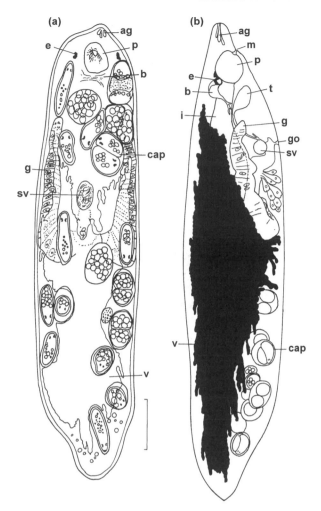

Figure 2.12 The graffillid *Paravortex gemellipara* from the digestive gland of the bivalve mollusc *Modiolus demissus*. (*a*) Whole animal drawn from life with slight coverslip flattening. (*b*) Animal viewed from the left side. ag = anterior glands; b = 'brain'; cap = capsule containing embryos; e = eye; g = germarium; go = genital opening; i = intestine; m = mouth; p = pharynx; sv = seminal vesicle; t = testis; v = vitellarium. Scale bar: 200 μm. Source: reproduced from Ball, 1916.

all available to the symbiote. This special kind of endosymbiosis, made possible by the peculiar digestive physiology of molluscs, has permitted the symbiote to greatly reduce its own output of digestive enzymes.

Other graffillids exploit gastropod molluscs in a similar way and the host enzymes sequestered by these endosymbiotes will be an ideal cocktail to deal with the range of particular foods ingested by their host.

Because of the metabolic dependency on the hosts for digestive enzymes, graffillids are unable to survive away from their hosts for more than a few days. This contrasts sharply with ectosymbiotes such as temnocephalans and bdellourid triclads, which produce the normal complement of digestive enzymes and can survive for several months and even breed away from their hosts (see above).

The bivalve *Scrobicularia plana*, the host of *P. scrobiculariae*, lives in black, deoxygenated mud near high water mark. It is open and feeding for a relatively short period at high water and is closed for the rest of the time. When feeding, the incurrent water also transports oxygen, and the mud around individual molluscs is oxidized to a light grey colour; when the shell valves are closed, oxygen availability to the tissues of the mollusc is reduced. *P. scrobiculariae* is most active and feeding at low tide when the oxygen availability is reduced, and it appears to have two adaptations to permit it to remain active in these circumstances. First, it has a physiologically active haemoglobin in the mesenchyme cells, in the 'brain' and around the extrinsic musculature of the pharynx (Phillips, 1978), and, secondly, there is evidence that glycolysis is of major importance in metabolism (Jennings and LeFlore, 1979). There is insufficient haemoglobin to act as an oxygen store, but it may allow the graffillid to compete successfully with host tissue in extracting what little oxygen there is available in the host's body (Jennings, 1981). The cockle *Cerastoderma edule* has a related endosymbiote, *Paravortex cardii*, but lives near low water mark near the surface of sand, and is submerged and well aerated for most of the time. The absence of haemoglobin and the reduced importance of glycolysis in *P. cardii* probably reflect the greater abundance of oxygen.

It is interesting that the kidney of *Scrobicularia plana* is parasitized by an adult fellodistomatid digenean parasite *Proctoeces subtenuis*, which is also found in the hind gut of sparid and labrid fishes. Digeneans from the mollusc contain haemoglobin but those from the fishes do not (Freeman and Llewellyn, 1958; see also section 19.2).

2.3.3 Fecampiids and crustaceans

The life cycle of *Fecampia erythrocephala* was investigated by Caullery and Mesnil (1903). The cylindrical parasite does not exceed 10–12 mm in length and between one and nine individuals were observed inside the haemocoel of small shore crabs (*Carcinus maenas*) less than 25 mm in carapace width (Fig. 2.13a). Young parasites have a pharynx and intestine, but the former disappears and only the intestine persists as a progressively diminishing space in the body (Fig. 2.13b). Each parasite is hermaphrodite but, before laying eggs, escapes from the crab, crawls beneath a stone and with secretion from glands opening through the ciliated epidermis secretes a cocoon around itself (Fig. 2.13c). Egg capsules, each containing two embryos (Fig. 2.13d), are then laid *via* a posterior genital opening and develop in the space between the spent adult and the wall of the cocoon. Each larva is about 220 μm long and develops a pair of eyes and a mouth, pharynx and intestine. It is presumably these ciliated larvae that penetrate the crab, after escaping from their cocoon, but how penetration is achieved is not known.

Kronborgia differs in important ways from *Fecampia*. Christensen and Kanneworff (1964) described the finding by a student of cylindrical red worms, 3–5 mm long, in the haemocoel of the tube-dwelling amphipod *Ampelisca macrocephala*. These worms lacked eyes, mouth, pharynx and intestine but possessed a distinctive gonopore at one end (Fig. 2.14a, c). When removed from the host they swam around actively in sea water using their epidermal cilia and remained alive for more than 1 week. While searching for more material they found quite different worms in the body cavity of another host individual. These were up to 40 mm long, white with no gonopore, but, like the red worms, they lacked eyes and gut (Fig. 2.14b). When removed from the host the white worms were fragile, sluggish and unable to swim. They perished within a few days in sea water, although some started to secrete

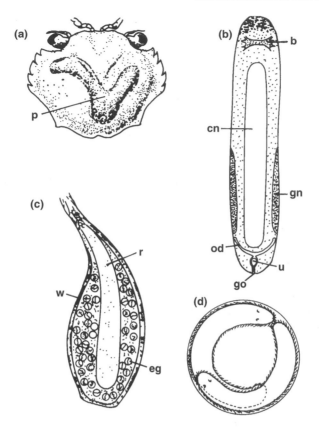

Figure 2.13 *Fecampia erythrocephala*. (*a*) Folded parasite (p) seen through the carapace of *Carcinus maenas*. (*b*) Diagrammatic representation of the main features of the adult parasite. (*c*) Cocoon. (*d*) Egg capsule containing two ciliated larvae. b = 'brain'; cn = central cavity; eg = egg capsules; gn = gonads; go = genital opening; od = oviduct; r = remains of adult; u = uterus; w = wall of cocoon. Source: reproduced from Caullery and Mesnil, 1903.

around themselves a long straight cocoon in which deposition of egg capsules was occasionally observed.

They eventually discarded the hypotheses that the red and white worms represented separate species and that they represented juveniles and adults respectively of a single species. The correct explanation, that the red worms were males and the white worms females of the same species, was surprising because most turbellarians are hermaphrodite, but was shown to be so by serial sectioning and anatomical study (Fig. 2.14*c–e*). Separate sexes (gonochorism or dioecism) is rare in platyhelminths, but we will meet it in some tapeworms (section 10.6), in schistosome digeneans (Chapter 18) and in some didymozoideans (section 19.1).

The finding in *Kronborgia* that the two sexes were often in separate host individuals implied that copulation occurs outside the host and this was confirmed later when the male was observed to enter the cocoons (Fig. 2.15). This parasite was sufficiently different from previously known fecampiids to be assigned as the type species of a new genus with the name *Kronborgia amphipodicola*.

Williams (1993) observed that some gutless females of *K. isopodicola*, a New Zealand fecampiid parasitizing amphipods, wrap themselves closely around the gut of their host, an attitude that may well enhance their ability to absorb

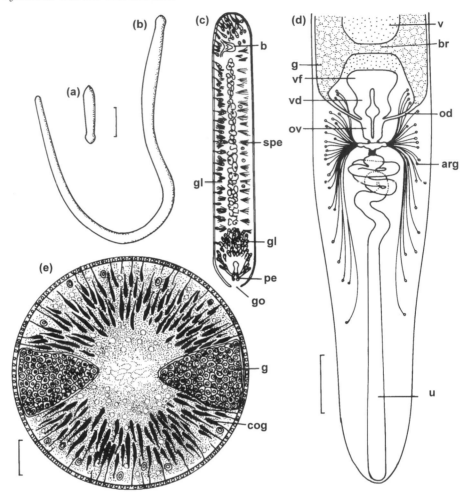

Figure 2.14 Morphology and anatomy of *Kronborgia* spp. (*a*) Male and (*b*) female of *K. amphipodicola*. (*c*) Schematic sagittal section of male *K. amphipodicola*. (*d*) Posterior end of female *K. pugettensis* showing the reproductive system. (*e*) Transverse section through the germarium of *K. caridicola*. arg = accessory reproductive glands; b = 'brain'; br = bridge connecting paired germaria; cog = cocoon gland; g = germarium; gl = gland; go = genital opening; od = oviduct; ot = ootype; ov = ovovitelline duct; pe = penis; spe = spermatozoa; u = uterus; v = vitellarium; vd = vitelline duct; vf = vitelline funnel. Scale bars: (*a*), (*b*) 2 mm; (*d*) 500 μm; (*e*) 1 mm. Source: (*a*)–(*c*) reproduced from Christensen and Kanneworff, 1964 with permission from the Marine Biological Laboratory, Helsingør, Denmark; (*d*) reproduced from Shinn and Christensen, 1985, with permission from Cambridge University Press; (*e*) reproduced from Kanneworff and Christensen, 1966, with permission from the Marine Biological Laboratory, Helsingør, Denmark.

nutrients, made available by the host's digestive processes, through their ciliated epidermis.

Christensen and Kanneworff (1965) delved further into the biology of *K. amphipodicola*. As expected, the mature worms escape from the host (Fig. 2.15). This is a traumatic event for the host, involving dislodgement of the gut so that the parasite can emerge at the site of the anus. During the emergence of the female the host becomes motionless, presumably under the influence of secretions from the worm, and dies soon afterwards. Escape of the female takes several minutes and host immobility may be essential to prevent damage to the delicate, partially emerged

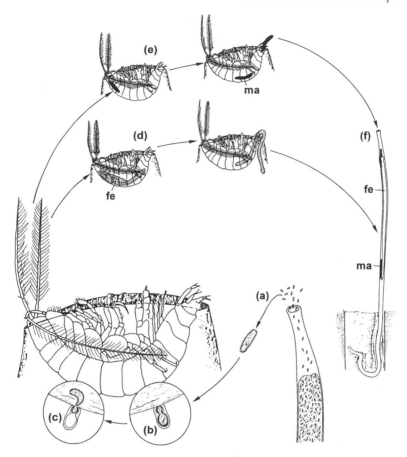

Figure 2.15 Life cycle of *Kronborgia amphipodicola*, parasitizing ampeliscid amphipods. Ciliated free-swimming larvae (*a*) escape from the cocoon, attach themselves to the host's carapace and secrete a cyst (*b*). The larva within perforates the host cuticle and enters the haemocoel (*c*). Females (*d*) and males (*e*) mature in the haemocoel and escape *via* the anus of the host. The female secretes a cocoon (*f*) within the host's tube. The male enters the cocoon and fertilizes the female. Egg capsules are deposited in the cocoon and the spent female dies. fe = female; ma = male. Source: based on a drawing by Olsen in Christensen and Kanneworff, 1965.

female worm. Immediately after leaving the host, the female begins to secrete a cocoon which is longer than its own body (Fig. 2.15), using glands opening through the ciliated epidermis on to the body surface (Fig. 2.14*e*). The worm attaches the base of the cocoon to the wall of the host tube and shapes the cocoon by moving backwards and forwards within it and by revolving around its body axis. The cocoon is completed in less than 30 min, and females 2–3 cm long usually produce cocoons 4–6 cm in length, projecting 2–3 cm out of the host tube into the water. One or more males enter the cocoon through the trumpet-shaped upper opening and in some unknown way inseminate the female. They have been observed ejecting sperm close to the female, and after releasing all their sperm the males leave the cocoon and die. Williams (1990) found spermatozoa of *K. isopodicola* between cells of the parenchyma of the female and between the interdigitating cell processes forming the wall of the ovovitelline duct. She suggested that insemination may take place *via* the epidermis, spermatozoa then migrating through the body to the female reproductive tract.

In the related *K. pugettensis*, Shinn and Christensen (1985) noted that the vitellarium in the female is extensive and diffuse, and that vitelline cells collected by a funnel are fed into two vitelline ducts, each of which receives an oviduct from one of the two germaria (Fig. 2.14d). They identified a central compartment, which they believe may be the ootype, where the egg capsules are assembled. Like *Fecampia*, each egg capsule contains two egg cells; as the capsules are made they are stored in a blind-ending uterus. They leave the body of the female through a rupture in the body wall and accumulate in the cocoon. Shinn and Christensen noted that the capsule does not develop the brown colour indicative of the tanning process so widespread in the egg capsules of other platyhelminths, and suggested that this may reflect the extra protection provided by the wall of the cocoon.

As the female of *K. amphipodicola* spawns it gradually moves upwards in the cocoon and its body becomes progressively shorter and thinner (in one case a 20–25 mm long female was reduced when spent to a length of only 7 mm and a thickness of 0.5 mm). The exhausted, dying female then crawls out of the cocoon through the upper opening, leaving behind a mass of capsules clinging together in a sticky matrix in the upper region of the cocoon.

The egg capsules of *K. amphipodicola* begin to hatch 50–60 days after spawning. The ciliated larvae are active swimmers for 2–3 days. If they fail to find a host at the end of this period, they become sluggish and no longer infective. The larvae of *K. isopodicola* have two rhabdomeric eyes, each with a remarkable curved reflector, consisting of an array of overlapping crystalline platelets, which may act to amplify and concentrate the light (Watson, Williams and Rohde, 1992). A strikingly similar arrangement has been reported in polystomatid monogeneans (section 7.1.5).

The mode of entry of the larvae of *K. amphipodicola* into the crustacean host is unique. After making contact, usually with the host's dorsal carapace, the larva attaches by the anterior end and encysts, a process that takes less than 5 min (Fig. 2.15). The cyst is thick-walled, shaped internally like an hour glass. The cyst cavity concentrates secretions, probably from anterior gland cells in the larva, on to the tiny circle of carapace enclosed by the cyst. A hole with a diameter of 4–8 μm appears in the carapace and the larva, complete with cilia, takes about 1 min to squeeze through this aperture into the host's haemocoel. Christensen and Kanneworff (1965) had no exact figures for the time needed to perforate the cuticle but observed that many larvae were swimming around in the haemocoel of small hosts on the morning after the same hosts had been exposed to free-swimming larvae on the previous afternoon. The stylet-bearing metacercariae of some digeneans (xiphidiocercariae) also penetrate their arthropod hosts from an external cyst (section 16.3.4(a)), but the larva of *K. amphipodicola* has no mechanical aid (stylet or teeth) with which to abrade the host cuticle and the larval secretions must be remarkably potent to soften crustacean cuticle so swiftly. The activity of the parasites declines as they grow.

Kanneworff and Christensen (1966) record showing the cocoon of *K. amphipodicola* to Professor Gunnar Thorson who stated that the Zoological Museum in Copenhagen housed a number of tubes which had not been assigned to any animal group but which reminded him of the cocoon of *K. amphipodicola*. Examination of this material revealed several distinct cocoon types, each type probably corresponding to an unknown parasitic fecampiid. One of these cocoon types was exceptionally large (Fig. 2.16), 0.5–0.75 m in length, and this was later identified as the work of a new species, *K. caridicola*, from the body cavity of North Atlantic shrimps. The females

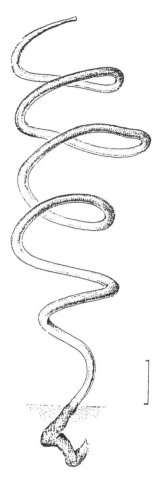

Figure 2.16 Cocoon of *Kronborgia caridicola*. The thickened, twisted lower end anchors the cocoon in the substrate. Source: reproduced from Kanneworff and Christensen, 1966 (drawn by P. H. Winther), with permission from the Marine Biological Laboratory, Helsingør, Denmark.

were correspondingly large, two individuals collected by Kanneworff and Christensen (1966) measuring 39 and 21 cm in length respectively. Christensen and Kanneworff (1967) went on to describe the cocoons of six other unknown fecampiids. One of these cocoon types was similar in length to that of *K. caridicola* but highly coiled; these coiled cocoons were attached to prosobranch shells, including one containing a hermit crab (Fig. 2.17), suggesting that hermit crabs might serve as hosts.

Ultrastructural evidence has been invoked to support a close relationship between fecampiids and the major groups of parasitic platyhelminths (Watson and Rohde, 1993), but DNA-sequencing evidence is said to indicate that fecampiids have no close relationship with any other platyhelminth taxon, including the rhabdocoels (Rohde *et al.*, 1994). That the fecampiids are highly specialized is without doubt. Their unusual features include a two-phase life cycle, with an endoparasitic (parasitoid) growth stage and a non-parasitic, non-feeding reproductive stage, the absence of a gut with absorption of nutrients through the ciliated cellular epidermis, and separate sexes with extreme sexual dimorphism. Their biology provides no clues that might be helpful in elucidating their obscure ancestry.

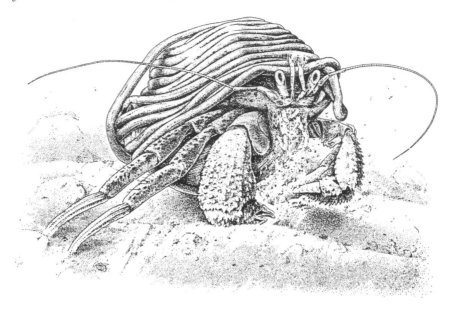

Figure 2.17 Fecampiid cocoon attached to a shell of *Natica clausa*, inhabited by the hermit crab *Pagurus trigonocheirus*. Source: reproduced from Christensen and Kanneworff, 1967 (drawn by K. Olsen), with permission from the Marine Biological Laboratory, Helsingør, Denmark.

2.3.4 Pterastericolid rhabdocoels and starfish

The pterastericolid rhabdocoels inhabit the alimentary systems of starfish (asteroid echinoderms). Jennings and Cannon (1985) studied the nutritional physiology of *Triloborhynchus astropectinis*, *Pterastericola australis* and *P. vivipara*, all of which inhabit the hepatic caeca of their respective starfish hosts. They found that they have a parasitic lifestyle reminiscent of that of graffillid endosymbiotes of molluscs (section 2.3.2). Like these graffillids, the pterastericolids have no morphologically distinct gastrodermal gland cells of their own and rely on the host for their supply of digestive enzymes. However, the echinoderm hosts do not provide a supply of enzyme-rich heterolysosomes as part of their digestive cycle, as in the bivalve molluscs, and the pterastericolids acquire their host enzymes by ingesting whole host caecal cells. These host cells not only contain enzymes but are also important storage sites for lipid and glycogen. Thus, all the dietary requirements of the flatworm and the enzymes necessary to deal with them are provided by the host's caecal cells.

According to Jennings and Cannon, the high lipid content of the food of pterastericolids is directly responsible for the high lipid content of the parasite's gastrodermis, a feature more typical of free-living than endosymbiotic flatworms (section 20.2). Jennings and Cannon also made the point that the well-known regenerative abilities of the asteroids are probably important in the toleration of these caecal-feeding flatworms by their hosts.

Another feature of pterastericolids that they share with some umagillid and graffillid endosymbiotes, is the presence of red pigment, mainly around the 'brain' and pharynx. Jennings and Cannon (1985) showed that this pigment in pterastericolids is haemoglobin and that, although it is converted to carboxyhaemoglobin by carbon monoxide, this has no apparent deleterious effect on the flatworm. Jennings and Cannon pointed out that, in many echinoderm organs, metabolism appears to be largely anaerobic, perhaps because of the lack of a vascular system for internal transport of oxygen. They suggested that oxygen availability may be low in the pyloric caeca of the host when the caecal epithelium is active, and may be further reduced by metabolic activity in the neighbouring gonads. In such a situation, the presence of haemoglobin may favour the parasite in the competitive struggle for oxygen with the host tissue, which lacks haemoglobin. The fact that parasite haemoglobin is not essential for survival *in vitro* suggests that its use may be restricted to times of severe oxygen stress.

Triloborhynchus astropectinis from the hepatic caeca of *Astropecten irregularis* has been described by Bashiruddin and Karling (1970; Fig. 2.18). It is about 1 mm long

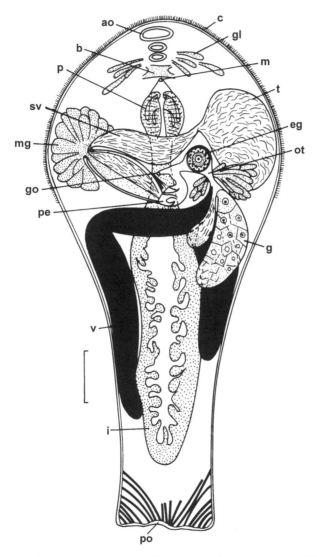

Figure 2.18 *Triloborhynchus astropectinis*, in ventral view, from the hepatic caeca of the starfish *Astropecten irregularis*. ao = apical organ; b = 'brain'; c = cilia; eg = egg; g = germarium; gl = glands; go = genital opening; i = intestine; m = mouth; mg = male accessory glands; ot = ootype; p = pharynx; pe = penis; po = posterior attachment organ; sv = seminal vesicle; t = testis; v = vitellarium. Scale bar: 100 μm. Source: modified from Bashiruddin and Karling, 1970.

with external ciliation restricted to the bright red anterior third of the body. The posterior end of the body is muscular and serves for attachment to the caecal epithelium. There is a muscular pharynx and a saccate intestine. When detached the animal can swim using its cilia, but it can also move in a leech-like manner using the posterior attachment organ and apical pockets anterior to the mouth. This is a striking convergence with the Monogenea, as is also the presence in *T. astropectinis* of a sclerotized male copulatory apparatus comprising a hollow tube with two blunt, finger-like processes and a sharp hook. Copulation has been observed and mutual insemination occurs *via* the hollow copulatory tube. It is presumed that eggs pass out in the host's faeces and lead to the infection of other starfish, but this has not been confirmed.

P. vivipara has a higher incidence per infected host than *P. australis* and *T. astropectinis*; Jennings and Cannon (1985) suggested that this might be related to the parasite's viviparous habit. Some fertilized eggs are laid, but others develop and hatch within the parenchyma of the parent and presumably reach the host's caecal lumen *via* the gut of the parent or by rupturing the body wall. We will meet this phenomenon of 'autoinfection' frequently in the parasitic platyhelminths.

An interesting feature of *Triloborhynchus psilastericola* is that immature individuals live in the coelom, between the branches of the hepatic caeca (Jespersen and Lützen, 1972). How they reach this site is unknown. These juveniles have a full covering of cilia but, after leaving the coelom and reaching maturity in the hepatic caeca, these cilia are lost, except for those carried on a ring-shaped, circumoral region.

2.3.5 The rhabdocoel *Acholades* and starfish

This relationship deserves special attention because it has a number of unique features that set it apart from other symbiotic relationships in the platyhelminths. *Acholades asteris* is the only known representative of the Acholadidae and parasitizes the tube feet of the starfish *Coscinasterias calamaria*. Its nutritional biology has been investigated by Jennings (1989). The organism has no mouth, pharynx or intestine and is a parasite, but it lives not in the lumen of the tube foot but embedded in its wall. The tube foot wall contains two layers of connective tissue and it is within the outermost of these two layers that the parasite is found, forming a bright yellow or orange bulge 4–5 mm in length near the base of the foot on the side opposite to the longitudinal nerve strand. This may have significance because the parasite does not interfere with the function of the tube foot in any detectable way.

The remarkable feature of the relationship is that the parasite conducts extracorporeal digestion of the host's connective tissue by the secretion of enzymes from gland cells in the epidermis and from subepidermal gland cells with ducts opening through or between the ciliated epidermal cells. These gland cells are particularly abundant at the end of the animal closer to the root of the tube foot; since the reproductive system of the parasite is at the opposite end and since the turbellarian reproductive system is typically posterior, this highly glandular zone is regarded as anterior. This anterior end also has deep epidermal invaginations which Jennings described as acting as miniature guts, drawing into their lumina strands of host connective tissue for digestion. Partially digested host tissue is then taken up by pinocytotic vesicles lying between the epidermal cell cilia; these vesicles fuse to form epidermal phagosomes where digestion is completed intracellularly. Food uptake occurs over the whole body surface but is especially active in the anterior region.

As Jennings has pointed out, this unique feeding method is fundamentally different from that of other gutless platyhelminths such as tapeworms because, in the latter, apart from a digestive contribution made by some membrane-bound surface enzymes, organic nutrients already liberated by the host's digestive activity are absorbed through the surface (Chapter 9). In *Acholades*, the epidermis and subepidermal glands constitute a 'gut', comparable in most functional respects with the endoderm (gastrodermis) of other turbellarians, secreting enzymes, taking up partly digested host tissue by pinocytosis and completing digestion intracellularly. It is unusual, but not unknown, for a digestive epithelium engaged in extensive pinocytosis to retain cilia, but Jennings suggested that their retention in *Acholades* reflects the need to circulate and mix the enzymes and to transport semi-digested broth from the predominantly digestive anterior region to the rest of the body surface for absorption. Jennings pointed out that, provided his interpretation of the polarity of the animal is correct, then the major gut-like invaginations of the anterior epidermis occupy the same position as the embryonic stomodaeal invaginations of other turbellarians; this may represent an alimentary system that has not developed along orthodox lines and is supplemented by modifications of the epidermis covering the rest of the body.

The larval stage and the life cycle of *Acholades asteris* are unknown.

2.4   TURBELLARIAN PARASITES OF FISHES

There is a trend among turbellarians to develop associations with fishes, especially with the skin and gills of these hosts (Table 2.1). There is uncertainty about the relatives of *Ichthyophaga* and udonellids, but *Micropharynx* (Fig. 2.19*a*) is undoubtedly a triclad and a rhabdocoel associated with fishes was placed provisionally in the graffillid genus *Paravortex* by Kent and Olson (1986). It seems likely that all of these relationships with fishes have evolved independently.

Records of host epithelial cells in the intestine of *Micropharynx parasitica* indicate a parasitic relationship with epidermis feeding (Ball and Kahn, 1976), and it is assumed that *Paravortex* (?) sp. also feeds on epithelial tissue (Kent and Olson, 1986). *Micropharynx* is attached to its host (*Raja* spp.) by a posterior attachment zone, which is the flat terminal area of a cylindrical feature produced by the rolling inwards of the posterolateral edges of the body (Fig. 2.19*b*, *c*). According to Ball and Kahn (1976), this attachment organ may be fixed deep in the host tissue and individuals often cannot be removed without damaging the organ. Nothing is known about the attachment mechanism. Kent and Olson (1986) described the

Table 2.1 Some turbellarian symbiotes associated with fishes

Symbiote	Host	Location	Reference
Micropharynx parasitica	*Raja radiata*	Dorsal body surface	Ball and Kahn, 1976
Paravortex (?) sp.	Various marine teleosts	Surface of skin and gills	Kent and Olson, 1986
Ichthyophaga subcutanea	*Cephalopholus pachycentron*	Within gill connective tissue	Menitskii, 1963
Udonella caligorum	Various marine teleosts	Attached to *Caligus* sp. on skin or gills	Sproston, 1946

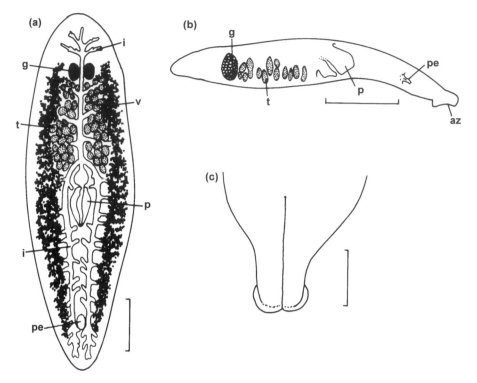

Figure 2.19   The triclad *Micropharynx parasitica* from the skin of the ray *Raja radiata*. (*a*) Whole animal. (*b*) Animal in side view. (*c*) Ventral view of folded posterior end used for attachment. az = adhesive zone; g = germarium; i = intestine; p = pharynx; pa = posterior attachment zone; pe = penis; t = testis; v = vitellarium. Scale bars: 1 mm. Source: modified from Ball and Kahn, 1976.

attachment of *Paravortex* (?) sp. to host epidermis by the protruding pharynx, but they made no reference to any other means of attachment.

Udonellids are found attached by a posterior attachment disc to the surfaces of copepods, which themselves are parasitic on the skin and gills of marine fishes (Fig. 2.20), but it seems likely that udonellids are themselves parasites and feed on their fish hosts. There is no evidence that udonellids ingest blood, even in those living on the gills, and Sproston (1946) believed that udonellids subsist on mucus and gill–skin epithelium dislodged during the feeding activities of the copepods. However, there is no confirmation that udonellids exploit the copepod's untidy feeding habits in this way. Nichols (1975) observed that *Udonella caligorum* 'oscillates' and sometimes everts the pharynx in the presence of freshly-hatched *Artemia* larvae and that pharynx eversion does not occur in the presence of flesh or scales from the fish host, but it seems unlikely that there would be sufficient animal food on the skin and gills of fishes to support a predatory rather than a parasitic way of life.

The pharynx (Fig. 2.21*a*, *b*) lies in a pharyngeal cavity and can be protruded through the mouth (Sproston, 1952; Ivanov, 1946). The distal opening of the pharynx is guarded by a circular lip and there are gland cells embedded in the pharynx wall with openings into the pharyngeal lumen proximal to the lip. The resemblance to the pharynx of skin-feeding monogeneans like *Entobdella soleae* (see Figs 4.2, 4.9*a*) is striking and lends some support to the suggestion that *Udonella* is an epidermis

Figure 2.20   A caligid copepod carrying on one of its two egg sacs a single adult *Udonella caligorum* and, on its body, bunches of udonellid eggs. Scale bar: 1 mm.

feeder. The homogeneous granular contents found in the gut by Ivanov (1952) are also consistent with a diet of fish epidermis. It is possible, indeed probable, that udonellids feed directly and independently on the skin in the manner of a monogenean fish-skin parasite (section 4.4).

There are two anterior adhesive sacs in *Udonella*, one on each side of the head region, and each contains a glandular cushion that supports a densely packed array of gland duct terminations from gland cells lying lateral to the pharynx (Fig. 2.21*b*). These cushions can be pushed out of the adhesive sacs and are capable of attaching the animal to the substrate (Nichols, 1975). The posterior adhesive disc (Fig. 2.21*c*) is also well supplied with glands and, according to Ivanov (1952), there is no doubt that attachment is achieved by secreting cement rather than by suction, and the organ does not contain hooks at any stage of development.

Although *Ichthyophaga subcutanea* is found on the gills, it is an endoparasite since it is enclosed in a cyst in the subcutaneous connective tissue of the gill lamellae (Menitskii, 1963). According to Menitskii, *Ichthyophaga* is a blood feeder; it has an anterior mouth, a pharynx housed in a pharyngeal cavity and an intestine lined by an apparently syncytial gastrodermis.

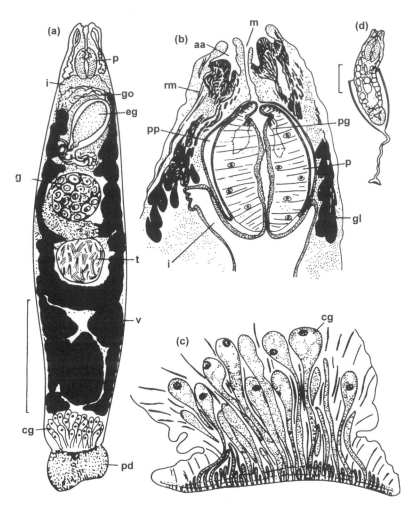

Figure 2.21 *Udonella caligorum.* (*a*) Adult. (*b*) Frontal section through the anterior region of the adult. (*c*) Sagittal section through the posterior attachment organ. (*d*) Larva emerging from egg. aa = anterior adhesive sac; cg = cement glands; eg = egg; g = germarium; gl = gland cells; go = genital opening; i = intestine; lp = pharyngeal lips; m = mouth; n = nucleus; p = pharynx; pd = posterior attachment disc; pg = pharyngeal gland duct; pp = cavity housing pharynx; rm = retractor muscle; t = testis; v = vitellarium. Scale bars: (*a*) 0.5 mm; (*d*) 100 μm. Source: (*a*), (*d*) redrawn from Sproston, 1946; (*b*), (*c*) redrawn from Ivanov, 1952.

   The life cycles of all of these organisms are imperfectly known. In *Micropharynx*, egg capsules have been found in preservative containing the parasites, but the development and fate of these capsules is not known. Since nearly all marine planarians are littoral organisms with no pelagic stage, Ball and Khan (1976) suggest that *Micropharynx* transfers from host to host during the prolonged bouts of copulation indulged in by their elasmobranch hosts. *Paravortex* (?) sp. leaves the host when it reaches a length of about 440 μm (Kent and Olson, 1986). As many as 160 young, each with a pair of pigment-shielded eyes, develop within the bodies of their parents. The ciliated young appear to be released by body rupture and presumably swim to a new host. According to Menitskii (1963), *Ichthyophaga* is ciliated and has a pair of eyes, but there is no information on the life cycle of the parasite.

The operculate, stalked eggs of udonellids are cemented to the copepods (Fig. 2.20), usually to their genital segments, and give rise to unciliated larvae (Nichols, 1975; Fig. 2.21*d*). The copepods are able to swim freely and presumably at some stage they leave their fish host and swim to a new one, transporting their udonellids with them. Thus, a phoretic relationship between udonellids and copepods seems likely to be of central importance and this is supported by reports of *Udonella* on free-swimming *Caligus* spp. (Baylis and Jones, 1933; Minchin, 1991). Prior to the evolution of a phoretic relationship between copepods and udonellids, new fish hosts may have been colonized by free-swimming ciliated udonellid larvae, as in modern monogeneans. The development of phoresy would render the free-swimming habit and epidermal cilia redundant. Similar phoretic relationships have developed independently between parasitic crustaceans and monogenean fish parasites (section 6.5.2).

There is little information on how udonellids spread to new copepods. Adults appear to prefer female copepods (Nichols, 1975; Timofeeva, 1977), which retain their egg sacs until hatching takes place, and Dawes (1946) claimed that eggs of *Udonella* and the copepods hatch synchronously. This raises the possibility that *Udonella* larvae attach themselves to the emerging nauplii. Nichols (1975) observed a synchronous hatching event and kept the animals under observation for 2 days, but no *Udonella* larvae established themselves on the copepod larvae. Timofeeva (1977) also failed to find any udonellid larvae attached to nauplii. Tufft (reported in Nichols, 1975) observed transfer of flatworms between copepods on starry flounder (*Platichthys stellatus*) and also reported that *Udonella* is capable of attaching itself temporarily to the fish's surface, later establishing itself on a new copepod.

2.5 CONCLUSION

The modern turbellarians provide examples not just of parasitism but of many other kinds of symbiotic association. In contrast, the monogeneans, cestodes and digeneans, with which the rest of this book is mostly concerned, are entirely parasitic, with a diversity so overwhelming that the parasitic turbellarians seem insignificant in comparison. However, there may be important lessons to be learned about the evolution of parasitism from the symbiotic turbellarians that the major parasitic groups cannot provide and, apart from some speculation in the next chapter on the origins of the major parasitic groups, consideration of the origins and evolution of parasitism in the platyhelminths, including the turbellarians, will be deferred until the concluding chapter. It is worth noting here that two host groups that figure significantly in the biology of the major parasitic platyhelminths are the molluscs and the fishes, both of which are favoured as partners by modern turbellarian symbiotes.

An ectoparasitic relationship between turbellarians and fishes has probably developed independently in *Micropharynx*, *Paravortex* (?) sp. and the udonellids. Convergence of some of their features with those of monogenean skin- and gill-parasites of fishes is striking. I am probably not alone in making the initial mistake of regarding *Micropharynx* as a monogenean when I encountered the animal for the first time, and there is a strong resemblance between the anterior adhesive pads, the pharynx and the posterior attachment organ of udonellids and the corresponding organs of some monogenean fish-skin parasites. This has led some parasitologists to regard udonellids as monogeneans but, after exhaustive study, Ivanov (1952) came to the conclusion that this was not so. Although this conclusion has enjoyed general support, there is no such consensus about their real affinities.

It is just such a turbellarian–fish association, albeit of great antiquity, that probably provided the starting point for the evolutionary expansion that led to modern monogeneans and cestodes and it is consideration of this expansion that will occupy most of the next few chapters.

CHAPTER 3

Origins of the major groups of parasites

3.1 INTRODUCTION

Llewellyn (1965) proposed that the proto-oncophoreans, which he regarded as ancestors of the monogeneans and cestodes, developed a parasitic relationship with early vertebrates (fishes) and that the proto-digeneans established themselves as parasites of molluscs. Such possible events imply separate origins from turbellarian stock for the proto-oncophoreans and the proto-digeneans, the proto-digenean–mollusc relationship being older (many millions of years?) than the oncophorean–vertebrate relationship (see Llewellyn, 1986). It is difficult to reconcile this view with a more recent proposal that the major parasitic groups of platyhelminths are a monophyletic assemblage. Comparative ultrastructural work is alleged to support monophyly (see review by Rohde, 1994a), as does also nucleotide sequencing data. In fact, on the basis of the latter evidence, Blair (1993) went so far as to state that the evidence for monophyly was so strong that he saw no need to consider dissenting views. This has led to the establishment of the taxon 'Neodermata' to embrace the monogeneans, cestodes and digeneans, together with the smaller group, the aspidogastreans, i.e. those parasitic platyhelminths that substitute during development a new body covering or neodermis for the ciliated epidermis of the larva. However, the following question, posed by Llewellyn in 1986, has not been answered convincingly by the advocates of monophyly: what were the ancestors of the oncophoreans doing while their sister-group the digeneans was developing its association with molluscs and before vertebrates became available as hosts to both digeneans and oncophoreans?

Whether the major parasitic groups are monophyletic or not, the free-living turbellarian ancestor or ancestors are usually regarded as rhabdocoel-like. This has been challenged by Rohde (1994a), but he was unable to identify a more likely candidate.

The following account of the possible origins and principal evolutionary trends of the main groups of parasitic platyhelminths is hypothetical and based mainly on the ideas of Llewellyn (1965, 1970, 1986). It accommodates the important biological features of these organisms but may have to be revised in the light of molecular studies.

3.2 THE MONOGENEAN–CESTODE LINEAGE (THE ONCOPHOREANS)

Llewellyn (1965, 1970) has suggested that the proto-monogeneans evolved from ancestors resembling modern rhabdocoels. Perhaps the proto-monogeneans were opportunistic browsers on the skin of the early fishes, using their epidermal cilia to swim from one host to another, as it seems the modern ectoparasitic rhabdocoel *Paravortex* (?) sp. probably does (section 2.4). By limiting their feeding activities to the host's epidermis, damage inflicted by the proto-monogeneans would have been minimal and transitory, because vertebrate epidermal cells have the ability to multiply and migrate, rapidly repairing superficial wounds. This reduced the risk of host infection by microorganisms and the parasites could afford to become permanently attached to their hosts rather than having to seek a host whenever they needed a meal.

Attachment may have been achieved at first by adhesive secretions, which are widely used for temporary attachment to substrates in modern turbellarians, especially those living between sand grains on the shore (Rieger *et al.*, 1991), but, eventually, so-called marginal hooklets, mounted on a posterior attachment organ or haptor, would have evolved as a means of permanent anchorage. These hooklets essentially provided attachment to host epidermal cells and, because of the rather delicate nature of these cells, there would have been a limitation on the sizes of the hooklets and the size of the parasite. However, by employing 16 hooklets deployed in a circle the load could have been spread and the size of the parasite maximized.

Curiously, these hooklets are not epidermal derivatives but are assembled by mesodermal oncoblast cells (Fig. 3.1); just how the 'blade' region of the hooklet is externalized is a challenging question. Their function, to provide multiple points of attachment to host epithelial layers, has been conserved throughout the monogeneans and is probably reflected in the relative uniformity of the size and shape of the hooklets (Fig. 3.2). However, they are not simply drag-anchors and are capable of performing 'gaffing' movements produced by tiny muscles attached to the 'blade' and root ('handle') of the hooklet and to a short projecting process or 'guard'.

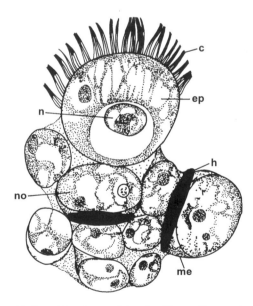

Fig. 3.1 Hook-secreting cells (oncoblasts) containing hooklet rudiments in the embryo of the monogenean *Entobdella soleae*. c = cilia; ep = epidermal cell; h = hooklet rudiment; me = cell membrane of oncoblast; n = nucleus of epidermal cell; no = nucleus of oncoblast. Source: reproduced from Lyons, 1966, with permission from Cambridge University Press.

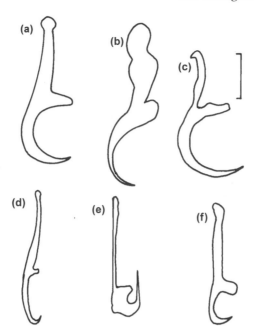

Fig. 3.2 (*a*)–(*e*) Single marginal hooklets from a range of monogeneans. Scale bar: 5 μm. (*f*) Hooklet of a pseudophyllidean cestode for comparison. Source: (*a*)–(*e*) redrawn from various sources; (*f*) redrawn from Llewellyn, 1965.

The acquisition of hooks is, arguably, the single most important development in these parasites. Hooks became the 'hallmark' of monogeneans and central to their subsequent evolutionary expansion and success. In fact, much of the subsequent evolutionary history of monogeneans concerns ways in which the primary attachment organs (marginal hooklets) have been supplemented by secondary structures (larger hooks or hamuli) or by tertiary structures (friction pads, muscular loculi, clamps).

The development of a haptor would have eliminated the uncertainty and risk involved in searching for a host every time a meal was required and also would have conserved energy. Initially, ciliary propulsion may have been used to find a mate but, sooner or later, these parasites would have adopted leech-like locomotion, achieved by extending and contracting the body and alternately attaching the anterior adhesive areas and the posterior haptor. The ciliated epidermis would then have served only for host-finding by the free-swimming larva, after which it could be jettisoned.

Strict host specificity, i.e. restriction of a symbiote to one or a few closely related species of hosts, is a phenomenon that we have already met in the symbiotic turbellarians. Even closely related hosts may differ in anatomical detail, in physiology and in behaviour; survival of a symbiote is dependent on close adaptation to these specific features of the host. Most modern monogeneans show a high degree of host specificity and it is highly probable that their ancestors have always done so. Thus, as the fish-like vertebrates speciated and diversified, host specificity would have imposed a parallel speciation and diversification on their monogenean parasites, creating a correspondence between host and parasite phylogeny. However, host specificity was not always strong enough to prevent monogeneans establishing themselves on ecologically available 'alien' hosts and there is a growing awareness that 'ecological transfer' ('host-switching' or 'capture') has probably been an important contributor to the evolutionary expansion and diversification of monogeneans.

The ectoparasitic proto-monogeneans diversified and colonized all the major groups of fishes, but they possessed a latent potential for endoparasitism that permitted two important lines of further development. First, an early monogenean may have colonized the gut of primitive fishes and initiated the process of fundamental evolutionary change that gave rise to the cestodes. Two major developments in this lineage were the loss of the gut and the exploitation of predator–prey relationships as a means of dispersal to new hosts, the prey organisms (invertebrates or vertebrates) serving as so called intermediate hosts and the predator being the definitive or final host. It is likely that the six hooklets retained by cestode larvae are a legacy of their monogenean ancestry (Fig. 3.2) and the term 'oncophoreans', coined for monogeneans and cestodes by Simmons (see Llewellyn, 1986), reflects this similarity. Secondly, and at a later date, more specialized monogeneans probably colonized the urinary bladder and this would have permitted these parasites and their descendants to survive the development of partly terrestrial habits in the amphibians. In Chapter 4 an in-depth study of the biology of *Entobdella soleae* will serve as an introduction to the monogeneans and in Chapters 5–7 the evolutionary expansion and diversification of the monogeneans will be considered. Cestode biology and evolution will be dealt with in Chapters 8–11.

3.3   THE DIGENEAN LINEAGE

The life cycle of a modern digenean involves more than one host, as does the life cycle of a cestode, but there is a fundamental difference between them that has been emphasized by Llewellyn (1965). In the digeneans there is rarely a predator–prey relationship between the two principal hosts, one of which is a mollusc and the other a vertebrate. In fact, the phases of parasitism involving the mollusc and the vertebrate are to some extent independent and the two hosts merely share the same environment, either permanently or temporarily. There are reasons, presented by Llewellyn (1965) and by Pearson (1972), for regarding the mollusc as the older of the two hosts. The proto-digenean may have been a ciliated flatworm that developed a symbiotic association with a mollusc.

The molluscs parasitized by modern digeneans are the gastropods (snails and slugs), lamellibranchs (bivalves) and scaphopods (tusk-shells). The amphineurans (chitons) and the cephalopods (squid, octopus) have not been reported as hosts. If, as indicated by Wright (1971), the ancestral amphineurans and cephalopods were the first groups to emerge from the main molluscan lineage, and if the development of the proto-digenean–mollusc association postdated these events and involved the ancestor of the modern gastropods, lamellibranchs and scaphopods, then this would explain the current distribution of digeneans among the molluscs.

In some marine sanguinicolids (e.g. *Aporocotyle simplex*) host-switching appears to have taken place between mollusc first intermediate hosts and polychaete worms (annelids; see Køie, 1982), and it is surprising that the sluggish, bottom-dwelling amphineurans have not been colonized in the same way. The cephalopods may have escaped the attentions of host-switching digeneans because of their active pelagic life style.

Modern digeneans are endoparasites of molluscs, inhabiting the haemocoel. This internal site is reached by penetration, effected by a ciliated larva or miracidium, which breaches either the surface covering of the mollusc or the gut wall if the egg is eaten by the host. However, Llewellyn (1986) postulated that the proto-digeneans began their association with molluscs as ectoparasites. Molluscs are covered by a layer of columnar or cuboidal epithelial cells, which are exposed except when the animal retracts within the shell. The surfaces of other invertebrates, such as annelids and crustaceans, are protected by a cuticle. Therefore, the molluscs, like the fishes that came later, would have offered excellent incentives for a potential ectoparasite, and it is possible that the proto-digeneans, like the proto-monogeneans, embarked on the road to parasitism in a similar way, beginning as opportunistic browsers and using their surface cilia to swim to new hosts.

Beneath the molluscan epidermis there are muscles, but in some regions, notably on the upper surface of the head and on the edge of the mantle, these layers are thin (Wright, 1971). An ectoparasitic proto-digenean feeding in these regions might have broken through into the haemocoelic spaces, where they would have been bathed by a highly nutritious fluid.

Pearson (1972) reached a similar position by a somewhat different route, favouring the establishment of the first proto-digeneans as ectocommensals in the mantle cavity of the mollusc. An alternative hypothesis proposed by James and Bowers (1967; in Pearson, 1972) envisaged the first proto-digeneans as gut-dwelling endocommensals, perhaps similar to *Paravortex* (section 2.3.2), which later broke through into the haemocoel, but it is not easy to see how or why such endocommensals later turned to penetration through the outer surface.

Some way of escape for the eggs would have been essential for the newly endoparasitic proto-digenean. It would probably have been no more difficult for the parasite to leave the mollusc through the thin areas of body wall than for the parasite to enter and, by abandoning the host before egg-laying, the adult parasites could have ensured egg release. The need to disperse eggs more widely may have led to the development of a unique muscular tail for propulsion.

So far, our hypothetical ancestral digenean is an endoparasite of molluscs with an oviparous, tailed, cercaria-like adult, which enjoys a brief, free-living existence when full of eggs (gravid). This life style has some similarity with that of modern fecampiid turbellarians (section 2.3.3). According to Jennings and Calow (1975), endoparasitism provides a continuous, superabundant and easily obtained food supply in a relatively predictable environment, and the consequences of this are that environmental limitations on fecundity are removed. Jennings and Calow pointed out that, given the evolutionary axiom that organisms will always reproduce as much as possible, high fecundity is likely to be an automatic consequence of adopting the endoparasitic way of life. Thus we would expect early digeneans living in nutrient-rich mollusc haemolymph to maximize their reproductive output, and they have achieved this in an unusual way. In the molluscan host they produce one or more generations by a unique process that has been variously interpreted as asexual (polyembryony, budding) or parthenogenesis (see Chapter 14 for further consideration of this phenomenon). The consequence of this process is the exponential increase in parasitic individuals in the mollusc, one invading miracidium giving rise ultimately to hundreds or thousands of free-swimming, tailed individuals.

Sooner or later fishes began to dominate the aquatic living space. Free-swimming adult digeneans could have interacted with them in two ways, differing in significance and potential. A few digenean adults may have established themselves as fish ectoparasites, but predation by fishes on the free-living adult digeneans may have provided the greatest impetus for change. Being eaten by fishes would not have been necessarily disadvantageous in the evolutionary sense, since eggs would be released by digestion from the bodies of the gravid adults, and, protected by digestion-resistant, tanned shells, would pass out and be dispersed in the fish's faeces. Eventually, some ingested adults may have become parasites, delaying their transit through and extending their stay in the gut of these fishes by virtue of persistence of parasitic adaptations from larval life. Llewellyn (1986) regarded the possession of suckers as being one such preadaptation, and maintained that these could have evolved during the ectoparasitic stage on molluscs, but Pearson (1972) considered that suckers could have evolved after recruitment of the vertebrate host.

Thus, a new vertebrate host would have been added to the digenean life cycle. A shift in sexual maturity would have followed so that the free-swimming stage linking the mollusc and the vertebrate host remained immature (cercaria), sexual development resuming in the vertebrate host.

Access to the gut of the vertebrate host would have been improved by utilizing the food chain. Prey organisms (invertebrates or vertebrates) eaten by the vertebrate definitive host would have been invaded by cercariae and become intermediate hosts with a similar function to those of cestodes.

It has been pointed out by Roberts and Janovy (1996) that digeneans have diversified extensively as parasites of all the major groups of vertebrates, with the exception of the Chondrichthyes (elasmobranchs: sharks and rays), in which they are relatively uncommon. They attributed this to the retention of high levels of urea in the tissues of elasmobranchs as a means of reducing the osmotic differential between their tissues and sea water. Such levels of urea are toxic to those digeneans that have been tested. Roberts and Janovy suggested that the osmoregulatory exploitation of urea by elasmobranchs may have evolved before digeneans were in a position to colonize these fishes and that high levels of urea may have since acted as a barrier to invasion by most digeneans. This contrasts with the situation in the cestodes, which have a rich and diverse fauna in sharks and rays (sections 10.6, 10.8) and either tolerate urea or degrade it. Monogeneans too have diversified as parasites of elasmobranchs; as ectoparasites they are not exposed externally to high concentrations of urea but must ingest substantial amounts, and some monogeneans have not been deterred from colonizing internal sites (see Table 7.1). Moreover, some monogeneans have exploited urea as a hatching stimulant; the relatively high concentrations of urea diffusing from the bodies of elasmobranchs provide a useful signal that a potential host is nearby (sections 5.2, 5.4).

The biology of the digenean *Fasciola hepatica* will be explored in detail in Chapter 13 and further evolutionary developments of the digenean life cycle will be considered in Chapters 15–18.

3.4 THE ASPIDOGASTREANS (= ASPIDOBOTHREANS, ASPIDOCOTYLEANS)

The most striking feature of this relatively small group of parasites is the great diversity of their life cycles. Molluscs figure in some of these and this has led parasitologists to regard them as relatives of the digeneans and to group them together as 'trematodes'. (This term originally encompassed the monogeneans, as well as the digeneans and aspidogastreans, but, in this wider sense, became redundant with the appreciation that monogeneans are closer to cestodes. Nevertheless, the term is still used in its narrow sense to describe the digeneans and aspidogastreans.) This relationship is supported by evidence from nucleotide sequencing (Blair *et al.*, 1993; Rohde, 1993). Llewellyn (1965) has suggested that proto-aspidogastreans may have been rhabdocoel-like turbellarians that developed an endoparasitic relationship with molluscs and this points to a common ancestry with the proto-digeneans. However, aspidogastreans gain access to the digestive or excretory systems of molluscs *via* their external orifices (see Chapter 19), not by tissue penetration as in the digeneans, and the two groups have followed dissimilar evolutionary pathways.

According to Llewellyn (1965), the proto-aspidogastreans in molluscan hosts may have inhabited sites from which their eggs could escape to the outside world, so that there would be no need for adults to leave the host as in the proto-digeneans. This direct life cycle with a single mollusc host persists today in some modern aspidogastreans, but vertebrates have been incorporated into the life cycles of others. In some aspidogastreans this association with vertebrates is casual, the parasite normally cycling between molluscs but capable of surviving in the gut of a vertebrate consuming an infected mollusc. But, casual or not, it is remarkable that the same parasite is able to survive in two such disparate environments as the pericardial sac of a mollusc and the intestine of a vertebrate. In time, the mollusc and the vertebrate may have become interchangeable, alternative hosts and later this evolutionary shift may have gone all the way, with the vertebrate becoming the obligatory and only host. But the evolutionary flexibility of the group extends further because some of these vertebrate parasites have recruited new invertebrate intermediate hosts (e.g. crustaceans), in much the same way as cestodes and digeneans acquired their intermediate hosts, i.e. by exploiting a predator–prey relationship.

These developments will be considered further in Chapter 19.

CHAPTER 4

The biology of the monogenean fish-skin parasite Entobdella soleae

4.1 INTRODUCTION

The monogenean *Phyllonella soleae* (= *Entobdella soleae*) was discovered on the skin of the common sole, now called *Solea solea*, by van Beneden and Hesse in 1864. The colour lithograph (see Frontispiece), published by Cunningham (1890) in his 'Treatise on the Common Sole' clearly shows two adult specimens of *E. soleae* in their typical location and attitude on the lower surface of the host. Like all monogeneans, *E. soleae* has a direct life cycle with no intermediate hosts (Fig. 4.1).

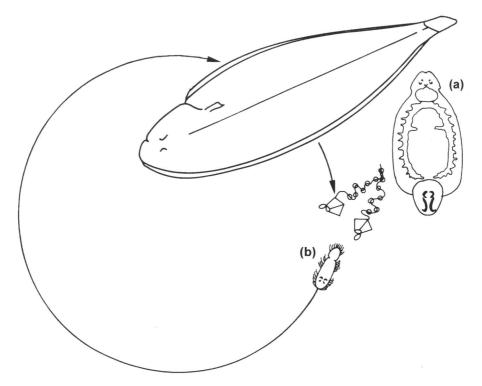

Figure 4.1 The life cycle of the monogenean *Entobdella soleae*. The adult parasite (*a*) inhabits the lower surface of the common sole (*Solea solea*). Ciliated larvae (*b*), hatching from tetrahedral eggs on the sea bed, invade the upper surface of the sole and migrate to the host's lower surface. Source: reproduced from Kearn, 1986, with permission from Academic Press.

Some anatomical studies were made by Little (1929), but when I started work on the parasite in 1960, virtually nothing was known about its general biology. It soon became clear to me that *E. soleae* was particularly suitable for experimental studies – the sole and its parasites survive well in aquaria, experimentally detached parasites readily attach themselves to a glass surface and survive for 2–3 days, continuing to behave with every indication of normality, and their semi-transparent bodies permit observation of internal events such as egg assembly.

4.2   PREVALENCE

Between 50% and 60% of sole of commercial sizes caught in the southern North Sea (UK) are infected with *E. soleae*, commonly with between one and five adults per fish (Kearn, James and Evans-Gowing, 1993). Adults vary in length depending on age (section 4.15), but flattened, preserved parasites range from about 4.5 mm to about 6.5 mm (Kearn, 1990). Young soles hatch from pelagic eggs and newly metamorphosed fishes are found among the breakers close to the beach (Wheeler, 1978). '0' group soles caught close to the beach at Lowestoft are uninfected, although fishes of this age are readily infected in the laboratory (Kearn, previously unpublished observations). In the wild, these young fishes presumably pick up their first parasites when they migrate into deeper water to grounds inhabited by larger infected soles.

4.3   ATTACHMENT AND LOCOMOTION

4.3.1   The haptor

The haptor of *E. soleae* is a posteriorly situated, muscular, saucer-shaped structure attached to the body by a stalk or peduncle (Fig. 4.2). Protruding from the ventral surface at the posterior margin of the haptor is a pair of formidable hooks (so-called anterior hamuli), seen particularly well with the scanning electron microscope (SEM; Fig. 4.3). Buried in the tissues of the haptor and extending in an anterior direction, each hook has a long girder-like root, which terminates approximately in the centre of the haptor. A second pair of hook-like structures (accessory sclerites) protrudes from the ventral surface at the centre of the haptor (Fig. 4.3) and there is a third much smaller pair of hooks (posterior hamuli) located near the hooked regions of the anterior hamuli (Fig. 4.2). In addition to the three medianly situated pairs of sclerites, there are 14 tiny marginal hooklets distributed around the edge of the haptor, which is bordered by a delicate marginal flap.

The haptor provides secure attachment to the host's skin and resists all attempts to dislodge it with artificial jets of water, which are much more forceful than any current experienced by the parasite in the natural world. A sharp blade or similar tool inserted between the haptor edge and the skin is required to dislodge the animal. The host's skin contains rows of overlapping flat scales, each with a spiny border, projecting in a posterior direction from the skin surface (Figs 4.4, 4.21) and the parasite adopts a characteristic attachment attitude in relation to these scales. The haptor is almost always orientated with its longitudinal axis parallel to the

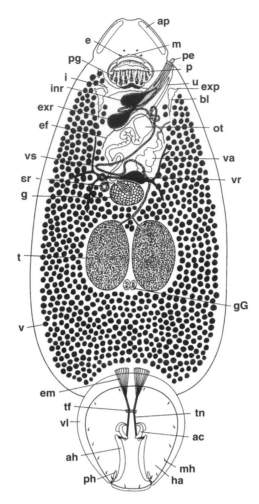

Figure 4.2 The anatomy of an adult specimen of *Entobdella soleae*, flattened in ventral view. ac = accessory sclerite; ah = anterior hamulus; ap = adhesive pad; bl = bladder; e = eye; ef = efferent duct from spermatophore gland; em = extrinsic muscle; exp = excretory pore; exr = external reservoir for spermatophore jelly; g = germarium; gG = glands of Goto; ha = haptor; i = intestine; inr = internal reservoir for spermatophore jelly; m = mouth; mh = marginal hooklet; ot = ootype; p = pharynx; pe = 'penis'; pg = pharyngeal gland cell; ph = posterior hamulus; sr = seminal receptacle; t = testis; tf = transverse fibres; tn = tendon; u = uterus; v = vitellarium; va = vagina; vl = marginal valve; vr = vitelline reservoir; vs = vas deferens. Scale bar: 1 mm.

longitudinal axis of the fish, with the haptor anteriorly located with respect to the fish and the body of the parasite projecting towards the host's tail. The posterior edge of the haptor is usually tucked beneath the projecting border of a scale (Fig. 4.4*a*).

The armament of hooks seems at first sight more than adequate for secure attachment but the ability of the parasite to attach to glass and the ease with which the haptor can be partially detached from the host skin simply by lifting the edge of the haptor, shows that suction also has an important part to play (Kearn, 1964). Low pressure (suction) between the haptor and the host skin is presumably maintained by the marginal flap (Fig. 4.3), which acts as a valve and prevents the influx of sea water; suction is destroyed by lifting the valve and breaking the marginal seal,

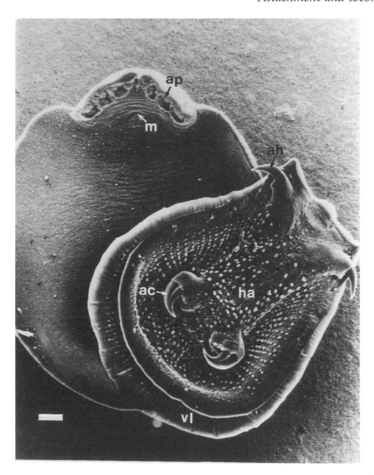

Figure 4.3 Scanning electron microscope (SEM) photograph of *Entobdella soleae* in ventral view. ac = accessory sclerite; ah = anterior hamulus; ap = adhesive pad; ha = haptor; m = mouth; vl = marginal valve. Scale bar: 50 μm. Source: photograph by Dr J. Rees reproduced with permission.

although the hooks of the anterior hamuli and probably those of the posterior hamuli continue to pin the posterior border of the haptor to the skin. A surprising discovery from experimental manipulation of this kind is that the hook-like accessory sclerites do not function as hooks but in reality act as semi-vertical props (Fig. 4.5), their curved feet pressing into the host's skin and creating curved pits that are evident after removal of the parasite (Fig. 4.6). The discovery that the anterior hamuli, with their powerful posterior hooks and girder-like roots, and the prop-like accessory sclerites are linked together by a muscle–tendon system, led to the suspicion that these sclerites were themselves involved in the generation of suction.

In the posterior region of the body of the parasite are two prominent muscles (Fig. 4.2). Each of these is linked to a long slender tendon, which runs through the peduncle into the haptor, passes round a notch at the proximal end of the prop and attaches to the anterior end of the girder-like root of the anterior hamulus. The spatial organization of one of the muscle–tendon–prop–anterior-hamulus systems is shown diagrammatically in a longitudinal section through the haptor and the posterior region of the body in Fig. 4.5. The pull exerted by contraction of the muscle

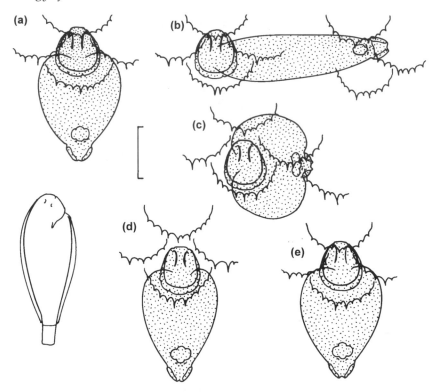

Figure 4.4   (*a*) Adult specimen of *Entobdella soleae* in its typical resting position and orientation relative to the scales of its host. (*b*)–(*e*) Consecutive stages in a single locomotory 'step'. Scale bar: 1 mm. The orientation of the parasite and the scales in relation to the body of the host can be determined by reference to the outline of a sole shown in the same orientation. Source: modified from Kearn, 1988a.

will tend to push the prop downwards into the skin. However, the curved foot of the prop and the underlying bony plates of the scales will ensure that penetration of the skin by the prop is minimal, and the net effect of the muscle contraction will be

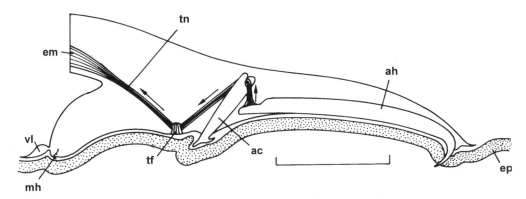

Figure 4.5   A diagrammatic parasagittal section through the haptor of *Entobdella soleae*; posterior hamulus omitted. The arrows show the direction of movement of the tendon when the extrinsic muscle contracts. ac = accessory sclerite; ah = anterior hamulus; em = extrinsic muscle; ep = host epidermis; mh = marginal hooklet; tf = transverse fibres; tn = tendon; vl = marginal valve. Source: modified from Kearn, 1971a.

Figure 4.6   'Footprint' on the skin of a sole at the site of attachment of the haptor of *Entobdella soleae*. Note spiny edges of host scales. Scale bar: 0.5 mm. Source: reproduced from Kearn, 1964, with permission from Cambridge University Press.

the lifting of the anterior ends of the girder-like roots of the anterior hamuli and hence the lifting of the roof of the central region of the haptor. With a good seal around the haptor margin and with the marginal hooklets functioning like tent pegs and preventing inward movement of the edge of the haptor, strong suction is likely to be generated.

The haptor of *E. soleae* may have a second way of generating suction using vertical (dorso-ventral) and circular intrinsic muscles. The evidence for this stems from observations made on the behaviour of parasites that have just moved to a new site on the host's skin (Kearn, 1988a). At the end of the locomotory 'step' (described below), the haptor is not yet lodged beneath the overhanging scale immediately behind it. This is achieved by slowly sliding the haptor back-wards across the skin until it can go no further (Fig. 4.4*d, e*). Clearly, the hooks are not used until this backward sliding is complete. Specimens of *E. soleae*, especially small ones, are also capable of attaching themselves to the underside of the surface film and the ability of the haptor to change its shape in this situation from a shallow saucer to a deeper cup is consistent with the generation of suction by intrinsic muscles.

The versatility of the monogenean haptor and its adaptation for attachment to other surfaces offered by their fish hosts will be considered in Chapters 5–7.

4.3.2   Locomotion, orientation and the anterior adhesive pads

Entobdella soleae has a second way of attaching itself to the host's skin. The parasite has two ventral pads, one on each side of the head (Fig. 4.3). These pads produce a remarkable adhesive that cements the parasite's head strongly to the wet slimy surface of the sole's skin and to other surfaces, such as glass. It is even more remarkable that the parasite can sever this cement bond instantly when the need arises. These pads are used for temporary attachment during the leech-like locomotory 'steps' of the parasite (Fig. 4.4). Such a step is usually preceded by 'searching' movements of the head region, in which the body alternately elongates and contracts, probing in different directions. Then, with the body fully extended, the adhesive pads are attached and the haptor is released (Fig. 4.4*b*). The body shortens and the haptor is reattached, immediately behind the head region (Fig. 4.4*c*). The adhesive pads then break their bond with the substrate and the body re-extends (Fig. 4.4*d*).

Whichever direction the parasite takes on the host, the haptor is locked in position relative to the body during the step (see Fig. 4.4) so that, when the haptor is relocated, its longitudinal axis is still parallel with the longitudinal axes of the fish scales. If parasites are detached experimentally and forced to reattach themselves to the host with the haptor abnormally orientated (rotated through 180°, see Kearn, 1988a), they reorientate within a few minutes and resume the typical orientation. They do this in darkness or on a freshly dead sole, suggesting that light and water currents play no part in the orientation of the parasite. There seems little doubt that *E. soleae* is able to detect the edges of the host's scales using sensilla (modified cilia) on their touch-sensitive body margins, and is able to use this information to establish the characteristic relationship between the haptor and the scales and to maintain this orientation and relationship during locomotion.

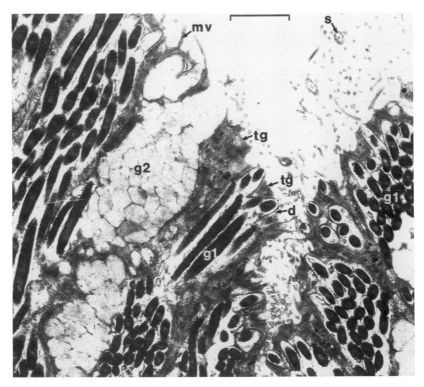

Figure 4.7 TEM photograph of a section through the surface of the adhesive pad of *Entobdella soleae*. d = desmosomal junction; g1 = gland ducts with rod-shaped secretory bodies and multiple apertures; g2 = gland duct with granular secretory bodies; mv = microvillus; s = sensillum; tg = tegument. Scale bar: 1 μm. Source: reproduced from El-Naggar and Kearn, 1983, with permission from the Australian Society for Parasitology.

Most of the surface of the adhesive pads of *Entobdella soleae* is taken up by gland duct openings (Fig. 4.7). There is a tegumental covering, with projecting microvilli, but there are so many gland ducts perforating it that the tegument is reduced to a network around the duct openings. Desmosomal junctions with the tegument hold the duct terminations in place. There are two kinds of duct opening and two distinct kinds of secretory body. The most abundant secretion consists of electron-dense, rod-shaped bodies and the multiple apertures of the ducts transporting this secretion resemble a pepper-pot. Each aperture is surrounded by tegument and permits the exit of a single rod-shaped body. The second kind of secretion consists of roughly spherical, electron-lucent granules transported in ducts each probably with only one aperture, like a salt-cellar.

There is evidence that the two kinds of secretion are extruded at the time of attachment of the pads and interact in some way to produce the cement that binds the parasite to its host (Kearn and Evans-Gowing, in press). This is reminiscent of the interaction between the two components of commercial epoxy resins.

There is another important question to ask and that is how the parasite achieves rapid detachment of the pads. Immediately after spontaneous detachment the pads have no adhering secretion (Rees, 1986), while 'padprints' are visible on a glass substrate when viewed by reflected light or with SEM (Kearn and Evans-Gowing, in press). Thus, separation appears to be achieved not by tearing the pads away from the substrate but by severing the adhesive bond cleanly and swiftly at the surface of the pad. If the only function of the granules and the rods is to provide raw material for the cement, then there is only one other possible instrument of detachment and that is the tegument covering the pads. Since the pad tegument forms a lattice surrounding all the duct openings, even the separate apertures of the rod-carrying ducts, its deployment is well suited for instantaneous universal severing of the adhesive bond, perhaps by release of a substance that dissolves the adjacent cement. Moreover, the pad tegument is isolated by cell membranes from the rest of the tegumental syncytium covering the parasite's body (El-Naggar and Kearn, 1983).

The haptor tegument has an important role in attachment in other monogeneans (section 6.4.1).

4.4 FEEDING

In *Entobdella soleae* the mouth lies on the ventral surface of the head region between the adhesive pads (Figs 4.2, 4.3). There are no hard mouth parts, no jaws or stylet. The mouth communicates immediately with a large cavity housing a conspicuous pharynx, the bulk of which consists of large, longitudinally arranged, uninucleate gland cells arranged in two tiers. Each gland cell has an opening into the lumen of the pharynx, mounted at the apex of a papilla (Fig. 4.8). The pharynx lumen communicates posteriorly with two gut caeca, each one following a lateral course, giving off blind-ending side branches and joining its fellow anterior to the haptor (Fig. 1.5). There are no gut diverticula in the haptor.

Many skin parasites such as leeches are blood feeders, but the colourless nature of the gut contents of *E. soleae* indicates a different diet; it is relatively easy to demonstrate that the parasite feeds on host epidermal cells, the delicate cells that constitute

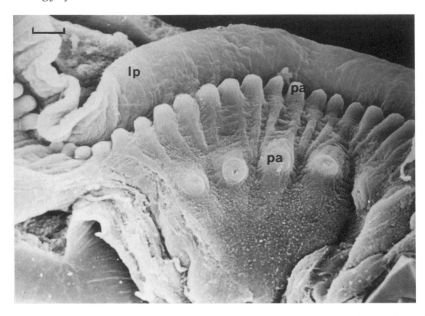

Figure 4.8   The inner surface of the pharynx of *Entobdella soleae*. The pharynx was dissected out of a living parasite, opened by a longitudinal cut, preserved and prepared for SEM. lp = pharyngeal lip; pa = papilla. Scale bar: 20 μm. Source: reproduced from Al-Sehaibani, 1990, with permission.

the outer layer of the sole's skin (Kearn, 1963a). Parasites can be induced to feed by isolating them from the host for 12 h or more and replacing the starved parasites on sole skin. Soon after re-establishment on the host, the pharynx is everted and protruded through the mouth (Fig. 4.9*a*). The protruding pharynx is circular, with the gland cell papillae projecting from its inner surface, and it is placed in contact with the host's skin. Detached hungry parasites will sometimes evert the pharynx and attach it to glass as shown in Fig. 4.10*b, c*. On host skin the circular pharynx maintains its feeding position for 5 min or so; peristaltic contractions of the pharyngeal muscles begin and pump colourless liquid (eroded epidermal tissue) backwards into the gut caeca. The pharynx is then detached and withdrawn into the head and at the feeding site a shallow circular pit, from which the superficial epidermis has been eroded, is evident (Fig. 4.9*b*). Histological sections through one of these feeding wounds show that only the epidermal cells are eroded, the dermis beneath, with its blood supply, seemingly undamaged (Fig. 4.9*c*). Experiments have shown that the secretion of the large pharyngeal gland cells readily digests the protein gelatin and it is likely that this secretion, ejected through the pharyngeal papillae on to the host's skin surface, plays a major part in eroding and digesting the epidermal cells.

If we consider the properties of fish epidermis, we begin to appreciate the great significance of this diet choice for the parasite–host relationship. Epidermis has an important role in wound healing – epidermal cells respond rapidly to skin damage by multiplying and migrating across exposed wound surfaces, sealing off the lesion and preventing invasion by microorganisms. Monogeneans like *E. soleae* exploit this property. The epidermis that they remove is rapidly replaced at relatively little cost to the host and with little likelihood of permanent damage. Since wild soles, which can reach 60 cm in length, carry on average

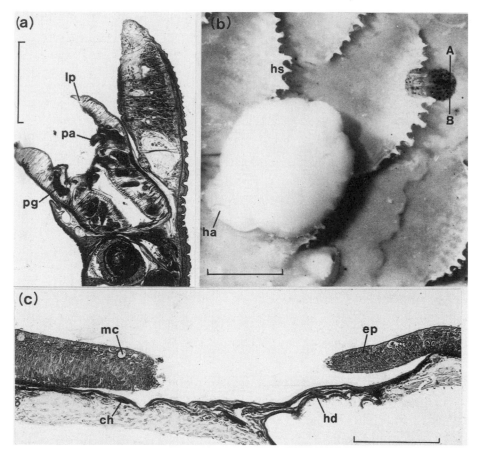

Figure 4.9 Feeding in *Entobdella*. (*a*) Sagittal section through the head region of *E. diadema* with everted pharynx. (*b*) *E. soleae* preserved in Bouin's fluid immediately after feeding, showing feeding wound at top right. (*c*) Section (AB) through the feeding wound shown in (*b*). ch = chromatophore; ep = host epidermis; ha = haptor; hd = host dermis; hs = host scale; lp = pharyngeal lip; mc = mucous cell; pa = papilla; pg = pharyngeal gland. Scale bars: (*a*) 250 μm; (*b*) 1 mm; (*c*) 150 μm. Source: reproduced from Kearn, 1963a, with permission from Cambridge University Press.

three (one to nine) adult parasites, mostly reaching about 6.5 mm in length, the damage inflicted on the host in its natural environment is negligible. However, latent pathogenicity is revealed by events in aquaria. In this situation, parasite eggs accumulate and soles are exposed to repeated infection by high densities of larvae. A single host may acquire 200 or more parasites and, unless these are removed, is likely to die. There has been no investigation of the causes of this host mortality, but perhaps replacement of epidermis by the host fails to keep pace with its removal by the parasites, leading to fatal infection of the host by microorganisms.

4.5   EGESTION

Entobdella soleae has no anus so the remnants of its meals must be ejected through the mouth. According to Hyman (1951, p. 206), egestion in turbellarians is facilitated by

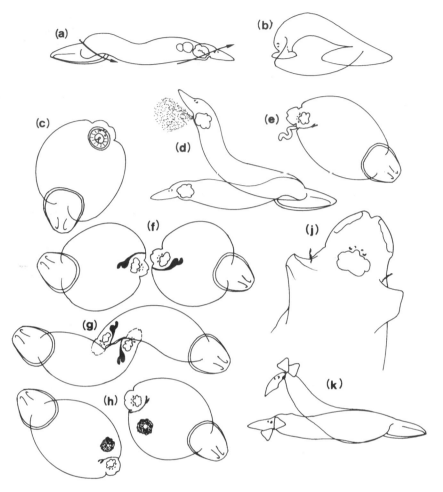

Figure 4.10   The biology of *Entobdella soleae*. Sketches of stills from a video recording made of living parasites attached to the bottom of a glass container of sea water. (*a*) Parasite performing body undulations (breathing movements) seen in side view; arrows indicate direction of water currents. (*b*), (*c*) Parasite attempting to feed on the bottom of the container, seen in side view and from above respectively. (*d*) Egestion. (*e*) Parasite ejecting a spermatophore. (*f*)–(*h*) Consecutive stages in mating – in (*h*) each parasite carries a spermatophore ventrally. (*j*) Anterior region of parasite ejecting material from the bladders. (*k*) Behaviour of parasite with eggs tethered to the body. Source: (*a*)–(*c*), (*e*)–(*k*) original sketches; (*d*) reproduced from Kearn *et al.*, 1996, with permission from Taylor & Francis.

swallowing water to flush out the gut and the same procedure is adopted by *E. soleae* (see Kearn *et al.*, 1996). An egestion episode begins with the head detached and the pharynx in the retracted position. The pharynx is then used as a peristaltic pump, just as it is when transferring newly eroded host epidermis to the intestine, to force ingested sea water into the gut lumen. Between four and eight of these swallowing events occur in rapid succession and lead to the distension of the gut caeca. This serves two functions. First, it stretches muscles in the walls of the gut diverticula and dorsoventral body muscles in the vicinity of the gut, so that they are in a position to contract, and, secondly, it provides a suspension medium for particulate waste material in the gut lumen. The parasite then raises its head and initiates a wave of body muscle contraction that serves to drive the ingested sea water and the material suspended in it out through the mouth (Fig. 4.10*d*). Thus, the pharynx has a secondary role as a peristaltic pump in egestion and the pharyngeal lip, which serves as a valve to prevent influx of sea water when the pharynx is everted and in the feeding position, probably has a rather different valvular role during egestion. With the pharynx in the retracted position, the lip is tucked inside the pharyngeal lumen and probably prevents efflux of sea water *via* the mouth while swallowing.

4.6 BREATHING

Although *E. soleae* is a parasite, it is influenced by the external environment because of its ectoparasitic habit. The external environment of the sole presents some special problems because sediment at the sea bottom often contains little oxygen as a result of bacterial decomposition. If the sea water around the parasite is stagnant, the parasite would rapidly use up the oxygen in its immediate environment. Adult specimens of *E. soleae* cope with this problem by undulating the body (Fig. 4.10*a*), thereby continually replenishing the oxygen-containing sea water around it. As shown by Kearn (1962), when oxygen concentration is low, these undulating breathing movements are relatively rapid (Fig. 4.11*a*) and of large amplitude. At the same time, the body becomes thin and flat, increasing the surface area for absorption of oxygen and decreasing diffusion distances to the tissues (Fig. 4.11*b*). When oxygen concentration increases, as it might do when the host becomes active, the rate of undulation and amplitude of the waves decrease and the body becomes thicker and smaller in surface area.

4.7 MATING

Entobdella soleae is hermaphrodite. There are two testes, communicating *via* a vas deferens with a protrusible penis-like structure, which also receives a copious accessory secretion from widely scattered gland cells (Fig. 4.2). The single germarium gives rise to an oviduct, which, after receiving the common vitelline duct from the vitelline reservoir, becomes the ovovitelline duct and follows a tortuous route to the ootype where the egg is assembled. The ootype in turn leads to the uterus, which opens into the distal end of the cavity housing the penis. This cavity has a single common genital aperture through which fully formed eggs and the penis emerge.

The vitelline follicles that produce the vitelline cells are widely dispersed in the body and are associated with the gut diverticula. Ducts from the follicles carry their products to a vitelline reservoir situated immediately in front of the germarium. The vitelline reservoir also receives a narrow communication from the coiled proximal storage region of the vagina, which is usually packed with spermatozoa. Distally the vagina becomes a very narrow straight tube which opens on the ventral surface of the body.

E. soleae has a novel way of inseminating a partner (Kearn, 1970). Two parasites exchange spermatophores, each comprising a central core of spermatozoa embedded in gelatinous material derived from the male accessory glands. Mating is brief

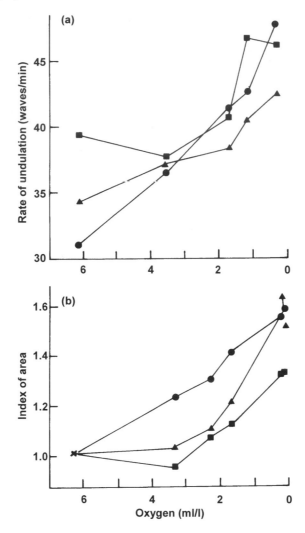

Figure 4.11 The relationship between (*a*) the rate of body undulation and (*b*) 'surface area' and the oxygen content of the medium in three individuals of *Entobdella soleae*. The product of maximum total length of the undulating parasite and body breadth was assumed to be proportional to surface area; the index of area in the graph is the calculated value of 'surface area' represented as a proportion of the 'surface area' of the animal in sea water containing 6.3 ml of oxygen per litre. Source: reproduced from Kearn, 1962, with permission from Cambridge University Press.

(Fig. 4.10*f–h*) and, after separation, each parasite usually carries a spermatophore externally (Fig. 4.10*h*), attached to the region of the ventral surface where the tiny vaginal opening is located. Because of the rapidity of the process the mechanics of mating are unknown, but the penis diameter greatly exceeds that of the vagina and it seems unlikely that intromission occurs. Soon after separation each mated partner undergoes body contractions that serve to draw the spermatozoa from the spermatophore into the vagina, where they are stored usually in an inactive state. Spermatozoa make two further moves. In some unknown way they are transferred from the vagina through the vitelline reservoir and into spherical seminal receptacles (usually two of these) opening from the ovovitelline duct. Thespermatozoa are active in these receptacles and later some of them gain access, again in an unknown way, to a chamber inside the germarium. Mature oocytes from the main body of the germarium enter this chamber, possibly by amoeboid motion, and it is here that fertilization takes place (Tappenden, Kearn and Evans-Gowing, 1993).

It would seem feasible for *E. soleae* to self-inseminate by inserting the large penis into its own spacious uterus, but this has never been observed in *E. soleae*. Immature specimens, kept in isolation from other parasites throughout their lives, fail to self-inseminate and are unable to make viable eggs (Kearn, James and Evans-Gowing, 1993). In fact the parasite seems unable to modulate its production of spermatozoa and spermatophore jelly and reaches the point where so much of this material accumulates that, in the absence of a receptive partner, it prefers to jettison a spermatophore (Fig. 4.10*e*) rather than inseminate itself by either the uterine or the vaginal route.

Self- or cross-insemination *via* the uterus does present a logistic problem for the parasite because in egg-laying adults egg traffic moves in a distal direction along the uterus and is more or less continual, while spermatophore assimilation requires transport in a proximal direction. Relatives of *E. soleae* belonging to the genus *Benedeniella* have solved this problem neatly by indulging in self-insemination *via* the uterus before the female system becomes mature and egg assembly starts (Kearn and Whittington, 1992a). This is made possible by the almost universal protandry (early maturation of the male reproductive system) that occurs in monogeneans.

Self-insemination in *Benedeniella* requires the male copulatory organ to turn through almost 180° in order to enter the opening of the uterus, and this manoeuvre is made easier by two new anatomical features (Fig. 4.12). First, the male copulatory organ of *Benedeniella* is a cirrus, not a penis, i.e. the organ extends by eversion like the finger of a glove rather than by protrusion, and secondly the common genital aperture is temporarily closed by a septum. As the cirrus unrolls, it progresses distally through the tube in which it lies, and when it reaches the septum is unable to leave the body and is directed into the uterus. The cirrus is so long that its tip may enter the ootype.

Adult specimens of *Entobdella soleae* have gland cells opening along their body margins (El-Naggar and Kearn, 1983); it is possible that these may be a source of chemical messengers (pheromones) that enable parasites to find each other rapidly for mating. It has already been noted that isolated parasites are unable to make viable eggs and an early meeting with a receptive partner is essential if egg output is to be maximized. Kearn, James and Evans-Gowing (1993) calculated that the chances of an early meeting taking place between two parasites wandering randomly, without pheromonal attraction, on a relatively small sole, are slim, even though the parasites are confined to the lower surface of the host. Since soles spend a lot of time resting, during which gill effluent from the lower gills is shunted out through the upper gill opening, the sole will have a relatively thin, stagnant layer of sea water beneath it, providing an ideal environment for the outward diffusion of pheromones. This may be the most significant factor in the preference of *E. soleae* for the host's lower surface. However, experimental proof of pheromonal attraction is still wanting.

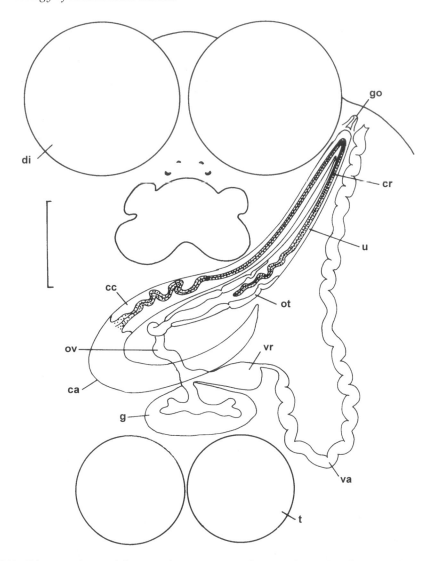

Figure 4.12 Diagram of part of the reproductive system of a juvenile *Benedeniella macrocolpa* in ventral view, showing penetration of the uterus (u) by the cirrus (cr). cc = cirrus canal; cs = cirrus sac; di = anterior attachment disc; g = germarium; go = common genital opening; ot = ootype; ov = ovovitelline duct; t = testis; va = vagina; vr = vitelline reservoir. Scale bar: 250 μm. Source: reproduced from Kearn and Whittington, 1992a, with permission from Cambridge University Press.

4.8   EGG ASSEMBLY AND EGG LAYING

Egg assembly is easy to observe because of the transparency of the animal, and lasts for 4–6 min (Kearn, 1985). A single fertilized oocyte is released from the germarium (Fig. 4.13) and, as it passes along the ovovitelline duct, numerous vitelline cells are released behind it and follow it in a string. Thus, the egg cell is the first cell to enter the distal tetrahedral chamber of the ootype and lodges in the distal corner of the ootype chamber, where the ootype lumen communicates with the uterus. Vigorous

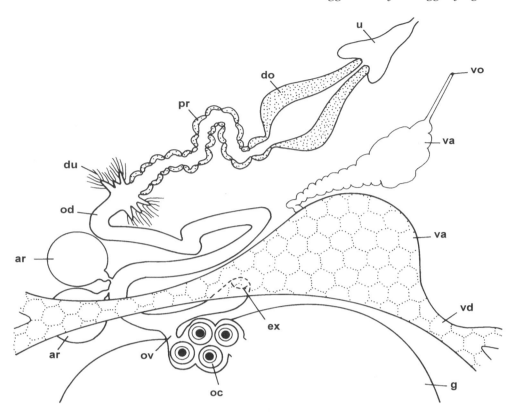

Figure 4.13 The egg assembly apparatus of *Entobdella soleae* in ventral view. do = distal tetrahedral chamber of ootype; du = ducts of ootype glands; ex = exit of common vitelline duct from the vitelline reservoir; g = germarium; oc = oocyte; od = oviduct; ov = ovovitelline duct; pr = proximal tubular region of ootype; sr = seminal receptacle; u = proximal region of uterus (distal region not shown); va = vagina; vd = transverse vitelline collecting duct; vo = vaginal opening; vr = vitelline reservoir. Source: reproduced from Kearn, 1985, with permission from the Australian Society for Parasitology.

peristaltic and antiperistaltic churning movements of the tetrahedral chamber then begin. Tappenden (1989) recorded as many as 100 contraction cycles during the assembly of a single egg, although the vigour of these movements decreases towards the end of the process. She observed that the egg cell maintains its distal corner position for the first 18–20 cycles. While the vitelline cells are squeezed backwards and forwards, they release their eggshell precursors in the form of droplets, and these coalesce adjacent to the wall of the ootype to form the eggshell. Without the ootype churning movements it seems unlikely that vitelline droplets released within the mass of vitelline cells would ever reach the wall of the ootype, and transport of vitelline droplets must be a vital if not the only function of these movements (Kearn, 1986).

Tappenden (1989) used a rapid preservation technique to 'freeze' the process of egg assembly at different stages in different individual parasites, and prepared sections for TEM study. She found that the cortical granules below the surface of the oocyte are not shed until the oocyte enters the ootype and that this material probably serves to hold the oocyte temporarily in place in the distal corner. Shell cannot be laid down in this corner until the oocyte moves out of the way. This delayed deposition of eggshell material in the distal corner creates the operculum or lid through which the larva will eventually escape. Tappenden suggested that cortical granule material sandwiched between the 'old' shell of the egg and the 'new' shell of the operculum may provide a thin layer of cement holding the operculum in place.

The ootype of *E. soleae* has a proximal tubular region as well as the distal tetrahedral chamber (Fig. 4.13), and vitelline droplets in this tube coalesce to form an appendage continuous with the eggshell. The ootype tube stretches during egg assembly and permanently increases the length of the egg appendage to about 880 μm; adhesive material, from gland cells (so-called Mehlis's glands) opening into the proximal end of the tube, is fashioned into eight to 11 more or less spherical adhesive droplets (Fig. 4.14), each about 15 μm in diameter and situated at intervals of about 60 μm along the appendage (Kearn, 1963b, 1985; Tappenden, 1989). According to Tappenden there is a second kind of Mehlis's gland, opening into the proximal end of the ootype tube. The secretion from these cells is released early in the egg assembly process and Tappenden suggested that it may serve to stimulate the vitelline cells to release their droplets and/or to initiate the tanning process (Chapter 1). There is evidence that tanning begins while the egg is still in the ootype, because Tappenden first encountered difficulty in cutting sections through vitelline material, a situation indicative of the commencement of the toughening process, towards the end of the period of shell deposition.

The churning movements of the ootype or egg mould ultimately cease and the egg adopts the tetrahedral shape of the ootype.

One or more eggs may be stored temporarily in the uterus, where their brown colour intensifies as tanning proceeds, but sooner or later they are forced out of the uterus by strong contractions. However, this ejection is usually only partial, the appendage remaining lodged in the uterus and tethering the egg capsule to the parasite (Fig. 4.14). When two or three partially ejected eggs are tethered in this way,

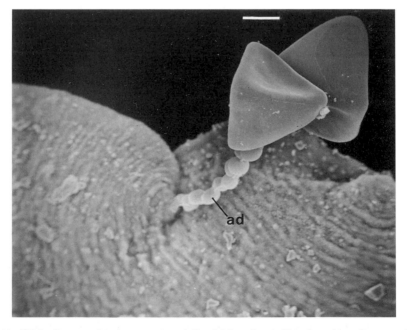

Figure 4.14 SEM photograph of two eggs of *Entobdella soleae* still tethered to the parasite by the appendages lodged in the uterus. ad = adhesive droplet. Scale bar: 50 μm.

the parasite lifts the anterior region of the body vigorously two or three times with the head deflected downwards (Fig. 4.10k). In this characteristic attitude, with the parasite attached at the bottom of an observation vessel, the tethered eggs are uppermost. This movement is repeated at intervals until the egg appendages are released. When the parasite is in its natural, inverted position on the lower surface of its flat-fish host, these movements are likely to be most effective in dislodging the eggs and attaching them by their sticky droplets to sand grains underlying the fish. In contrast with the secretion from the anterior adhesive pads of the parasite, the egg stalk adhesive fails to attach to sole skin, so that all the eggs leave the host and are implanted in the sedimentary material at the sea bottom.

Adult parasites measuring about 5 mm in length produce about 30 eggs per day (1.25 eggs per hour) at 12°C, but growth continues after sexual maturity is reached and 6 mm adults may produce more than 60 eggs per day (2.5 eggs/h; Kearn, 1985). Most of the tetrahedral eggs have a side length of 160–170 μm and there is evidence of a small but significant increase in egg size as parasites increase in size.

4.9 EGG DEVELOPMENT AND MODE OF HATCHING

Eggs attached to sand grains at the sea bottom undergo a period of development of about 1 month at 12°C. During this period the eggs are exposed to predation and, since sediment at the sea bottom is often anoxic, they may also be subjected to relatively low oxygen concentrations. Macdonald (quoted by Kearn, 1975) showed that eggs swallowed without undergoing shell damage by crustaceans that might be natural predators passed through the gut intact and, after leaving the predator's body, continued to develop and hatch. This underlines the potential survival value of the chemically and physically resistant tanned eggshell. The tetrahedral shape, with its relatively high surface area–volume ratio and short diffusion distances from surface to centre, may be an important factor in the supply of oxygen to an actively growing embryo in an anoxic environment (Kearn, 1986).

The growing embryo progressively uses up the vitelline cells. Ultimately, the fully developed, cylindrical but folded larva or oncomiracidium fills the tetrahedral chamber (Fig. 4.15). The brown shell is translucent and the four, fully developed, pigment-shielded eyes readily identify the position of the head of the larva, which occupies the opercular corner of the tetrahedron (Kearn, 1975). The marginal hooklets are also fully developed at this stage, but the edges of the circular haptor are folded inwards and the three pairs of median sclerites are not yet fully developed. Ciliated epidermal cells are present at three levels on the larva: on the head, in the midbody region and on the haptor.

The imminence of hatching is indicated by the commencement of activity of the body cilia. This ciliary motion imparts rotation to the oncomiracidium, such that the head slowly revolves in the opercular corner at a speed of 6 s (2–8 s) per revolution at 17–20°C (Kearn, 1975). Rotation continues until the operculum suddenly breaks loose from the rest of the shell. The larva uses its muscles to squeeze through the narrow opercular opening but the cilia also continue to beat and the traction that they generate increases in strength as more and more of the body emerges from the shell. The anchorage provided by the adhesive droplets is important at this stage because, if the egg capsule is freed by severing the appendage, the oncomiracidium usually fails to escape and swims around dragging the

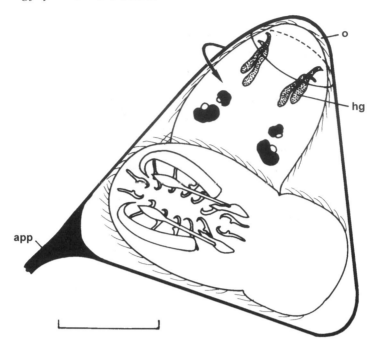

Figure 4.15 Egg of *Entobdella soleae* showing the position of the fully developed oncomiracidium relative to the operculum (o). Only the root of the egg appendage (app) is shown. hg = hatching glands. The arrow shows the direction of rotation of the hatching larva. Scale bar: 50 μm. Source: modified from Kearn, 1975.

shell behind it. Oncomiracidia that are successful in hatching take at least 4 min to escape from the egg but, as we shall see later, other monogeneans hatch much more quickly (Chapters 5, 6).

The rotation of the head of the larva in the opercular corner during hatching is likely to be of special significance. The unhatched larva possesses two pairs of small gland cells on each side of the head region (Fig. 4.15) and during rotation the openings of these glands are adjacent to the opercular discontinuity. After hatching, these cells are no longer visible and are presumably exhausted; the implication is that they provide a hatching fluid, which is smeared on to the opercular seal, thereby dissolving the cement holding the operculum in place.

By exposing intact and experimentally punctured eggs to digestive enzymes it has been shown that the opercular cement is readily dissolved by proteases, but only if these enzymes have access to the inside of the opercular seal (Kearn, 1975). This suggests that the opercular cement is protein-based, that the hatching fluid has proteolytic properties and that the opercular cement either does not reach the outer surface of the egg or, if it does so, is resistant to proteases where it is exposed.

4.10 THE ONCOMIRACIDIUM AND HOST-FINDING

4.10.1 General

The anatomy of the oncomiracidium is shown in Fig. 4.16. Free-swimming larvae keep the haptor folded so that the hooks are not exposed and follow a spiral path at

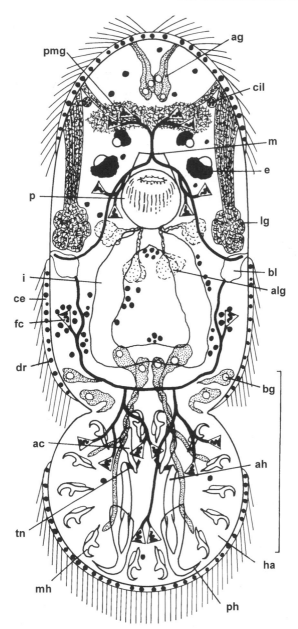

Figure 4.16 The oncomiracidium of *Entobdella soleae*, in ventral view. ac = accessory sclerite; ag = anterior median head glands; ah = anterior hamulus; alg = glands associated with the alimentary canal; bg = body glands; bl = bladder; ce = ciliated epidermis; cil = ciliary eye (?); dr = refringent droplet; e = rhabdomeric eye; fc = flame cell; ha = haptor; i = intestine; lg = lateral head glands; m = mouth; mh = marginal hooklet; p = pharynx; ph = posterior hamulus; pmg = posterior median head glands; tn = tendon. Scale bar: 100 μm. Source: modified from Kearn, 1974a.

a speed of about 5 mm/s at 20°C (Kearn, 1967a). They do not rotate about their longitudinal axes as they swim, as they do when escaping from the egg.

The intriguing question is how this tiny larva succeeds in locating its host – many undoubtedly fail to do so, but sufficient numbers are successful to maintain the natural population at its low level. The larva seems to have little to enhance its chances of survival; it is only $250\,\mu$m in length and, although the sole is not a fast swimmer, the oncomiracidium is unable to match the host's swimming speed. Moreover, the larva has a gut that seems capable of functioning but it cannot feed until it reaches a sole and is unlikely to survive for more than 24 h without feeding.

4.10.2 Rhythmical hatching

A partial answer to this puzzle emerged as a result of a chance observation. I left a batch of fully developed eggs of *E. soleae* near a window and observed them at intervals over a period of a few days. I noticed that the dish contained plenty of free-swimming larvae in the mornings, but, having removed these larvae, I had to wait until the following morning to find further significant numbers of larvae. I began to suspect that the eggs had a hatching rhythm and went on to confirm this using an experimental environment with a constant temperature and an artificial light regime of alternating 12 h light and dark periods (LD 12:12; Kearn, 1973). In these circumstances, eggs kept in clean sea water uncontaminated by the host usually hatch during the first 2–3 h following 'dawn' and rarely at other times (Fig. 4.17). Eggs exposed to natural illumination cycles show a similar dawn hatching pattern. Hatching appears to be controlled indirectly by way of the photoperiod rather than by the direct stimulatory effect of light, because rhythmic hatching persists when eggs are kept in continuous light or continuous darkness, provided

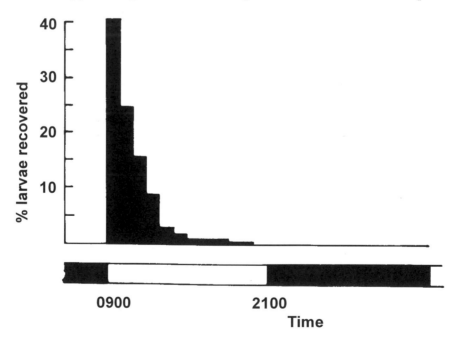

Figure 4.17 The daily pattern of hatching of eggs of *Entobdella soleae* cultured in sea water uncontaminated by the host at 14°C. The histogram summarizes observations made over a period of 15 days, with fluorescent lighting turned on and off abruptly (LD 12:12). White panel = period of illumination; black panel = period of darkness. Larvae collected in each of the 24 daily 1 h periods have been added together and given as hatching percentages. Source: reproduced from Kearn, 1986, with permission from Academic Press.

that they are exposed to LD 12:12 until hatching begins (Kearn, 1973). Thus, the parasite has an endogenous circadian hatching rhythm, entrained by the pattern of light and darkness to which the eggs are exposed during incubation.

The possible significance of this rhythmicity emerges when the behaviour of the host is considered. Soles are normally active at night, swimming about at or near the bottom and feeding mainly on small crustaceans and worms which they locate with the sensory filaments beneath the head (Wheeler, 1978). At dawn, soles become inactive, partially burying themselves in the sediment and resting throughout the hours of daylight, except on dull days or in heavily sedimented water when they may be moderately active at any time. Fishermen are familiar with the activity pattern of soles, making bigger catches at night when the active flat-fish are more readily scooped into the trawl. The oncomiracidia will have a better chance of making contact with a resting sole than with an active one and, since the period of host inactivity usually begins at dawn and persists throughout the hours of daylight, natural selection will favour emergence of larvae at the beginning of this period, i.e. soon after dawn.

4.10.3 Hatching in response to sole mucus

E. soleae has another fascinating adaptation that will enhance its chances of locating its host. As we have seen above, in the absence of contamination by the host, *E. soleae* rations out the larvae hatching from the eggs, small numbers of larvae hatching early each morning. However, if a drop of sole skin mucus is added to fully developed eggs during the dawn spontaneous hatching period, hatching is greatly enhanced (Fig. 4.18). Moreover, hatching can be induced within a few minutes by the addition of mucus at any time (Kearn, 1974b). Thus, if a sole settles on or near the eggs of *E. soleae* during the hours of daylight or during the night, mucus from the fish is likely to activate the larvae and, provided that the sole remains stationary for a few minutes, there is every chance that the larvae will make contact with the host.

The discovery that the eggs of *E. soleae* hatch when treated with sole mucus appeared at first to be at odds with statements in the literature to the effect that monogenean oncomiracidia are not immediately infective after hatching and may not be ready to attach to the host for periods varying from minutes to hours. Such a non-infective period in *E. soleae* would destroy any advantages gained by rapid hatching stimulated by sole mucus. However, it is easy to demonstrate that the oncomiracidia of *E. soleae* are immediately infective after hatching and, indeed, do not necessarily have to swim before establishing themselves on the host. When fully developed eggs are placed on the skin of a sole, some of the oncomiracidia attach themselves to the skin by the head region as soon as the head emerges through the opercular opening of the egg, the haptor being attached when escape from the egg is completed (Kearn, 1981).

The identity of the chemical clue (or clues) in sole mucus that stimulates hatching in *E. soleae* remains elusive. Some evidence accumulated to suggest that urea is effective (Kearn and Macdonald, 1976) – it is a potent hatching stimulus in other monogeneans (see Chapters 5, 6) – but later tests with purer grades of urea were less effective and there is a possibility that some contaminant may account for the earlier success (see Kearn, 1986). Whatever the nature of the stimulus, it is not specific to soles and mucus from a range of fishes, many unrelated to soles, is effective (Kearn,

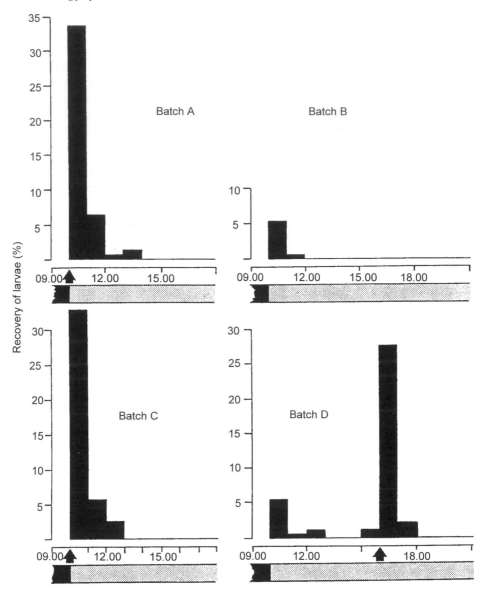

Figure 4.18 Effects of sole body mucus on hatching in four batches of eggs (150–250 eggs per batch) of *Entobdella soleae*. Mucus was added at the times indicated by the arrows. No mucus was added to batch B (control). Eggs exposed to LD 12:12 at 13°C. Number of larvae collected is expressed as a percentage of the total number of eggs in the batch. Stippled panel = light period; black panel = darkness. Source: reproduced from Kearn, 1974b, with permission from Cambridge University Press.

1974b). This may be of little consequence for *E. soleae*, if soles are the most common inhabitants of areas of the sea bottom where the eggs of the parasite occur.

4.10.4 The eyes and responses of oncomiracidia to light

There is no evidence that oncomiracidia are able to track down their hosts by visual means. The four eyes of the larva are relatively simple in construction (Fig. 4.19).

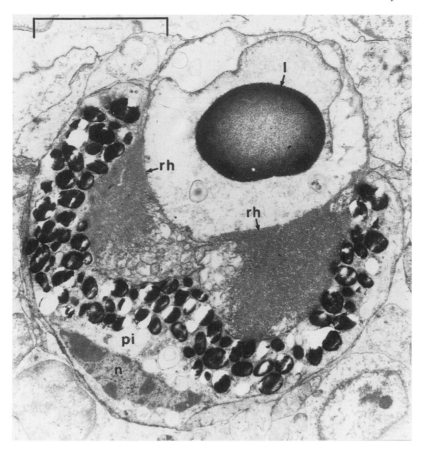

Figure 4.19   TEM photograph of a section through one of the large, posterior, pigment-shielded eyes of the oncomiracidium of *Entobdella soleae*. l = lens; n = nucleus of pigment cell; pi = pigment cell; rh = rhabdomere. Scale bar: 5 μm. Source: reproduced from Kearn and Baker, 1973, copyright © by Springer-Verlag 1973, with permission from Springer-Verlag.

Each of the two large posterior eyes (Fig. 4.16) has two light receptors or rhabdomeres, each consisting of tightly packed microvilli and arising from a retinular (nerve) cell outside the eye. Each of the two small anterior eyes has only one rhabdomere. An important feature of these eyes is that they are directional. The rhabdomeres lie within the concavity of a cup-shaped cell containing pigment that is opaque to light, so that light can only reach the rhabdomere(s) through the aperture of the cup, which accommodates a crystalline lens. The large posterior eyes can receive light only from an anterolateral direction, while the smaller anterior eyes can receive light only from a posterolateral direction. So, as the larva travels through the water in an open spiral, the eyes are ideally arranged to compare light intensities in front of and behind the animal and to control the direction of swimming in relation to light direction.

The eyes probably have an indirect role in host finding by controlling the direction of swimming. In the laboratory, larvae hatching spontaneously without stimulation by host skin mucus swim vertically when illuminated from above, as they would be in the sea, but the direction of these vertical movements changes abruptly at intervals (Kearn, 1980). Each larva is photopositive for a time, swimming upwards

towards the light, and then suddenly becomes photonegative, turning through 180°
and swimming vertically downwards until it reaches the bottom. In the sea,
a descending larva may make contact with and attach itself to a resting sole. In
the absence of a sole there is no advantage in staying on the bottom and the
larva switches to a photopositive response, swimming upwards for a time before
making a new descent. If there are no soles about, these upward and downward
movements may persist for a long time, but, since in the sea the water column is
likely to be continually disturbed by horizontal water movements, it is most unlikely
that a descending larva will make contact with the same spot on the bottom
more than once, so each fresh descent will offer a new opportunity to locate a
sole. Since soles usually rest during the daylight hours, this combination of a
restricted dawn hatching period and a behavioural response to the direction of
illumination provides an excellent strategy for promoting contact between oncomiracidium and host.

The pigment-shielded eyes may perform another task. They may respond to the
light filtering through the eggshell and, by monitoring day length, control
rhythmical hatching. However, there is a second candidate for the role of day-
length monitor. In a TEM study, Lyons (1972) found two unusual cells, one on
each side of the head region of the oncomiracidium. Most of each cell is occupied
by a large vacuole containing coiled lamellate bodies, which appear to be extensions of cilia rooted in the wall of the vacuole (Fig. 4.20). I have located these cells
in the living oncomiracidium viewed with the phase contrast microscope (Fig. 4.16).

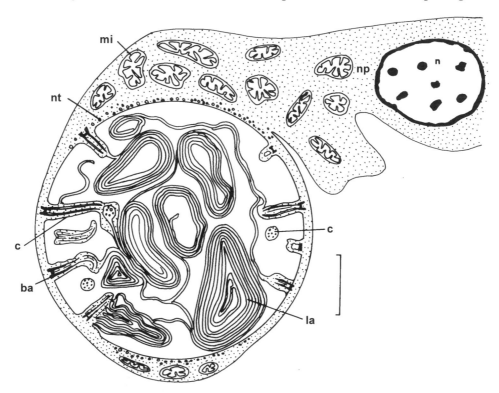

Figure 4.20   Diagram of the ciliary eye (?) of the oncomiracidium of *Entobdella soleae*. ba = basal body;
c = cilium; la = lamellate body; mi = mitochondrion; n = nucleus; np = nerve process; nt = neurotubule.
Scale bar: 1 μm. Source: modified from Lyons, 1972.

Lyons was struck by the similarity between these cells and the cilia-based photoreceptors of molluscs, and she went as far as to suggest that these 'ciliary eyes'(?) may be 'a long term light receptor, setting, for instance, some hypothetical diurnal or nocturnal activity phase'. Rhythmical hatching may be this 'activity phase'.

4.10.5   Responses of oncomiracidia to host skin

There is no evidence to indicate that the oncomiracidia of *E. soleae* are able to follow scent trails to their hosts, but it is likely that the larva uses its chemical sense to identify a sole when it makes contact with it. The scales of soles and some other fishes are readily removed with a pair of forceps and usually carry a small patch of skin (epidermis and dermis; Fig. 4.21). Given the choice of scales from a variety of fishes, the oncomiracidia of *E. soleae* show a strong preference for those of its natural host (Kearn, 1967a). The larva first attaches itself to the patch of skin on the scale by the adhesive areas on the head region and then by the haptor, which is unfolded, exposing the marginal hooklets and other sclerites. Remarkably, within half a minute or so, the larva jettisons its ciliated cells – with their cilia still beating these cells swim away – and the larva is committed to its parasitic life. It is interesting that larvae respond in precisely the same way to a piece of Agar jelly that has been in contact with sole skin, indicating that chemical substances produced by the sole skin and soaked up by the Agar jelly provide the clues used by the parasite for host identification.

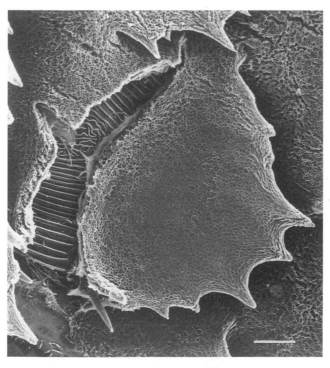

Figure 4.21   SEM photograph of a scale of *Solea solea*, partly withdrawn from its socket in the dermis. Scale bar: 100 μm. Source: photograph by Dr Ian Whittington reproduced with permission.

Evidence from a further experiment suggests that these clues come from the epidermis rather than from the dermis (Kearn, 1967a). Removal of a sole scale involves fracturing the skin; this stimulates the epidermal cells to adopt their wound healing role and multiply–migrate down the bony lamina of the scale away from the less responsive dermal tissue. By bisecting the scale, a sample of epidermis without dermis can be obtained, and larvae respond to this precisely as they do to whole skin by attaching themselves and jettisoning their ciliated cells. The conclusion is that the chemical clues used by *E. soleae* for host identification are provided by host epidermis, which, as we established above, is the parasite's food.

4.10.6　Responses of oncomiracidia to gravity

The sensory versatility of the oncomiracidium of *E. soleae* is underlined by another observation. When hatching was induced in total darkness, most of the hatched larvae gathered at the top of the vessel, indicating that the oncomiracidia have a gravity sense (Kearn, 1980). There is one situation in which eggs are likely to find themselves where a geonegative response would have survival value. The egg appendages of *E. soleae* adhere readily to sand grains and, because of the attached ballast, the eggs are likely to become buried in the shifting sediment. In fact eggs soon disappear below the surface when sandy sediment in a vessel is disturbed by gentle currents. Light may not reach larvae hatching from buried eggs, especially if a sole is resting on top of them, and a geonegative response is required to extricate them from the sand. When eggs were buried experimentally at depths of 0.5–1 cm in sand from an area of the sea bed inhabited by soles, some larvae found their way to the surface (Kearn, 1980).

4.10.7　Hatching in related species

The discovery of the dawn hatching rhythm of *E. soleae* has stimulated the study of hatching in related monogeneans. *E. hippoglossi* lives on the skin of the halibut (*Hippoglossus hippoglossus*) and is one of the largest monogeneans, reaching 2 cm in length. In spite of its close anatomical similarity to *E. soleae*, its hatching pattern is different, most larvae emerging during the first few hours of darkness (Fig. 4.22), and hatching is not stimulated by halibut mucus (Kearn, 1974c). Little is known about the behaviour of the halibut, but the dusk hatching pattern of *E. hippoglossi* suggests that the host rests at night. It is the largest flat-fish and an active predator, taking fish, cephalopod molluscs and some crustaceans, and it is likely that these are captured during the day. Unlike most flat-fishes it forages in midwater (Wheeler, 1978). Reduced host contact with the bottom might explain the lack of a hatching response to host mucus in *E. hippoglossi*.

E. diadema is less closely related to *E. soleae* and *E. hippoglossi*; it is a skin parasite of an elasmobranch host, the stingray *Dasyatis pastinaca*. I have had only one opportunity to experiment with the eggs of *E. diadema*, which were exposed to an LD 12:12 light regime (Kearn, 1982). It was found that hatching could be induced during the 12 h period of illumination by switching off the light. More surprising was the finding, using infrared illumination and an infrared–visible-light converter, that hatching occurred with great speed, larvae emerging between 3 and 5 s after extinguishing the light, and that freshly hatched larvae were strongly geonegative. This

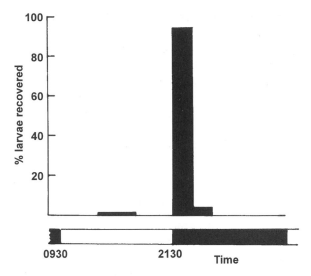

Figure 4.22 The daily pattern of hatching of eggs of *Entobdella hippoglossi* at 7°C. Details as in Fig. 4.17, except that hatching percentages are calculated for 12 daily 2 h periods over a total period of 7 days. Source: reproduced from Kearn, 1986, with permission from Academic Press.

indicates that, unlike *E. soleae*, the opercular cement is pre-weakened in readiness for the thrust of the activated larva.

It is conceivable that hatching in *E. diadema* is stimulated in the natural environment by shadows cast by the host. Stingrays prefer shallow water and sandy bottoms, where there is likely to be little turbidity and good illumination (Wheeler, 1978). Parasite eggs on the bottom may be stimulated to hatch when a stingray settles on them, the emerging larvae being in a position to attach themselves directly to the host. Since the freshly hatched larvae are geonegative, cruising hosts may also become infected, provided that they are moving slowly enough and that they are not too far from the bottom. A host-finding strategy involving a shadow-hatching response seems well suited to a parasite of a large, diurnally active flat-fish with the habit of swimming intermittently and resting for brief periods on the bottom. This strategy would be most effective if the distribution of available hosts is aggregated rather than dispersed, because, in the latter situation, a disproportionately large number of larvae might respond to shadows cast by non-host organisms. Unfortunately, little is known about the daily behaviour pattern and distribution of stingrays and we still do not know how the eggs of the parasite respond when exposed to the gradual fading of light at dusk.

4.11   POST-LARVAL MIGRATION

Adult specimens of *E. soleae* (with a few exceptions – see below) are found on the lower surface of their flat-fish host, *Solea solea*. However, freshly attached and developing parasites are more frequently encountered on the upper surface, both on freshly caught soles and on heavily infected soles kept in an aquarium (Kearn, 1963c). The possibility that the upper skin is more attractive to oncomiracidia has been tested by offering them a mixture of detached upper and lower scales, but the larvae showed no preference, attaching themselves equally readily to scales from the upper and lower surface (Kearn, 1967a). Another possible explanation relates to the habits of the host. The sole spends much of its time resting on the sea bottom or partly covered by sand, and it is reasonable to suppose that the lower surface of the fish is less accessible to oncomiracidia than the upper surface. Free-swimming larvae will mostly approach from above, and less frequently a larva hatching from an egg beneath the sole will attach itself to the lower surface.

This distribution, with newly invaded larvae on top and adults beneath, implies either that larvae migrate from the upper surface to the lower surface or that only those larvae gaining access to the lower surface survive to maturity. That some or all of the larvae alighting on the upper surface are able to find their way to the lower surface has been confirmed experimentally (Kearn, 1984). Uninfected soles can be obtained by brief immersion in fresh water, which is rapidly lethal to the parasites but not to the fish, and a selected area of the upper surface of such a treated sole can then be infected with oncomiracidia. After 40 days most of the larvae were established on the lower surface.

There is another aspect to this migration which has its roots in an observation made on an aquarium host (Kearn, 1963c). A sole from one of the tanks in the Public Aquarium at the Plymouth Laboratory of the Marine Biological Association of the UK was found to be heavily infected. Closer examination revealed that the parasites on the upper surface (mostly small ones) were concentrated on the head of the fish while those on the lower surface were more widely scattered. Since it is the head region of resting, buried soles that tends to be exposed above the sand, more small parasites would be expected to alight here. However, the tank containing this particular sole served as a colour-change display and had a stony bottom with no sediment. The implication is that post-larvae migrate forwards on the upper surface before transferring themselves to the lower surface. This was confirmed later by experimentally infecting freshwater-treated soles and studying the distribution of larvae after 9 days (Kearn, 1984). This work also indicated that larvae may not move directly forwards but may follow a zigzag path along the diagonal rows of scales.

4.12   TRANSFER OF POST-LARVAE AND ADULTS FROM HOST TO HOST

It is not uncommon to find one or two adult parasites on the upper surfaces of heavily infected soles in aquaria. Since soles in aquaria commonly lie for a time on top of one another, it is conceivable that adult parasites can transfer themselves from the lower surface of one host to the upper surface of the other. That this does indeed take place was demonstrated by placing uninfected soles in aquaria with infected fishes; within a day or so, parasites had become established on the upper surfaces of the uninfected fishes (Kearn, 1988b). Even more intriguing, adult parasites experimentally implanted on the upper surfaces of soles were found to behave like newly-invaded larvae, migrating forwards on the host and usually returning to the lower surface. Whether contact between wild soles occurs during their day to day activities is unknown, but prolonged contact between mating soles does occur and may provide a significant opportunity for transfer. Gyrodactylid monogeneans have abandoned ciliated larvae as a means of spreading to new hosts and rely largely on host contacts for transfer (see Chapter 5).

The persistence into adulthood of the ability to migrate from the upper to the lower surface suggests that transfer of parasites from host to host is not just an abnormal consequence of confining soles in aquaria. So, skin-parasitic monogeneans like *E. soleae* may have two ways of infecting new hosts, by way of eggs and ciliated larvae and by direct transfer of juveniles and adults. It is particularly interesting that an undescribed *Entobdella* parasitizing stingrays in Australia appears to have taken adult transfer one step further. When sub-adults are experimentally detached from the host, they swim vigorously head first by undulating the body, and this provides a means of transfer to neighbouring hosts without the necessity for any contact between them (Kearn and Whittington, 1991).

The modern distribution of *Entobdella* spp. spans elasmobranchs and teleosts, groups that are only distantly related, and can best be explained by a host-switching event from flat elasmobranchs to flat teleosts sharing the same benthic environment. It has been suggested that monogeneans like *Entobdella* that spread by transfer of adults from fish to fish may speciate more readily by host-switching than parasites relying entirely on oncomiracidia for dispersal (Kearn, 1994). In this context, host-switching appears to have been particularly prevalent in the gyrodactylids, which have abandoned free-swimming larvae as a means of infecting new hosts (section 5.5.1). In experimental infections of 'alien' hosts with *E. soleae*, oncomiracidia were less successful at establishing themselves on rays than adult parasites, which survived for 2–8 days, but the opposite situation occurred with the sole *Buglossidium luteum*, which became infected with oncomiracidia but could support adults for no more than 24 h (Kearn, 1967a). In spite of the success of oncomiracidial invasion, the parasites did not reach sexual maturity on *B. luteum*. The factors that determine whether oncomiracidia or adults are able to survive on alien hosts are unknown, but these experiments indicate that some hosts create problems for attachment and others for feeding.

4.13 FINDING THE WAY ABOUT

There are two sources of directional information that could be used by post-larvae or adults to find their way forward on the host: (1) antero-posterior water currents generated by gill ventilation and host swimming; (2) the backwardly projecting host scales. The scales provide visual and tactile clues, but the use by the parasite of the former as a guide to direction can be rejected because forward migration persists in total darkness (Kearn, previously unpublished observation). There is some evidence that adult parasites have a tactile 'awareness' of the host's scales (section 4.3.2) and an extension of this tactile sense and exploitation of the scales as 'signposts' may provide the basis for the anterior migration.

4.14 WHY MIGRATE?

Two separate questions require an answer: why migrate forwards and why migrate to the lower surface?

If migrating larvae do indeed travel diagonally forwards on the host as suggested above, they will soon reach the dorsal or ventral fins, which run continuously along the edges of the body (see Frontispiece and Fig. 4.1). However, larvae do not appear in significant numbers on the ventral surface adjacent to these fins, suggesting that the fins provide an effective barrier (Kearn, 1984). There are no fins along the anterior and ventral borders of the head and migrating larvae and adults probably have free access to the ventral surface in this region.

Adult parasites readily feed on the upper skin and a nutritional reason for the preference for the lower surface seems unlikely. Adult parasites on the upper surface would be vulnerable to predators and eggs laid by these adults might be less likely to adhere to sand grains and might therefore be swept away by currents from areas inhabited by sole (Kearn, 1984). Perhaps the most significant advantages are related to mating. Finding a mate is easier with the breeding population restricted to the lower surface rather than scattered over the whole fish, and a pheromonal system for location of a mating partner, if such a system exists, is unlikely to be effective on the exposed upper surface of the fish but is ideally suited for parasites on the lower surface (see above, section 4.7).

4.15   GROWTH AND DEVELOPMENT

In the oncomiracidium, the seven pairs of peripherally-arranged marginal hooklets are fully developed and probably play their main role at this time as larval attachment sclerites (Fig. 4.16). They are able to make individual gaffing movements and movements of the edge of the haptor may also serve to engage them. There are three pairs of central sclerites, but embryological evidence suggests that the two anterior-most or accessory sclerites are derived from a persistently-growing, centrally-placed, eighth pair of marginal hooklets (Kearn, 1963b). In the oncomiracidium these accessory sclerites are already different in shape and significantly larger than the other marginal hooklets. The other two pairs of central sclerites or hamuli, like the accessory sclerites, will continue to grow. The hooked regions of the posterior hamuli are well developed and probably supplement the marginal hooklets in the early post-larva, but the roots of these hooks are only partially developed (Kearn, 1963b). At this stage, the anterior hamuli are rod-shaped with no hooks. As the parasite increases in size, the growth of the anterior hamuli keeps pace with the growth of the accessory sclerites. This reflects the close functional relationship between them in juveniles and adults, and eventually they dwarf the non-growing marginal hooklets. Thus, there is an ontogenetic shift in the attachment apparatus, the marginal hooklets with the aid of the posterior hamuli being sufficient for the attachment of small post-larvae but requiring replacement by the suction-generating anterior hamulus–accessory-sclerite apparatus to cope with the stresses and strains of large size.

The male reproductive system reaches maturity before the female system (protandry) and the protandrous males appear able to donate and receive spermatophores (Kearn, 1970), presumably storing any spermatozoa received until the female organs become functional. *E. soleae* begins to assemble eggs when about 85 days old at 12°C and can survive on soles in aquaria for up to

182 days (Kearn, 1990). The parasite continues to grow throughout life, so the larger the parasite the older it is. The anterior hamuli (and possibly the other median sclerites) also grow persistently and larger adults produce more eggs (section 4.8).

CHAPTER 5

Other monogeneans parasitizing the skin of fishes

5.1 *ENOPLOCOTYLE*

During a visit to Lisbon, Portugal in 1970, I had the opportunity to examine a moray eel, *Muraena helena*. I knew that a monogenean, *Enoplocotyle minima*, had been collected in 1912 by Tagliani from the skin of eels kept in the Naples Aquarium, but there had been no further reports of this parasite. I took skin scrapings from my eel and was fortunate to find several specimens of *E. minima*, including adults (Kearn and Vasconcelos, 1979). Later, during a visit to Japan, I collected a second and previously undescribed species, *E. kidakoi*, from the skin of a related eel, *Gymnothorax kidako* (see Kearn, 1993).

Enoplocotyle species are interesting because they are small – flattened, preserved, adult specimens of *E. minima* and *E. kidakoi* measure about 350 μm and 500 μm in length respectively (the oncomiracidium of *Entobdella soleae* is 250 μm in length). Because the haptor is armed only with marginal hooklets, 14 peripherally arranged and two in the centre (Fig. 5.1*a*), *Enoplocotyle* must look today much like the first monogeneans that established themselves on the skin of ancestral fish-like vertebrates. Whether in fact *Enoplocotyle* is a direct descendant of these early monogeneans, persisting relatively unchanged during the enormous length of time during which the fishes evolved and diversified, is debatable and, as we shall see shortly, the parasite may be secondarily simplified. Perhaps being small and living on an inactive host inhabiting holes in reefs, the marginal hooklets are sufficient to cope with the relatively minor stresses and strains of attachment and require no supplementation by larger hooks (hamuli) as in *Entobdella soleae*.

Another reason for interest in *Enoplocotyle* is that the larva of *E. kidakoi* is unciliated (Fig. 5.1*b*). The eggs are sausage-shaped with a short appendage bearing an adhesive droplet, which attaches the eggs firmly to the bottom of observation vessels (Kearn, 1993). After incubation for about 9 days at 20°C, the eggs contain larvae that appear to be fully developed. These eggs fail to hatch spontaneously, although the larvae inside remain alive for at least another 30 days. However, the operculum of 40-day-old eggs, but not that of 9-day-old eggs, becomes detached with the slightest pressure and, when a piece of fresh host skin is brought into contact with eggs between 18 and 40 days old, the larvae hatch promptly by extending the body and pushing off the operculum. Attempts to induce hatching by applying fresh skin mucus from an eel and by placing a piece of skin near the eggs (2 mm away) were unsuccessful (Kearn, 1993). Thus, a chemical trigger from host skin appears to be involved, but the need for contact with host tissue suggests

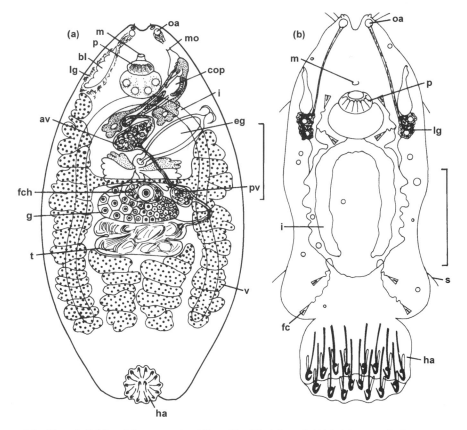

Figure 5.1 The adult (*a*) and the oncomiracidium (*b*) of *Enoplocotyle kidakoi*, in ventral view. av = anterior seminal vesicle; bl = bladder; cop = copulatory apparatus; eg = egg inside ootype; fc = flame cell; fch = fertilization chamber; g = germarium; ha = haptor with marginal hooklets; i = intestine; lg = lateral head glands; m = mouth; mo = opening of male genital pocket; oa = opening of adhesive sac; p = pharynx; pv = posterior seminal vesicle; s = sensillum; t = testis; v = vitellarium. Scale bars: (*a*) 100 μm; (*b*) 25 μm. Source: reproduced from Kearn, 1993, with permission from the Zoological Society of London.

that this chemical does not diffuse well, is released in low concentrations by the host or possibly needs to be combined with physical disturbance of the eggs to induce hatching.

Moray eels will make frequent contact with the walls of their submarine caverns. The eggs of their enoplocotylid skin parasites will readily become attached to the walls and, when fully developed, will equally readily be stimulated to hatch by contact with the host. The ciliated epidermis of the oncomiracidium is redundant and has been lost.

5.2 THE ACANTHOCOTYLIDS

Acanthocotylids are parasites of agnathan fishes (hagfishes) and rays (elasmobranchs). They are similar in size to *Entobdella soleae*, and *Acanthocotyle lobianchi* from the skin of *Raja* spp. feeds on epidermis in the same way (Kearn, 1963a). However, there is an important difference in the haptor. *Entobdella* copes with increased strain on the haptor imposed by growth of the parasite by supplementing

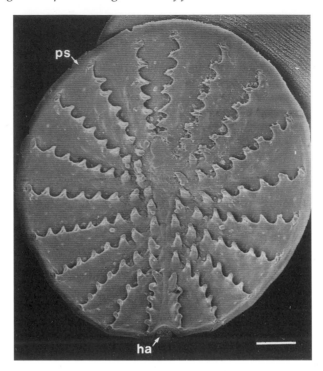

Figure 5.2   SEM photomicrograph of the haptor (ha) and pseudohaptor (ps) of *Acanthocotyle lobianchi*. Scale bar: 100 μm. Source: reproduced from Kearn, 1994 (original photograph by R. Evans-Gowing), with permission from the Australian Society for Parasitology.

its marginal hooklets with two pairs of hamuli. In addition, growth of the central pair of marginal hooklets (hooklets I, according to the numbering scheme proposed by Llewellyn, 1963) is persistent and, as accessory sclerites, they become associated with the anterior hamuli and generate suction. In *Acanthocotyle*, the larval haptor, with 16 (14 peripheral and two central) marginal hooklets, does not grow and the haptor and the hooklets are the same size in the adult as in the oncomiracidium (Fig. 5.2; Kearn, 1967b). There are no hamuli and, in the adult, marginal hooklets I are similar to all the other hooklets. Supplementation of the marginal hooklets is achieved in acanthocotylids by the separate development of a disc-shaped pseudohaptor, immediately anterior to the true haptor (Fig. 5.2). In *A. lobianchi*, this pseudohaptor is derived during post-larval development from the posterior region of the body, and acquires radial rows of hooked sclerites (Fig. 5.3). The pseudohaptor is muscular and generates suction, but how this is achieved is not clear.

The eggs of *A. lobianchi* are similar in shape to those of *Enoplocotyle* and the resemblance extends to the appendages, which are furnished with cement, and the oncomiracidium, which is unciliated (Fig. 5.4). Macdonald (1974) found that the eggs of *A. lobianchi* contain fully developed larvae after 15 days at 13°C, and, although they fail to hatch spontaneously, they are able to survive for as long as 83 days, probably by utilizing lipid reserves in the form of droplets in the oncomiracidium. The eggs of *A. lobianchi* hatch with much greater rapidity than those of *Entobdella soleae* when treated with host skin mucus (Fig. 5.4). The opercular cement is weakened ready for hatching, as in *Enoplocotyle*, and on receipt of the chemical

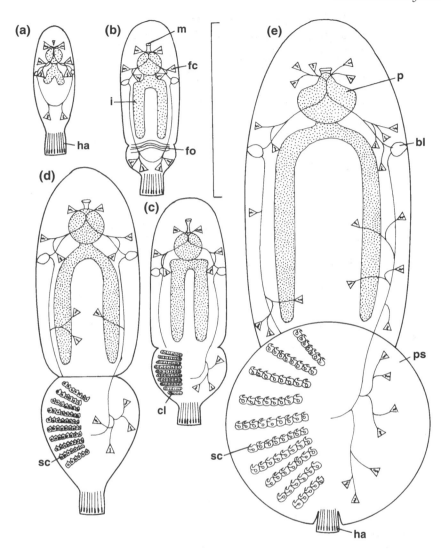

Figure 5.3 (a)–(e) Successive stages in the larval development of *Acanthocotyle lobianchi*. Folds (fo) on the ventral surface of the body indicate the position of the developing pseudohaptor (ps). Rows of prominent cells (cl) appear beneath the folds and each cell produces a single sclerite (sc). bl = bladder; fc = flame cell; ha = haptor with marginal hooklets; i = intestine; m = mouth; p = pharynx. Scale bar: 200 μm. Source: reproduced from Kearn, 1967b, with permission from Cambridge University Press.

stimulus the larva immediately extends the body, at least doubling its length, and dislodges the operculum.

Eggs of *A. lobianchi* attached to sand grains at the sea bottom are likely to be stimulated to hatch by mucus from a ray settling on the eggs. Freshly hatched larvae will have a good chance of making direct contact with the host and cilia are no longer required. It can be predicted that, unlike *E. soleae*, it will be more common for *A. lobianchi* to attach to the ray's lower (ventral) surface; this has been confirmed by collecting and measuring parasites from freshly caught *Raja clavata* (Fig. 5.5; Kearn, 1967b). *A. lobianchi* is essentially a parasite of the ventral surface of rays throughout its life, although some larvae may reach the upper surface when the host partially

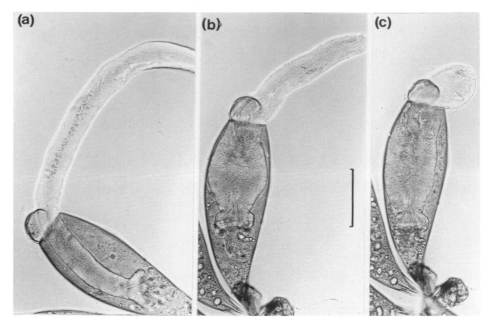

Figure 5.4 Hatching in *Acanthocotyle lobianchi*. Successive positions of the living larva photographed with electronic flash. (*a*) Immediately after stimulation of the unhatched egg with host mucus. The larva responds by rapidly extending the body, thereby pushing off the operculum. (*b*), (*c*) Taken successively as the larva withdrew into the shell after hatching. Scale bar: 50 μm. Source: reproduced from Kearn, 1986, with permission from Academic Press.

buries itself and throws sand containing eggs on to its upper surface. However, it is interesting that *A. elegans* is invariably found on the upper surface of the same host, *R. clavata* (see Kearn, 1967b), suggesting that this parasite embarks on a migration from the lower to the upper surface of its host, in the opposite direction to *E. soleae*.

There is good evidence that the chemical hatching factor for *A. lobianchi* is urea. Not only is commercial urea a potent hatching stimulant but the effectiveness of ray mucus is destroyed by incubation with urease (Kearn and Macdonald, 1976). The recruitment of urea as a hatching signal is probably related to the habit of elasmobranch fishes of maintaining a high urea concentration in the blood as part of their osmoregulatory strategy, with the consequence that relatively large amounts of urea are present also in their skin mucus. Urea appears to be ineffective as a hatching stimulant for *E. soleae* (section 4.10.3) and for *Enoplocotyle kidakoi* (see above).

There is a third acanthocotylid on *R. clavata*, namely *A. greeni* (Fig. 5.6a), which is found on the lower surface. There is a development in this parasite that we will meet in other monogeneans and in other parasitic platyhelminths, and that is the retention of eggs by the parent. In *A. greeni*, the eggs are tethered externally in the same way as in *E. soleae*, with the ends of their appendages held within the uterus. The difference is that in *E. soleae* only two or three eggs are retained in this way and are released within a period of an hour or so, while in *A. greeni* as many as 80 eggs are retained and the bunch is probably not released until some of the eggs are fully developed. Macdonald and Llewellyn (1980) discovered that the eggs of *A. greeni* hatch only in response to host mucus and suggested that the prolonged egg

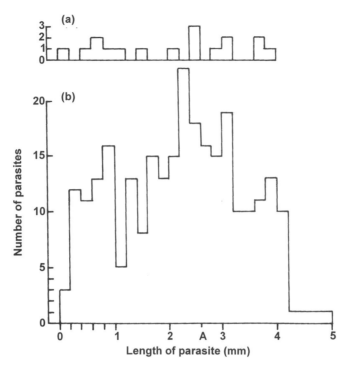

Figure 5.5 The distribution of *Acanthocotyle lobianchi* on (*a*) the dorsal surface and (*b*) the ventral surface of 14 rays (*Raja montagui*), freshly caught near Plymouth, UK in April 1965. A = onset of sexual maturity. Source: redrawn from Kearn, 1967b.

retention practised by the parasite provides an opportunity for the first-made eggs in the bunch to hatch and infect the host of their parent (autoinfection). If the egg bunch were then released, or actively implanted in the sediment by an arm-like body extension on which the uterus opens, the younger eggs could continue their development on the sea floor and be available to infect other rays settling on them. Macdonald and Llewellyn predicted that this dual reproductive strategy would lead to over-dispersion, a feature not uncommon in parasite populations in which a few hosts have high parasite burdens while most have low burdens. However, there are too few observations on the distribution of *A. greeni* to test this suggestion.

When *R. clavata* is resting on the bottom, any slight movement of the host is likely to dislodge prematurely an external egg bundle like that of *A. greeni*, by frictional contact with the sediment. Water currents may pose a similar threat when the host is active. There is, however, a shield-like extension of the body margin covering the uterine arm and the egg bundle, and this may alleviate this danger. *Acanthocotyle pugetensis* also retains an egg bundle and presumably follows a similar reproductive strategy to that of *A. greeni*, but in the former the egg bunch is stored internally in an enlarged uterus (Fig. 5.6*b*).

Other features of the reproductive system of *Acanthocotyle* spp. are intriguing. The terminal male organ is little more than a papilla, which is not especially muscular and lacks glands; there is also no vagina. Macdonald and Llewellyn (1980) observed ejection by *A. lobianchi* of clusters of spermatozoa mixed with a fluid from an accessory male gland; they regarded these bodies as spermatophores.

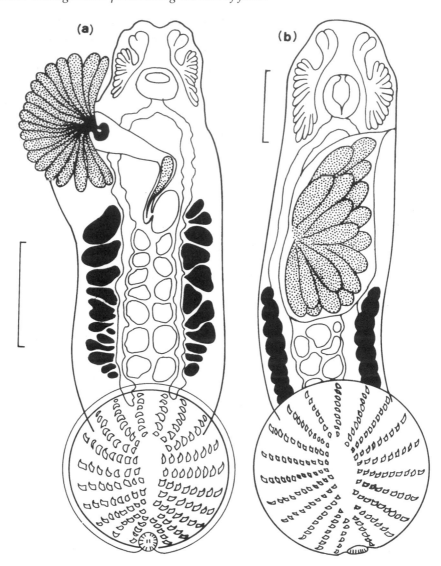

Figure 5.6   (*a*) *Acanthocotyle greeni* and (*b*) *A. pugetensis*. Eggs stippled. Scale bars: (*a*) 0.5 mm; (*b*) 0.2 mm. Source: (*a*) modified from Macdonald and Llewellyn, 1980; (*b*) reproduced from Bonham and Guberlet, 1938.

In the absence of any piercing organ that might introduce spermatozoa by hypodermic impregnation, the only access for spermatozoa is *via* the uterus and Macdonald and Llewellyn reported finding spermatozoa in this site in *A. lobianchi*. All of these features led Macdonald and Llewellyn to suggest that sperm transfer is accomplished by using the flexible uterine arm to pick up spermatophores, but they did not observe action of this kind and I have never seen mating take place in spite of long periods of continuous observation. It seems mechanically impossible for the uterine arm to pick up and assimilate a spermatophore when an egg bundle is blocking the uterus, so perhaps *Acanthocotyle* assimilates spermatozoa *via* the uterus prior to the commencement of egg assembly, as in *Benedeniella* (section 4.7).

There are several points of similarity between *Enoplocotyle* and the acanthocoty-lids, but a major difference between them is the absence of a pseudohaptor in the former. As mentioned above, *Enoplocotyle* could have ancient origins, having evolved before monogeneans acquired a pseudohaptor or other supplementations such as hamuli. Alternatively, *Enoplocotyle* may be a progenetic[*] acanthocotylid, i.e. an acanthocotylid that achieves sexual maturity while an early post-larva, before development of the pseudohaptor. Progenesis is another phenomenon that we will meet again among the parasitic platyhelminths.

5.3   THE CAPSALIDS AND MONOCOTYLIDS

The capsalids, typified by *Entobdella soleae*, are diverse, parasitizing teleost and elasmobranch fishes. *E. soleae* relies principally on marginal hooklets for initial attachment but, as the parasite grows, the phased introduction of two pairs of larger, ventrally directed hamuli occurs and suction generated by lifting the roof of the saucer-shaped haptor becomes important (section 4.3.1). Lifting is effected by mus-cles in the body which are linked by long tendons to the girder-like anterior hamuli embedded in the roof of the saucer; the accessory sclerites, which are derived from the central pair of marginal hooklets, provide a supporting role (Fig. 5.7*a*). Most capsalids have disc-shaped haptors armed with two pairs of ventrally-directed hamuli and a pair of accessory sclerites, but in *Benedenia* and its relatives (e.g. *Neobenedenia*, *Metabenedeniella*) suction is generated by directly lifting the roof of the saucer rather than by way of the hamuli (Fig. 5.7*b*).

There is evidence that *Entobdella* uses intrinsic haptor muscles to generate suction independently of the extrinsic muscle system (section 4.3.1). There has been a tendency to develop this intrinsic muscle system in many capsalids, the haptors of which are subdivided by partitions to produce numerous separate loculi, each with intrinsic suction capability. *Pseudobenedenia* (Fig. 5.7*c*) has followed this route, while retaining its extrinsic muscle–tendon system and full complement of sclerites. How-ever, *Capsala martinieri* has abandoned entirely the extrinsic muscle system in favour of muscular loculi and has lost all its median sclerites (Fig. 5.7*d*).

Capsalids include some of the largest monogeneans, *C. martinieri* from the skin of the sunfish *Mola mola* having a circular body exceeding 2 cm in diameter. This large body is attached to the host by a haptor 8 mm in diameter, i.e. bigger than a whole adult specimen of *E. soleae*. Haptors with such large diameters may be more efficient and may provide greater security for attachment to relatively inflexible fish skin if they opt for multiple independent loculi rather than a single suctorial unit.

There is evidence that some capsalids have followed a completely different route leading to loss of the median sclerites. A single specimen of an undescribed species of *Trimusculotrema* was found by Kearn and Whittington (in Kearn, 1994) on the

---

* There is much confusion over the terms 'progenesis' and 'neoteny'. They are regarded by some as synonymous, while others recognize a difference between precocious sexual devel-opment of a juvenile (progenesis) and the retention of juvenile characters in the sexually mature adult (neoteny; see Pearson, 1972; Williams and Jones, 1994, p. 65). In practice, it is often difficult to discriminate between these two situations and the term 'progenesis' will be used in this book in the sense of precocious development of the reproductive system, which may lead to the achievement of sexual maturity in juveniles or adults at an earlier stage than in its relatives.

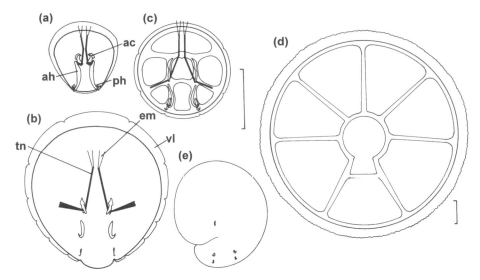

Figure 5.7 Haptors of some capsalid monogeneans. (*a*) *Entobdella soleae*. (*b*) *Benedenia seriolae*. (*c*) *Pseudobenedenia nototheniae*. (*d*) *Capsala martinieri*. (*e*) *Trimusculotrema* sp. ac = accessory sclerite; ah = anterior hamulus; em = extrinsic adductor muscle; ph = posterior hamulus; tn = tendon; vl = marginal valve. Scale bars: 1 mm; (*a*)–(*c*) and (*e*) drawn to the same scale. Source: (*a*)–(*c*), (*e*) reproduced from Kearn, 1994, with permission from the Australian Society for Parasitology; (*c*) modified from an original drawing by Williams, Ellis and Spaull, 1973; (*d*) reproduced from Kearn, 1976a, with permission from Elsevier Science.

epidermis of a stingray. This parasite had reduced median sclerites and no loculi or marginal valve, features usually associated with suction generation (Fig. 5.7*e*), and attempts to dislodge the parasite from glass showed that the circular haptor was attached by cement. Thus, monogenean cements appear to be suitable for long-term attachment of the haptor to living epidermal surfaces, as well as for temporary attachment of the head region to similar surfaces during locomotion (section 4.3.2). The development of cement for haptoral attachment may have led to loss of suctorial ability and reduction of the sclerites. Cement is also used by microbothriid monogeneans for semi-permanent attachment of the haptor to hard surfaces (see below).

The habitats of some skin-parasitic capsalids are remarkably restricted. *Benedenia lutjani* shows a preference for the pelvic fins of *Lutjanus carponotatus* and *Metabenedeniella parva* is exclusively found on the dorsal fin of *Diagramma pictum* (see Whittington and Kearn, 1993 and Horton and Whittington, 1994 respectively). How the parasites find their way to these specialized habitats is unknown but, having done so, the chances of meeting another individual for sperm exchange will be high.

In contrast with the high degree of host specificity of many capsalids, *Neobenedenia melleni* has been recorded from more than 100 species of teleosts, belonging to more than 30 families and five different orders (Whittington and Horton, 1996).

Evidence that the immune system of fishes may be capable of responding against skin-parasitic monogeneans comes from experimental work by Bondad-Reantaso *et al.* (1995) on *N. girellae* from the Japanese flounder (*Paralichthys olivaceus*; see also work on gyrodactylids, section 5.5.1). Fewer and smaller parasites were recovered from fishes which had experienced a primary infection. Unlike gyrodactylids, the work on *N. girellae* did not indicate the involvement of humoral antibody and the implication is that skin-mediated immunity may have an important role.

Figure 5.8   SEM photomicrograph of the haptor of *Dendromonocotyle* sp. Scale bar: 250 μm. Photograph by R. Evans-Gowing from a specimen donated by I. Whittington. Source: reproduced from Kearn, 1994, with permission from the Australian Society for Parasitology.

Monocotylid monogeneans are elasmobranch parasites that provide a striking example of convergence with some capsalids. Stingray skin parasites of the genus *Dendromonocotyle* have disc-shaped multiloculate haptors that have lost or reduced the single pair of hamuli characteristic of most monocotylids (Fig. 5.8). As well as 14 peripheral marginal hooklets, the free edges of the partitions between the loculi are provided with tiny, non-slip sclerites. However, most monocotylids inhabit the gill cavities, nasal fossae and other internal sites (see Chapter 6 and Table 7.1).

Predation may be a significant threat to parasites living on the exposed outer surface of fishes and some features of capsalid, monocotylid and other monogeneans may have evolved in response to selection pressures exerted by predators (see also section 4.14). Body pigmentation is not uncommon in skin parasites and, at least from the human viewpoint, can provide very effective camouflage. 'Cleaning' symbioses, in which fishes permit 'cleaner' organisms (smaller fishes, crustaceans) to search their bodies and presumably to remove parasites, are well known in coral reef communities and may be common elsewhere (Kearn, 1994). Although there is not yet any field evidence that cleaners prey on monogeneans, there are reports of predation on gyrodactylids and microbothriids (see below) in aquaria (see Kearn, 1994), and Cowell *et al.* (1993) demonstrated experimentally that cleaner fishes (gobies) are particularly effective at removing *Neobenedenia melleni*.

The capsalid *Capsala martinieri* and the monocotylid *Dendromonocotyle kuhlii* contain brown–black body pigment and are difficult to detect on the skin of their hosts (Kearn, 1994), in spite of the large size of these parasites. The pigment in

D. kuhlii lies in the gut and may be derived from ingested host epidermal cells, but because the gut branches overlie the reproductive organs they provide a dorsal shield of pigment and this suggests that the pigment is not merely a dietary residue waiting for disposal. It is interesting that the stingray *Dasyatis americana*, a host for *Dendromonocotyle* sp., is regularly cleaned by a wrasse, *Thalassoma bifasciatum*, but Snelson *et al.* (1990) found monogeneans on the upper surface of a ray only 2 days after a cleaning episode. Although the stomach contents of wrasse contained no identifiable remains of *Dendromonocotyle*, this is not conclusive evidence that they are not occasionally taken. In *C. martinieri*, much of the pigment lies in the parenchyma.

Kearn (1994) has suggested that other skin-parasitic monogeneans may achieve inconspicuousness in a different way. Like many free-living pelagic organisms they may achieve virtual invisibility from predators by adopting transparency. Comparison between the capsalids *Entobdella soleae* and *Benedenia seriolae*, the former living on the lower surface of a flat-fish and unlikely to be subject to predation and the latter being exposed on the flanks of a round-bodied teleost (*Seriola quinqueradiata*), reveals a striking difference in transparency. The vitellaria, which fill most of the bodies of these parasites, are highly transparent in *B. seriolae*, imparting a glass-like quality to the body, while those of *E. soleae* are opaque, rendering the animal yellowish-white and much less difficult to detect, even on the host's white underside (see Frontispiece).

5.4 THE MICROBOTHRIIDS

Microbothriids are skin parasites of elasmobranch fishes. Hooks are absent in adults and this feature has led some to doubt that microbothriids are monogeneans. However, these doubts have been dispelled by the finding of six spicules, which are most probably vestigial marginal hooklets, in the larva of *Leptocotyle minor* (see Kearn and Gowing, 1990) and by studies of attachment that have revealed explanations for the loss of functional hooks in the adult.

Leptocotyle minor attaches itself to a single denticle embedded in the skin of the common dogfish (*Scyliorhinus canicula*; Fig. 5.9), and since the denticles have no covering of skin and are extremely hard, hooks would be useless (Kearn, 1965). The relatively small, hookless haptor attaches itself to the denticle not by suction but by cement secreted by gland cells in the posterior region of the body. Parasites can be induced to attach to glass and, when removed, leave behind at the haptor site a patch of cement. *Leptocotyle* also has adhesive glands in the head region and these enable the parasite to move in leech-like manner from denticle to denticle, although locomotion is retarded by the length of time needed to strip the haptor away from the denticle (Kearn, 1965).

The common dogfish is a small, bottom-living shark found on sand, fine gravel and mud and feeding largely on bottom-dwelling invertebrates (Wheeler, 1978). The eggs of *L. minor* (Fig. 5.10a) are roughly ovoid in shape and do not hatch spontaneously (Whittington, 1987a). The ciliated oncomiracidium completes its development in 16–18 days at 13–14°C and can survive within the egg for as long as 88 days. Rapid hatching is triggered by washings from dogfish skin and experimentally by urea, a situation which is somewhat surprising since the relatively active, free-swimming habits of the dogfish seem to preclude contact between eggs and host

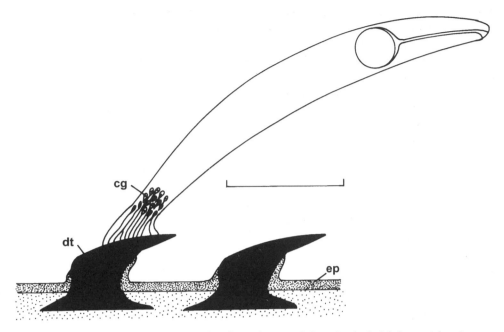

Figure 5.9 Diagram illustrating the mode of attachment of the microbothriid *Leptocotyle minor* to a denticle in the skin of its dogfish host, *Scyliorhinus canicula*. cg = cement gland; dt = denticle; ep = host epidermis. Scale bar: 500 μm. Source: reproduced from Kearn, 1994, with permission from the Australian Society for Parasitology (redrawn from an original diagram in Kearn, 1965).

for long enough to permit hatching and host invasion. However, the eggs have a very long and exceedingly fine appendage (approaching 3 mm in length on an egg capsule about 120 μm long), which enhances the chances of eggs being swept up into suspension by passing hosts and greatly retards the sinking of suspended eggs (Whittington, 1987a). Dogfish foraging in groups near the bottom are likely to

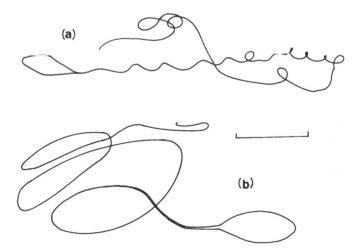

Figure 5.10 The egg of (*a*) the microbothriid skin parasite *Leptocotyle minor* and (*b*) the polyopisthocotylean gill parasite *Hexabothrium appendiculatum* from the dogfish, *Scyliorhinus canicula*. Scale bar: 200 μm. Source: reproduced from Whittington, 1987a, with permission from Cambridge University Press.

swim through clouds of suspended eggs, which will be stimulated to hatch by the host's skin secretions and may even become entangled on denticles.

The ability of the oncomiracidia to swim and, therefore, to remain in suspension is probably important to maximize opportunities for host contact. Whittington (1987b) found that the larvae swim erratically, changing speed and direction abruptly, and show no response to shadows or to changes in light direction. They also show a readiness to attach to any suitable surface that is available, and all of these behavioural features seem well suited to promote contact and attachment to foraging dogfishes.

Leptocotyle minor shares its dogfish host with an unrelated monogenean gill parasite, a polyopisthocotylean *Hexabothrium appendiculatum* (section 6.5.3). These two parasites provide a remarkable example of convergent evolution. Their eggs (Fig. 5.10) resemble each other in shape, have similar appendages and hatch rapidly in response to urea; there are similarities too in the behaviour of their larvae (Whittington, 1987a, b). The exploitation of urea as a hatching signal seems to have evolved at least three times in monogenean parasites of elasmobranchs, namely in the acanthocotylids, in the microbothriids and in the hexabothriid polyopisthocotyleans.

5.5   THE GYRODACTYLIDS

5.5.1   General biology

The gyrodactylids are small epidermis feeders that attach themselves by hooks (one pair of hamuli and 16 peripherally situated marginal hooklets; Fig. 5.11), mostly without complementary suction. In the acanthocotylids we have already met a tendency towards egg retention with possible autoinfection. In the gyrodactylids this practice has gone to extraordinary lengths and has led to a reproductive strategy so successful that a high proportion of teleost fishes are colonized by them.

As long ago as 1832, von Nordmann observed hooks inside the body of a *Gyrodactylus* but regarded them as part of the gut, and it was von Siebold in 1849 who recognized that these hooks are part of an embryo within the parent and that *Gyrodactylus* is viviparous (see review by Harris, 1993). The single embryos are retained inside the uterus (or possibly the ootype) until they are virtually adult. There is no longer a need for an eggshell and, since nutrients can be supplied to the 'naked' embryo from the gut, the vitellarium is redundant (Fig. 5.11*b*), although isolated syncytial bodies are regarded as persistent vitellaria by Cable, Harris and Tinsley (1996). The infrapopulations (an infrapopulation is defined as all individuals in or on an individual host; see Margolis *et al.*, 1982) of these viviparous gyrodactylids increase by autoinfection and the skin-inhabiting parasites are entirely responsible for infecting new hosts, either by direct transfer when hosts make contact or by using the substrate as a staging post between hosts (see review by Kearn, 1994). Gyrodactylid adults cannot swim, unlike *Entobdella*, in which the sub-adults of at least one species can do so (section 4.12).

In ancestral monogeneans, spread of adults or sub-adults from host to host by contagion may well have occurred side by side with infection of new hosts by freely deposited eggs and larvae, as in *Entobdella soleae*. Many monogeneans that have colonized gills and other internal microhabitats have probably been forced to

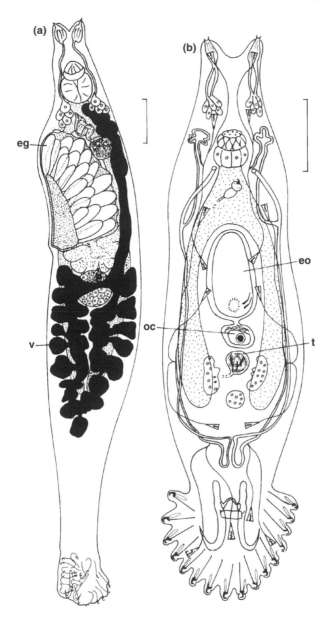

Figure 5.11 (*a*) An oviparous gyrodactylid, *Phanerothecium harrisi* from *Plecostomus plecostomus* and (*b*) a viviparous gyrodactylid *Gyrodactylus pungitii* from *Gasterosteus pungitius*. eg = egg; eo = embryo; oc = oocyte; t = testis; v = vitellarium. Scale bars: 100 μm. Source: redrawn from (*a*) Kritsky and Boeger, 1991 and (*b*) Malmberg, 1970.

abandon transfer by contagion, but in the viviparous gyrodactylids it is the only option. Some gyrodactylids have colonized the gills, but they must leave the gill cavity to transfer to new hosts.

It was not until 1983 that an oviparous gyrodactylid, *Oogyrodactylus farlowellae*, was discovered by Harris on an Amazonian catfish. Since then, the rich Amazonian fauna has yielded others, and their reproductive features foreshadow those of their viviparous relatives. They retain eggs *in utero* (Fig. 5.11*a*), where unciliated larvae develop inside them. According to Boeger, Kritsky and Belmont-Jégu (1994), the embryonated eggs are attached by adhesive droplets to the skin or bony plates of their catfish hosts. Thus, they may have already abandoned freely disseminated eggs and free-swimming larvae as a means of infecting new hosts in favour of spreading by contagion. The attachment of eggs to the host of the parent parasite will boost the infrapopulation, increasing opportunities for transfer from host to host. It is but a short step to loss of the eggshell and viviparity.

The problem with the viviparous gyrodactylids is that the embryo must spend a long time inside the uterus (or ootype?) of the parent in order to grow to full size. This places severe restraint on reproductive output. Some gyrodactylids go some way towards alleviating this problem either by giving birth early before the embryo is fully grown (as in *Macrogyrodactylus*; see Khalil, 1970) or by supporting two embryos side by side (as in *Gyrodactylus gemini*; Ferraz, Shinn and Sommerville, 1994), but they also have a unique development that goes much further towards countering the loss of fecundity – a second embryo begins to develop inside the first before the latter is born. Thus, when the first offspring is born it already contains a well developed embryo of its own (Fig. 5.12).

The viviparous gyrodactylids divert resources from their own reproductive system into the development of their embryos. This extends to a delay in the

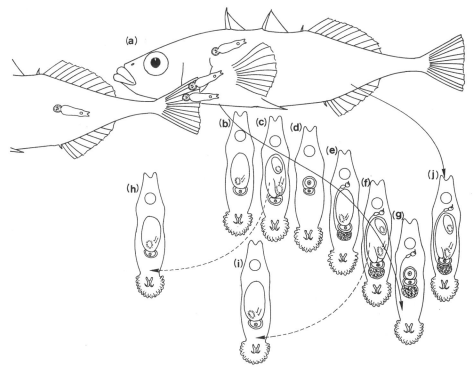

Figure 5.12   The life cycle of *Gyrodactylus* sp. (*a*) Parasites transfer from fish to fish when hosts make contact with each other. (*b*)–(*g*) Successive stages in the life cycle of a newborn individual: (*c*) gives birth to first daughter (*h*); (*d*) oocyte enters empty ootype–uterus (?); (*e*) male reproductive system develops; (*f*) gives birth to second daughter (*i*). Mating may take place between late stages – (*e*), (*f*) or (*g*) – and another adult (*j*). Source: reproduced from Kearn, 1994, with permission from the Australian Society for Parasitology.

appearance of the male reproductive system, creating a situation of protogyny in contrast with the pronounced protandry already encountered in this book and typical of oviparous monogeneans. In fact, the male system does not appear until after the first birth and the commencement of development of the second oocyte (Fig. 5.12).

There is good evidence that the first-born daughter is produced asexually by mitosis, but the origins of the second daughter, which develops in the parent after the birth of the first, seem to be more complex (see Harris, 1993). Meiotic chromosomes have been identified in the second oocyte and there have been reports of spermatozoa associated with and even inside it, but the source of these spermatozoa and their role remain unknown. Harris emphasized that these intracellular spermatozoa were too few to be detected with the light microscope and that the situation differs from the normal results of cross-insemination, in which spermatozoa are readily visible in the seminal receptacle. It is also known that sex is not an essential requirement for production of the second daughter because development of this daughter can commence, perhaps *via* some kind of meiotic parthenogenesis, in isolated parasites with no opportunity for cross-insemination, and, as far as we know, no spermatozoa of their own. Sex is thought to be restricted to the third-born and any subsequent daughters.

The relative importance of the asexual and sexual phases is related to the population age structure and mortality of individual species (Harris, 1993). For example, in *G. gasterostei* and *G. pungitii* on sticklebacks, mortality is high and parasites are unlikely to reach ages and densities that favour sexual reproduction; no inseminated individuals have been found and sex is probably rare. In contrast, in *G. salaris*, mortality is low, with 10–15% of individuals surviving to reproduce sexually; inseminated individuals occur even in light infections and sex is a normal part of their biology.

The special brand of viviparity (termed hyperviviparity by Cohen, 1977) displayed by the gyrodactylids, combined with autoinfection, leads to a rapid increase in numbers of parasites on a single host. This ties in with the reliance on contagion for transmission, since it will ensure that parasites are sufficiently numerous and widespread on the host to take advantage of any host to host contacts. Furthermore, according to Scott (1985a), the lethargic behaviour and abnormal swimming of heavily infected guppies attracts other fishes and increases contacts. However, there is a down-side because large numbers of parasites pose a threat to host survival, which, in turn, threatens survival of the parasites.

One way in which natural selection may have reduced the impact of high infection intensity on the host is by promoting a reduction in size of individual parasites. Most gyrodactylids are small, with lengths falling within the range 0.4–0.8 mm. This may explain why most gyrodactylids have a supplementary pair of hamuli, while other monogeneans of similar size subsist with just marginal hooklets. Hamuli possessed by relatively large ancestral gyrodactylids may have been retained in modern descendants in spite of a reduction in body size, except in *Isancistrum* and *Anacanthocotyle*, which have lost the hamuli (see Llewellyn, 1984 and Kritsky and Fritts, 1970 respectively).

A second way in which the effects of large numbers of gyrodactylids is modulated is *via* the host's antiparasite defences. Guppies (*Poecilia reticulata*) may be infected with two separate species of *Gyrodactylus*, namely *G. turnbulli* and *G. bullatarudis* (see

Harris, 1986), and the relationships between these parasites and their host have been extensively studied. Figure 5.13(*a*) shows the typical course of a primary infection of a previously uninfected ('naive') guppy with *G. turnbulli* (incorrectly identified by Scott (1985b) as *G. bullatarudis*, according to Harris, 1986). The infrapopulation rises to a peak over a few days and then declines, often reaching zero. The decline and rapid elimination of a secondary (challenge) infection, administered immediately after recovery from the primary (Fig. 5.13*b*), is consistent with the expression of a host immune response generated by the primary infection. Scott (1985b) found that many fishes did not regain their full susceptibility until 4–6 weeks post-recovery, and even at 6 weeks some fishes were only partially susceptible to reinfection. According to Madhavi and Anderson (1985) susceptibility of the guppy is variable and genetic factors are involved in its determination. Richards and Chubb (1996) found that the primary response of guppies provided some protection against challenge from either *G. turnbulli* or *G. bullatarudis*, irrespective of the species of parasite involved in the initial infection.

Although the immunological basis of the host reaction to gyrodactylids is well established, how this resistance operates is still unknown. According to Lester (1972) the epidermis of uninfected sticklebacks sheds sheets of solidified mucus or 'cuticle' at 1- or 2-day intervals. Sticklebacks infected with *G. alexanderi* lose any parasites attached to this 'cuticle' and an increase in density of the material was regarded as a reaction to *Gyrodactylus*. However, its role in the control of parasite numbers has

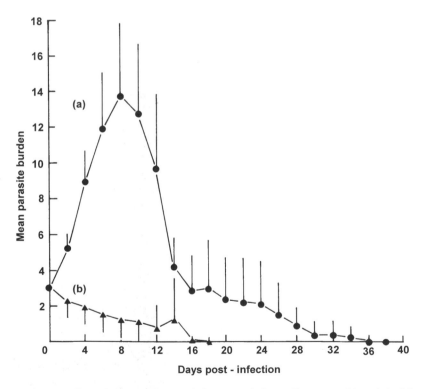

Figure 5.13   Mean number of *Gyrodactylus turnbulli* (not *G. bullatarudis*, as stated in original figure) per guppy (*Poecilia reticulata*) during (*a*) an initial infection and (*b*) a challenge infection immediately after recovery. Bars represent one standard error. Source: reproduced from Scott, 1985b, with permission from Blackwell Science.

been questioned by Scott and Anderson (1984), who did not observe continual shedding of 'cuticle' by guppies infected with *G. turnbulli* (identified as *G. bullatarudis*), and were unable to offer an explanation for the dramatic decline in parasite infrapopulations. Richards and Chubb (1996) also found no cuticular shedding in guppies infected with *G. turnbulli* or *G. bullatarudis*, but reported a thickening of the epidermis of the caudal peduncle in some fishes infected with *G. turnbulli*.

By measuring the length of survival of *G. stellatus* in serum and in mucus collected from the host *Pleuronectes vetulus* at stages during the rise and fall of parasite infrapopulations in the laboratory, Moore, Kaattari and Olson (1994) found evidence to suggest that both serum and mucus contain factors involved in resistance to the parasite (cf. work on capsalids, section 5.3). In general, parasite survival was shorter in serum and mucus samples collected from flat-fishes at the later recovery stages of infection. Mucus from infected fishes contained proteins antigenically similar to fish serum proteins, as demonstrated by a gel diffusion immunoassay using a rabbit antiserum against fish whole serum. Precipitation reactions were strongest in mucus collected during the later stages of infection and not detected in the mucus of uninfected fishes, indicating that the precipitation bands observed were associated with the gyrodactylid infection.

Intraperitoneal injection of fishes with a preparation of whole, killed *G. stellatus* generated serum with a significant effect on parasite survival, but no such effect was demonstrated with mucus. This is consistent with the notion that fish have a secretory immune system that is separate from the systemic immune system (references in Moore, Kaattari and Olson, 1994), and that the route of administration of the antigen determines whether or not the epidermal defence system responds. In this respect, it is interesting that Richards and Chubb (1996) found evidence to suggest that the host response to gyrodactylids by guppies is largely localized to areas of heavy infection.

Surprisingly little is known about the behavioural factors influencing the transfer of gyrodactylids from one individual host to another. Bakke *et al.* (1992) reported that detached specimens of *G. salaris* and specimens attached to dead hosts increase their 'searching' activity and transmission rate and reduce their host-preference selectivity. The regions of the head near the openings of the anterior adhesive glands are likely to be important in sensory discrimination between hosts. A multiciliate 'spike' sensillum located close to these openings on each side was reported by Lyons (1969b) and may have chemosensory or mechanoreceptive functions. Two pairs of cilia-based structures that may also have a part to play in host-to-host transfer were found below the body surface near the 'spikes' by Watson and Rohde (1994). One of these pairs closely resembles the cilia-based organs presumed to be photoreceptors in the oncomiracidium of *E. soleae* (section 4.10.4).

In *G. salaris*, parasites that had not yet given birth for the first time were less likely to transfer to a new host than worms that had already given birth at least once (Harris, 1993; Harris, Jansen and Bakke, 1994). Harris *et al.* saw this as an adaptive strategy, since the asexually produced first daughter, being genetically identical with its parent, may be better suited for survival on the host of its parent than on a new host. Also, parents with large first embryos have already committed considerable resources and it may be safer to remain on the old host, which has supported this investment, rather than transfer to a new host, which may fail to sustain this support (Harris, Jansen and Bakke, 1994). The fitness of sexually produced offspring with new gene combinations may be suboptimal relative to that of the parent when infecting the same host and such offspring may be better able to colonize a new host successfully. However, host defences introduce a new dimension. On overcrowded hosts, defences against the parasite may be approaching their maximum expression and sexually produced offspring may be able to survive longer, thereby increasing their chances of encountering a new host for transfer.

A feature of *Gyrodactylus* spp. that has caused much consternation among parasitologists is the conservatism of their morphology – species are distinguished mostly on the basis of morphometric differences and subtle differences in shapes of the sclerites. In spite of this, more than 400 species of *Gyrodactylus* are known, according to Harris (1993). Recent evidence on host specificity, host range and morphological variation creates difficulties in determining the limits of species within this genus.

There are different views on the degree of host specificity enjoyed by viviparous gyrodactylids. Bychowsky (1957) considered them to be the least specific of the monogeneans, while Malmberg (1970) regarded them as narrowly host specific. Bakke *et al.* (1992) made an analysis of host specificity using data from well-known species that had been used in field and laboratory studies; they came to the conclusion that gyrodactylids are less host-specific than commonly thought, with 30% of the 76 species being recorded from a single host species. The most detailed analysis is available for *G. salaris* (see Bakke *et al.*, 1992), and this reproduces on several salmonid fishes including the salmon (*Salmo salar*). The parasite can also survive for up to 8 days on the eel, *Anguilla anguilla*, but it fails to feed and cannot reproduce. *G. salaris* is also notable for its high degree of morphological variation and Harris (1993) has suggested that frequent indulgence in sexual reproduction in this species may have generated the genetic diversity to account for this variation and also its wide host range.

In view of these findings and the practice of basing descriptions of *Gyrodactylus* spp. on specimens collected from a single host, there may be a need to reconsider the validity of many species attributed to the genus.

In ovigerous monogeneans such as the polyopisthocotyleans (see Chapter 6), there is evidence for substantial parallel evolution between parasites and their hosts (phylogenetic radiation; see review by Kearn, 1994). However, according to Harris (1993), the complexities of the distribution patterns of gyrodactylids are not consistent with a high degree of phylogenetic radiation, and host-switching is likely to have played a significant role. In fact, the suggestion has been put forward that transfer of parasites by contagion may favour speciation by host-switching (Kearn, 1994; see also section 6.5.4).

In gyrodactylids there are some spectacular examples of discontinuous species distributions that can best be explained by host-switching. For example, the only monogeneans so far recorded from an invertebrate are gyrodactylids of the genus *Isancistrum* on squid of the genus *Haloteuthis*, and species of *Gyrodactylus* are ectoparasites of both anuran and urodele amphibians (see Kearn, 1994). There are less spectacular but equally interesting examples. Phylogenetic radiation is indicated by the primary distribution of species belonging to the so-called *G. wegeneri* group on cyprinid teleosts, but host-switching has also taken place from cyprinids to sticklebacks, percids and cottids sharing the same habitat (see Harris, 1993). In this example, it is especially interesting that some of the recipient hosts are predators feeding on cyprinids, and predator–prey contacts may provide important opportunities for host-switching. Bakke *et al.* (1992) also regard interspecific fish shoals and contact between scavengers and dead fishes as providing other important opportunities for transfer. As an example of the second of these possibilities there is a record by Mo (in Bakke *et al.*, 1992) of the salmon parasite *G. salaris* on the flat-fish *Platichthys flesus*.

As Harris (1993) has pointed out, the complexity of population genetics in organisms with mixed reproductive strategies like *Gyrodactylus* is considerable, and there is much to be learned about the relative contributions to speciation of the asexual and sexual phases of reproduction and how these contributions are influenced by environmental factors. These are matters of interest to a wide range of biologists, not just to parasitologists.

Houde and Torio (1992) claimed that male guppies that had been infected with *G. turnbulli* were significantly less attractive to females than uninfected males. They attributed this to discrimination by the females on the basis of the brightness of the males' orange spots, which were paler in the infected individuals. As predicted by Hamilton and Zuk (1982), this may ensure that the offspring inherit genes for resistance to parasites, but it may also reduce the chances of females contracting the parasitic infection and suffering the pathological consequences of such an infection. Houde and Torio concluded that more information is needed on natural patterns of transmission and heritability of resistance for a meaningful appraisal of these possibilities (see also sections 7.1.4(d), 17.7.1).

5.5.2 *Gyrodactylus salaris* and salmon

Gyrodactylus salaris was first reported as a pathogen of wild juvenile salmon (*Salmo salar*) in the Lakselva River in northern Norway in 1975 (see reviews by Dolmen, 1987 and Johnsen and Jensen, 1991). The parasite caused heavy mortality of salmon parr (young river fish from 0+ to 3 years in age), with several thousand parasites on individual fishes; most of the parr population was wiped out over a 2-year period and subsequently the parasite appeared in other river systems and salmon hatcheries. As a consequence of the decline in parr density, catches of adult salmon in the rivers and in the sea have fallen.

This devastating epidemic posed questions about the source of the pathogen and the reasons for its virulence – few monogeneans are pathogenic, except in overcrowded aquarium situations (section 4.4). Some authorities regard *G. salaris* as endemic to the Norwegian fauna and consider that the heightened susceptibility of the salmon is the consequence of deleterious effects on the fish of environmental factors, or unfavourable changes created by man in the genetic make-up of the hosts. Others believe that *G. salaris* is an exotic species, introduced on fish stocks from elsewhere in Scandinavia.

In suitable river sites, salmon parr populations are dense and individuals may be territorial (Bakke *et al.*, 1992). Within a population, *G. salaris* may spread by physical contact during agonistic interactions. The parasite can endure brackish water, but sea water is lethal, so adult salmon arriving from the sea for spawning in the rivers are uninfected. Adult fishes readily pick up parasites as they pass upstream through populations of infected parr and also when they make contact with precocious male river fishes that take part in spawning with the adults. Since the adults cover large distances, they are potentially able to spread the infection through the river system, and fishes (smolts) heading for the sea for the first time may likewise spread the infection downstream. Eels, which can sustain a non-reproducing infection for several days, and flounders, which are potentially able to carry infections through brackish water separating one river system from another, may also be important in the distribution of the disease. Newly hatched salmon (alevins) can also be infected, but the pathogenicity of the parasite at this stage and the significance of this age group as hosts are unknown (Bakke *et al.*, 1992).

One approach to control has been to eradicate salmon in infected rivers with rotenone, followed by restocking with disease-free fish.

5.6   THE WAY FORWARD

The gill chambers of fishes are readily accessible to skin parasites and it is not surprising that several different kinds of monogeneans have established themselves in these chambers. In the next chapter we will examine gill morphology, possible selection pressures that have led to their colonization and some of the adaptations adopted by gill dwellers.

CHAPTER 6

Monogeneans parasitizing the gill chambers of fishes

6.1 FISH GILLS

We are perhaps most familiar with the arrangement of gills in a bony fish. If the operculum or gill cover is raised, we see beneath it the first of usually four gills (Figs 6.1*a*, 6.2*a*). Each gill has a bony arch and projecting from it are many flattened and tapering primary gill lamellae (Figs 6.1*b*, 6.2*a*). Each primary lamella carries two rows of closely-spaced, flap-like secondary gill lamellae (Figs 6.1*b*, 6.9), projecting in opposite directions from its two flat surfaces. The gill owes its red colour when fresh to these secondary lamellae, which are thin-walled, contain blood and are the main sites of gaseous exchange. Blood flow inside the secondary lamellae is opposite to

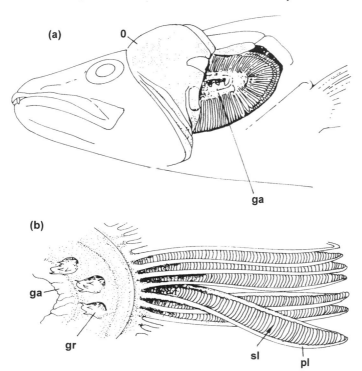

Figure 6.1 (*a*) Gills of the left side, as seen after raising the operculum of *Lucioperca*. (*b*) Part of a gill. ga = gill arch; gr = gill raker; o = operculum; pl = primary gill lamella; sl = secondary gill lamella. Source: (*a*) reproduced from Rauther, 1937; (*b*) based on Schmidt, 1942.

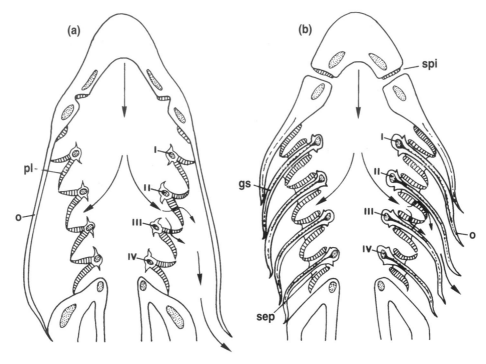

Figure 6.2 Diagrammatic sections in a horizontal plane through the buccal and gill chambers of (*a*) a teleost and (*b*) an elasmobranch fish. The opercula (o) are shown closed on the left and open on the right of each diagram. I–IV = gill arches; gs = gill septum; pl = primary lamella bearing secondary lamellae; sep = septal canal; spi = spiracle. Arrows indicate paths of gill ventilating currents. Source: based on Bertin, 1958.

the water flow in the interlamellar space, so creating the advantages of a counter-current exchange system; this is an important adaptation for maximizing gaseous exchange.

There is a basic similarity between the respiratory apparatus of teleost and elasmobranch fishes (Fig. 6.2). Elasmobranchs also have primary gill lamellae projecting from a gill arch and carrying rows of blood-filled secondary lamellae. In teleosts and elasmobranchs, each gill (holobranch) has two rows of primary gill lamellae and each of these rows is known respectively as the anterior and posterior hemibranch (half-gill). The main difference between teleosts and elasmobranchs concerns the length of the gill septum running between the hemibranchs of each gill. In elasmobranchs this septum is long, projecting beyond the distal ends of the primary lamellae and forming a flap-valve for the gill slit immediately posterior to it (Fig. 6.2*b*). Thus the septum forms a barrier between the two hemibranchs of each gill. In teleosts there is a single flap-valve or operculum covering all the gill slits on one side and the gill septa are greatly reduced, so that in many teleosts there is no barrier between the two hemibranchs (Fig. 6.2*a*).

Physiological studies of teleosts have shown that their gills are ventilated by a double pumping system, a buccal cavity force pump and an opercular cavity suction pump; these pumps complement each other so that water flow through the gills is more or less continuous (see Hughes, 1984). During normal respiration, the tips of the primary gill lamellae of adjacent gill arches touch, so that water passing between the gill arches from the buccal cavity is forced to pass between the secondary gill lamellae in order to enter the opercular cavity (Fig. 6.2*a*). The water current strikes one hemibranch of each gill on its anterior face and the other on its posterior face. As we shall see later (section 6.5.2), the continuous nature of the gill-ventilating current and the difference in the direction of the flow in different hemibranchs have important consequences for gill-parasitic monogeneans.

There is no gill septum to impede water flow as it passes between the secondary gill lamellae of many teleosts, although in some (e.g. the clupeomorphs) the gill septum is well developed proximally and absent from only the distal third of the gill. In the elasmobranchs the septum is a complete barrier between hemibranchs and might be thought to seriously impede water flow through the gill. However, between the secondary lamellae and the gill septum there is a septal channel, which permits water to flow distally (Fig. 6.2*b*).

6.2   COLONIZATION OF THE BUCCAL AND BRANCHIAL CAVITIES

This has taken place so many times in the monogeneans that these cavities must offer significant improvements in life style for skin parasites. Avoidance of predation by 'cleaner' organisms (sections 4.14, 5.3) might be an important incentive, but a more important advantage may be related to opportunities for cross-insemination, which will be greater in the more compact living space.

No special challenge faces monogenean colonizers from the outer skin surface of fishes, since the buccal and branchial cavities offer extensive, relatively flat areas of epidermis. Ambient water currents are not necessarily stronger than they would be on the skin surface of a rapidly moving fish, although gill-ventilating currents are likely to be more or less continuous. This situation is reflected in the lack of obvious specialization in many monogeneans living in these habitats.

Many gyrodactylids inhabit the gill cavity and they are not only similar to their skin-parasitic relatives but must move freely from gills to skin and *vice versa* because of their dependence on skin contact between hosts for dispersion. Some of the larger monocotylids live on the flat surfaces within the gill pouches of elasmobranchs and have no obvious adaptations for this life style (Fig. 6.3). There are capsalid relatives of *Entobdella* that occupy similar niches in teleosts, again with little obvious change. For example, *Trochopus plectropomi* has a haptor that is small enough to attach to the relatively wide and flat inner border of the host's primary lamella, where there are no secondary lamellae (Fig. 6.4*a*). The disc-shaped haptors of capsalids are flexible and are able to mould to curved surfaces. Some of them attach themselves to the borders of the primary lamellae by folding the haptor, either longitudinally, as in *Benedenia rohdei*, or transversely, as in *T. pini* (Fig. 6.4*b, d*). A versatile capsalid of the genus *Benedenia* is able to attach itself to the gill arches, gill rakers or pharyngeal tooth pads of its teleost host (Whittington and Kearn, 1991).

Few of these invaders of the gill cavity are blood feeders, suggesting that ready access to host blood was not a significant factor in the colonization of the gill cavity. One group of monogeneans, the polyopisthocotyleans, are exclusively blood feeders, but there is no reason to suppose that their origins were any different from those of other gill-dwelling monogeneans, i.e. they may have evolved from skin-feeding gill parasites (see also below).

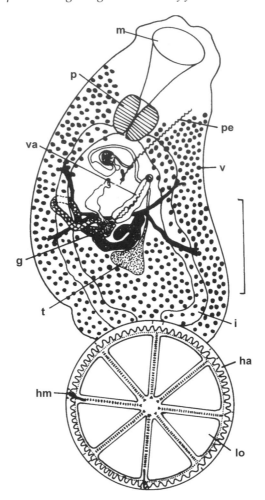

Figure 6.3 *Monocotyle spiremae* from the non-respiratory surfaces in the gill chamber of *Himantura uarnak*. g = germarium; ha = haptor; hm = hamulus; i = intestine; lo = loculus; m = mouth; p = pharynx; pe = penis; t = testis; v = vitellarium; va = vagina. Scale bar: 1 mm. Source: modified from Measures, Beverley-Burton and Williams, 1990.

The secondary gill lamellae are small, closely spaced and delicate; attachment to them by relatively large multicellular organisms like monogeneans threatens the host by inflicting serious damage and/or by interfering with gaseous exchange. Nevertheless, many monogeneans have colonized these respiratory surfaces, a move that has often involved important changes in the haptor (see below).

6.3   THE MONOCOTYLIDS

Relatively minor changes in the haptor have permitted some monocotylids to colonize the interlamellar spaces between the secondary gill lamellae of their elasmobranch hosts. *Horricauda* has a typical monocotylid haptor with ventral loculi (Fig. 6.5). Juveniles are small enough to fit into the interlamellar space and

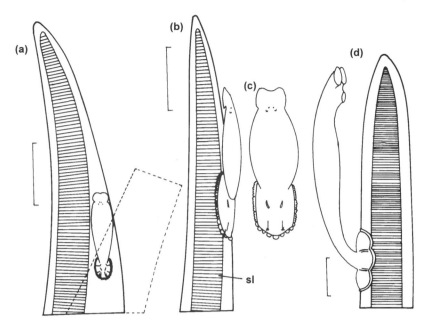

Figure 6.4 Attachment attitudes of some capsalid parasites on primary gill lamellae. (*a*) *Trochopus plectropomi*. (*b*) *Benedenia rohdei*. (*c*) *B. rohdei* in dorsal view with haptor unfolded. (*d*) *Trochopus pini*. sl = secondary gill lamellae. Scale bars: 1 mm. Source: (*a*)–(*c*) reproduced from Whittington and Kearn, 1991, with permission from CAB International; (*d*) reproduced from Kearn, 1971b, with permission from Cambridge University Press.

their haptors have acquired six anteriorly directed, dorsal spikes. Thus, each parasite secures itself in its narrow niche by using loculi, a single pair of hamuli and 14 marginal hooklets to attach to the secondary lamella in contact with the ventral surface of the haptor, and it seems likely that the six spikes impale the lamella in contact with the dorsal surface of the haptor. Adult specimens of *H. rhinobatidis* abandon these interlamellar spaces and relocate in the septal canals (Kearn, 1978).

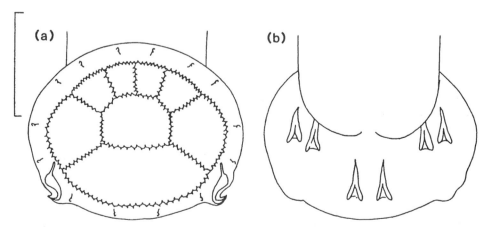

Figure 6.5 The haptor of *Horricauda rhinobatidis*, (*a*) in ventral view, and (*b*) in dorsal view. Scale bar: 100 μm. Source: reproduced from Kearn, 1994, with permission from the Australian Society for Parasitology (modified from Kearn, 1978).

6.4   THE DACTYLOGYROIDEANS

The dactylogyroideans have speciated explosively in parallel with the teleost expansion and rival the gyrodactylids and the polyopisthocotyleans as teleost parasites. One group, the dactylogyrines, has made a major contribution to this expansion, especially on the diverse cyprinid freshwater fishes, and in countries where cyprinids are harvested for food they may create problems in fish farms. Another group, the amphibdellids, has the distinction of parasitizing elasmobranchs (electric rays), an association that seems likely to have arisen by host-switching.

6.4.1   Attachment, locomotion and reproduction

The sizes of most dactylogyroideans are limited by the need for the haptor to fit into the interlamellar niche between adjacent secondary gill lamellae. Like the capsalids, dactylogyroideans have two pairs of hamuli but in the latter one of these pairs points in a dorsal direction.

In *Amphibdelloides* on electric rays and in *Tetraonchus* on a teleost host (the pike *Esox lucius*), the hamuli are linked functionally in two lateral pairs, each pair comprising a dorsal and ventral hamulus, which are tethered to a supporting bar and bound together by fibres in such a way that the hamuli can counter-rotate (Fig. 6.6). The consequence of this counter-rotation is that the sharp points of the hooks will impale the adjacent secondary gill lamellae, the two dorsal hamuli impaling one of them and the two ventral hamuli the other (Fig. 6.7). Each lateral pair of hooks is operated as a unit by a single extrinsic muscle, which is linked *via* a long tendon and fibrous loops to the hamuli. In *Amphibdelloides* the system is relatively simple, with the tendon passing through a loop attached to the root of the ventral hamulus and inserting on the root of the dorsal hamulus (Fig. 6.6*a*); contraction of the muscle will draw the roots of the two hamuli together, causing the hooks to rotate outwards into the gill lamellae. In *Tetraonchus*, there is an extra loop attached to the root of the dorsal hamulus (Fig. 6.6*b*); this 'pulley' system probably has greater mechanical advantage than that of *Amphibdelloides* (see Kearn, 1966).

In spite of their biological interest, little attention has been paid to the attachment mechanisms of other dactylogyroideans, but what we do know suggests that they are likely to be mechanically diverse. In *Haliotrema balisticus*, the functional linkage between the four hamuli is different, the two ventral hamuli being linked by tendons (Fig. 6.6*c*) and in *Chauhanellus australis* the hooks are operated by intrinsic muscles (Kearn, 1994). In *C. australis*, hook protraction in detached specimens is extremely rapid – faster than the eye can follow – and this may be related to the fact that the hook-operating muscles are striated. Striated muscle is rare in the platyhelminths but occurs in a few other situations, such as in the scoleces of some cestodes (section 10.6) and the tails of digenean cercariae (sections 13.5, 15.3).

Some dactylogyroideans supplement their hamulus apparatus with accessory structures, such as dorsal and ventral friction pads in diplectanids (Fig. 6.8*a*, *b*) and spines in *Chauhanellus youngi* and *Rhamnocercus* spp. (Fig. 6.8*c*, *d*). In *Furcohaptor* the haptor has become laterally elongated and these lateral extensions with the reduced hooks at their ends encircle and grasp the whole primary lamella (Fig. 6.9).

Fourteen or 16 marginal hooklets are typically present in adult dactylogyroideans. Most of them are on the ventral surface, but two pairs are usually dorsally situated. The contribution of the hooklets to attachment seems minor compared

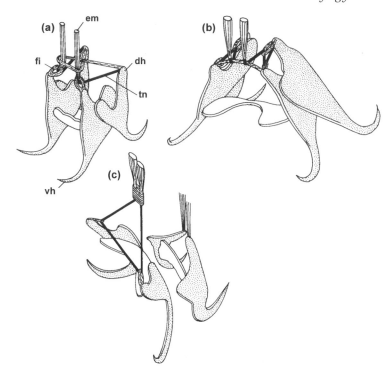

Figure 6.6 The adhesive mechanisms of some dactylogyroidean gill parasites. (*a*) *Amphibdelloides maccallumi*. (*b*) *Tetraonchus monenteron*. (*c*) *Haliotrema balisticus*. dh = dorsal hamulus; em = extrinsus muscle; fi = fibrous loop; tn = tendon; vh = ventral hamulus. Source: based on original diagrams from (*a*) Llewellyn, 1960, (*b*) Kearn, 1966 and (*c*) Kearn, 1971c; reproduced from Kearn, 1994, with permission from the Australian Society for Parasitology.

with that of the relatively massive hamuli and it is difficult to understand why they persist. However, in *C. australis*, the marginal hooklets are especially active immediately after the haptor moves to a new interlamellar niche (Kearn, 1994) and, by pinning the haptor to the lamellae, they may facilitate penetration of the lamellae by the hamuli. Each marginal hooklet is mounted on a mobile papilla. In *Hamatopeduncularia* these papillae have become independently mobile tentacles that slowly extend and provide numerous widespread attachment points (Fig. 6.10). This development has permitted some *Hamatopeduncularia* spp. to leave the restrictive niches between secondary lamellae and attach themselves in exposed sites such as the edges of the primary lamellae (see Kearn and Whittington, 1994). Here there is no limitation on haptor size and some, like *H. major*, are relatively large. However, there is evidence that their attachment is reinforced by an adhesive secretion.

A second and probably independent development of cement for adhesion to gills has occurred in the calceostomatines. Their haptors have reverted to mobile, flattened discs (Fig. 6.11) and it has been shown that the ventral side of this disc in *Neocalceostomoides brisbanensis* is strongly adhesive (Kearn, Whittington and Evans-Gowing, 1995). The hamuli are reduced to one pair and these and the slender marginal hooklets are all ventral. Cytons within the haptor may be the source of the cement; their secretory granules are diverted into and widely dispersed throughout the ventral haptor tegument and are probably released on the ventral surface by

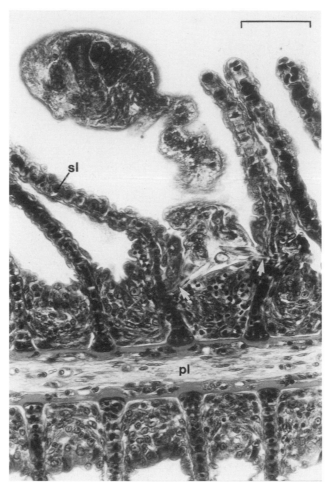

Figure 6.7   Photomicrograph of a section through *Tetraonchus monenteron* attached between the secondary gill lamellae (sl) of the pike, *Esox lucius*. White arrows indicate positions of the hamuli. pl = primary gill lamella. Scale bar: 50 μm.

exocytosis. The cements produced by monogeneans are remarkable enough for their ability to bond to the wet, slimy, body surface of fishes (see sections 4.3.2, 5.3), but even more so for their ability to maintain this bond in the face of the continuous gill-ventilating current.

The dactylogyrines have placed a different kind of emphasis on marginal hooklets. The ventral hamuli are reduced to vestiges and the ventral and dorsal marginal hooklets, which are enlarged by handle extensions, supplement the dorsal hamuli (Fig. 6.12). In *Neodactylogyrus crucifer*, the ventral marginal hooklets have an accessory, supporting, x-shaped bar (Fig. 6.12a), and in *Dactylogyrus crassus* a pair of dorsally orientated marginals is enlarged and creates a pincer-like arrangement with the dorsal hamuli (Fig. 6.12b).

Few dactylogyroideans are sedentary, in spite of a possible threat of dislodgement by the continuous gill-ventilating current and the high energy-costs of locomotion. Locomotory 'steps' in a dactylogyroidean like *Tetraonchus* have been observed on excised gills (Kearn, 1987a) and are similar to those of *E. soleae* (section 4.3.2). After

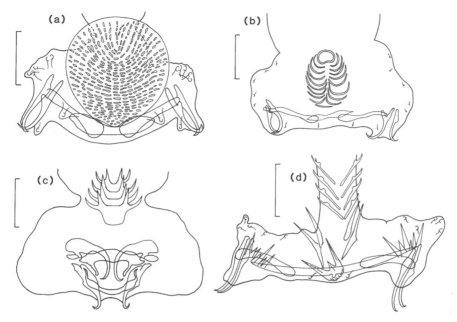

Figure 6.8 Dactylogyroideans with accessory attachment structures. (*a*) *Diplectanum aculeatum*. (*b*) *Lamellodiscus parisi*. (*c*) *Chauhanellus youngi*. (*d*) *Rhamnocercus rhamnocercus*. Squamodiscs (sq) in (*a*) and (*b*) and spines in (*d*) occur on both sides of the haptor but are shown only on one side; marginal hooklets are omitted in (*c*). Scale bars: 50 μm. Source: modified from original drawings in (*a*) Bychowsky, 1957, (*b*) Oliver, 1969, (*c*) Kearn and Whittington, 1994 and (*d*) Monaco, Wood and Mizelle, 1954; reproduced from Kearn, 1994, with permission from the Australian Society for Parasitology.

extending the body, the sides of the head are attached to the free borders of two adjacent secondary lamellae (Fig. 6.13) – on each side of the head there are three eversible sacs that receive the ducts from adhesive glands. Thus the head forms a bridge between two lamellae and the haptor is inserted into the space beneath the bridge. Sensilla on the haptor (see Kearn and Gowing, 1989) probably play a part in the positioning of the haptor and in the subsequent initiation of marginal hooklet movements and hamulus protraction. At least two of these sensilla are mobile and Kearn and Gowing suggested that the movement might enhance interaction with chemical signals. Alternatively, cessation of ciliary movement on touching host tissue may have a role in contact reception (see also section 17.4).

An important advantage in preserving mobility is the promotion of cross-insemination and generation of genetic flexibility. The need to avoid interspecific matings has led to a bewildering array of copulatory sclerites in dactylogyroideans, illustrated in Fig. 6.14 by just a few of the male copulatory sclerites of dactylogyrines (see also below). The functional morphology of few of these organs has been considered, but the study by Llewellyn (1960) of the male copulatory apparatus of *Amphibdelloides* illustrates their high level of sophistication. In *Amphibdelloides* there are two opposable accessory sclerites acting like pincers (Fig. 6.15). These are operated by special muscles and grip a sclerotized vaginal pad on the co-copulant, thereby countering the gill-ventilating current, which will tend to separate the mating partners. The penis is a long, narrow, sclerotized tube, which passes through a perforation in one of the two accessory sclerites and is ensheathed by longitudinally running muscle fibres linking the expanded base of the tube to the accessory sclerite. Llewellyn (1960) has likened the penis to the flexible cable-release

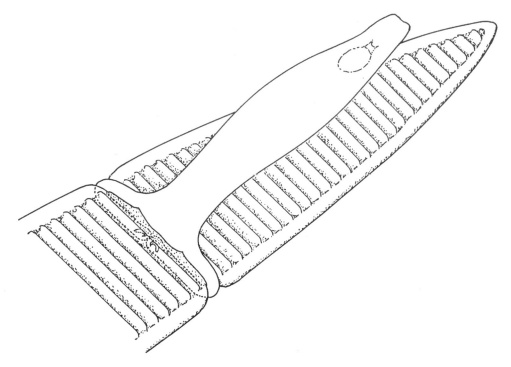

Figure 6.9 *Furcohaptor cynoglossi* attached to the primary gill lamella of *Cynoglossus macrostomus*. Source: reproduced from Bijukumar and Kearn, 1996, with permission from Kluwer Academic Publishers.

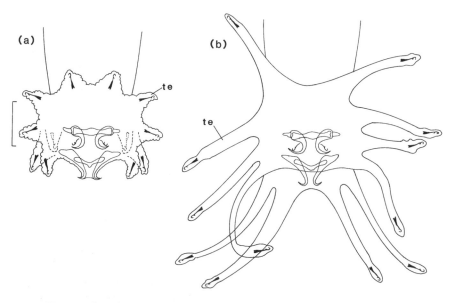

Figure 6.10 *Hamatopeduncularia pearsoni* with haptoral tentacles (te) retracted (*a*) and extended (*b*). Scale bar: 50 μm. Source: reproduced from Kearn and Whittington, 1994, with permission from the Australian Society for Parasitology.

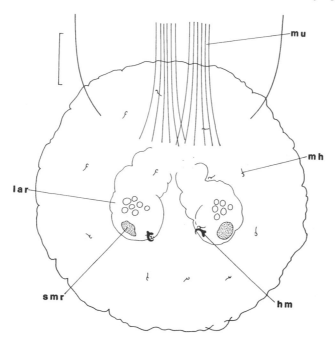

Figure 6.11 Attached haptor of *Neocalceostomoides brisbanensis*. hm = hamulus; lar = large gland reservoir; mh = marginal hooklet; mu = muscle; smr = small gland reservoir. Scale bar: 100 µm. Source: reproduced from Kearn, Whittington and Evans-Gowing, 1995, with permission from the Australian Society for Parasitology.

system of a camera; when the muscle sheath contracts the penis tube will protrude and enter the vagina.

Some dactylogyroideans have developed ways of promoting cross-insemination without having to resort to extensive locomotion, an activity that may incur increased risk of dislodgement by gill-ventilating currents. *Chauhanellus australis*, for example, can greatly extend the body to reach parasites on adjacent primary lamellae or hemibranchs (Kearn and Whittington, 1994), and the diplectanid *Telegamatrix ramalingami* has an extensile copulatory tentacle bearing at its distal extremity the vaginal opening and the male copulatory apparatus (Fig. 6.16).

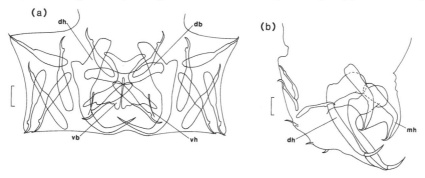

Figure 6.12 (*a*) Haptor of *Neodactylogyrus crucifer*. (*b*) Haptor of *Neodactylogyrus sparsus* in dorsolateral view, showing the relationship between the dorsal hamuli (dh) and the enlarged marginal hooklets (mh). db = dorsal bar; vb = ventral bar; vh = ventral hamulus. Scale bars: 10 µm. Source: based on original drawings by Gusev (1985); reproduced from Kearn, 1994, with permission from the Australian Society for Parasitology.

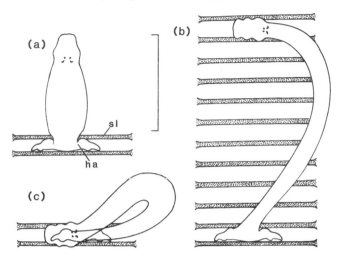

Figure 6.13   (*a*)–(*c*) Successive stages in locomotion of *Tetraonchus monenteron* from one site to another on the same side of the primary lamella of its host, the pike, *Esox lucius*. ha = haptor (hooks not shown); sl = free border of secondary gill lamella. Scale bar: 0.5 mm. Source: reproduced from Kearn, 1987a, with permission from the American Society of Parasitologists.

Occasional change of location on the gill may be important to avoid local host inflammatory reactions, which could lead to complete encapsulation of the haptor by host tissue. According to Bychowsky (1957, p. 75), haptor encapsulation may be the first step in a process of parasite elimination by the host, the next step being the shedding of the host nodule with the parasite attached to it.

A few dactylogyroideans have turned this situation to their own advantage. Overgrowth of the haptor by host tissue, provided that it poses no threat during the reproductive life of the parasite, will provide secure attachment without expenditure of energy by the parasite. This has been achieved in *Ancylodiscoides parasiluri* (Fig. 6.17) and has led to significant reduction in the size of the haptor and its armature. Similar developments occur in polyopisthocotyleans (see below).

The diplectanid *Lamellodiscus acanthopagri* interacts with the host's inflammatory response in an entirely different way. Autoinfection is promoted not by retention of eggs by the parasite, as in *Acanthocotyle* or *Phanerothecium* (see Figs 5.6 and 5.11*a* respectively), but by attaching the eggs individually to the gills (Roubal, 1994a). This is achieved in *L. acanthopagri* by the egg appendage, which is not long and flexible, as it is in many of its relatives, but short, curved and rigid, terminating in a sharp point (Fig. 6.18). This thorn-like appendage is hooked through the host epithelial tissue lying between the secondary gill lamellae and initiates the development of a host inflammatory nodule. This eventually forces the egg out of the gill tissue and leads to its expulsion from the host. The timing of larval development is such that most eggs hatch before they are shed; the emergent ciliated larvae either attach themselves adjacent to the eggs (autoinfection) or escape into the water and infect other fishes. A few eggs may be shed before they hatch and contribute to infection of other hosts.

6.4.2   Diversity and speciation

The dactylogyroideans have outstripped most other groups of monogenean gill parasites in terms of numbers of species. Part of this wealth of species can be

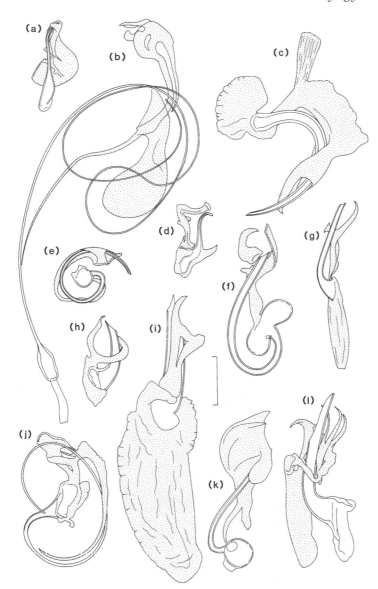

Figure 6.14 Diversity of male copulatory sclerites in *Dactylogyrus* and *Neodactylogyrus* spp. (*a*) *N. goktschaicus*. (*b*) *D. auriculatus*. (*c*) *N. cristatus*. (*d*) *D. molnari*. (*e*) *N. spirocirrus*. (*f*) *N. crucifer*. (*g*) *D. amphibothrium*. (*h*) *D. colonus*. (*i*) *D. crassus*. (*j*) *D. fallax*. (*k*) *N. alatus*. (*l*) *N. mongolicus*. All drawn to the same scale; scale bar: 20 μm. Source: based on original drawings by Gusev (1985); reproduced from Kearn, 1994, with permission from the Australian Society for Parasitology.

attributed to the diversity of the teleostean hosts that provide a platform for their evolution, but an extra dimension is provided by a marked tendency for individual teleost species to harbour up to ten or more congeneric gill parasites (species clusters; see Gusev, 1995).

Gusev (1995) has summarized some of the possible factors responsible for the creation of these species clusters on freshwater hosts. He regarded the isolation of river systems by rising sea levels in the late Tertiary and Quaternary periods as

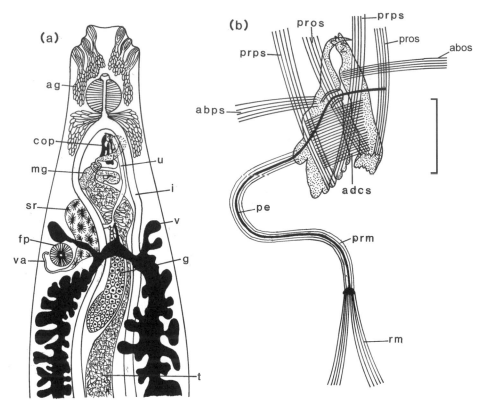

Figure 6.15 Genitalia of *Amphibdelloides maccallumi*. (*a*) Anterior region of the body, showing the position of the male copulatory apparatus and the vagina. (*b*) The male copulatory apparatus and the muscles that operate it. abos = abductor muscle of opposable sclerite; abps = abductor muscle of penis-bearing sclerite; adcs = adductor of copulatory sclerites; ag = anterior head glands; cop = copulatory apparatus; fp = fibrous pad; g = germarium; i = intestine; mg = male accessory gland; pe = sclerotized penis tube; prm = protractor muscle sheath of penis; pros = protractors of opposable sclerite; prps = protractors of penis-bearing sclerite; rm = retractor muscle of penis; sr = seminal receptacle; t = testis; u = uterus; v = vitellarium; va = vagina. Scale bar: 50 μm. Source: reproduced from Llewellyn, 1960, with permission from Cambridge University Press.

central to this development, since each isolating episode subdivided infected host populations and provided the opportunity for the evolution of new taxa of parasites and hosts. In the intervening periods of sea water regression, the isolated river systems became reunited, their faunas intermingled, and related or conspecific fishes from different river systems exchanged their monogeneans, so that individual fish species acquired clusters of divergent parasites. This process was reinforced by the strict host specificity of the parasites and the availability of different microhabitats on the gills of each host.

Thus, although the species clusters give an impression of sympatry, Gusev believed that many of them are allopatric in origin. However, Euzet and Combes (1980) envisaged a mechanism whereby sympatric speciation could take place *in situ* on the gills of the host. They assumed that the morphology of the male copulatory sclerites and the shape of the corresponding vagina were genetically linked, and that a mutation would alter both of these structures in such a way that they retained compatibility with each other but were no longer compatible with the sclerites of the parent species. The mutant would be able to self-inseminate but be unable to mate

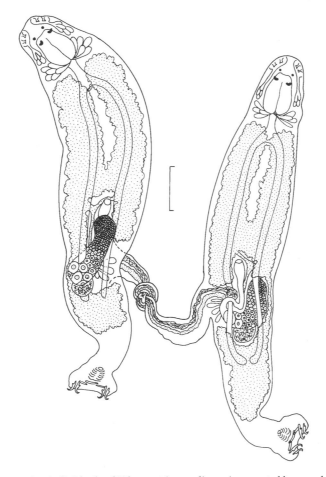

Figure 6.16   Two mating individuals of *Telegamatrix ramalingami* connected by copulatory tentacles, each bearing the vaginal opening and male copulatory apparatus. Scale bar: 100 μm. Source: based on an original drawing by Bychowsky and Nagibina (1976); reproduced from Kearn, 1994, with permission from the Australian Society for Parasitology.

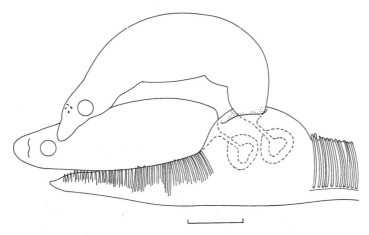

Figure 6.17   *Ancylodiscoides parasiluri* attached to a primary gill lamella. Scale bar: 1 mm (approx.). Source: based on an original drawing by Gusev and Strelkov (1960); reproduced from Kearn, 1994, with permission from the Australian Society for Parasitology.

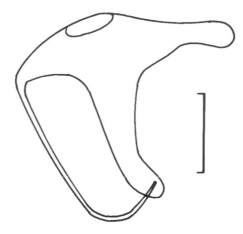

Figure 6.18   The egg of *Lamellodiscus acanthopagri*. Scale bar: 30 μm. Source: redrawn from Roubal, 1994b.

with individuals of the parent species. Provided that the mutant and its offspring could utilize a different ecological niche from that of its parent species, it would survive and flourish. There is no evidence yet that speciation does occur in this way, but one prediction of the hypothesis, that the most significant morphological differences between congeneric species sharing the same host would concern their reproductive sclerites, is borne out by observation (Euzet and Combes, 1980). Moreover, partitioning of the gills by congeneric species is one aspect of niche diversity that has been identified (Euzet and Combes, 1980).

6.5   THE POLYOPISTHOCOTYLEANS

6.5.1   Overview

Polyopisthocotyleans are almost exclusively blood feeders, a habit that was recognized by Llewellyn in 1954. This diet is reflected in some unique features, which, together with some other distinctions, set them apart from the other monogeneans (Monopisthocotylea). There are indications from sequencing work with 18S ribosomal DNA that monogeneans may not be monophyletic, (see Blair, 1993; Rohde *et al*, 1993, 1995).

In polyopisthocotyleans the terminal or subterminal mouth opens into an oral cavity that is either suctorial, as, for example, in the polystomatid parasites of amphibians and the hexabothriid gill parasites of elasmobranchs, or contains two buccal suckers, as in the teleost gill parasites (mazocraeideans; Fig. 6.19).

The pharynx is relatively small and muscular and the intestinal lining comprises digestive cells, separated and supported by a syncytial connective tissue (Halton, 1975). The digestive cells are responsible for the uptake and intracellular digestion of host blood proteins. The haematin-rich residues from this digestive process accumulate in vacuoles and it is these cells, with their dark brown haematin contents, that give polyopisthocotyleans their distinctive pigmented appearance. The cells release their haematin into the gut lumen from time to time, the sloughing of whole haematin cells being a rare event (Halton, 1976). The haematin residue is egested by a route that bypasses the pharynx; according to Llewellyn (1972), there is a narrow

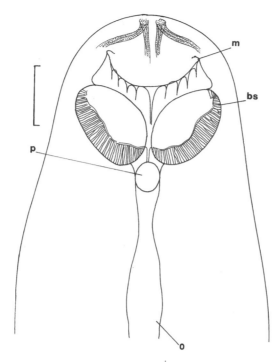

Figure 6.19 The anterior region of the blood-feeding ·polyopisthocotylean, *Pricea multae*. Scale bar: 100 μm. Source: based on an original drawing by Rohde, 1976; reproduced from Kearn, 1994, with permission from the Australian Society for Parasitology.

canal linking the oesophagus with the buccal cavity or prepharynx and haematin has been observed leaving *via* this canal.

Polyopisthocotyleans also possess another unique feature, a genitointestinal canal linking the oviduct with the intestine. A relationship with the blood-feeding habit is indicated by the suggestion that this tube provides a route for symbiotic bacteria to reach the eggs and hence the next generation of blood-feeding parasites (see Llewellyn, 1972). Such bacteria are concerned with blood digestion and are characteristically found in the gut of haematophagous animals. However, Morris and Halton (1975) found no bacteria in the guts of fresh specimens of the polyopisthocotylean *Diclidophora merlangi*.

The advantages of blood-feeding versus skin-feeding are not obvious. It might be supposed that the ensured supply of blood would favour a sedentary life style, with corresponding energy savings, and, indeed, many polyopisthocotyleans appear to be sedentary (see below). However, given the likelihood that gill epithelium can regenerate and that monogeneans have extensible and flexible bodies, there is no reason to believe that skin feeders need to move about to obtain sufficient food. It might, therefore, be more profitable to look elsewhere for reasons for adopting a sedentary life style, in particular to interactions with gill-ventilating currents (see below), to mating patterns and to possible interactions with local host defences. As far as adopting a diet of blood is concerned, there may be nutritional advantages.

Each of the major modern groups of lower vertebrates possesses its own special group of blood-feeding polyopisthocotyleans (Table 6.1). The inference from this is that the polyopisthocotyleans were well-established as inhabitants of the buccal and

Table 6.1   Some groups of vertebrates and their polyopisthocotylean monogeneans (source: based on Llewellyn, 1963)

Host group	Parasite group
Holocephalans	Chimaericolids
Elasmobranchs (Chondrichthyans)	Hexabothriids
Chondrosteans	Diclybothriids
Teleosteans	Mazocraeideans
Urodeles	Sphyranurids
Dipnoans } Anurans } Chelonians }	Polystomatids

gill cavities before the ancestral vertebrate stock diverged. Perhaps their ancestors were skin parasites in which muscular suckers developed at the sites of some of their 16 marginal hooklets. These suckers would have been equally suitable for attachment to non-respiratory surfaces in the buccal or gill cavities. In the chimaericolids and the hexabothriids (Fig. 6.20*a*), sclerites have become associated with the suckers, and in the modern mazocraeideans this trend has continued and has led to the modification of each sucker to form two opposable, sclerite-supported jaws, which are used in a unique way to clamp the parasite to the secondary gill lamellae

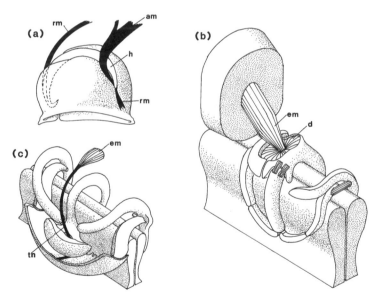

Figure 6.20   Suckers and clamps of polyopisthocotylean monogeneans. (*a*) Diagram of a sucker of a hexabothriid monogenean. Suction is generated by contraction of the adductor muscle (am), which lifts the roof of the sucker by lifting the hook (*h*) attached by its hooked end to the lining of the gill cavity. (*b*) Stereogram of a suction clamp of *Diclidophora* grasping two secondary gill lamellae. d = diaphragm which, when lifted by the extrinsic muscle (em), creates suction and draws the two jaws of the clamp together. (*c*) Stereogram of *Plectanocotyle* clamp grasping two secondary lamellae. The extrinsic muscle (em) gives rise to a tendon (tn) which passes through a hole and fairlead in the centrally positioned sclerite of one jaw and attaches to the other jaw. Contraction of the muscle draws the jaws together. The parasite tissue enclosing the clamps is omitted. Source: (*a*) based on an original drawing by Euzet and Maillard (1976); reproduced from Kearn, 1994, with permission from the Australian Society for Parasitology; (*b*), (*c*) reproduced from Llewellyn, 1958 and 1956a respectively, with permission from Cambridge University Press.

(Fig. 6.20*b*, *c*). Most modern polystomatid polyopisthocotyleans have no supporting sclerites associated with their haptoral suckers and may have descended directly from early polyopisthocotyleans with unarmed suckers (but see below).

Concinnocotyla australensis (Fig. 6.21) is a polystomatid monogenean of special interest because it lives in the buccal and gill cavities of a dipnoan, the Australian lungfish (*Neoceratodus forsteri*), while most of its relatives parasitize amphibians (see Chapter 7). This adds a little parasitological support to current views indicating that

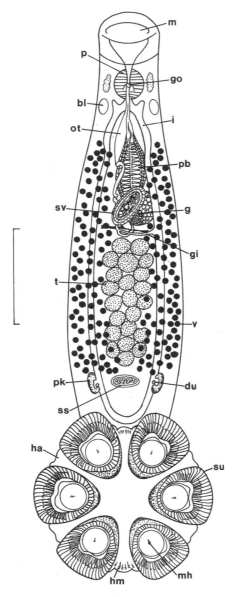

Figure 6.21 *Concinnocotyla australensis*, ventral view. bl = bladder; du = duct connecting gut pocket to dorsal surface; g = germarium; gi = genitointestinal canal; go = common genital pore; ha = haptor; hm = hamulus; i = intestine; m = mouth; mh = marginal hooklet; p = pharynx; pb = penis bulb; pk = pocket connected to gut; ss = sperm-filled sac; su = haptor sucker; sv = seminal vesicle; t = testis; v = vitellarium. Scale bar: 500 μm. Source: redrawn from Pichelin, Whittington and Pearson, 1991.

tetrapods (amphibians) evolved from dipnoan ancestors (Hedges, Hass and Maxson, 1993).

C. australensis has many unusual features, including the occurrence of supporting sclerites (rings and ribs) in the wall of each haptoral sucker, the presence of a duct on each side connecting the posterior end of the intestine to the outside and the absence of gut haematin, the presence of which usually indicates blood feeding (Pichelin, Whittington and Pearson, 1991; see also section 6.5.3). Dipnoans are highly specialized fishes and the Australian lungfish has been isolated for a very long time; it is conceivable that all the unusual features of *C. australensis* arose as special adaptations for life on lungfishes, after the separation of the early tetrapods. Alternatively, if the sucker sclerites of *C. australensis* are homologous with the sucker–clamp sclerites of other polyopisthocotyleans, then it would suggest that the first polyopisthocotyleans had sclerite-supported suckers and that some of these sclerites were bequeathed to their chimaericolid, hexabothriid and mazocraeidean descendants. This would also mean that sucker sclerites were totally lost in the antecedents of polystomatid parasites of amphibians.

If *C. australensis* is a skin feeder, as seems likely, it may be an indication that ancestral polyopisthocotyleans inhabiting the gill chamber were skin feeders.

6.5.2 Attachment and symmetry

In the ancestors of the modern hexabothriids, a single curved sclerite developed in a groove in the roof of each sucker (Fig. 6.20a); extrinsic muscles attached to this sclerite generate suction in a manner similar to *Entobdella soleae* (section 4.3.1). In other inhabitants of the non-respiratory surfaces, suction came to be generated by the action of an extrinsic muscle on a diaphragm supported by a sclerotized ring (Fig. 6.20b); by converting this to a clamping action rather than an open sucker, *Diclidophora* spp. were able to grip the secondary gill lamellae.

The parasites with open, sclerite-supported suckers may have resembled the modern mazocraeidean *Cyclocotyla* (= *Choricotyle*), which inhabits non-respiratory host surfaces (see Kearn, 1994). *Cyclocotyla* and its relatives are also sufficiently versatile to attach to the carapaces of copepod and isopod gill parasites on which they are presumably transported to new hosts. Such phoretic relationships must have evolved independently in udonellid turbellarians (section 2.4) and in a capsalid gill parasite (*Tristomum biparasiticum*; see Goto, 1894).

Other polyopisthocotylean gill parasites have abandoned suction, if they ever employed it, as a means of clamp operation. In *Plectanocotyle* and *Kuhnia* the clamps are closed by an extrinsic muscle–tendon–fairlead system (Fig. 6.20c; Llewellyn, 1956a, 1957a) and in *Anthocotyle* by intrinsic muscles running directly between the jaws of the clamp (Llewellyn, 1970). *Caballeraxine chainanica* reduces the strain on its clamps by knotting the stalk joining the haptor to the body around the primary gill lamella (Fig. 6.22).

The mazocraeidean *Heterobothrium elongatum* (Fig. 6.23l) undergoes an amazing change as it develops. The haptor in some unknown way burrows backwards into the subcutaneous gill tissue; the adult is so deeply embedded that a long tubular stalk joins the haptor to the body proper, which emerges from the gill surface. In the adult the stalk threads its way through the gill arch from one end to the other, with the haptor sometimes emerging from the proximal end of this canal and apparently serving as a plug preventing withdrawal of the parasite (Williams and Lethbridge,

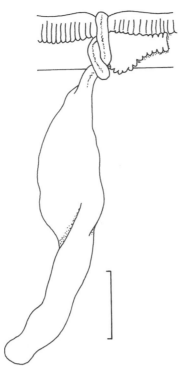

Figure 6.22 *Caballeraxine chainanica* attached to the primary lamella of *Velifer multispinosus*. Scale bar: 0.5 mm (approx.). Source: redrawn from Mamaev and Kurochkin, 1976; reproduced from Kearn, 1994, with permission from the Australian Society for Parasitology.

1990). Whether this tissue-burrowing habit has any function other than for attachment remains to be investigated, but a similar habit has evolved in the chimaericolid *Callorhynchicola multitesticulatus* (see Llewellyn and Simmons, 1984). In both parasites there is a strong inflammatory response by the host and in *C. multitesticulatus* this response has been exploited for attachment by the parasite. Annular flanges on the stalk between the haptor and the body engage a sleeve of fibrous tissue generated by the host reaction, thereby preventing withdrawal of the haptor.

There is considerable variation in the numbers, distribution and relative sizes of the clamps in mazocraeideans (Fig. 6.23). Ancestral mazocraeideans may have possessed four pairs, as in *Kuhnia* (Fig. 6.23*a*), sometimes mounted on peduncles, as in *Diclidophora* (Fig. 6.23*b*), but in *Plectanocotyle* there are only three pairs (Fig. 6.23*c*) and in *Microcotyle* many pairs (Fig. 6.23*f*). In *Anthocotyle* the anteriormost pair is greatly enlarged (Fig. 6.23*e*), permitting the parasite to grasp the edge of a primary gill lamella, like some of the capsalids (section 6.2). Many mazocraeideans are asymmetrical, *Grubea* having four clamps only on one side of the haptor and a minute fifth clamp on the other side (Fig. 6.23*g*, *h*). In *Gastrocotyle* there are more than 30 clamps on one side only (Fig. 6.23*i*) and, although *Axine* has multiple clamps on both sides of the haptor, the parasite becomes asymmetrical by differential growth, which results in all the clamps becoming aligned in a single lateral row (Fig. 6.23*j*).

An explanation for this asymmetry has been offered by Llewellyn (1956b). Because of the arrangement of teleost gills (Fig. 6.2*a*), polyopisthocotyleans attached

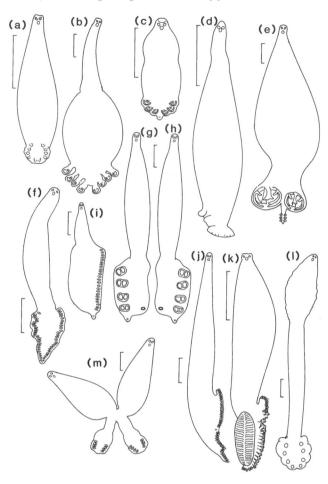

Figure 6.23   Clamp disposition and body shapes of some mazocraeidean polyopisthocotyleans. (*a*) *Kuhnia scombri*. (*b*) *Diclidophora merlangi*. (*c*) *Plectanocotyle gurnardi*. (*d*) *Lethacotyle fijiensis*. (*e*) *Anthocotyle merluccii*. (*f*) *Microcotyle donavini*. (*g*), (*h*) *Grubea cochlear* showing 'left-footed' and 'right-footed' individuals. (*i*) *Gastrocotyle trachuri*. (*j*) *Axine belones*. (*k*) *Rhinecotyle crepitacula*. (*l*) *Heterobothrium elongatum*. (*m*) *Diplozoon paradoxum*. Scale bars: 1 mm. Source: reproduced from Kearn, 1994, with permission from the Australian Society for Parasitology; (*d*), (*f*), (*k*), (*l*) from original drawings by Manter and Prince (1953), Euzet and Marc (1963), Euzet and Wahl (1970) and Williams and Lethbridge (1990) respectively.

to the secondary gill lamellae will be exposed to strong, unilateral, gill-ventilating currents. The strongest water flow impinges on the upstream side of the parasite and it is here that the clamps develop, rather than on the relatively sheltered downstream side (Fig. 6.24). This asymmetrical development gives maximum security for the parasite while minimizing interference with the host's respiratory function. However, the gill-ventilating current strikes some hemibranchs on one side and others on the opposite side, so that the direction of asymmetry of the parasite (whether 'right-or left-footed') will depend on its location on the gill. A logical extension of this argument is that these gill parasites have become sedentary – there is no evidence that once adapted to a particular primary lamella they can reverse their body asymmetry, or even move to an alternative site requiring a similar direction of symmetry. They do not possess the conspicuous anterior adhesive glands characteristic of mobile monogeneans.

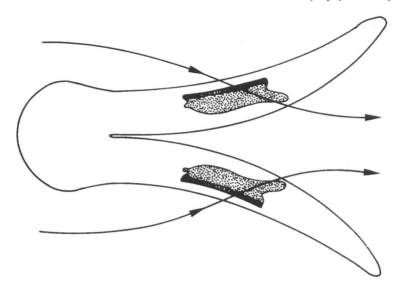

Figure 6.24 Diagram illustrating the relationship between direction of asymmetry ('left-' or 'right-footed' individuals) in *Gastrocotyle trachuri* and flow of water (arrows) across the hemibranchs of the gill of the host *Trachurus trachurus*. The parasite on the upper hemibranch has clamps only on its right side, while the one parasite on the other hemibranch has clamps only on its left side. Source: redrawn from Llewellyn, 1957b.

This account explains the asymmetry of parasites like *Grubea*, *Gastrocotyle* and *Axine* but does not account for the symmetry of *Diclidophora*, *Plectanocotyle* and *Kuhnia*. Symmetrical *Kuhnia scombri* inhabits the gills of the same host, the mackerel (*Scomber scombrus*), as asymmetrical *Grubea cochlear*. Each individual specimen of *Diclidophora* maintains its body symmetry by attaching to both sides of the same primary gill lamella (*D. merlangi* spans more than one lamella), thereby exposing both sides of the body to equal flow (Llewellyn, 1956b). There is evidence that *Plectanocotyle* is not sedentary (Llewellyn, 1966) and therefore, like dactylogyroideans, is likely to be exposed to currents from any direction as it moves about. *K. scombri* lives within the shelter of the gill septum while *Grubea* attaches more distally (Llewellyn, 1966), so the currents to which *Kuhnia* is exposed may be too weak to generate asymmetry.

In spite of the sedentary way of life adopted by most mazocraeideans, there is rarely any evidence of a host reaction at the sites of their clamps. *Heterobothrium* is an exception, as is the chimaericolid *Callorhynchicola*, but in these rare situations host inflammatory responses are associated with tissue penetration by the parasites. How most mazocraeideans escape the host's reactions is not known, but perhaps clamps are less provocative than hooks. Hooks have a greatly reduced role in adult mazocraeideans. In some species a pair of marginal hooklets and one or two pairs of hamulus-like hooks persist on a small posterior lappet but in others, e.g. *Diclidophora* spp., there are no hooks in the adult.

6.5.3 Reproduction and the life cycle

An interesting feature of the biology of mazocraeidean gill parasites is their non-random distribution on the gill arches. For example, *Diclidophora merlangi* is more commonly encountered on the outermost gill arch (I), while *D. luscae* occurs more frequently on arches II and III (Llewellyn, 1956b). As Llewellyn has pointed out, it is unlikely that oncomiracidia are sufficiently manoeuvrable or discriminatory to settle selectively, and any post-oncomiracidial migration, if it occurs, must take place early because most of the adults are sedentary. If the assumption is made that the parasites are unable to identify or seek out a particular gill arch and that infective larvae are swept involuntarily over the gills by the ventilation current, then two other factors may have a strong influence on gill distribution. First, individual gill arches may differ in the volume of water passing through them and hence the number of larvae carried to them and, secondly, attached larvae may have a greater chance of survival on some gill arches than on others.

An advantage of restricted gill distribution is that it provides opportunities for cross-insemination. This is well demonstrated in a study of *D. merlangi* by Macdonald and Caley (1975). They confirmed that the parasite is more common on gill arch I and that the outer hemibranchs (of all gills) are more frequently parasitized than the inner hemibranchs. Most fishes carried one to three adults, and in situations where more than one adult occupied the same gill arch they most frequently (on 74% of the arches) occupied neighbouring sites on the same hemibranch. However, when juveniles occupied the same gill arch as an adult, they occupied adjacent sites in only 31% of the cases, suggesting that there may be limited movement by juveniles and perhaps some attraction between parasites.

Macdonald and Caley observed copulation frequently between detached adult specimens of *D. merlangi*. In these circumstances, the event was unilateral, the spined penis of one individual attaching to the anterior region of a second adult. There is no vagina, and spermatozoa either penetrate the tegument or enter through the wounds inflicted by the penis hooks. Spermatozoa appear to reach the seminal receptacle *via* the intercellular spaces.

There is no vagina in *Gastrocotyle trachuri* but the male copulatory organ is better equipped for hypodermic impregnation since there is a rigid, central penis tube. Llewellyn (1983) offered a functional explanation for this apparatus (Fig. 6.25)

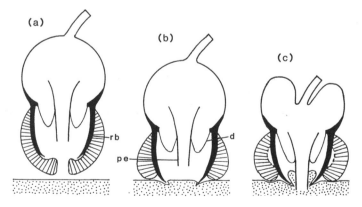

Figure 6.25 Diagrams illustrating a possible mechanism for hypodermic impregnation in *Gastrocotyle trachuri*. (*a*) Apparatus at rest. (*b*) Distal aperture widens; tips of ribs (rb) press against co-copulant; muscular tissue of the distal bulb forms a seal between partners. (*c*) Radial muscles of the distal chamber contract and generate a suction pressure, causing a blister of co-copulant tissue to be sucked into the distal chamber; at the same time the diaphragm (d) and penis tube (pe) are drawn into the distal chamber so that the penis tube penetrates the tegumental blister. Source: reproduced from Llewellyn, 1983 (fig. 11, p. 443), with permission from The Royal Society.

and proposed that spermatozoa injected into the vitelline system would have an uninterrupted route to the oviduct. He regarded the non-localized injection of spermatozoa into the vitelline system as a persistent primitive feature. He suggested that, in other polyopisthocotyleans, particular vitelline ducts became favoured impregnation sites, leading to acquisition of external openings and specialization as vaginae. Since these favoured ducts could be dorsal, ventral, paired or unpaired, this may have produced the considerable variation in male copulatory organs and in vaginal architecture that is seen within closely related groups of mazocraeideans.

Staying together long enough to exchange spermatozoa is difficult in a strong gill-ventilating current. In freshwater mazocraeideans of the genus *Diplozoon*, a permanent solution has been found to this problem: two juveniles meet on the gills and then undergo complete tissue fusion in such a way that their reproductive ducts communicate in life-long copulation (Fig. 6.23*m*). Fusion of two individuals is known to occur in only one other platyhelminth parasite, the didymozoidean *Diplotrema* (section 19.1).

The eggs and larvae of polyopisthocotylean gill parasites display a range of special adaptations to enhance their chances of reaching new hosts. Convergence between *Hexabothrium appendiculatum* and the unrelated microbothriid *Leptocotyle minor* extends to egg morphology, rapid hatching in response to host-derived urea and oncomiracidial behaviour (section 5.4). There is convergence too between *Plectanocotyle gurnardi* (see Whittington and Kearn, 1989) and *Entobdella diadema* (section 4.10.7) in that eggs of both parasites hatch rapidly when a shadow is cast over them by the host. Shadows also influence the behaviour of the oncomiracidium of *P. gurnardi*. They spend much of their time conserving energy, attached by the anterior end to the substrate, but are readily reactivated by shadows and by mechanical disturbance, stimuli that are likely to be provided by their bottom-dwelling gurnard hosts.

Diclidophora luscae stores up eggs inside an enlarged uterus, but this habit is not related to ovoviviparity, because the eggs are released before any significant embryonic development has occurred (Whittington and Kearn, 1988). Egg retention serves to produce an egg bundle with an average of 60 eggs entangled together by long abopercular appendages (Fig. 6.26). The eggs also have opercular appendages resembling grappling hooks. Egg bundles sink more rapidly than separate eggs, so that they are less likely to be swept away as they descend and more likely to be anchored by their hooked appendages in areas inhabited by their bottom-living teleost host (the pouting, *Trisopterus luscus*). Oncomiracidia hatch rapidly in large numbers when egg bundles are mechanically disturbed, either experimentally or presumably by movements of foraging hosts. Temporary egg storage may also permit release of eggs on a once-a-day basis. This could be timed so that egg bundles are deposited in sites that are visited by hosts at a particular time of day or night for feeding or resting.

The polystomatid *Concinnocotyla australensis* has adopted the unusual practice of cementing its eggs to the tooth plates of its lungfish host (Whittington and Pichelin, 1991). Large numbers of eggs attached by the ends of their appendages carpet the teeth and some of these complete their development and hatch, liberating an unciliated larva. After hatching, larvae of *C. australensis* migrate to the gills and it is presumed that adults return to the oral cavity to attach their eggs to the tooth plates. How new hosts are infected is unknown.

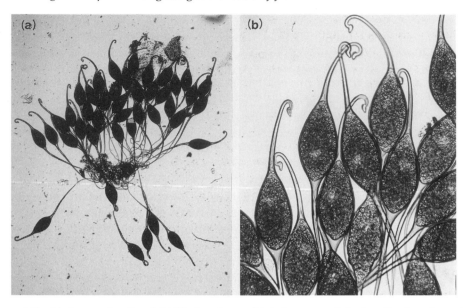

Figure 6.26   (*a*) An egg bundle of *Diclidophora luscae*. (*b*) A few eggs of the bundle viewed at a high magnification showing the hook-like opercular appendages. Source: reproduced from Kearn, 1986, with permission from Academic Press.

An important study by Llewellyn (1962) of the population dynamics of *Gastro-cotyle trachuri* and *Pseudaxine trachuri* from the gills of the scad (*Trachurus trachurus*) has shed light on adaptations of the life cycles of gill parasites to seasonal and longer term changes in the biology of their hosts. Llewellyn found that the parasites infect 3–4-month-old scad when they descend to the sea bottom. Sexual maturity of the parasites may be reached as early as 1 month post-infection and their life span is normally no longer than 1 year. At the beginning of their second year, scad pick up a largely new infection, but 2- and 3-year-old fishes are poorly parasitized, not because of an age-immunity but because of a post-spawning migration of fishes away from coastal waters where the parasite is common. The parasites have also adapted to a seasonal change in the feeding habits of scad by ceasing to produce larvae before the summer disappearance of scad from the sea bottom in pursuit of pelagic food. This change in the host's behaviour may be controlled by hormones and one way in which the parasite could 'anticipate' coming events in the life of the host is by responding to host hormones ingested with its blood meals. Links between host and parasite reproduction become particularly significant in mono-geneans parasitizing amphibians and will be considered in the next chapter.

6.5.4   Host-switching and speciation

The question of whether or not parasite phylogeny provides information about host relationships is frequently raised. It has already been suggested (section 5.5.1) that the level of contribution of host-switching to the evolution of monogeneans may differ from group to group and that monogeneans that spread from host to host by contagion, like the gyrodactylids, may be more likely to switch hosts than mono-geneans that disperse only by freely deposited eggs and oncomiracidia, like the polyopisthocotyleans. Llewellyn (1957b) has emphasized that each major group of aquatic vertebrates has its characteristic group of polyopisthocotylean parasites (see Table 6.1), arguing strongly for evolutionary radiation of these parasites keeping in step with radiation of their hosts.

In spite of this, an apparent example of host-switching in action was described for a polyopisthocotylean by Llewellyn, Macdonald and Green (1980). They reported an influx of the gadoid marine fish *Trisopterus esmarki* into Plymouth waters from the west and north, bringing with it the gill parasite *Diclidophora esmarkii*. This brought the parasite into contact with *T. minutus minutus* and *T. luscus*; no parasites had previously been recorded on the former in the Plymouth area and *T. luscus* was known to be infected with *Diclidophora luscae*. Parasites then appeared on *T. minutus minutus* for the first time and were identified by Llewellyn *et al.* as *Diclidophora esmarkii*. They interpreted this as a host-switching event, but found no switching from *T. esmarki* to *T. luscus*.

Another dimension has been added to this story by the work of Tirard *et al.* (1992). They found *D. luscae* on *T. luscus* from the North Sea, Bay of Biscay and Gulf of Cadiz (Atlantic). *T. minutus minutus* occurred only in the North Sea and Bay of Biscay, and fishes from the latter site harboured a *Diclidophora* sp. *T. minutus capelanus* is restricted to the Mediterranean and is also infected with *Diclidophora* sp. They analysed and compared the genetic structure of parasite and host populations.

D. luscae and the parasites from *T. m. minutus* were found to differ at six loci and, in the Bay of Biscay, total reproductive separation existed between the parasites from these two hosts. Surprisingly, differences were found at five loci between the parasite population of *T. m. minutus* (Atlantic) and *T. m. capelanus* (Mediterranean), whereas only one locus differed between the Atlantic parasites of *T. luscus* and the parasites of *T. m. capelanus* in the Mediterranean. Thus, parasites of fishes regarded as related sub-species (*T. m. minutus* and *T. m. capelanus*) are genetically further apart than parasites of one of these sub-species (*T. m. capelanus*) and parasites from a supposedly different host species (*T. luscus*).

Since *T. esmarki* is absent from the Bay of Biscay, Tirard *et al.* rejected the possibility that the parasite on *T. m. minutus* came from *T. esmarki*; they regarded it as a new species (*Diclidophora minuti*) specific to *T. m. minutus*. Their statement that its distribution may exceptionally reach as far north as the Channel and their observation that *T. m. minutus* caught in northern Scotland (North Sea) were uninfected, in spite of the presence of *T. esmarki* frequently infected with *D. esmarkii*, inevitably raise the possibility that the parasites found on *T. m. minutus* by Llewellyn *et al.* were *D. minuti*.

Genetic analyses of the hosts support the parasitological indications, namely that there are two taxonomic entities: (1) *T. minutus* found from the North Sea to the Bay of Biscay (Atlantic) and (2) *T. luscus* found in the North Sea, the Atlantic and the Mediterranean. Morphoanatomical characteristics that led earlier ichthyologists to distance *T. luscus* from *T. m. minutus* and *T. m. capelanus* must be ascribed to convergences between *T. minutus* (Atlantic) and *T. luscus* (Mediterranean). The small genetic divergence between the Atlantic and Mediterranean populations of *T. luscus* and their parasites reflects the relatively recent colonization of the Mediterranean by Ice Age immigrants moving south from the Atlantic.

This careful work has revealed a high degree of correspondence between parasite and host evolution, so much so that it has revealed with great clarity past errors in interpretation of morphoanatomical data and should perhaps give us more confidence in placing reliance on parasites as biological markers.

6.6 THE WAY FORWARD

The buccal and gill cavities of fishes are essentially extensions of the external surface and their parasites essentially ectoparasites. However, a surprisingly large number of monogeneans have penetrated further into the bodies of their hosts. This has imposed new demands on their biology and in the next chapter we will consider some of these endoparasites, especially those that have colonized the urinary bladder and the gut. These events assume added importance because they probably include the distant origin of the tapeworms.

CHAPTER 7

Endoparasitic monogeneans

The adaptability of monogeneans is well illustrated by the diversity of internal sites that they have colonized (Table 7.1). Two of these trends deserve special attention. One is the colonization of the bladder of amphibia by polystomatid polyopisthocotyleans and the other is the invasion of the gut of fishes.

7.1   THE POLYSTOMATIDS

7.1.1   The colonization of tetrapods

The emergence of amphibia from fish ancestors offered a significant challenge to monogenean parasites. The polystomatids are not the only monogeneans that answered this challenge, but they are undoubtedly the most successful, having diversified extensively among the anurans (frogs). They have also established a foothold in chelonian reptiles (section 7.1.6). The success of polystomatids may be due partly to the preservation of a relatively unspecialized haptor with the following features: 16 marginal hooklets, one or two pairs of large hooks and six suckers, each of which is unarmed, except for a single marginal hooklet (see section 6.5.1 and Fig. 7.1). These suckers are effective for attachment to a variety of host surfaces. Their blood-feeding habits may also have favoured colonization of amphibians by allowing the parasites direct access to host hormones, which provide valuable information about seasonal breeding cycles and dramatic changes such as metamorphosis (see also section 6.5.3). The parasite's survival depends on appropriate responses to these host events, especially in those amphibians that are semi-terrestrial in life style.

Using the available fossil evidence and deductive reasoning, Pough *et al.* (1990) have attempted to reconstruct the possible evolutionary path leading to the land-dwelling tetrapods from their fish ancestors, which they considered to be osteolepiform fishes (but see section 6.5.1). They proposed that all the morphological features that were advantageous for life on land were originally functional in the aquatic habitat. In particular, lungs would be advantageous for fishes living in warm, shallow, deoxygenated pools, and limbs would be useful for supporting the body in shallow water, for raising the head above the water to breathe and for moving along the bottom in search of prey. The next step, the adoption of a terrestrial way of life, would be unlikely to take place until these morphological preadaptations were completed in the aquatic environment. Several theories have been advanced to explain why some of these vertebrates became terrestrial. One that is reasonably convincing proposes increasing competence and agility in exploiting the terrestrial environment for shelter from aquatic predators and

Table 7.1 Site diversity in endoparasitic monogeneans (excluding buccal and gill chambers)

Group	Parasite	Host	Site	Source
Dactylogyroideans	*Amphibdella torpedinis* (juveniles)	*Torpedo marmorata*	Heart	Llewellyn, 1960
?	*Paraquadriacanthus nasalis*	*Clarias lazera*	Nasal cavities	Ergens, 1988
?	*Acolpenteron catostomi*	*Catostomus commersoni*	Ureters	Sproston, 1946
	Enterogyrus spp.	Cichlid fishes	Stomach	Pariselle, Lambert and Euzet, 1991
Gyrodactylids	*Gyrodactylus cryptarum*	*Gadus callarias*	Preopercular sensory canals of the head	Malmberg, 1970
Capsalids	*Nasicola klawei*	*Neothunnus macropterus*	Nasal cavities	Yamaguti, 1968
?	*Montchadskyella intestinale*	*Paristiopterus gallipavo*	Intestine	Bychowsky, Korotaeva and Nagibina, 1970
		Zanclistius elevatus	Stomach	
Monocotylids	*Calicotyle kroyeri*	*Raja* spp.	Cloaca and rectal gland	Kearn, 1987b
	Gymnocalicotyle inermis	*Pristiophorus cirratus*	Oviducts	Yamaguti, 1963
	Dictyocotyle coeliaca	*Raja* spp.	Body cavity	Yamaguti, 1963
	Empruthotrema raiae	*Raja* spp.	Nasal fossae	Kearn, 1976b
Polystomatids	*Polystoma integerrimum*	*Rana temporaria*	Bladder	Sproston, 1946
	Protopolystoma xenopodis	*Xenopus* spp.	Kidneys (juveniles) Bladder	Tinsley and Owen, 1975
	Pseudodiplorchis americanus	*Scaphiopus couchii*	Lungs (juveniles) Bladder	Tinsley and Earle, 1983

for the acquisition of food such as terrestrial invertebrates. Thus, fully aquatic tetrapods may have been around for a long time before any excursions on to the land took place.

Gills may have been lost before the tetrapods moved on to land and the buccal cavity of terrestrial tetrapods would have been a very different place. Some polystomatids found a more hospitable alternative environment inside the urinary bladder, which offered accessibility to blood vessels for feeding and ready access to the outside world *via* the cloaca.

7.1.2 *Protopolystoma xenopodis*

Protopolystoma xenopodis (= *P. xenopi*; Fig. 7.1) is a modern polystomatid that parasitizes the urinary bladder of a fully aquatic anuran, the clawed toad (*Xenopus laevis*), and its life style may be similar to that of the early polystomatid parasites. The parasite lays eggs throughout the year, embryonic development and hatching taking place after the eggs have left the host and ciliated oncomiracidia gaining direct entry to the bladders of new hosts (Tinsley and Owen, 1975).

7.1.3 *Eupolystoma* spp.

Retention of eggs, which we have already met (section 5.2), is the single most important feature that has enabled polystomatids to cope with the amphibious habits of their hosts (Tinsley, 1983). For example, *Eupolystoma anterorchis* uses its uterus to store up to 300 eggs, which are assembled when the host (*Bufo pardalis*) is on land (Tinsley, 1978). In *Eupolystoma* spp., space for this enlarged uterus has been created by moving the germarium backwards and displacing the vitelline and testicular follicles laterally (Fig. 7.2). The stored eggs of *E. anterorchis* complete their development *in utero* and when the host returns to water the eggs are laid, hatch immediately and enter the bladders of new hosts. Visits to water by the toads are infrequent and often brief, but large numbers of toads may congregate in newly-formed pools after rainfall.

Several features of the reproductive biology of *E. anterorchis* have attracted the attention of Tinsley and are worth noting. He observed that, in parasites with many ready-to-hatch eggs in the uterus, there were few recently assembled eggs, indicating that egg assembly is not continuous and may be modulated by some kind of feedback from the gravid uterus. Also, it seems likely that unhatched oncomiracidia are able to survive for appreciable lengths of time *in utero*.

Tinsley suspected that egg-laying is induced by lowering of urine concentrations as a result of hydration of toads returning to water after a long absence. Indeed, gravid parasites taken from their hosts and placed in water did lay their eggs, but attempts to induce egg-laying *in vivo* by placing dehydrated infected toads in water failed, suggesting that factors other than osmotic ones are involved.

As soon as the fully developed eggs of *E. anterorchis* enter a hypo-osmotic medium they hatch explosively, suggesting that high internal pressure induced by osmotic uptake of water is responsible for hatching. In these circumstances, there is no call for a strong, protective eggshell and the shell of *E. anterorchis* is thin, with little tanning, judging by its faint yellow colour. Moreover, the shell is flexible, stretches as the embryo grows and hatches by splitting – there is no operculum.

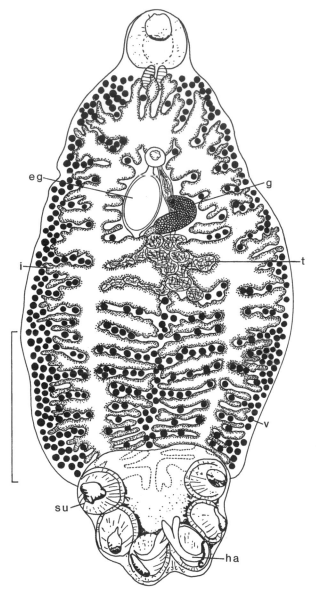

Figure 7.1 *Protopolystoma xenopodis* (=*Polystoma xenopi*), ventral view. eg = egg; g = germarium; hm = hamulus; i = intestine; su = sucker; t = testis; v = vitellarium. Scale bar: 0.5 mm. Source: redrawn from Price, 1943.

Thus, special adaptations promoting survival of *E. anterorchis*, in a host which visits water intermittently, include egg storage with *in utero* development, possible modulation of egg assembly, enhanced longevity of the larvae, mass deposition of the ready-to-hatch eggs, reduction of the eggshell and rapid hatching. Transmission is likely to occur at every opportunity, not just during the period when the host is in water for spawning.

The prolonged storage of eggs in the uterus, as practised by *E. anterorchis*, is but a short step away from hatching of some of these unlaid eggs and autoinfection, a

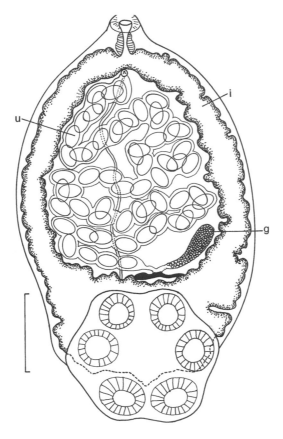

Figure 7.2 *Eupolystoma alluaudi*, ventral view. g = germarium; i = intestine; u = uterus containing eggs. Scale bar: 0.5 mm. Source: redrawn from Euzet and Combes, 1967.

phenomenon that we have already encountered in skin- and gill-parasitic mono-geneans (sections 5.2, 6.4.1). *E. alluaudi* from *Bufo regularis* has taken this step. Some of the fully developed eggs retained in the large uterus have the option of hatching without leaving the host, thereby augmenting the population of parasites in this host; other fully developed eggs pass out in the urine and contribute to the infection of new hosts. These two separate options are accomplished in *E. alluaudi* by two morphologically distinct larvae: larvae responsible for autoinfection are unciliated and incapable of swimming, while larvae involved in infecting new hosts are ciliated and can swim (Fournier and Combes, 1979).

The distribution of *E. alluaudi* in the toad population is non-random (over-dispersed). Most toads are uninfected (prevalence 5%), while the few that are infected often harbour large numbers of parasites: infection intensities may reach several hundred parasites and there is a record of 2000 parasites in a single toad (Combes, Bourgat and Salami-Cadoux, 1973). Over-dispersion is a common feature of parasites (sections 5.2, 7.1.4(c)) and indicates that all hosts do not have the same chance of becoming infected. There are several possible reasons for this inequality of infection; autoinfection is one of them and we will meet others later.

A potentially serious problem with autoinfection is that the infrapopulation of the parasite could reach a lethal level for the host. While studying *Bufo regularis* harbouring populations of *E. alluaudi* of different sizes, Fournier and Combes (1979) made a remarkable discovery that has an important bearing on this problem. In toads with less than five parasites in the bladder, only non-ciliated crawling larvae were produced; these hatched *in utero* where the shells remained until ejected later and the larvae established themselves in the bladder. In toads with more than 1000 parasites, ciliated larvae were exclusively produced from eggs released into the bladder and carried out with the urine. In toads with intermediate populations both unciliated and ciliated larvae were produced. Thus, there seems to be some kind of control by the parasite of the proportions of the two kinds of larvae produced, and this control is related to the density of parasites in the bladder. Nothing is known about the mechanism of this remarkable density-dependent control of the parasite population.

In the dry season, *B. regularis* is terrestrial and produces almost no urine (Salami-Cadoux, 1975). During this period autoinfection can proceed. In the rainy season, the toads visit water for spawning, urine begins to flow and this provides the opportunity for eggs laid by the parasites to enter the water. The fully developed eggs hatch immediately and, according to Combes, Bourgat and Salami-Cadoux (1976), transmission occurs during the act of mating.

7.1.4 *Pseudodiplorchis americanus* and desert toads

(a) The biology of the host

In order to appreciate the remarkable adaptations of *P. americanus* it is necessary to describe the main features of the host's biology. The desert toads (*Scaphiopus*) of North America, like *B. regularis*, maintain only a tenuous link with water, but they differ in that most of their terrestrial life is spent in an inactive state, buried at depths of nearly a metre in the desert sand. Their emergence from the sand and their contact with water for spawning take place once a year following torrential annual rains (Fig. 7.3). The toads are strictly nocturnal and enter the temporary rain pools only after dark (9 pm) and leave before dawn (4 am). A succession of rain storms may provide opportunities for two, occasionally three, extra spawning assemblies on other nights. Female toads spawn only once and therefore attend only one spawning assembly while males will participate in all the assemblies. Following spawning the toads feed on a variety of desert invertebrates during favourable conditions lasting little more than 8 weeks but their activity during this period is actually much less, since they forage only on nights when the desert surface is dampened by rain and spend the rest of the time in shallow burrows (Tinsley, 1990a). It is during these limited forays on fewer than 20 nights a year that the toads must accumulate sufficient food reserves (mainly fat) to sustain them during the long hibernation period (September to July in Arizona).

In spite of this extreme life style, polystomes such as *Pseudodiplorchis americanus* survive and flourish in these desert toads, and the work of Tinsley and colleagues on the biology of these parasites has provided one of the most exciting and interesting of parasitological studies (Tinsley, 1983; Tinsley and Earle, 1983; Tinsley and Jackson, 1986, 1988 and references below).

(b) The general biology of Pseudodiplorchis americanus

In the biology of *P. americanus*, ovoviviparity has an essential role, eggs being assembled and stored in an enlarged uterus and their contained embryos complet-

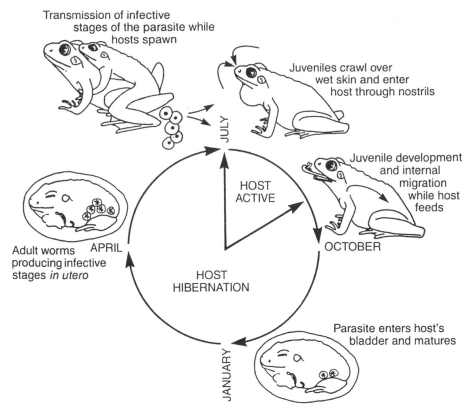

Figure 7.3 The life cycle events of *Pseudodiplorchis americanus* in relation to the annual cycle of activity of its host *Scaphiopus couchii*. Source: reproduced from Tinsley, 1989, with permission from Elsevier Trends Journals.

ing their development during the subterranean phase of the toad's life. However, the huge uterus is accommodated in a different way from that of *Eupolystoma*. In *Eupolystoma*, space for the uterus is created by a posterior migration of the germarium and by the lateral displacement of vitelline and testicular follicles, while in *P. americanus* the germarium retains its anterior position with the enlarged uterus behind it (Fig. 7.4). The testes of *P. americanus* are consolidated and laterally displaced and, for reasons that will become apparent, the vitellarium is reduced to a few follicles near the germarium. The implication is that ovoviviparity has arisen independently in *Eupolystoma* and *Pseudodiplorchis*. Another difference between these two polystomes is that there is no evidence for autoinfection in *Pseudodiplorchis*.

With the advent of the summer rains and the brief opportunity for nocturnal spawning, *P. americanus* is ready to release its eggs immediately and, like those of *Eupolystoma*, they hatch as soon as they enter water. The oncomiracidia of *P. americanus*, unlike those of *Eupolystoma*, enter the nostrils of the toads, spend time in the lungs and eventually migrate to the bladder *via* the stomach and the intestine (Fig. 7.3). The reasons for such a circuitous route are obscure; the hazards the migrants face are formidable and include desiccation on the exposed skin of the toad, exposure to surfactants and air in the lungs and to acidity, bile, digestive enzymes and physical obstruction by food boluses in the gut (Cable and Tinsley,

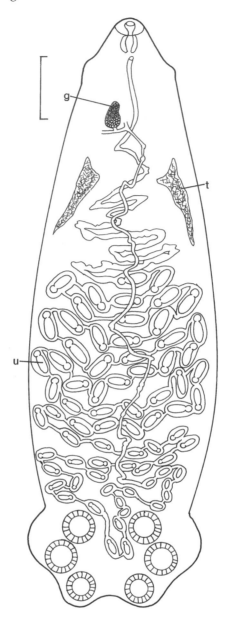

Figure 7.4   *Pseudodiplorchis americanus.* g = germarium; t = testis; u = uterus containing developing eggs. Scale bar: 1 mm. Source: from Rodgers and Kuntz, 1940.

1992a). Migrating parasites may also experience host immune responses and changes both in the ambient oxygen concentration and the osmotic environment. During hibernation host urine becomes concentrated, exposing the parasites to considerable osmotic stress. No other monogenean experiences such a wide range of physiological conditions during its life time.

The newly formed eggs of *P. americanus* contain only two or three vitelline cells (Tocque, 1990, in Cable and Tinsley, 1991). An irregular electron-dense 'shell' enclosing early embryos may be equivalent to the eggshell of other monogeneans and may be derived from these vitelline cells, but this is soon replaced by a membranous sac produced by the uterus (Cable and Tinsley, 1991). Growth of the embryo is substantial and the sac stretches to accommodate the increase in size, but the vitelline cells are so few that they are unlikely to make any significant contribution to the nutrition of the growing embryo and it seems likely that food from the intestine of the parent reaches the embryo *via* the wall of the uterus. Cytoplasmic connections between the larval tegument and the sac were interpreted as performing a placenta-like function in the nourishment of the embryo by Cable and Tinsley (1991). Thus, the substantial changes in the eggshell reflect the demands of the new provisioning route for the embryos and the need for shell permeability and extensibility; a protective tanned shell is no longer a requirement for eggs that hatch as soon as they leave the host and, in a situation where there is a premium on accommodating as many eggs as possible *in utero*, a rigid tanned shell is likely to be less advantageous than a flexible one (Cable and Tinsley, 1991). The functions of the vitellarium have been superseded and the organ is greatly reduced.

So narrow is the infection window for *P. americanus* that a very precise stimulus to induce parasite oviposition is necessary. During hibernation the host accumulates urea in the liver and blood (Tinsley, 1990b) and, when the host enters water, a parasite in the toad's bladder would be exposed to major osmotic and chemical changes, which could provide oviposition stimuli for the parasite. However, Tinsley found no evidence that phenomena linked to rehydration stimulate oviposition; this is perhaps not surprising because such cues signal only immersion in water and not the presence of other toads. A much better cue for this purpose, and one which the parasite has adopted, is linked with sexual activity of the hosts, but the manifestations of this stimulus differ in male and female toads. In females, parasites release eggs more or less exactly as the female starts to oviposit and cease to do so as host spawning stops, while in males, release of parasite eggs occurs during intense sexual excitement, whether this takes place during mating, while calling to attract females or while interacting with other males. The significance of this difference is that in a mating assembly all female toads spawn, so a stimulus linked with spawning is infallible. However, many males fail to mate, so a stimulus linked with sexual arousal rather than gamete emission ensures that their parasites can contribute to transmission, whether their hosts mate or not. The nature of this stimulus is unknown, but host sex hormones may be involved directly or indirectly.

The oncomiracidia are larger than those of other monogeneans, reaching 600 μm in length (the larva of *Entobdella soleae* is 250 μm long). They also live longer, retaining their swimming ability for over 48 h at 25–27°C. They migrate across the skin of the toad using their haptor and the anterior adhesive areas but, unlike *E. soleae*, for example, they do not lose their ciliated cells immediately after attachment to the host. Thus, larvae of *P. americanus* are not necessarily committed to the first host to which they attach. Cable and Tinsley (1992b) observed dislodgement by the host's forelimb of larvae approaching the host's nostrils; such larvae may be able to resume swimming and locate a new host.

The larvae have a remarkable tolerance to drying, surviving for up to 1 h at 32°C and 45% relative humidity. The cilia survive this desiccation; locomotion following reimmersion is not impaired. Cable and Tinsley (1992b) found no obvious ultrastructural features to explain the ability of the larvae to resist drying, but the tegument appears to be exceptionally tough, persisting for several days after death. The larvae are also able to push their way through the water–air interface, unlike many other oncomiracidia, which are readily trapped by surface tension. These properties permit the larvae to survive periodic submergence and exposure on the heads of mating toads and to accomplish the water–air crossing that is necessary to reach the nostrils. Also, provided that the toad remains moist, late invaders may be able to complete their migration on toads that have left the mating assemblage; this is important because during the last hour of such assemblages free swimming larvae are especially abundant.

Vesicles that accumulate in the tegument during the 28-day developmental period in the lungs seem to be involved in protection of the parasite during migration through the gut to the bladder (Cable and Tinsley, 1992a). Most larvae set out on this migration while the host is actively feeding and, although the journey takes less than 30 min, the need for protection from the harsh processes of digestion is underlined by the rapid death within 2 min of unprepared juveniles transferred experimentally to the stomach from the lungs. An unknown trigger is required to prime the parasites for the journey and a significant contributor to their survival is the mass release of tegumental vesicles on to the parasite's surface.

(c) Population dynamics

The work of Tinsley and colleagues has given us a detailed insight into the complex interactions between parasite, host and environmental factors. Since infection takes the form of an exceedingly brief annual pulse and since the total reproductive output of a parasite is accumulated *in utero* during hibernation, a great deal can be learned about the population dynamics of the parasite.

Tinsley and Jackson (1988) sampled toads from an area of desert of about 30 km², and recorded a prevalence of 50%, with infected toads harbouring a mean of about five adult parasites. This population of adult parasites will contribute to transmission of the parasite at the next spawning assembly. In a study conducted in 1983 in the San Simon valley in Arizona, Tocque and Tinsley (1991a) found a strong positive correlation between reproductive output (number of larvae *in utero*) and body length of the parasite (Fig. 7.5a). Moreover some toads contained three distinct size classes of parasites and, since invasion occurs only in July in Arizona, these size classes were considered to represent different yearly cohorts (Fig. 7.5b). Most of the adult parasites in the suprapopulation (defined as all individuals in all stages of development within all hosts in the ecosystem – see Margolis *et al.*, 1982) were aged 1 and 2 years post-invasion, with 19% of parasites older; some individuals may survive for a fourth year. The annual reproductive output increases dramatically with age, with first-, second- and third-year parasites producing means of 5, 43 and 96 larvae respectively. The maximum reproductive output, achieved by a parasite that was probably 4 years old, was 350 larvae. Thus, a relatively small number of older parasites make the greatest contribution to transmission but reproductive output, even in the highest producers, is exceptionally low for a parasite (*Entobdella soleae* adults of medium size produce about 50 eggs per day – section 4.8).

There is a reduction of reproductive output when infrapopulations of *P. americanus* become dense (Tocque and Tinsley, 1991a); this density-dependent phenomenon increases in significance as the parasites age and may reflect increased demand for resources by large parasites maintaining many larvae *in*

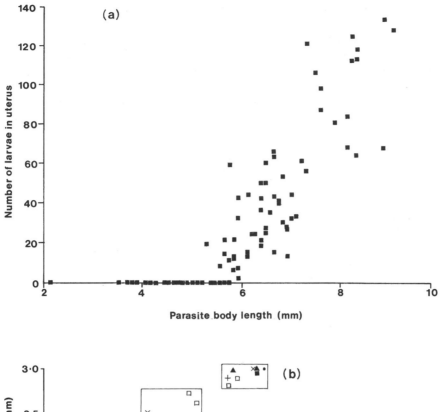

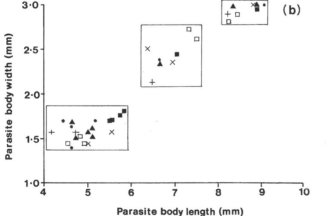

Figure 7.5 (*a*) Relationship between reproductive output and body length of adult *Pseudodiplorchis americanus* collected prior to host spawning in Arizona in July 1988. (*b*) Age cohorts of *P. americanus* distinguished by the length and width of parasites from a sample of toads infected with three distinct parasite size classes (parasites from each host represented by a different symbol). The body lengths of the largest parasites in first (up to 6 mm) and second (up to 7.5 mm) year groups were used to divide the whole population of parasites into different age cohorts. Source: redrawn from Tocque and Tinsley, 1991a.

utero and hence increased competition between larger individuals for one or more of these resources.

Tocque and Tinsley (1991b) demonstrated that desert temperature cycles greatly affect the life cycle of *P. americanus*. Maximum growth and reproductive development occur at 25°C and are reduced at both higher and lower temperatures. At 16°C reproductive development is inhibited. The maximum annual period in which development can occur in *P. americanus* is probably the 4–5.5 months during which temperatures are above 20°C, and this period includes the time when toads are active after the summer rains. The first month spent in the respiratory tract is taken up with pre-adult development, so first-year parasites have only 3–4.5 months for the reproductive preparation that will contribute to transmission in the following summer. Even at the optimum temperature of 25°C, this period is barely sufficient for the accumulation of larvae by first-year adults. Moreover, weather patterns may affect the length of this preparation period. If the rains are late in one year and early in the next, this period may be too short for first-year parasites to produce larvae, and only parasites older than one year will be able to contribute to transmission at the host breeding event of the following summer.

According to Tocque and Tinsley (1991b), the San Simon Valley is at the northerly limit of distribution of *S. couchii* and the long duration of cool winter temperatures in this area will limit the reproductive preparation of first-year parasites. Different temperature patterns elsewhere in the toad's range may favour the completion of reproductive development by first-year parasites. Different patterns of rainfall also occur in other parts of the toad's range, permitting extended toad breeding periods in some areas and, in other areas, no breeding when rainfall is poor.

In spite of the narrowness of the infection window and the extremely low reproductive capacity of *P. americanus*, transmission is highly efficient, with oncomiracidia having a greater than 30% chance of establishing themselves in a new host (Tinsley and Jackson, 1988).

By focusing on the events in a breeding assembly there is more to be learned about parasite–host interactions. The distribution of parasites in toads just emerged from hibernation is over-dispersed (Tinsley, 1989; see also section 7.1.3), but separation of the data for male and female toads reveals that heavy parasitic burdens (15–30 parasites per host) occur in males (Fig. 7.6), and there are four times as many adult parasites in males than in females. Such sex differences in parasite burdens are well known (section 20.3) and are usually attributed to behavioural and/or physiological–immunological differences between the two sexes (Zuk and McKean, 1996). There is, however, little direct evidence to support specific causal relationships and Tinsley's study outlined below (Tinsley, 1989) is outstanding in that it identifies the ecological mechanisms leading to sex differences in the *Pseudodiplorchis–Scaphiopus* association.

The infection window at each nocturnal spawning assembly has a maximum width of 7 h (9 pm to 4 am). Since the distances travelled by toads to the temporary rain pools vary, the host population steadily increases during this 7 h window. As toads arrive the eggs of their parasites are released and hatch, so that there is a build-up of oncomiracidia as time progresses. Consequently the infection pattern in the 7 h window is highly asymmetrical, with 10% of the night's total invasions occurring up to midnight (during the first 3 h of the assembly) and more than 30% occurring during the final hour of the assembly before dawn. Females generally spawn within about 3 h of entering the pools and then leave. Therefore, females arriving early and leaving before midnight will pick up fewer larvae than females arriving at 1 am and leaving at dawn. Males also arrive at different times, but tend to remain in the assembly until it disperses at dawn. Hence they experience longer and more intense exposure than most females. This behavioural difference is partly responsible for sex differences in parasite burdens.

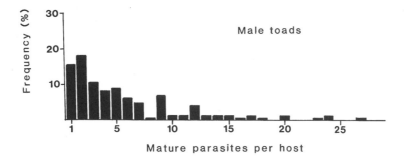

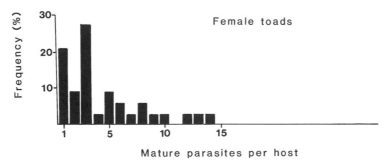

Figure 7.6 Distribution of infection levels of adult *Pseudodiplorchis americanus* (parasites containing infective stages *in utero* and therefore 1 year or more post-infection) in male toads ($n = 297$; prevalence 48%) and female toads ($n = 126$; prevalence 26%). Data from *Scaphiopus couchii* populations in the San Simon valley, Sonoran Desert, Arizona, USA. Source: reproduced from Tinsley, 1990b, with permission from Taylor & Francis.

Males are also exposed to more parasites because they visit all the assemblies permitted by the rainfall pattern, while females visit only one assembly. Tinsley and Jackson (1988) found that among males there was an approximate doubling of mean parasite burden at each of three separate exposures, leading to a prevalence of 100% and a mean intensity of more than 100 new parasites per infected toad. Among females about 40 parasites per toad were recorded.

There is a third behavioural factor influencing the distribution of parasites between the sexes. Males always outnumber females in the assemblies and many males will fail to mate. Tinsley found that males that mate have fewer parasites than those that do not and suggested that relative mobility may be significant. Unpaired males move about as they search for females and interact with other males, while mating males are relatively sedentary. Unpaired males may therefore make contact with more oncomiracidia, since they swim through a greater volume of water. Thus males are likely to be exposed to more larvae than females for two reasons: because they spend longer in the pools and because many of them are more active and will meet more larvae.

Tinsley tested these ideas in the field by experimentally eliminating these two differences between the sexes. He confined mating pairs of infected toads for a fixed time period in enclosures in newly formed pools, thereby ensuring that both sexes experienced the same exposure times and activity patterns. The result was that females acquired slightly more larvae than males, a probable consequence of the fact that mating females lie lower in the water and larvae have shorter distances to travel to the nostrils. This bias towards females during mating is greatly outweighed by the vulnerability of males before and after mating as a result of their behaviour.

Although transmission is initially highly successful with all male toads becoming infected, there is a catastrophic fall in infrapopulations during hibernation. Prevalence falls to about 50% in the males, with a mean intensity of five parasites in those that retain their infection. Death of heavily infected toads might be a contributory factor, but Tinsley found little evidence for this. Parasites reduced fat reserves and erythrocyte concentrations in hibernating hosts, but serious effects occur only in toads in the poorest nutritional condition and there is evidence that toads are long-lived (up to 15 years). There is no evidence that hormonal differences between the sexes have any influence on post-invasion reduction in parasite numbers and, although competition between parasites may play a part, this does not explain the loss of all parasites by half the male toads. The inference is that host immunity is a major factor in these post-invasion losses.

In this context, it is interesting that Tocque and Tinsley (1992) found evidence that parasites in the bladders of hibernating toads do not egest waste material from the gut, even though older parasites may ingest as much as 1.9–5.3 μl of blood per week. They pointed out that, since hibernating toads do not urinate, egested material from the parasites would accumulate and might provoke the host immune system. Parasites may avoid such provocation by not egesting.

(d)   Test of the Hamilton–Zuk hypothesis

This hypothesis (1982) has stimulated much debate and research at the interface between parasitology and evolutionary biology and proposes that secondary sexual characters are used by females as indicators of the genetic ability of the male to resist disease, since these characters and the displays in which they are used depend on healthy, vigorous males (see also sections 5.5.1, 17.7.1). One prediction of the theory is that, within a species, successful males should harbour fewer parasites than unsuccessful ones. Tinsley (1990a) was quick to appreciate that the *Pseudodiplorchis–Scaphiopus* system was highly suitable for testing this prediction and favourable features include the following: *P. americanus* is the only common parasite of *S. couchii*; during hibernation the host starves, and toad spawning and parasite transmission take place on the first night after rainfall, before the host replenishes its food reserves; mate selection is determined by female choice and the important male secondary sexual character directing this choice is male vocalization, an energetically costly procedure; females enter the mating arena only once and complete spawning and mate selection in a matter of hours; all toads entering 'first night' assemblies are comparable with regard to the effects of parasite burden on a chronically-starved host.

Parasites carried by male toads when they enter a 'first night' assembly could affect mating performance. Assuming that the ability to eliminate parasites is heritable and that the parasite is pathogenic (they are competing during hibernation for the host's food reserves), toads with few or no parasites should perform better in mating assemblies and be more 'desirable' to females. However, Tinsley found no correlation between mating success and parasite burden. Although infection can prejudice survival, it is only one of several inter-related factors, including host feeding success and choice of hibernation site. Selection for disease resistance occurs before the hosts enter the mating arena.

7.1.5 *Polystoma* spp.

According to Gallien (1935), the presence of a polystomatid monogenean in the European common frog, *Rana temporaria*, was first mentioned by Rösel von Rosenhof in 1758 and this elevates *Polystoma integerrimum* (Fig. 7.7) to the status of first monogenean to which written reference was made. During the 19th century, it attracted the attention of many eminent parasitologists, including Rudolphi, Baer, von Siebold, Leuckart and van Beneden. Even in some modern textbooks it takes pride of place as a type example of a monogenean, in spite of the fact that its biology has many atypical features.

The common frog spends most of the year on land among damp herbage. Opportunities for oviposition by the parasite are therefore restricted to 1–2 weeks in the spring, during which the frogs visit water for spawning. *P. integerrimum* copes with this brief oviposition opportunity in an entirely different way from *Pseudodiplorchis americanus*. Egg assembly is switched off during the long terrestrial phase of the host's annual cycle, but the adult parasites in the bladders of the terrestrial frogs occupy themselves by accumulating resources for the period of intense egg assembly which will take place during the brief aquatic breeding phase of the host.

Tinsley (1983, 1990b) quoted an average figure of 2300 for the total mean egg output of an adult parasite over a period of a few days, with a maximum of up to 4000, implying a maximum egg assembly rate of almost 2 eggs per minute. This exceeds that of any other monogenean; *Entobdella soleae* assembles up to 2.5 eggs per hour, but during much of this period the formed egg rests in the ootype, or the ootype is empty, and the actual time required to assemble the eggshell is 4–6 min (section 4.8). Nevertheless, it is still difficult to understand how *P. integerrimum* is able to assemble an egg in approximately 30 s. Even more remarkable is the fact that this high rate of egg production is achieved at frog breeding temperatures as low as 5–9°C (Combes, 1972). However, in terms of annual output the fecundity of *P. integerrimum* is very low.

The eggs laid by adult parasites in the bladders of breeding frogs are undeveloped. After leaving the frog, the eggs require 42 days to complete their development at 10°C, and at 5°C development may exceed 80 days. Ciliated oncomiracidia emerge from the egg (Fig. 7.8), but by this time the adult frogs have left the water and there is no contact with and no infection of adult frogs. The oncomiracidia infect the gills of the tadpoles. Llewellyn (1957b) observed invasion of tadpoles by many individual larvae and identified a consistent behaviour pattern. After making chance contact with a tadpole, the larva uses its cilia to glide over the tadpole's body, eventually reaching a position about 1 mm anterior to the spiracle. From this position the larva travels directly to the edge of the spiracle and, at a slack moment in the rhythmical outflow of water from the spiracle, folds itself ventrally through 180° and glides swiftly into the gill chamber.

Larvae on the gills continue to develop slowly until host metamorphosis, when they migrate to the bladder of the young frogs. Gallien (1935) claimed that this migration took place by way of the gut of the tadpole as in *Pseudodiplorchis*, but Combes (1968) showed that in tadpoles in which the gut had been disrupted experimentally, migrating parasites still turned up in the bladder after metamorphosis. He went on to establish that migration occurs at night *via* the ventral skin of the immobile tadpole (Fig. 7.8*d*) and takes about 1 min. The parasites use the haptor and the anterior adhesive glands during locomotion and the glands become

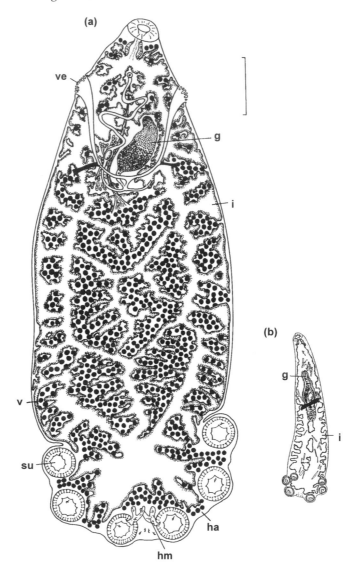

Figure 7.7 *Polystoma integerrimum*: (*a*) adult from the bladder of adult *Rana temporaria*; (*b*) progenetic parasite from the gills of *Rana temporaria* tadpole. Drawn to the same scale; scale bar: 1 mm. g = germarium; ha = haptor; hm = hamulus; i = intestine; su = sucker; v = vitellarium; vo = vaginal openings. Source: redrawn from Combes, 1968.

noticeably distended and refringent prior to migration. In the related parasite *Polystoma pelobatis* from the frog *Pelobates cultripes*, Combes noted that some parasites appeared capable of anticipating metamorphosis, since they left the gills and established themselves on the growing stumps of the front legs, which at this time stretch the wall of the branchial cavity. He believed that a physicochemical change such as reduced availability of oxygen, rather than a morphological change, provided the stimulus for this first phase of migration.

The tadpole gill parasite has additional potentiality. If infection of the tadpole takes place early enough, the parasite becomes precociously sexually mature

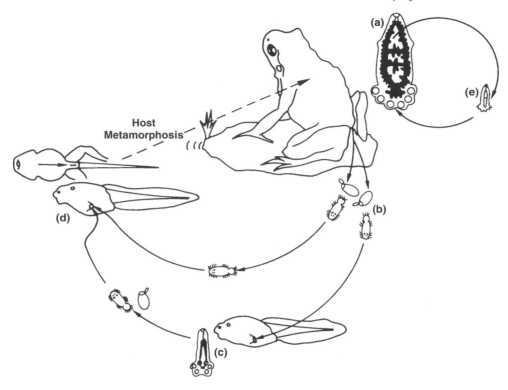

Figure 7.8   The life cycle of *Polystoma integerrimum*. The adult (*a*) inhabits the bladder of the common frog *Rana temporaria*. Ciliated larvae (*b*) emerging from eggs laid by adults in the frog's bladder invade tadpole gills. Larvae encountering the gills of young tadpoles undergo progenesis (*c*). Larvae from bladder parasites or from eggs laid by progenetic parasites encountering the gills of older tadpoles (*d*) grow slowly and migrate at night over the ventral skin of metamorphosing tadpoles to the bladder. Ovoviviparity in bladder parasites leads to infection of the same frog (*e*). Source: reproduced from Kearn, 1986, with permission from Academic Press.

(progenesis; see footnote, section 5.2; Fig. 7.7*b*). These progenetic parasites produce a late generation of eggs and ciliated larvae (Fig. 7.8*c*). Gallien found that some of these parasites were mature after 20 days and laid approximately 10–12 eggs per day; he calculated that these miniature adults produced between 400 and 500 eggs during an egg assembly period of about 40 days.

Combes (1968) showed that the age of the tadpole determined whether or not invading oncomiracidia underwent accelerated sexual development. On young tadpoles precocious development and miniature adults are common, but the proportion of invading larvae undergoing progenesis decreases as the tadpoles age and this phenomenon does not occur on old tadpoles. Late oncomiracidia hatching from eggs laid by progenetic parasites invariably meet old tadpoles, do not themselves undergo progenesis and, after a brief sojourn on the tadpole gills, migrate to the bladder.

Work in more northerly latitudes (e.g. by Gallien in the north of France) suggested that this progenetic generation is an essential part of the life cycle of *P. integerrimum*, but Combes (1968), working in the Pyrenees, where *R. temporaria* is at the southern limit of its range and usually confined to altitudes of 1000–2400 m, found that progenetic parasites were rare on wild tadpoles. However, he was able to produce them experimentally by infecting Pyrenean tadpoles in the laboratory and this led him to seek an ecological explanation for their rarity in the wild. He found that, at any given temperature, frog eggs developed more quickly than polystomatid eggs and that the difference in time between the completion of development of host and parasite eggs increased from 4 days at 25°C to more than 50 days at 5°C (Fig. 7.9). Since host and parasite lay their eggs at almost the same time and since temperatures in the mountain study area are low, then, provided that eggs of host and parasite experience the same conditions, the oncomiracidia will hatch when the tadpoles are well developed and progenesis is unlikely to occur.

Another factor reduces even further the chances of progenetic development. Frog spawn usually floats while the eggs of *Polystoma* sink. During the day the sun warms the surface layer of the Pyrenean pools, creating a thermocline. This accelerates development of the floating frog spawn, while development of the parasite's eggs is retarded in the colder deeper water. This leads to a further separation in time between oncomiracidia and tadpoles. Combes recorded a temperature difference of 10°C in 30 cm of water and observed that spawning occurred in water at least 20 cm deep; thus, parasite eggs are unlikely to be deposited in shallow warm water at the edges of the pools. However, tadpoles tend to gather in shallow, warmer water, where their development is further accelerated.

Progenetic parasites were not entirely absent from tadpoles in the Pyrenean pools and Combes showed that localized temperature differences created by the presence of shade or snow on one side of the pool, coupled with movements of some tadpoles from the cool to the warm side of the pool, produced a favourable situation for progenetic development.

The conclusion that can be drawn from the story so far is that frogs can be infected only once in their lives, at the tadpole stage. Parasites in the frogs cannot contribute to the life cycle until they are 3 years old and thereafter only once a year. However, the bladders of adult frogs are known to contain parasites of different sizes, and this led Gallien (1935) to suggest that there is individual variation in growth rates. Combes (1968) disputed this and suspected that some pathway existed for infecting frogs after metamorphosis. He showed experimentally that oncomiracidia are unable to enter the bladders of frogs *via* the cloaca and that parasites in young, newly-metamorphosed frogs cannot transfer to and establish themselves in cannib-

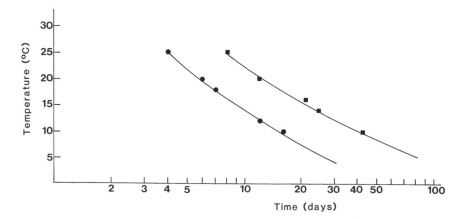

Figure 7.9   Relationship between temperature and the duration of development of the eggs of *Polystoma integerrimum* (squares) and those of *Rana temporaria* (circles). Source: from an original figure by Combes (1968); reproduced from Kearn, 1986, with permission from Academic Press.

alistic adult frogs. He discovered that the answer lies in autoinfection by ovovivi-parity, adult parasites having the ability to retain a single egg in the uterus at the end of the laying season (Fig. 7.8e). The larva hatching from this egg becomes established in the bladder of the same frog, effectively doubling the host's burden of parasites. This phenomenon offsets to some degree the disadvantages of the parasite's reproductive strategy: concentration of infection on the stage in the host's life cycle that suffers the highest mortality, i.e. the tadpole stage, and the long period (3 years) during which the parasite makes no reproductive contribution.

Correspondence between the reproductive activities of parasite and host is striking in polystomatid parasites of anurans. The severe seasonal limitation of egg production in *P. integerrimum* coincides not with high environmental temperatures but with the brief spawning period of the host. Although there are latitudinal differences in the dates of the host spawning period, egg laying in *P. integerrimum* corresponds in general with these dates (Fig. 7.10). There is also correspondence between parasite and host in the periods of recession and regeneration of the

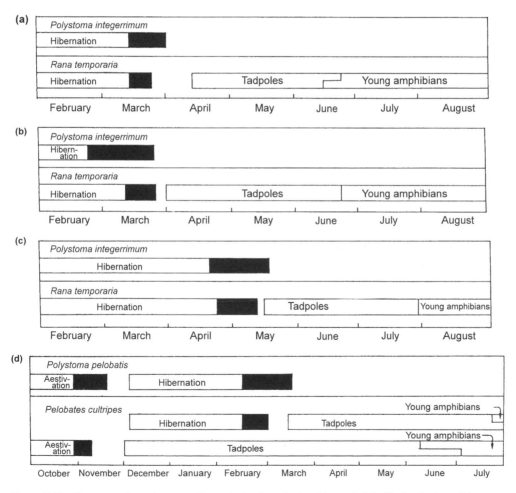

Figure 7.10 Correspondence between the periods of egg laying (shown in black) of polystomatid mono-geneans and their frog hosts. (*a*) *Polystoma integerrimum* in the Pyrenees. (*b*) *P. integerrimum* in the Paris region, France. (*c*) *P. integerrimum* in the region of Leningrad (now St Petersburg, Russia). (*d*) *P. pelobatis* and the frog *Pelobates cultripes*, in the south of France (Roussillon). Source: based on Combes, 1968.

reproductive systems, and sexual maturity in parasite and host is achieved at the same age (3 years). The frog *P. cultripes* may have an egg-laying period in the autumn as well as in the spring and Combes (1968) observed that its parasite, *Polystoma pelobatis*, also has two egg-laying periods corresponding with those of the host (Fig. 7.10*d*).

The tree frog *Hyla versicolor*, which is parasitized by *Polystoma nearcticum*, joins mating assemblies of less than 4h duration; according to Tinsley (1991), egg assembly of the parasite switches on and off abruptly at the beginning and end of these brief periods and appears to be linked with host sexual excitement. These changes in parasite reproductive activity are paralleled by changes in the expression of FMRF-amide-related peptide (neuropeptide), but not the classical neurotransmitter serotonin (5-HT), in neurons innervating the ootype (Armstrong *et al.*, 1997). This is

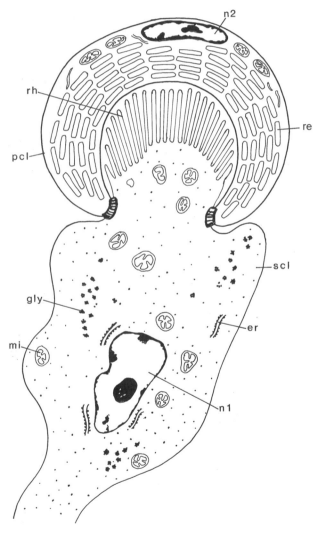

Figure 7.11   Diagram of a median section through one of the eyes of the oncomiracidium of *Polystoma integerrimum*. er = endoplasmic reticulum; gly = glycogen; mi = mitochondrion; n1 = nucleus of sensory cell; n2 = nucleus of platelet cell; pcl = platelet cell; re = reflecting platelet; rh = rhabdomere; scl = sensory cell. Source: redrawn from Fournier and Combes, 1978.

evidence that regulatory peptides may be involved in egg assembly in flatworm parasites.

This reproductive synchrony between *H. versicolor* and *P. nearcticum* and the relationship between egg laying and host sexual activity in *Pseudodiplorchis americanus* point strongly to the possibility that the reproductive activities of polystomatids are controlled in some way by hormonal changes in their hosts, and the blood feeding habits of these parasites provide ready access to these hormones. Some attempts have been made to investigate this possibility experimentally but none of these attempts has provided an unequivocal answer (see Kearn, 1986).

A memorable experience for me was the opportunity to see living oncomiracidia of *Pseudodiplorchis americanus* in Professor Richard Tinsley's laboratory, but it was not just their large size that left a lasting impression. The four eyes of these larvae were visible as intense pinpoints of light, quite unlike the black eyespots of the larvae of *Entobdella soleae*. Gallien (1935) also remarked, with reference to larvae of *Polystoma integerrimum*, that 'les deux paires d'yeux...brillent comme des miroirs'. Fournier and Combes (1978) studied the ultrastructure of the eyes of *P. integerrimum* and showed that there is indeed a mirror behind the single rhabdomeric sensory cell (Fig. 7.11). Light reflected by this mirror accounts for the brilliance of the eyes, like the tapetum of a cat.

Fournier and Combes found that the cup-shaped mirror cell occupies the position of the pigment cup in other monogeneans and that the mirror consists of concentric rows of platelets separated by 120 nm gaps. They suggested that this composite structure amplifies the light by constructive interference between rays reflected from the platelets and focuses light on the rhabdomere; focusing is achieved in *E. soleae* by a crystalline lens (see Fig. 4.19). Similar reflecting eyes are present in the larva of the turbellarian *Fecampia isopodicola* (section 2.3.3).

It is difficult to envisage the function of these reflective, light-concentrating eyes because, according to Macdonald and Combes (1978), the oncomiracidia of *P. integerrimum* hatch only during daylight, when tadpoles are in shallow and presumably well lit water. The oncomiracidia of *P. americanus* hatch and infect their hosts at night (section 7.1.4), and the larvae may be able to use dim moonlight or starlight for orientation.

7.1.6 Other polystomatids of tetrapods and the sphyranurids

The polystomatids are well established not only in the anuran amphibians but also in the oral cavities and bladders of freshwater chelonians (turtles; Table 6.1). Sphyranurids are related to polystomatids but possess only one pair of suckers and have been reported only from urodele amphibians (newts and salamanders; Table 6.1). If it is assumed that coevolution has taken place between these polyopisthocotyleans and their vertebrate hosts, then chelonians and anurans would be expected to have closer affinities with each other than either group has with the urodeles (Llewellyn, 1963). An alternative advocated by Tinsley (1981) is that host-switching has taken place, chelonians acquiring their polystomatids by transfer from primitive pipid anurans sharing their aquatic environment. The polystomatids of reptiles feed on epithelial cells and mucus and Allen and Tinsley (1989) regarded this not as a primitive feature but as secondary development from an ancestral blood feeding habit, perhaps induced by the thickness and toughness of the epithelial linings of their habitats and the relative inaccessibility of blood vessels.

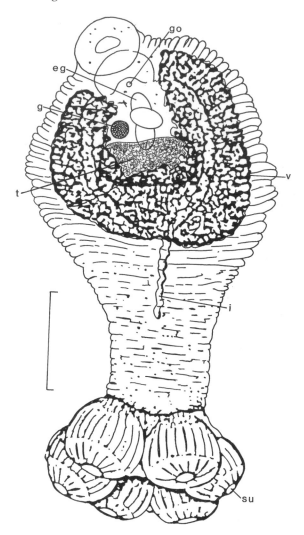

Figure 7.12 *Oculotrema hippopotami,* ventral view. eg = egg; g = germarium; go = genital opening; i = intestine; su = haptoral sucker; t = testis; v = vitellarium. Scale bar: 1 mm (approx.). Source: reproduced from Stunkard, 1924.

Kearn (1994) has suggested that the reduction of haptoral suckers to one pair in the sphyranurids may be an adaptation to their special life style. They are found on the skin and gills of fully aquatic mud puppies (*Necturus*), where they move about rapidly like leeches and feed largely on epithelial cells (Wright and Macallum, 1887). Manipulation and co-ordination of three pairs of suckers is not compatible with such an active life style and one way in which reduction of suckers could have been achieved from a polystomatid ancestor is by progenesis, with sexual maturity being achieved before the inception of the second and third pairs of suckers.

In 1924, Stunkard published a description of a polystomatid monogenean which had been given to him by Professor Hindle of the School of Medicine in Cairo, Egypt. The vessel, containing five specimens, bore the label 'from the eye of a hippopotamus' and Stunkard presumed that the parasites came from a hippopotamus in the Cairo Zoological Gardens. He attributed the parasite to a new genus and named it *Oculotrema hippopotami* (Fig. 7.12). However, the provenance of the material was regarded by some with scepticism and, according to Thurston and Laws (1965), Baer relegated the parasite to a footnote, with comments to the effect that this was either an accidental infection or a case of faulty labelling. This scepticism was laid to rest when Thurston and Laws (1965) found that the eyes of *Hippopotamus amphibius* in Uganda were indeed infected with the parasite and Thurston (1968) reported a prevalence of 90% and a mean intensity of eight parasites.

Thurston and Laws found no evidence of blood feeding – haemoglobin is present in the body of the parasite but appears to be of parasite origin (Thurston, 1970). Hence, the parasite is likely to feed on epidermal tissue like the reptilian polystomatids. There seems little doubt that the hippopotamus acquired its monogenean by host-switching; chelonians, already parasitized by skin-feeding polystomatids, may have been the source of this transfer.

7.2   INVASION OF THE GUT OF VERTEBRATES

The gut of vertebrates presents a substantial challenge for potential parasites since its function is to hydrolyse most biological macromolecules. The problems are forcibly underlined by the work of Tinsley and Jackson (1986), who showed that *Pseudodiplorchis americanus* requires special protection, probably in the form of tegumental secretions, while travelling through the gut of its toad host from the lungs to the bladder (see above). If protection is required for this brief journey of less than 30 min, then it is even more remarkable that some monogeneans survive permanently in the gut of fishes and that gut parasitism has evolved on more than one occasion.

7.2.1   Monogeneans dwelling in the gut of fishes

The dactylogyroideans *Neodiplectanotrema* and *Paradiplectanotrema* appear to inhabit the oesophagus (Gerasev, Gayevskaya and Kovaleva, 1987) and are unlikely to be exposed to digestive fluid but other dactylogyroideans, *Enterogyrus* spp., live in the stomach (see, for example, Pariselle, Lambert and Euzet, 1991). In *E. foratus* and *E. coronatus*, the posterior region of the haptor is cylindrical with terminally situated hamuli, and has invasive properties, penetrating deeply into the stomach wall (Fig. 7.13). A parapet produced by an inflammatory host reaction surrounds the parasite where it emerges from the stomach wall, and the bulbous anterior region of the haptor is anchored to this parapet by five pairs of marginal hooklets arranged in a circle. *Montchadskyella intestinale* (Fig. 7.14) also inhabits the stomach and the anterior part of the intestine (Bychowsky, Korotaeva and Nagibina, 1970). Its affinities are uncertain; Bychowsky *et al.* tentatively suggested an affiliation with the monocotylids and Kearn (1994) has made an equally tentative suggestion that it may be related to capsalids like *Encotyllabe* that live on the pharyngeal tooth pads of teleost fishes (see Kearn and Whittington, 1992b).

The ability of monogeneans to assemble eggshells of tanned protein may have been an important prerequisite for colonization of the host's gut. The eggshells of most monogeneans are chemically resistant and physically strong, providing protection for embryos as they pass through the gut to the outside world (section 1.4.5).

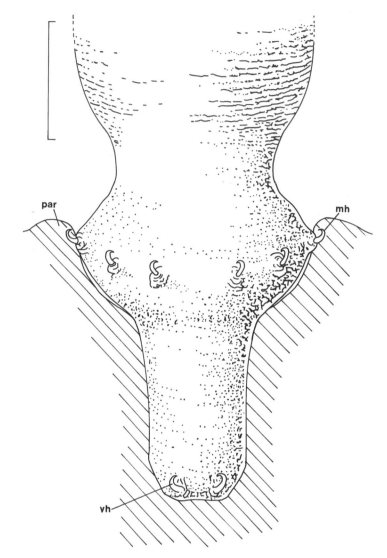

Figure 7.13 Diagram of the haptor of *Enterogyrus foratus* or *E. coronatus* embedded in the stomach wall of the host. mh = marginal hooklet; par = 'parapet' – swollen ring of inflamed stomach tissue; vh = ventral hamulus. Scale bar: 100 μm. Source: reproduced from Pariselle, Lambert and Euzet, 1991, reprinted by permission of Kluwer Academic Publishers.

All of these monogenean gut parasites retain their own gut and presumably feed on host epithelia like their ectoparasitic ancestors. It is not yet known whether they gain an additional nutritional advantage by transtegumental absorption of components of the predigested food of the host, but the highly folded body of *M. intestinale* (Fig. 7.14) would provide an increased surface area for nutrient absorption.

7.2.2 The gyrocotylideans

The invasions of the gut described in the last section may have taken place relatively recently and demonstrate that specialization of monogeneans for ectoparasitism

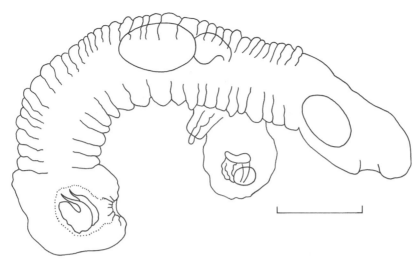

Figure 7.14 *Montchadskyella intestinale* in side view. The large outgrowth from the ventral surface carries the male copulatory sclerites and the uterine opening; the small outgrowth carries the vaginal opening. Scale bar: 200 μm. Source: redrawn from an original illustration by Bychowsky, Korotaeva and Nagibina (1970); reproduced from Kearn, 1994, with permission from the Australian Society for Parasitology.

does not preclude colonization of such a harsh environment. The colonization event that gave rise ultimately to the cestodes was much more remote and probably involved a primitive, unspecialized, ectoparasitic proto-monogenean. Llewellyn (1965) has argued that the emergence of the main groups of modern monogeneans took place before the Ordovician and that the ancestors of the cestodes must have separated from the proto-monogenean stock even earlier. We seem to have been fortunate in that some of these early colonizers of the vertebrate gut left descendants that survived with little change to the present day and these 'missing links' between the proto-monogeneans and the cestodes are the gyrocotylideans, found in the intestines of chimaeroid (holocephalan) fishes.

Some of the features of gyrocotylideans reflect their proto-monogenean ancestry, but they also lack a gut (Fig. 7.15) and this tapeworm feature suggests that loss of the gut was a very early event in the endoparasites of the proto-monogenean–cestode lineage. All nutrients must be absorbed through the tegument in gyrocotylideans and folding of the body margins in some species (Fig. 7.15) will increase the absorptive surface. However, much greater surface amplification is provided by tegumental microvilli (Fig. 7.16). These are densely packed like the microtriches of a tapeworm, but lack the thickened tips of microtriches (Lyons, 1969c).

Gyrocotylideans emerged from the proto-monogenean–cestode lineage before the marginal hooklets were reduced to the cestode number of six. They retain 10 hooklets (Fig. 7.17*f*), which, according to Lynch (1945), persist in post-larval stages (Fig. 7.17*c–e*) and are functional in the smaller juvenile specimens. During development the haptor is functionally replaced and engulfed by a funnel-shaped pseudohaptor (Figs 7.15, 7.17*e*), a development that resembles that of acanthocotylid monogeneans (section 5.2), but the hooklets are still to be found in the dorsal wall of the funnel of some fully mature specimens. In fact, Lynch stated that they could be seen clearly in one of his whole mount preparations of *Gyrocotyle urna* 32 mm in length.

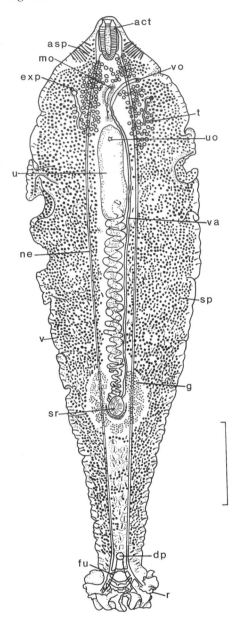

Figure 7.15   *Gyrocotyle urna*, in dorsal view. act = acetabulum (anterior sucker); asp = acetabular spines; dp = dorsal pore of funnel; exp = excretory pore; fu = funnel; g = germarium; mo = opening of male system; ne = longitudinal nerve; r = rosette organ; sp = spine; sr = seminal receptacle; t = testis; u = uterus; uo = opening of uterus; va = vagina; v = vitellarium; vo = vaginal opening. Scale bar: 0.5 cm. Source: redrawn from Lynch, 1945.

The funnel has foliaceous margins (the so-called rosette organ) and the dorsal wall of the funnel is perforated by a canal (Fig. 7.15). The inner surfaces of the rosette are abundantly supplied with glands (Lyons, 1969c) and the secretion of these glands appears to have adhesive properties. According to Allison (1980), the organ is highly extensile, spreads out over the host's mucosal surface and is attached firmly to the villi by adhesive, features reminiscent of the haptor of the calceostomatid monoge-

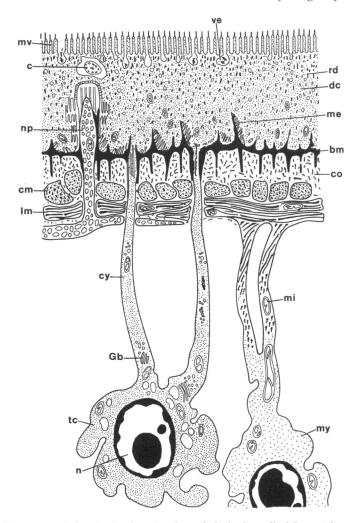

Figure 7.16   Diagrammatic longitudinal section through the body wall of *Gyrocotyle urna*. bm = basement lamina; c = cilium; cm = circular muscle; co = interstitial connective tissue; cy = cytoplasmic connection between cyton and surface syncytium; dc = syncytial surface layer of tegumentary cytoplasm; Gb = Golgi body; lm = longitudinal muscle; me = basal plasma membrane; mi = mitochondrion; mv = microvilli; my = myocyton; n = nucleus; np = nerve process; rd = rod bodies; tc = tegumentary cell body; ve = vesicle. Source: reproduced from Lyons, 1969c, copyright © by Springer-Verlag 1969, with permission from Springer-Verlag.

nean *Neocalceostomoides brisbanensis*, which attaches itself to surfaces in the gill chamber by secreting an adhesive (section 6.4.1). However, the funnel of gyrocotylideans also appears to generate suction, which is released when the dorsal canal opens.

Many aspects of the biology of gyrocotylideans remain tantalizingly enigmatic, in spite of the intense interest and attention that their evolutionary position has generated. In particular, it is still not known whether gyrocotylideans retain a monogenean-like, direct life cycle or have acquired an intermediate host as in the cestodes. *Gyrocotyle* lays amber-coloured, presumably tanned eggs, which are enclosed in a layer of jelly (Fig. 7.17*a*) and liberate a ciliated, free-swimming larva (Fig. 7.17*b*). This so-called lycophore larva has no gut, a haptor with 10 hooklets

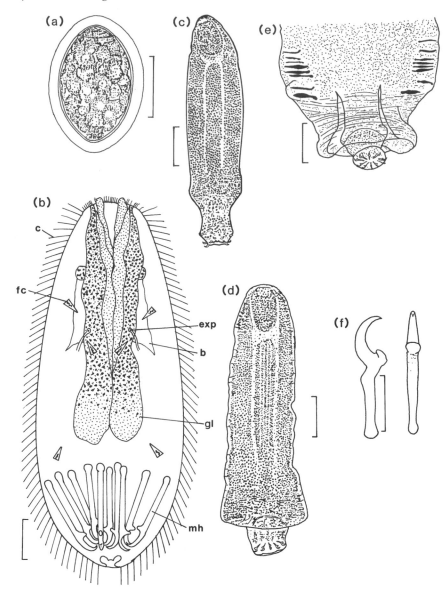

Figure 7.17 Development of *Gyrocotyle*. (*a*) Egg of *G. urna*. (*b*) Larva of *G. urna*. (*c*)–(*e*) Consecutive stages in the development of the post-larva of *G. fimbriata* in the intestine of the ratfish host. The swelling in the early post-larva (*c*), anterior to the haptor, is probably the developing pseudohaptor. In (*e*) the early rosette is a sleeve around the haptor. (*f*) Two views of a marginal hooklet of *G. fimbriata*. b = 'Brain'; c = cilia; exp = excretory pore; fc = flame cell; gl = gland cell; mh = marginal hooklet. Scale bars: (*a*) 50 μm; (*b*) 10 μm; (*c*) 100 μm; (*d*) 100 μm; (*e*) 100 μm; (*f*) 10 μm. Source: (*a*), (*c*)–(*f*) reproduced from Lynch, 1945; (*b*) redrawn from Xylander, 1986.

(hence the alternative name of decacanth larva), prominent glands opening at the anterior end of the body and a pair of organs interpreted as cilia-based eyes in the head region (Xylander, 1984). Llewellyn (1970, 1986) found very small specimens of *G. urna* attached to the tips of villi in the intestine of the ratfish *Chimaera monstrosa*. These specimens were little bigger than the free-swimming larvae and, apart from the lack of surface cilia and anterior glands, were indistinguishable from them. He interpreted this as evidence for a direct, monogenean-like life cycle, perhaps similar to that of stomach-inhabiting monogeneans of the genus *Enterogyrus*. Eggs of *Enterogyrus* pass out with the host's faeces and liberate a typical ciliated oncomiracidium, which establishes itself first on the head of a new host; the post-oncomiracidium then migrates *via* the gill arches and the oesophagus to the stomach (Bender, in Cone, Gratzek and Hoffman, 1987).

A different viewpoint on the life cycle was taken by Xylander (1986). He emphasized the fact that young specimens of *C. monstrosa* remain uninfected with *G. urna* until after the absorption of the yolk sac. In other words, the appearance of *G. urna* in the host coincides with the start of host feeding. Xylander claimed this as evidence that *Gyrocotyle* larvae enter an intermediate host which is ingested by the ratfish, so conveying the parasite to the gut. However, host behaviour will undoubtedly change when feeding begins and this change could well expose the host to direct invasion by larvae for the first time. Moreover, all attempts to infect potential intermediate hosts, such as amphipod crustaceans, which are eaten by ratfishes, have failed.

There appear to have been no attempts to infect whole fishes with larvae of *Gyrocotyle*. Chimaeroids are generally regarded as difficult to keep alive, but Simmons (1974) pointed out that *Hydrolagus colliei* can be procured and husbanded and therefore provides a vehicle for experimental attempts at infection. So far, this challenge has not been taken up. Xylander (1986) exposed the surface of a piece of skin cut from *C. monstrosa* to larvae of *G. urna*, but detected no behavioural response. Gills or buccal and branchial linings do not appear to have been exposed experimentally to larvae.

Another feature of the larvae that has direct relevance to the life cycle is their propensity for tissue penetration. Larvae have been observed to penetrate the cut surfaces, but not the intact surfaces, of detached pieces of host intestinal mucosa (Manter, 1951), but whether this experiment reflects real parasite–host interactions remains to be determined. However, there is little doubt that the larvae do have penetrative ability, which they use in a surprising way – they have been found embedded inside the bodies of larger specimens from the host's intestine (Lynch, 1945; Simmons, 1974). There is the possibility that these 'hyperparasites' come from eggs hatching prematurely in the uterus of the parent, the larvae dispersing into its parenchyma (as in the turbellarian *Paravortex scrobiculariae* – section 2.3.2), but the presence of these hyperparasites in immature *Gyrocotyle* indicates external penetration by free larvae. This may be a way for vulnerable larvae to escape the attentions of the host's immune system until they have had time to develop protection from this system.

The population dynamics of gyrocotylideans is of exceptional interest. A positive correlation between parasite size and host size suggests that the parasites are long-lived and that most infections are established early in the life of the host. According to Williams, Colin and Halvorsen (1987), gyrocotylideans generally have a very high prevalence with a very low intensity, with a strong tendency towards under-dispersion (variance to mean ratio less than 1). The probability of infection seems to be high since, with the exception of the smallest size group of fishes, prevalence reaches 90–100% but in spite of this most fishes contain only two adults. Williams *et al.* proposed four mechanisms that might be responsible for limiting the infra-population size: (1) a reduction in the probability of infection caused by changes in host biology; (2) a dynamic equilibrium with establishment of new parasites keeping pace with loss of already established individuals; (3) lethal effects on the host of infrapopulations greater than two; (4) failure of new parasites to establish because of competitive or immunological effects generated by resident parasites. They argued that existing data offer most support to the fourth hypothesis, i.e. that there is some kind of homeostatic control of infrapopulation size by the resident parasites.

7.3   THE WAY FORWARD

The fascination of the gyrocotylideans is that they most probably encapsulate the characteristics of the ancient intestinal parasites that gave rise to the enormously diverse tapeworms. They still preserve some features of their monogenean origins, such as the persistent haptor with 10 hooklets, the paired anterior excretory pores (Figs 7.15, 7.17*b*) and, possibly, a direct life cycle, but have taken a single major step on the road to the tapeworms by losing their own gut.

We will take up the evolutionary story of the tapeworms in Chapter 10, but first we will digress in the next two chapters to consider the way in which the stages in the more complex, indirect life cycles of tapeworms were unravelled by the scientists of the last century and to explore the biology of one of them in more detail.

CHAPTER 8

Unravelling the cestode life cycle

8.1 INTRODUCTION

We saw in the last chapter that, at this late stage in the 20th century, the life cycle of gyrocotylideans still eludes us. This is not because we are unaware of the appropriate questions to ask or unable to formulate experiments to answer these questions but because the investigation is fraught with practical difficulties. It is surprising that, no more than 150 years ago, no life cycle of a tapeworm was fully understood, but the problem with the elucidation of the tapeworm life cycle was quite different – material was abundant and accessible but deep-seated preconceptions and prejudices had to be overcome and the experimental approach to biological problems was in its infancy. Thus, the story of the unravelling of the tapeworm life cycle encapsulates important developments in biology and provides a most rewarding as well as interesting study. Grove (1990) gives a comprehensive account of this topic and much that follows is derived from his authoritative description.

8.2 THE ANATOMY OF ADULT *TAENIA*

Some adult tapeworms are so large (Fig. 8.1; *Taenia solium* in man is 2–7 m in length) and the detached body units or proglottides so conspicuous in the faeces that it is not surprising that they have been known since written records were first made. (With regard to anatomical terms for tapeworms the recommendations of Arme (1984) are followed here. A proglottis is a portion of the cestode body containing one set of reproductive organs or in a few species two sets side by side – see Mehlhorn *et al.*, 1981.) Moreover, the origins of such large creatures must have taxed many an enquiring mind, since the whole worm is clearly not ingested with the food and its ability to survive in the dark, corrosive environment of the small intestine, where other organisms are rapidly destroyed, must also have been deeply puzzling (the latter problem is still occupying modern physiologists, see section 9.3.2(e)).

From the earliest of times, biological phenomena that proved difficult to explain, such as the appearance of maggots in putrefying flesh, had been attributed to a process of 'spontaneous' or 'equivocal generation', during which organisms were supposed to materialize by a mysterious process of reorganization from putrefying material. The great appeal of this concept as an explanation for the otherwise inexplicable appearance of large tapeworms in the intestine can be appreciated. Parasitic worms in general were widely believed to originate from poorly digested food or from corrupt juices emanating from diseased organs. Although some early

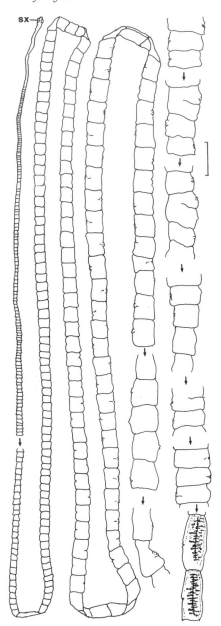

Figure 8.1 Several portions of an adult tapeworm (*Taenia saginata*) from man, showing the scolex (sx) and the gradual increase in size of the proglottides. Note branched uterus full of eggs in terminal gravid proglottides. Scale bar: 1 cm. Source: reproduced from Stiles, 1906.

writers, e.g. Aristotle, regarded tapeworms as animals, others did not, and some believed them to be a transformed strip of the intestinal lining. Individual proglottides were regarded by some observers as distinct worms, called 'cucurbitini' after their resemblance to pumpkin seeds, and some Arab writers believed that the whole tapeworm was not an organism but an elongated membranous sac produced by the intestine to hold the cucurbitini. Another Arab view saw the whole tapeworm and the cucurbitini as distinct organisms.

The tapeworm body (= strobila*) is broad and flat at one end and thread-like at the other (Fig. 8.1) and it is understandable that, in the days before microscopes, some observers regarded the thread-like end as the tail. It was an Englishman, Edward Tyson, in 1683, who opened the intestine of a dog and wrote the following:

> I observed this Worm alive in the *Ilion*; not lying streight, but in many places winding and doubling. Having taken notice how the Joynts were, I traced it up, by carefully opening the Intestine, to the smallest Extream; where I expected the *head* to be; and which did ly towards the *Duodenum*; whereas the broader end was downward towards the *Rectum*; and this broad end was free, and did nothing adhere; whereas that smaller extream did so firmly stick, and had fasten'd itself to the inward coat of the Intestine, that it was not without some trouble, by gently raising it with my Nail, that I freed it from it's adhesion.

He observed the head (= scolex) with a microscope and noted the presence of hooks of two sizes (Fig. 8.2). Concerning the head he made the following statement: 'I could not upon my strictest Enquiry and with extraordinary Glasses too, inform myself of any orifice here, which we may suppose to be the mouth'. He mistakenly thought that the conspicuous lateral apertures in each proglottis, which are now known to be genital apertures (Fig. 8.3), were mouths.

This observant man had other important things to say. He appreciated that the descent of faeces would be apt to carry away the long and sometimes convoluted body of the worm and proposed that the hooks, by attaching the scolex to the wall of the intestine, would counteract this. He also laid to rest two ideas: that tapeworms were not living organisms (he saw them move) and that the worm was a sac derived from the intestine and enveloping the cucurbitini.

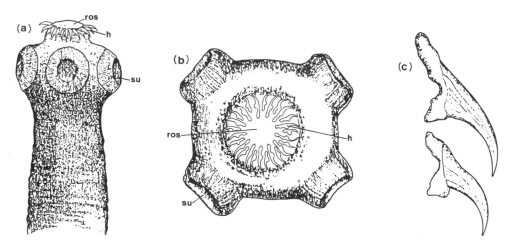

Figure 8.2 The scolex of *Taenia solium*. (*a*) In side view. (*b*) Viewed from the apical surface. (*c*) A large (anterior) and smaller (posterior) scolex hook. h = hook; ros = rostellum; su = sucker. Source: reproduced from Leuckart, 1886.

---

* Use of this term has been criticized by Mehlhorn *et al.* (1981), on the grounds that it is borrowed from coelenterates, where it has a different meaning. While recognizing the problem, the term is in common use and will not be avoided here.

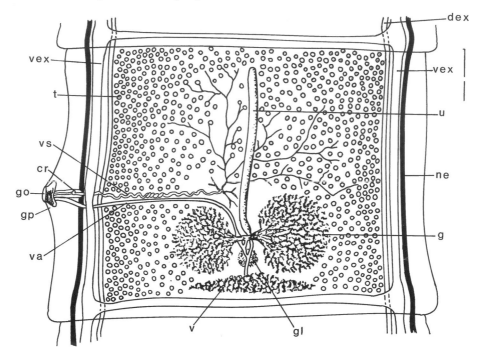

Figure 8.3 A mature proglottis of *Taenia saginata*. cr = cirrus; dex = dorsal excretory canal; g = germarium; gl = glands associated with female reproductive system; go = common genital opening; gp = genital papilla; ne = longitudinal nerve; t = testis; u = uterus; v = vitellarium; va = vagina; vex = ventral excretory canals connected by transverse canal; sd = vas deferens. Scale bar: 1 mm. Source: reproduced from Leuckart, 1886 and Stiles, 1906.

Tyson was prepared to believe that adult tapeworms developed from eggs entering the intestine from the outside world, but was mystified by the absence of any free-living creatures resembling tapeworms that could have been the source of these eggs. Consequently he was inclined to fall back on spontaneous generation as an explanation. Since he regarded the genital pore in each proglottis as a mouth, he assumed that material expelled from this pore was ingested host intestinal contents. As Grove (1990) has pointed out, if he had examined this with a lens he might have discovered eggs, which would have provided a clue to the origin of the worms.

Progress in the understanding of the anatomy of tapeworms took some steps forward and some backwards. Andry (1700, in Grove, 1990) regarded the genital pores in the proglottides as respiratory openings and mistook the branched uterus in each proglottis (Fig. 8.1) for a tracheal system, like that of an insect. The hooks on the scolex were thought by some to be teeth and the suckers on the scolex (Fig. 8.2) were variously regarded as eyes, nasal openings or mouths. The discovery of eggs (Fig. 8.4) by Andry was important and Goeze (1782, in Grove, 1990) demonstrated by inserting a horse hair into the genital pore that this aperture is not a mouth or a respiratory opening but leads to the germarium (Fig. 8.3).

During the next few decades, it was established that the tapeworm has no gut and that each proglottis contains a complete set of reproductive organs, the anatomy of which was elucidated (Fig. 8.3). However, the ignorance of tapeworm biology that existed as late as 1854 is underlined by a patent taken out in that year in the United

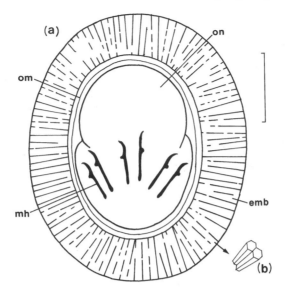

Figure 8.4   (*a*) Egg of *Taenia saginata*, showing (*b*) prismatic nature of embryophore (emb). mh = marginal hooklet; om = oncospheral membrane; on = oncosphere (= larva). Scale bar: 10 μm. Source: based on drawings by Silverman, 1954a and Silverman and Griffiths, 1955.

States by a Dr Alpheus Myers of Logansport, Indiana (see Pappas, 1987). The patent concerned a device for trapping tapeworms and consisted of two cylinders of gold or platinum, one housed inside the other (Fig. 8.5). Holes in the two cylinders were held in correspondence by a spring-loaded catch, so that free access from the outside of the device to the inside of the inner cylinder occurred as long as this correspondence was maintained. It was necessary to prime the inner cylinder with bait – Dr Myers recommended cheese. According to the text of the patent, the patient should be deprived of nourishment for several days, so that 'the worm becomes very hungry and will draw himself into the stomach by an instinct which belongs to all species of worms...'. A cord was then attached to the baited trap, which was swallowed. In the stomach the hungry worm inserted its head through the holes in the cylinders and seized the bait. The feeding activities of the worm then dislodged the catch and activated the spring-loaded mechanism, which led to a displacement of one cylinder relative to the other, hence trapping the worm. Motion of the cord was said to alert the patient that the trap had been sprung, and it remained only to withdraw the trap and the dependent worm by means of the cord.

8.3   THE LIFE CYCLE OF *TAENIA SOLIUM*

In 1835 von Siebold (in Grove, 1990) discovered that tapeworm eggs contain an embryo (oncosphere or hexacanth) with six hooklets (Fig. 8.4). Von Siebold (1857) remarked on the similarity of these embryos in tapeworms differing widely in their scolex morphology. He also extracted them from the eggs and commented as follows on the activity of one of them: 'It draws its round body together, and enlarges and contracts its transverse diameter, and by this operation protrudes, first in front and then at the sides, the six little hooks...'.

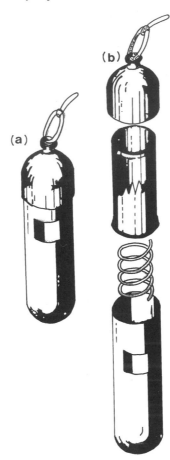

Figure 8.5 A device designed to remove a human tapeworm *via* the mouth, (*a*) assembled and (*b*) disassembled. Source: redrawn from Pappas, 1987.

The discovery of tapeworm eggs naturally raised the possibility that adult tapeworms in the intestine developed from eggs ingested by the host. A very early attempt to test this experimentally was made by Pallas in 1760 (see Grove, 1990). He inserted eggs of a dog tapeworm (now known as *Dipylidium caninum*, sections 8.5, 11.6) into the body cavity of a puppy and claimed to have found small tapeworms a month later. This has never been successfully repeated. Attempts to raise adult worms by feeding eggs directly to their hosts failed, and this gave strength to the supporters of spontaneous generation, since it appeared to contradict the doctrine of '*omne vivum ex ova*' (that all life comes from eggs). Thus, at the beginning of the 19th century, the concept of spontaneous generation was still very much alive, although the finding that tapeworms produce eggs and important discoveries concerning the biology of digeneans (see Chapter 12), sowed seeds that would ultimately lead to its downfall.

In 1790 a piece of work of great importance was conducted. A Dane, Abildgaard, noticed a similarity between tapeworms (*Schistocephalus solidus*, see section 10.5.4) from the intestines of fish-eating birds and parasites from the body cavity of sticklebacks (see Grove, 1990). The main difference between them was the lack of reproductive organs in the latter, which led Abildgaard to wonder whether the stickleback parasites were immature. Significantly, he did not let it rest there but decided to test it experimentally. He collected sticklebacks and fed these to two ducks. After 3 days he killed the ducks and after opening their intestines found tapeworms in one of them. These were alive and more active than the similar worms taken from the sticklebacks.

In spite of the tremendous parasitological importance of this observation, which, according to Grove, was the first successful experiment designed to elucidate the life cycle and transmission of an internal parasite, and the implications for spontaneous generation, it went virtually unnoticed.

Taeniid tapeworms now took the stage in life cycle investigations. The so-called cystic worms (bladderworms; Fig. 8.6) had been known for a long time. In a Greek play, *The Knights* by Aristophanes, there is reference to the custom of examining the tongue of a pig to detect the presence of such cysts, which reach 5–10 mm in size, and in the 17th and 18th centuries several workers recognized their animal nature. Goeze in 1784 pointed out that within the cyst was a head resembling that of an adult *Taenia* from the human gut, but he continued to regard the cyst and the tapeworm as separate organisms. When Gmelin revised the *Systema Naturae* of Linnaeus in 1790 (see Grove, 1990), he recognized the similarity between the cysts and adult worms and named the cysts *Taenia cellulosae*, but Zeder (1803, in Grove, 1990) created a new genus *Cysticercus* for these parasites and a few years later the cysts were named *Cysticercus cellulosae* by Rudolphi. Another view held by von Siebold was that cystic worms were tapeworms that had strayed from their normal habitat and consequently had undergone abnormal dropsical development.

Cystic worms were also discovered in other animals, for example *Cysticercus pisiformis* in the rabbit and *C. fasciolaris* in the mouse. It was an enlightened German, Friedrich Küchenmeister, in the 1850s, who performed the critical experiments and fed cysts from these common animals to their natural predators, namely foxes and cats. At autopsy he found young tapeworms in the intestines of the carnivores and he concluded that the cysts were not abnormal stray worms, as von Siebold

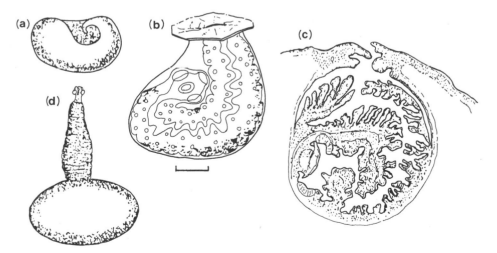

Figure 8.6 Bladder worms (cysticerci) of *Taenia solium* from the pig. (*a*) Cysticercus with invaginated scolex. (*b*) The invaginated scolex at a higher magnification. (*c*) Histological section through the invaginated scolex of a cysticercus. (*d*) Cysticercus with head evaginated. Scale bar (*b*): 0.5 mm. Source: reproduced from Leuckart, 1886.

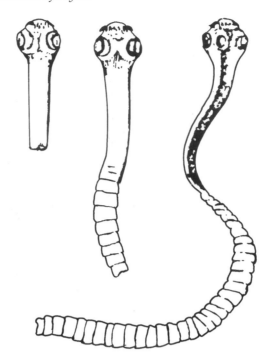

Figure 8.7 Early development of *Taenia multiceps* from the intestine of a dog. The scar at the posterior end indicates the site of detachment from the bladderworm vesicle. Source: reproduced from von Siebold, 1857.

proposed, but tapeworm larvae from which adult worms developed. Further studies showed that when a cyst reaches the intestine the scolex everts (it is inside out within the cyst; Fig. 8.6) and attaches to the intestine wall. Proglottides develop from a proliferative zone behind the scolex (Fig. 8.7); as they are pushed backwards by newly formed proglottides they increase in size and a set of reproductive organs develops in each of them (Fig. 8.3). The segment continues to grow after sexual maturity (Fig. 8.1), providing accommodation for an enormous branched uterus in which the eggs accumulate. When full of eggs, the so-called gravid proglottis detaches from the worm and, while still alive and capable of movement, passes out with the faeces (Fig. 8.8).

As mentioned above, the similarity between *C. cellulosae* in pigs and *Taenia solium* in humans had not gone unnoticed. Connections between the occurrence of the adult worms in humans and the consumption of pork, especially when eaten raw or lightly cooked, were also well known. *T. solium* was particularly common in Germany, where raw pork and fresh sausage were habitually consumed, and Leuckart (1886) noted that the frequent infection of expatriate Germans was probably due to the persistence of this custom. Von Siebold (1857) refers to the common occurrence of tapeworms in persons employed in slaughterhouses and in kitchens. Küchenmeister, referring to a man who voided 33 *Taenia* at once, remarked that he was 'the sweetheart of a butcher's daughter, and often got bladder-worms to swallow from his love' (see Leuckart, 1886). Tapeworms often put in an appearance after the consumption of raw flesh on medical advice. In contrast, Carthusian monks, who ate fish and consumed no meat, were never troubled with tapeworms (von Siebold, 1857).

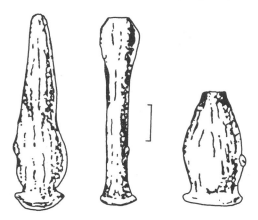

Figure 8.8 Shapes adopted by living proglottides of *Taenia saginata*. Scale bar: 0.5 cm. Source: reproduced from Leuckart, 1886.

Because of the difficulties with experiments on humans, German investigators fed cysts from pork to dogs. Von Siebold claimed to have raised adult worms, but similar experiments by Küchenmeister failed and he believed that von Siebold's worms originated from another pig bladderworm, *C. tenuicollis* (see Grove, 1990). Küchenmeister decided that the only way to solve the problem was to feed cysts to humans, and he sought permission for a bizarre experiment. He proposed to feed cysts to a criminal condemned to death and to examine the intestines of the corpse after execution.

The story is told graphically by Grove (1990). An opportunity to conduct this experiment presented itself in 1854, but permission was received rather late and only *C. pisiformis* and *C. tenuicollis* were available. These were fed to the convict unknown to him in lukewarm soup. However, a few days before the execution Küchenmeister's wife found *C. cellulosae* in roast pork obtained for their evening meal from a nearby restaurant. After much pleading at the restaurant, Küchenmeister obtained some uncooked meat from the same carcass and the following morning bladder worms from this meat were exchanged for pieces of fat in blood sausage offered to the prisoner for breakfast. A further 61 cysts were administered over the next two and a half days, either in sausage or in soup. After the execution, several young tapeworms were recovered and the distinctive hooks of the double crown on the scolex (Fig. 8.2) confirmed that some of the worms were *T. solium*, hence demonstrating that *C. cellulosae* and *T. solium* were one and the same species.

Not surprisingly, the ethics of this experiment were severely criticized, but Küchenmeister attempted to mollify his critics by stressing the harmlessness of the experiment and the fact that, should a convict be pardoned, the worms could be easily expelled.

A longer interval was clearly desirable between administration of the cysts and collection of the worms, in order to permit the worms to complete their development. Leuckart and others devised some rather less controversial experiments. In 1856, Leuckart gave four cysts in lukewarm milk to a young healthy volunteer who had been told about the development of the tapeworm (see Leuckart, 1886). Two and a half months later, tapeworm proglottides appeared in his faeces and after a further month, two worms about 2 m long were voided after treatment with an anthelminthic. Only one of these worms had a head but this was unmistakably that of *T. solium*.

In 1859, Küchenmeister repeated his controversial experiment with a much longer time interval. According to Grove, a total of 40 bladderworms was consumed by the prisoner, 127 and 72 days prior to decapitation. Half of these cysts produced worms; many of them were sexually mature and one was 1.5 m long.

Later, cysts were fed to other mammals, including rabbits, cats, dogs, pigs, sheep and monkeys, but all of these experiments failed and it became accepted that man is the only host of *T. solium*. The term 'cysticercus' has been adopted as an alternative name for a bladderworm and 'cysticercus cellulosae' is still in use for the encysted stage of *T. solium*.

The complete elucidation of the tapeworm life cycle demanded the demonstration by experiment that eggs eaten by pigs gave rise to bladderworms in the muscles. Pork containing cysts was known as measly pork and this condition had been attributed formerly to bad food (decomposing corn or acorns) or to some kind of contagious disease spreading from one pig to another. Feeding tapeworm eggs to

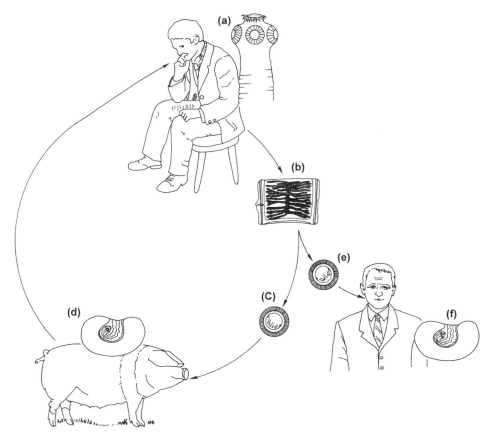

Figure 8.9 The life cycle of *Taenia solium*. Adult worm (*a*) in human small intestine. Detached, gravid proglottides (*b*) leave in faeces, contract and scatter eggs (*c*). Eggs eaten by pig hatch in intestine; oncosphere larva enters a blood vessel in intestine wall and is carried to muscle where it grows into a bladder worm (cysticercus) (*d*). Humans ingest pea-sized cysticerci in undercooked pork, scolex everts, sheds the cysticercus vesicle and tapeworm grows to maturity. Eggs (*e*) eaten by humans respond as in pigs and give rise to cysticerci (*f*) in muscle or nervous tissue (section 11.5.2).

mammalian hosts was not a new concept – such experiments had been performed at the end of the 18th century, but these early investigators were looking for adult worms in the intestine, not cysts in the muscles. In fact, in the year before Küchenmeister conducted his first notorious experiment, he obtained small cysts (*Coenurus cerebralis*, section 11.5.3) from the brain of sheep that had been fed 16 days previously with gravid proglottides of the dog tapeworm *Taenia multiceps*. The adult tapeworms had been obtained by feeding cysts from sheep to the dog, so that the whole life cycle had been demonstrated. However, it was van Beneden in 1853 who first established that eggs of *T. solium* fed to pigs produced *C. cellulosae*. This was confirmed in 1855 by Küchenmeister and Haubner, whose work was commissioned by the Government of Saxony, and they went on to show that *C. cellulosae* does not develop when eggs of *T. solium* are fed to dogs and sheep. It is particularly unfortunate for us that eggs of *T. solium* hatch in the human intestine and give rise to cysticerci in the tissues; this unpleasant twist in the life cycle will be considered later.

The life cycle of *T. solium* is summarized in Fig. 8.9 and has been encapsulated in a most amusing and appealing way in the verses of Walter Garstang (1954). I quote these verses in full here, with permission from Blackwell Publishers.

The Onchosphere

The Onchosphere or Hexacanth was not designed for frolic,
His part may be described as coldly diabolic:
He's born amid some gruesome things, but this should count for virtue,
That steadily, 'gainst fearful odds, he plies his task – to hurt you!

He's very small, a mere pin's head, beset with six small hooklets,
Is whirled about by wind and rain through puddles, fields and brooklets;
But if a pig should swallow him, as many porkers do,
He's made a start with no mistake: he's on the road to you!

Again I say don't blame the brat – he hasn't any head!
It isn't any fault of his – he wasn't painted red!
But once inside, he burrows through, and gropes his way about,
Then swells and sprouts a head at last, though this is inside out!

He's now a *Cysticercus* in the muscles of a pig,
With just a sporting chance of getting out to grow up big.
If you'll consent to eat your pork half-raw or underdone,
His troubles will be over, and a Tapeworm will have won:
He'll cast his anchors out, and on your best digested food
Will thrive, and bud an endless chain to raise a countless brood.

8.4   THE LIFE CYCLE OF *TAENIA SAGINATA*

During the 18th century accounts of the morphology of human tapeworms showed inconsistencies. As early as 1700, Andry observed a tapeworm scolex without hooks and there were conflicting and confusing descriptions of large fat tapeworms and thin, transparent ones (see Grove, 1990). Over the years a view gradually emerged that human taeniid tapeworms had two distinct forms or varieties, one with a scolex armed with hooks and a thin, flat, transparent body and the other without scolex hooks, a larger fatter body and more uterine branches. Regional variations in the occurrence of the two forms were also recognized. For example, Rudolphi observed only *Taenia* with hooks in Berlin, while the tapeworms studied by Bremser in Vienna had no hooks.

All the clues pointed to the existence of two distinct species of *Taenia* in humans, but no one was prepared to commit themselves to this conclusion. In fact, a popular view was that all young taeniid tapeworms had hooks and that these were shed like the hair of the human male with advancing age, and two major textbooks published in the 1840s by Dujardin and Diesing did not acknowledge the occurrence of two distinct species. This was all taking place more than 150 years after Tyson first described the *Taenia* scolex.

It fell to Küchenmeister to take the final step in 1852 and declare that *T. solium* with an armed scolex (Fig. 8.2) was distinct from *T. mediocanellata* (later to be called *T. saginata*) with an unarmed scolex (Fig. 8.10). (This tapeworm was placed in the genus *Taeniarhynchus* by Weinland (1858; in Rausch, 1994) on the grounds that it lacks the armed apical 'rostellum' found on the scolex of *Taenia*. However, the rostellum and its hooks appear during development of *Taenia saginata* and then regress (Goeze, 1782, in Rausch, 1994). Because of this Rausch rejects this distinction and regards *Taeniarhynchus* as a synonym of *Taenia*.) With the demonstration that the pig serves as the intermediate host for *T. solium*, the search for the intermediate host

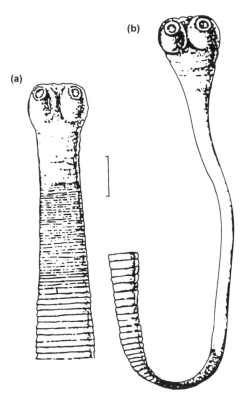

Figure 8.10 Scolex of *Taenia saginata* (*a*) contracted and (*b*) extended. Scale bar: 1 mm. Source: reproduced from Leuckart, 1886.

of *T. saginata* began. There were several clues. Sickly children were frequently prescribed raw beef to increase their strength. In the St Petersburg region of Russia it was noted that *T. saginata* infection followed this treatment. As predicted from their religious abstention from eating pork, European Jews were free of *T. solium*, but not from *T. saginata*, and the inhabitants of Abyssinia (Ethiopia), where raw beef was consumed by almost everyone, were almost universally infected with *T. saginata*. In fact, according to Leuckart (1886), Abyssinians considered themselves abnormal if they did not pass tapeworm proglottides, and regarded the worms as beneficial to their health in that a somewhat fluid and uniform stool was maintained in their presence.

In 1861, Leuckart fed mature proglottides of *T. saginata* to a 4-week old calf. The animal died but contained large numbers of cysts, which he interpreted as young bladderworms. A second experiment yielded some fully developed cysts (*Cysticercus bovis*), their longest diameter being 3–5 mm. Between the four suckers Leuckart reported a small apical rostellum which he believed carried hook rudiments, but there were no hooks in the fully formed bladderworm. Studies by other investigators established that pigs, sheep and rabbits are not susceptible to the parasite.

No attempt was made to complete the life cycle study by feeding *C. bovis* to condemned humans, and it was some time before voluntary ingestion of the cysts provided the final proof. Perhaps the most reliable study was performed by Perroncito in 1877 (see Grove, 1990), in which cooked and uncooked *C. bovis* were consumed. Some 54 days after eating a bladderworm, proglottides were passed and on day 57 a *T. saginata* 4.27 m long with 866 progottides was expelled after treatment. It was estimated that 13 or 14 proglottides were passed per day. Leuckart (1886) stressed the long life of this tapeworm, reporting that one of his students had harboured two worms for more than 5 years.

It was the experimental method that finally triumphed in the disentanglement of the lives of these parasitic platyhelminths and Küchenmeister in particular had done much to establish experimentation firmly in the forefront in the field of parasitology research. In addition, there was an important spin-off from these investigations in relation to spontaneous generation. As Grove has pointed out, none of the tapeworm studies was specifically designed to refute this doctrine, the investigators being concerned with clarifying the relationships between cystic worms and adult tapeworms, but by rationalizing the origins of tapeworms and cystic worms the demise of the doctrine of spontaneous generation of tapeworms was assured.

8.5   THE MYSTERY OF *DIPYLIDIUM CANINUM* (SEE ALSO SECTION 11.6)

In 1858, van Beneden published an account of some experiments designed to elucidate the life cycle of *Tenia serrata* (now called *Taenia pisiformis*). He removed cysticerci of *T. serrata* from rabbits and fed 32 of them to a male puppy, which up to that time had received no solid animal food. The sister of the puppy was treated in the same way as her brother but received no cysticerci. On dissecting the dogs, van Beneden recovered *T. serrata* only from the male dog, but, to his surprise, found that both dogs contained a distinctly different tapeworm, which he called *Tenia canina*.

The following quotation from van Beneden (1858) illustrates his dilemma. '*Nous avons trouvé ces Ténias dans deux chiens de sept semaines, dont l'un avait avalé des Cysticerques pysiformes, et l'autre n'avait pris, que nous sachions, aucune nourriture animale. Ils avaient tous les deux couru librement dans le jardin. Où vivent les Cysticerques de* Tenia canina?' It was not until 1889 that the solution to this puzzle was found, revealing that tapeworms have another route to carnivorous definitive hosts. Grassi and Rovelli (in Grove, 1990) showed that the dog flea acts as the intermediate host (see Fig. 11.19). Adult fleas are blood feeders and do not ingest tapeworm eggs, but the free-living flea larva feeds on detritus and is likely to pick up eggs of *D. caninum* from the floor of the kennel. The infective stages of the tapeworm (cysticercoids) persist after the metamorphosis of the flea and Grassi and Rovelli noted that the itch induced by the adult flea caused the dog to paw and gnaw at its skin, then bite and swallow the vector.

8.6   THE WAY FORWARD

Thus it was established that insects could serve as intermediate hosts for tapeworms and it was the discovery at a later date of the life cycle of another insect-propagated species that led to significant advances in our knowledge of tapeworm biology. *Hymenolepis diminuta* is a parasite of the rat and is acquired by consumption of tiny flour beetles (*Tribolium* spp.) containing cysticercoids. The ease with which this life cycle can be maintained in such convenient laboratory animals was soon appreciated and exploited by parasitologists with physiological, biochemical and immunological interests. Their findings will be considered in the next chapter.

CHAPTER 9

The biology of Hymenolepis diminuta

9.1 THE LIFE CYCLE

Adult *Hymenolepis diminuta* inhabit the small intestine of their rat definitive host (Fig. 9.1). Tapeworm eggs voided in the faeces of the rat are ingested by flour beetles (*Tribolium* spp.; *Tenebrio molitor*). The hexacanth larva is released from the egg in the beetle's gut, penetrates the gut wall and develops into a cysticercoid (essentially an encysted scolex) in the haemocoel. In order to reach the rat's small intestine, the

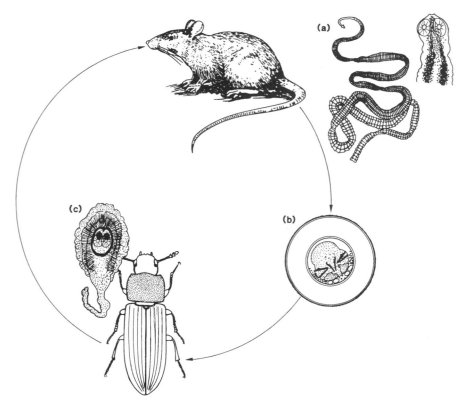

Figure 9.1 The life cycle of *Hymenolepis diminuta*. The adult tapeworm (*a*) lives in the small intestine of the rat and releases gravid proglottides into the faeces. Eggs (*b*) shed by the proglottides are eaten by flour beetles, hatch in the midgut and the tapeworm larva (oncosphere or hexacanth) enters the haemocoel; oncospheres grow into cysticercoids (*c*). The rat is infected by eating beetles containing cysticercoids.

beetle and the cysticercoid must be eaten by the rat, and the cysticercoid has to run the gauntlet of the rodent's grinding molar teeth and has to survive passage through the stomach, with its very low pH and proteolytic environment. The way in which tapeworms overcome these difficulties and arrive alive in the intestine will be considered later (section 9.7).

9.2   THE INTESTINE AS A PLACE TO LIVE

9.2.1   The structure of the intestine wall

The rat small intestine has an average length of 100 cm. The duodenum is followed by the jejunum and the ileum (see reviews by Mettrick, 1980 and Moog, 1981; Fig. 9.2). These three regions merge into one another and, although there are

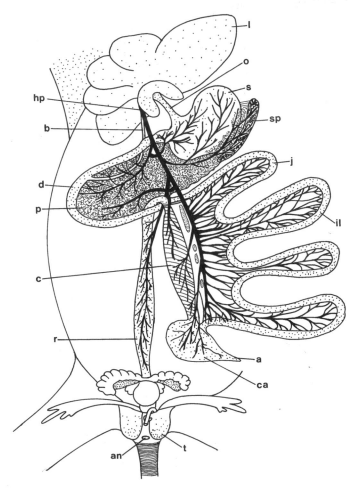

Figure 9.2   Diagram showing the abdominal digestive system of the rat. a = appendix; an = anus; b = bile duct; c = colon; ca = caecum; d = duodenum; hp = hepatic portal vein; il = ileum; j = jejunum; l = liver; o = oesophagus; p = pancreas; r = rectum; s = stomach; sp = spleen; t = testis. Source: reproduced from Wells, 1964, by permission of Heinemann Educational, a division of Reed Educational and Professional Publishing Ltd.

histological differences between them, there is a basic similarity in the arrangement of layers in their walls.

It is convenient to subdivide the intestine wall into the inner mucosa (adjacent to the lumen), middle submucosa and outer muscularis externa (Fig. 9.3). The mucosa consists of an outer muscularis mucosae comprising longitudinal and circular muscle, a middle lamina propria comprising blood and lymph vessels embedded in connective tissue, and an inner monolayer of epithelial cells lining the lumen. Finger-like villi, consisting of lamina propria covered by a single layer of columnar epithelial cells, project into the lumen and the exposed apical surfaces of these cells have a dense array of projecting microvilli (so-called brush border; Fig. 9.5b). Between the villi, tubular crypts penetrate into the lamina propria as far as the

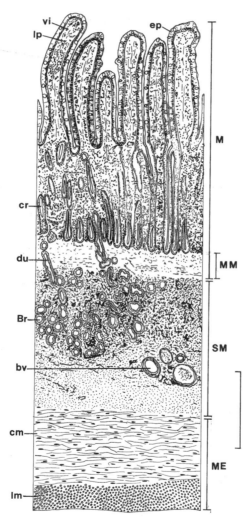

Figure 9.3 Diagrammatic interpretation of part of a transverse section through the wall of the mammalian duodenum. Br = Brunner's gland; bv = blood vessel; cm = circular muscle; cr = crypt of Lieberkühn; du = duct of Brunner's gland; ep = epithelium; lm = longitudinal muscle; lp = lamina propria; M = mucosa; ME = muscularis externa; MM = muscularis mucosae; SM = submucosa; vi = villus. External serosa not shown. Scale bar: 0.5 mm. Source: drawn from a photograph of a section through dog intestine in Reith and Ross, 1965.

muscularis mucosae. The crypts are lined by a single layer of epithelial cells, which are more nearly cuboidal in shape, and this layer is continuous with the epithelial covering of the villus.

The single continuous layer of epithelial cells lining the crypts and covering the villi appears static in histological sections like that shown in Fig. 9.3, but this impression is misleading. The cells lining the crypts undergo rapid proliferation – an unusual feature in the adult body – and migrate distally, moving out of the crypts on to the surface of the villus shaft. Some of them differentiate here into mucus-secreting goblet cells, but most of them become columnar epithelial cells or enterocytes (see Fig. 9.5*b*), which combine digestive and absorptive functions. The reason for this cell proliferation and migration is that the enterocytes have a limited life and are shed continuously from the tips of the villi into the intestinal lumen. A range of enzymes produced by the enterocytes and incorporated into the apical membrane covering their brush border has a vital part to play in digestion of carbohydrate and protein (Alpers, 1994).

The same apical membrane must be traversed by nutrients and other materials destined for absorption; tight junctions between the membranes of adjacent epithelial cells (see Fig. 9.5*b*) effectively prevent molecules from using the intercellular route to the core of the villus.

The contents of the intestine are not static: peristaltic movements are strong, especially in the proximal intestine, and both the villi and the microvilli pulsate.

9.2.2 Digestion in the intestine

The protein digestion that takes place in the intestine involves a remarkable interplay between enzymes from the pancreas and from the intestine itself. The pancreas lies in the mesentery supporting the duodenal loop (Fig. 9.2) and its secretion empties *via* several drainage ducts into the bile duct, and thence to the duodenum. Autodigestion of the pancreas is prevented because the proteolytic enzymes are secreted as inert zymogens or proenzymes, which are not activated until they enter the duodenum. Here, a brush border enzyme, enterokinase, converts the inert endopeptidase trypsinogen to its active form, trypsin, which then activates more trypsinogen and the other pancreatic proenzymes (see Alpers, 1994). Thus a cascade of activation reactions involving the pancreatic proteases is initiated by the release of trypsin by enterokinase. The resulting mixture of enzymes includes endopeptidases (trypsin, chymotrypsin, elastase) and exopeptidases (carboxypeptidases A and B). Their activity produces amino acids and a range of peptides; many of the latter are hydrolysed further by peptidases of the enterocyte brush border. There is evidence that some di- and tripeptides are absorbed intact.

Starch digestion is initiated by pancreatic α-amylase, which is found not only in the intestinal lumen but also attached to the apical surface of enterocytes. Earlier work suggesting that such an attachment conferred a kinetic advantage on pancreatic amylase in the digestion of starch (so-called membrane or contact digestion; section 9.3.2(c)) has not been confirmed (see Alpers, 1994). Digestion of starch initiated by α-amylase and the digestion of ingested disaccharides is completed by disaccharidases associated with the enterocyte brush border.

Bile is secreted by the parenchymal cells of the liver and, since the rat has no gall bladder, it is released continually into the duodenum. The functions of bile, which are mediated primarily by the bile salts, include the activation of pancreatic lipase and the lowering of surface tension; the latter leads to the emulsification of fats and the enhancement of lipase activity.

Recently, a picture has emerged of a flexible and adaptable digestive process in which changes in diet lead to adjustments in the proportions of enzymes secreted, so maximizing food utilization. For example, on a high carbohydrate diet the production of amylase by the pancreas increases at the expense of the proteases, while the reverse occurs on high protein diets.

9.2.3   Oxygen in the intestinal lumen

Our conception of the availability of oxygen to parasites inhabiting the small intestine has undergone radical changes. The contents of the small intestine used to be regarded as anoxic except for the 'paramucosal lumen', i.e. an ill-defined zone close to the gut lining (mucosa) where 'worms...have access to larger amounts of oxygen' (Read, 1950). The main support for this concept came from comparative measurements made with oxygen electrodes. However, the reliability of such measurements has been challenged (see Podesta and Mettrick, 1974) and the anoxia concept would only be true if the intestinal contents were stagnant or semi-solid. In fact, considerable quantities of fluid enter the proximal small intestine and thorough mixing is provided by peristalsis and by movements of the villi. Using a perfusion technique, Podesta and Mettrick showed that oxygen diffuses readily from the mucosa (P_{O_2} of arterial blood 100 mmHg) into the bulk aqueous phase of the luminal contents, and their evidence suggests that the contents are aerobic with a P_{O_2} of 40–50 mmHg.

9.2.4   pH and carbon dioxide in the intestinal lumen

Another misconception about the intestine is that the acidity of the stomach contents (chyme) is neutralized on entering the duodenum by bile and by alkaline juices from the pancreas and intestine. In fact, high luminal acidity is maintained for some time, with a marked pH gradient down the length of the small intestine. Both trypsin and chymotrypsin operate best at about pH 8, so their activity is below optimum following the passage of chyme into the proximal duodenum. Between meals the anterior two thirds of the small intestine retain slightly acidic contents, but these become alkaline (pH 7.3) towards the end of the ileum. The reaction between acid chyme and the bicarbonate-rich fluids entering the duodenum generates a luminal P_{CO_2} that may exceed 600 mmHg (see review by Befus and Podesta, 1976). There are few organisms that can survive the seemingly lethal combination of exceptionally high P_{CO_2} and low pH to which tapeworms are subjected (section 9.3.3).

9.2.5   The immune response in the intestinal lumen

Another old concept that has now been laid to rest is that the intestine is incapable of mounting an immune response against inhabitants of the lumen. The gut lumen used to be regarded as an extension of the outside world, with its inhabitants living essentially outside the body (see review by Befus and Podesta, 1976). It was generally supposed that effective host responses could only be mounted against parasites of the blood and other tissues, and that, unless a gut parasite had a phase of tissue penetration in its life cycle, or caused damage to the mucosa, it was likely to be ignored by the host. The assumption was made that the undamaged intestine was unable to take up antigenic macromolecules and that, even if this were possible, the intestinal environment was far too harsh to permit the survival of antibodies and cellular components of the immune system. It is surprising that this attitude persisted for so long, because early in the century it was known that the intestine could develop resistance to cholera and shigellosis in the absence of systemic immunity, although local antibody was not implicated in this resistance until the late 1940s. A significant finding that added impetus to work on intestinal immunity was the discovery that one of the five classes of serum immunoglobulins, IgA, was particularly abundant in the secretions of the gut mucosa.

We now know that there is an elaborate gut-associated lymphoid system composed of lymphoid cells. These are not only dispersed throughout the gut intestinal epithelium and lamina propria but also aggregated to form Peyer's patches in the intestine wall, the appendix and mesenteric lymph nodes. Many other immunologically important cells, for example basophils, eosinophils, macrophages and mast cells, are dispersed through the intestine wall.

The main immunological events that may be triggered by parasites in the lumen are summarized in Fig. 9.4. Potentially immunogenic quantities of macromolecules are absorbed intact from the gut lumen and this sampling is achieved by pinocytotic activities of the epithelial covering of structures such as Peyer's patches. Cells in the patches become sensitized by interaction with parasite antigen and undergo a complex journey *via* the lymphatic vessels and blood system returning to the intestine wall. Antigen may also enter the circulation directly and sensitize lymphoid

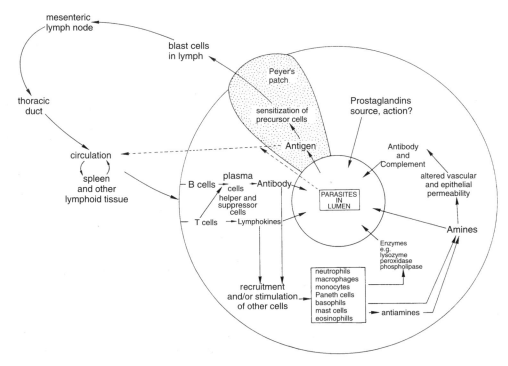

Figure 9.4   A simplified model of the immunological events which may lead to expulsion of parasites from the intestine. See text for explanation. Source: modified from Befus and Podesta, 1976.

cells in other sites. In the intestine wall, the sensitized cells differentiate into effector cells, whose products may act directly on the parasite or recruit other components of the immune system. These components in their turn may attack the parasite or undertake a regulatory role. The following list includes some of the substances that are thought to affect gut parasites directly: antibody, lymphokines, enzymes, amines, complement (a subset of special proteins that are activated by antibody), prostaglandin.

9.3   THE ADULT TAPEWORM

9.3.1   General histological features

According to Lumsden and Specian (1980) an unusual feature of tapeworm histology is the prevalence of syncytia – multinucleated cytoplasmic structures formed by fusion of mononucleated cells. Syncytial organization occurs in the tegument, the musculature and the ducts of the excretory and reproductive systems. Lumsden and Specian regarded this phenomenon as advantageous for continually growing tapeworms because it permits sustained growth without restructuring of previously differentiated tissue and efficient transfer of substances throughout a large tissue volume. Another feature, which according to Lumsden and Specian is unusual, is the high degree of coupling, mainly by 'close' or 'gap' junctions, between cytological elements, such as, for example, between the tegument and muscle.

9.3.2   The tegument

(a)   General features

Intuitively, we would expect any organism living in the small intestine to have an effective shield, an inert covering to protect its tissues from attack by enzymes, surfactants and effectors of the immune system. It is therefore not surprising that the thin outer covering of tapeworms was originally regarded as a dead cuticle secreted by the cells beneath it. However, the results of early histochemical and physiological studies were strikingly inconsistent with this concept; in particular, several enzymes, including phosphatases, were found to be associated with the 'cuticle' (see Read, 1966). These are characteristic of living cells and not of a dead cuticular covering. Although hair-like projections from the tapeworm surface were seen by microscopists, their nature was not immediately appreciated and the light microscope was unable to resolve structural details of the layer. The advent of the electron microscope in the 1960s changed all of this and it soon emerged that the 'cuticle' was in reality a layer of living cytoplasm, demarcated at its free and basal surfaces by plasma membranes. The cytoplasmic layer was a continuous sheet (a syncytium), not subdivided into cellular units and communicating by means of cytoplasmic connections (internuncial processes) with nucleated cell bodies (tegumentary cytons) lying in the parenchyma beneath the surface layer (Fig. 9.5a).

The surface layer of *H. diminuta* betrays its cellular nature by the presence of abundant disc-shaped inclusions and mitochondria (see review by Lumsden and Specian, 1980). Granular endoplasmic reticulum and Golgi complexes are confined to the cytons, and the disc-shaped inclusions are assembled by these Golgi bodies

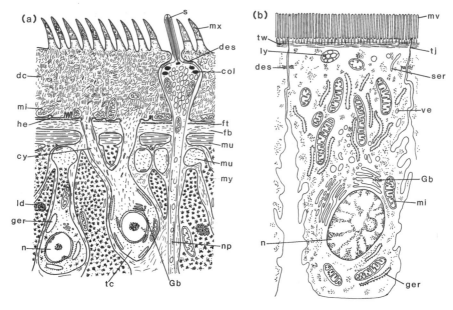

Figure 9.5 Diagrams of sections through (*a*) the tegument of *Hymenolepis diminuta* and (*b*) a columnar epithelial cell from the mammalian small intestine. col = electron-dense collar; cy = cytoplasmic connection; dc = distal cytoplasm; des = desmosome; fb = fibrillar layer of basement lamina; ft = felt-like layer of basement lamina; Gb = Golgi body; ger = granular endoplasmic reticulum; he = hemidesmosome; ld = lipid; ly = lysosome; mi = mitochondrion; mu = superficial muscle fibre; mv = microvillus; mx = microthrix; my = myocyton containing glycogen particles; n = nucleus; np = sensory nerve process; s = sensillum; ser = smooth endoplasmic reticulum; tc = tegumentary cyton; tj = tight junction; tw = terminal web; ve = vesicle. Glycocalyx not shown. Source: (*a*) reproduced from Lumsden and Specian, 1980, with permission from Academic Press; (*b*) redrawn from Moog, 1981.

and exported to the surface cytoplasm *via* the microtubule-lined internuncial processes.

The hair-like processes seen by the microscopists are dense microvilli. Lumsden (in Read, 1966) noted the general similarity between cestode surface microvilli and those possessed by vertebrate cells that are concerned with absorption, such as the epithelial lining of the intestine (Fig. 9.5*b*). This added further weight to the growing awareness of the importance of the tegument in the absorptive processes of the tapeworm. In *H. diminuta* each microvillus has a diameter of about 0.14 μm and a length of 0.9 μm. The distal tip forms a tapering, electron-dense cap, which is separated from the rest of the shaft by a multilaminate base-plate (Fig. 9.5*a*). This cap appears to be a unique feature and has led to use of the term 'microthrix' (plural 'microtriches') for the special tapeworm microvillus. Intimately associated with the outermost leaflet of the trilaminate membrane covering the tegumentary surface is a secreted macromolecular layer called the glycocalyx.

The tegumentary covering of the tapeworm and the epithelial covering of the rat intestinal villus share many of the same functions and operate in the same microenvironment, and this is reflected in many common features, both at the cellular (Fig. 9.5) and molecular levels. The syncytial organization of the tapeworm tegument and the binding together of host epithelial cells by tight junctions are two different ways to create a barrier and achieve control over the inward and outward passage of chemical substances. We shall see below that the similarities between the brush borders (apical microtriches/microvilli) extend to the molecular level in that some molecules embedded in the apical membrane have an enzymatic function.

(b) Nutrient uptake

Hymenolepis diminuta is unable to digest starch but readily absorbs glucose. Glucose uptake involves active transport rather than simple diffusion (see reviews by Read, Rothman and Simmons, 1963 and Read, 1966). Diffusion is the movement of solute molecules due to their kinetic energy. Although molecular movement is random, where there is a concentration gradient there will be a net movement away from the region of high concentration towards the region of low concentration. The most significant characteristic of active transport is that a substance is moved against the concentration gradient and this 'uphill' transport requires the expenditure of energy. *H. diminuta* is capable of accumulating glucose against a concentration gradient and this uptake is inhibited by substances that inhibit energy metabolism. The uptake of D-glucose is also inhibited by chemically similar monosaccharides such as D-galactose and D-allose, but not by D-fructose. This is interpreted as competition for the uptake locus by chemically similar substances and is another feature of active transport.

Only glucose and galactose, from a great many sugars that have been tested, appear to be metabolized to any appreciable extent by *H. diminuta*. Since galactose has little dietary significance and since the worm is unable to absorb disaccharide sugars, carbohydrate metabolism is fuelled principally by glucose and carbohydrate is stored as glycogen.

By utilizing active uptake, the worm ensures that glucose uptake can continue even when the glucose concentration in the gut is low (i.e. between meals). In this situation, diffusion might even lead to loss of glucose by the worm. It should be noted that the worm is in competition for glucose with the mucosa, which also has an active uptake.

H. diminuta is also incapable of digesting protein, although Walski *et al.* (1995) claimed that the tegument of some related hymenolepidids can take up horseradish peroxidase by pinocytosis (see also section 10.5.4). Amino acids are readily absorbed *via* six separate kinds of uptake locus (see Bryant and Behm, 1989). These are not absolutely specific, but generally each one shows an affinity for a particular amino acid or group of amino acids as follows: dicarboxylic amino acids, basic amino acids, glycine, serine, leucine, phenylalanine. Uptake is by active transport or by facilitated diffusion, the latter involving a specific transporting mechanism but one incurring no expenditure of energy.

Uptake of purines and pyrimidines is essential because most helminths are incapable of synthesizing these compounds *de novo*. The process is complex, with three uptake loci, two of which involve binding of the purine or pyrimidine to an activator site, thereby activating the transport mechanism and facilitating passage of the compound across the membrane (see Bryant and Behm, 1989).

Tapeworms also absorb many other substances, including fatty acids, glycerol and vitamins. It seems that *H. diminuta* relies on the host's habit of eating its faeces (coprophagy) to provide some essential vitamins. When coprophagy is prevented, these vitamins are absent from the rat's intestine and the tapeworms are adversely affected (Platzer and Roberts, 1969).

Thus, nutrients are acquired by adult tapeworms in the form of low-molecular-weight organic compounds, which can be absorbed by the tegument, in some cases (hexose sugars, some amino acids, nucleosides) against a concentration gradient.

It was already known from the work of Chandler (in Roberts, 1983) that growth of *H. diminuta* is impaired when rats are fed on a carbohydrate-free diet and that no such impairment occurs on a protein-free diet. Chandler assumed that the tapeworms are able to synthesize all the amino acids that they require and/or are able to absorb nitrogen compounds from the mucosa. However, the amount of nitrogen recoverable from the rat small intestine may exceed the amount of nitrogen ingested, even on a diet containing protein (see Read, 1971), and this indicates that the lumen of the small intestine receives a significant amount of endogenous nitrogen from the body, as well as exogenous nitrogen in the food. According to Read, this endogenous nitrogen comes from more than one source: from digestive enzymes and other proteins secreted into the gut; from intestinal epithelial cells shed into the lumen; from amino acids leaking into the lumen from the mucosa. Thus, the tapeworm parasitizes the endogenous pool of nitrogen from the body proper, as well as ingested exogenous nitrogen, and because the endogenous source is still available on starvation, can continue to grow on a protein-free diet. Because there is no endogenous source of carbohydrate, *H. diminuta* is unable to grow if this is denied in the host's food.

(c) Surface enzymes

Ultrastructural cytochemical studies established that phosphatases are localized in the external surface membrane of *H. diminuta* and are not removed by washing (see Pappas, 1980). The widely accepted notion among physiologists that phosphatases (phosphohydrolases) are involved in the absorption of sugars was naturally extended by parasitologists to tapeworms as an explanation for the role of the enzymes associated with the cestode tegument. However, Phifer (in Read, 1966) presented direct evidence that phosphatase is not involved in glucose absorption by *H. diminuta*. The concentrations of phlorizin that are needed to inhibit glucose uptake by the worm have a negligible effect on phosphatase activity. Conversely, Phifer found that phosphatase activity is inhibited by molybdate concentrations that have no effect on glucose uptake. What then is the role of the tapeworm's surface phosphatase?

After demonstrating the impermeability of *H. diminuta* to fructose and fructose diphosphate, Arme and Read (1970) observed that fructose accumulates in the external medium when the worm is incubated with fructose diphosphate. This confirmed earlier indications that the phosphatase responsible for splitting fructose diphosphate is located at the surface and led to the suggestion that this enzyme has a digestive role. A phosphate ester such as glucose-6-phosphate (G6P) would be hydrolysed and the liberated glucose absorbed by the worm. This curious idea that tapeworms have a digestive enzyme on their outer surface has a parallel in the hydrolytic enzymes embedded in the apical membranes of the columnar epithelial cells lining the host's gut. This runs counter to the popular view of a degenerate parasite, with the tapeworm having lost all digestive ability with the loss of its own gut.

Dike and Read (1971) incubated worms in the presence of G6P and glucose oxidase. The latter oxidizes free glucose to glucuronic acid, to which worms are impermeable. They found that glucose oxidase had no effect on the 'apparent' uptake of G6P by the worm and concluded that none of the glucose liberated during

G6P hydrolysis diffuses away from the worms. This is highly significant since it suggests that there is a close spatial relationship between the surface phosphatase and the glucose uptake locus, so that the glucose is 'snapped up' by the locus as soon as it is liberated. Bearing in mind the competitive nature of nutrient absorption by worm and host, this arrangement will have a strong kinetic advantage for the worm.

Other enzymes are now known to be present on the tapeworm surface and these include monacyl hydrolase (responsible for the hydrolysis of monoglycerides to free fatty acids and glycerol), ATPase and an RNAase (see Pappas, 1983).

The glycocalyx seems likely to play an important part in the functioning of these surface enzymes. Many of them require divalent cations for maximum activity and these are strongly adsorbed by the glycocalyx, which has a net negative charge (see Pappas, 1983). Pappas has also suggested that the glycocalyx may act as a barrier to the outward diffusion and loss of hydrolysis products liberated by the worm's surface enzymes.

The glycocalyx also absorbs organic substances including proteins. Evidence that the activity of pancreatic amylase was enhanced by adsorption to the glycocalyx of the mammalian intestinal cell, a phenomenon known as 'contact digestion', stimulated a search for a similar phenomenon in cestodes. Enhancement of the digestive ability of mammalian pancreatic amylase by the presence of tapeworms was claimed by two laboratories (Taylor and Thomas, 1968 and Read, 1973; in Pappas, 1983), but the significance of this was dubious because the tapeworm cannot absorb the di- and trisaccharides produced by amylase digestion. It was assumed that the host would intercede at this point by liberating monosaccharides from the di- and trisaccharides, thereby producing by 'enzyme sharing' an overall increase in glucose available to both parasite and host. However, Mead and Roberts (1972; in Pappas, 1983) found no evidence of enhancement of amylase activity by *H. diminuta in vivo*, and the results of Thomas and Turner (1980; in Pappas, 1983) suggested that the *in vitro* stimulation of enzyme activity was an artefact of the amylase assay procedure. Hence contact digestion involving pancreatic amylase in *H. diminuta* is not proven.

(d) Secretory mechanisms

Oaks and Holy (1994) have found evidence to suggest that *H. diminuta* has two distinct tegumental secretory mechanisms. First, Golgi-derived vesicles (the disc-shaped bodies mentioned above, section 9.3.2(a)) travel from the cytons to the bases of the microtriches, where their boundary membranes fuse with the apical membrane of the tegument, expelling their contents from the parasite. Secondly, tiny microvesicles formed from the tegumental apical membrane are shed from the surface between the microtriches. The functions of these two mechanisms are not understood, but the first of these may provide surface membrane destined for the microtriches. The corresponding membrane covering the microvilli of the intestinal epithelial cells is known to have a limited life (see Moog, 1981), and it is not unreasonable to suppose that the apical membrane of the tapeworm microthrix requires regular replacement.

(e) Avoiding damage by host enzymes

Hymenolepis diminuta lives in an environment that specializes in degrading biological macromolecules by way of potent proteolytic and lipolytic enzymes. The parasite has no cuticle to protect it and the idea that such parasites secrete defensive 'antienzymes' is attractive. Pappas and Read (1972a, b) added fuel to this idea by demonstrating that the living tapeworm inactivates trypsin and α- and β-chymotrypsin, and Ruff and Read (1973) reported inhibition of pancreatic lipase. Lipase inhibition is reversible and appears to involve weak bonding to the surface of the worm (see Pappas, 1980), while the proteases are irreversibly inhibited, suggesting a different type of interaction, possibly by way of a secreted inhibitor. However, Schroeder, Pappas and Means (1981) found no significant difference between the molecular weights of active and inactivated trypsin; in other words, if there is an inhibitor molecule secreted by the worm, it is very small. They also discovered that, although the inactivated enzyme could not interact in the normal way with the protein substrate, it was still able to interact with, and split up, synthetically made, short peptide molecules. This evidence indicates that the worm does not secrete an antienzyme and that the enzyme's active site is not destroyed by interaction with the tapeworm. It seems that the trypsin molecule undergoes a small structural or conformational change that prevents it from associating with large protein molecules in the normal way.

Some doubt as to whether the results of these *in vitro* experiments reflected *in vivo* events was introduced by Pappas (1978), who found no significant difference between the tryptic and total proteolytic activities of infected and uninfected rats. However, he pointed out that the amounts of enzyme inactivated by *H. diminuta* are small and, although significant for the worm because they affect its immediate microenvironment, are insignificant compared with the large quantities of active enzyme in the intestine (Pappas, 1983). The worm also profits nutritionally from host tryptic activity and it would be disadvantageous to immobilize large quantities of the enzyme.

9.3.3   Metabolism

During the first days in the rat's intestine the young tapeworm uses a fermentative process (anaerobic glycolysis) to generate energy (see review by Tielens, 1994). This is an inefficient process taking place in the cytosol and creating a mere two molecules of ATP per molecule of glucose, with lactate as an end product (Fig. 9.6). However, a change takes place between day 6 and day 12. The developing strobila switches over to a more energy-productive fermentation called malate dismutation. Glucose is degraded to phosphoenolpyruvate (PEP) *via* the same glycolytic pathway, but PEP is then carboxylated by phosphoenolpyruvate carboxykinase (PEPCK) to form oxaloacetate, which is then reduced to malate. Malate then enters the mitochondrion; one portion of it is oxidized to acetate and the other portion is reduced with the aid of fumarate reductase to succinate, with the production of five molecules of ATP – a significant gain compared to the production of lactate. The generation of precursors of other important carbon compounds (see Bryant, 1994) may be an additional advantage. Curiously, the anterior region near the scolex retains its predisposition to glycolysis with lactate production, while the posterior part of the animal favours malate dismutation (see Tielens, 1994).

This raises an intriguing question that has not yet been answered satisfactorily. If enough oxygen is available in the intestinal lumen (see above) and the tapeworm possesses mitochondria and has all the necessary biochemical machinery, then why not adopt aerobic energy metabolism, as practised by the free-living stages of

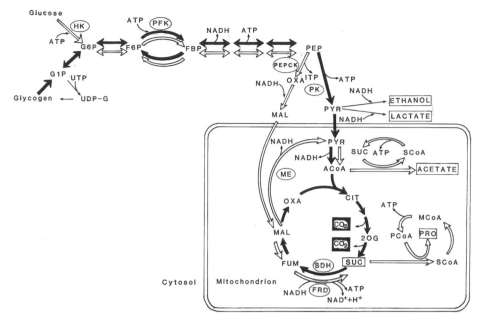

Figure 9.6 Generalized pathways of carbohydrate degradation by parasitic helminths. Black arrows = pathways of aerobic metabolism of free-living and larval stages; white arrows = fermentation pathways of adults. ACoA = acetyl-CoA; CIT = citrate; FBP = fructose-1,6-biphosphate; F6P = fructose-6-phosphate; FRD = fumarate reductase; FUM = fumarate; G1P = glucose-1-phosphate; G6P = glucose-6-phosphate; HK = hexokinase; MAL = malate; MCoA = methyl malonyl-CoA; ME = malic enzyme; 2OG = 2-oxoglutarate; OXA = oxaloacetate; PEP = phosphoenolpyruvate; PEPCK = phosphoenolpyruvate carboxykinase; PFK = phosphofructokinase; PK = pyruvate kinase; PRO = propionate; PCoA = propionyl-CoA; PYR = pyruvate; SDH = succinate dehydrogenase; SUC = succinate; SCoA = succinyl-CoA; UDP-G = UDP-glucose. Source: reproduced from Tielens, 1994, with permission from Elsevier Trends Journals.

parasitic platyhelminths and presumably by ectoparasites like *Entobdella soleae*? Aerobic metabolism involves conversion of glucose in the cytosol to pyruvate, which then enters the mitochondrion where it is oxidized completely to carbon dioxide and water, thereby maximizing the release of energy.

In addition to the energy gain, there is another special problem, which is exacerbated by following anaerobic metabolic pathways. High levels of CO_2 are generated in the intestine by interaction between acid chyme from the stomach and bicarbonate-rich fluids from the pancreas and elsewhere (see section 9.2.4 and Befus and Podesta, 1976). This will lead to diffusion of CO_2 into the tissues of *H. diminuta*, shifting the balance of the reaction:

$$CO_2 + H_2O \rightarrow HCO_3^- + H^+.$$

This will elevate the level of H^+ ions in the worm. Anaerobic metabolism adds to this problem because additional H^+ ions result from the acidic end products. There seems little doubt that *H. diminuta* possesses mechanisms to excrete H^+ ions. The acidity of the parasitized intestine is significantly increased (see Podesta and Mettrick, 1974). Provided that the tapeworm occupies a significant proportion of the luminal volume, H^+ excretion by the worm may contribute to this. However, it is still not clear why the tapeworm fails to resort to aerobic metabolism, unless there are times when levels of oxygen in the intestine are low.

9.3.4 Immunity

In the rat, the normal host, single specimens of *Hymenolepis diminuta* grow to maturity in 17 days and live as long as the rat. The longevity of the worm seems limited only by the life span of the host. Read (1967) transplanted worms from one rat to another younger rat at roughly annual intervals; two of these worms were still alive after 14 years and producing viable eggs.

Other mammals vary in their susceptibility. Colleagues of Clark Read failed to become infected and although Read succeeded in infecting himself twice, the infections persisted for only a few weeks and led him to conclude that the American male is not a highly satisfactory host. However, mice are readily infected experimentally and the course of their infection is revealing.

Over 90% of cysticercoids administered to 6-week-old mice become established and grow (for references see review by Hopkins, 1980). This success rate is comparable with rats. However, the worms never achieve sexual maturity. They grow more slowly than in the rat and from days 8–10 onwards, when the worms are 4–30 cm long, loss of worms and destrobilation (loss of the strobila leaving a scolex and neck between 0.5 and 2 mm in length) begin to occur. However, if these stunted worms are transplanted surgically either to a rat or to a previously uninfected ('naive') mouse, normal growth is resumed within hours. This suggests that the poor growth in the mouse has an immunological rather than a physiological basis and there are other indications that this is so. For example, treatment of mice with cortisone acetate, which is known to suppress the immune system, delays rejection and permits worms to grow until they stretch from the stomach into the caecum and reach maturity on day 17, as they do in rats. Also, mice less than 5 weeks old at the time of infection take, on average, about 4 days longer to reject their worms than mice 5–7 weeks old; this correlates well with important developmental changes in the lymphocyte populations in Peyer's patches (section 9.2.5) and the intestinal epithelium.

The best evidence for an immune response in the mouse comes from secondary (challenge) infections given to mice after anthelminthic termination of the primary infection. Mice that had received a primary infection of eight worms when 7 weeks old were challenged by surgical transplantation of single 7-day-old worms from donor mice. Nearly all transplanted worms became established and in naive control mice continued to grow during the next 6 days. However, transplants to immunized mice ceased to grow in 2–3 days and were all rejected by day 6. Worms were transplanted to the mice at intervals for 6 months and the immunized mice showed no weakening of resistance. Thus, there is real immunological memory, not just a transient change in the reactivity of the intestine wall, and this memory persists for at least 6 months.

The minimum duration of a primary infection required to evoke a response to a challenge infection is surprisingly short. Mice given a three-worm primary infection terminated by drugs on day 3 were strongly resistant, even though the primary worms were only 2 mm long and weighed a mere 10 μg at the time of termination. This demonstrates how little tapeworm tissue is necessary to evoke a response and suggests that the scolex and/or neck region are of primary immunogenic importance. However, as Hopkins (1980) pointed out, this does not mean that antigenic stimulation ceases on day 3, and he favoured the hypothesis that the whole tegument is a continuing source of antigenic material, in particular from the glycocalyx, which undergoes rapid turnover.

The observation that a single specimen of *H. diminuta* in the rat can survive as long as its host (see above) suggests that there is no immune interaction between the worm and its normal host. However, observations on larger infections reveal evidence of immunogenicity and a host response. In a five-worm infection, the mean dry weight per worm decreases from day 19 but stabilizes later (Andreassen, 1981). This stunting is also a feature of larger infections, but with 15 or 30 worms per rat a gradual loss of worms also occurs. In infections of 40–100 worms destrobilation takes place, usually about day 11, followed by expulsion of the destrobilated worms, although 5–10 worms per rat often survive. That this phenomenon reflects a host immune response rather than competition for some limited resource such as carbohydrate (crowding effect) receives support from the progress of secondary infections.

Andreassen and Hopkins (1980) established a 50-worm primary infection and removed it after 87 days. An uninfected control was treated at the same time with an anthelminthic. Eight days after termination of the primary, a 50-worm challenge infection was given and the control rats received a 50-worm primary. The rats were killed 7 days later and the growth of the challenge worms was found to be only 0.5% of that of the control worms, demonstrating that a strong initial resistance had developed. The same protocol was repeated, except that the period between removal of the primary worms and administration of the challenge was extended to 23 days. In a third experiment, in which the primary infection was removed by anthelminthic after 48 days, the interval between removal of the primary infection and administering the challenge was increased further to 41 days. The respective growths of worms in the secondary infections in these two experiments were 3% and 25% of that of the control worms, i.e. there was a 50-fold weakening of resistance over a period of less than 5 weeks. Even primary infections of five worms, which would have survived indefinitely if not expelled by drugs, depressed growth of secondary infections given 8 days later by over 80%.

Thus, even in the normal host, the rat, an immune response is generated, but a high level of parasites is needed to induce rejection and immunological memory is short. In an abnormal (poor) host, the mouse, rejection is induced by low levels of parasites and immunological memory is long (at least 6 months). Andreassen and Hopkins suggested that the rapid waning of memory in the rat indicates that the protective mechanism evoked in the intestine results in the limitation of worm load (if too high this could threaten the survival of the host and its parasites) and not the prevention of reinfection after the loss of a previous infection.

An interesting feature of the rat is that when a heavy secondary infection (50 worms) is superimposed on a light primary infection (five worms), the secondary worms are stunted while the co-existing primary worms are apparently unaffected. This implies that the primary worms can evade the immune response in some way. We will meet this so-called concomitant immunity again in schistosomes (section 18.3.11).

At the time when *H. diminuta* in mice are about to destrobilate, large swollen opaque areas appear in the tegument of the neck region (references in Hopkins, 1980). Transmission electron microscopy reveals signs of degeneration in the tegumental cytoplasm. Since these areas increase in number in multiple, more strongly rejected infections and are more frequent and severe immediately prior to rejection, it is possible that they represent immune damage. Hopkins (1980) pointed out that this is the region of the worm where new proglottides are made; massive cell proliferation is occurring and new tegument is appearing and expanding. He postulated that new tegument is more vulnerable to immune attack than older tegument on the scolex and strobila. This idea is consistent with the fact that destrobilation occurs in the neck region and that the surviving scolex and the shed strobila are apparently undamaged. Although the scolex may remain attached for some time, it cannot regrow until it is transplanted to a naive host, and this may be because new tissue generated by it is rapidly destroyed.

Dark areas are also found on the neck of *H. diminuta* from rats, but disappear as the affected area grows and moves down the strobila. Thus, similar immune attack appears to occur in the rat but is weaker, so that the new tegument survives long enough to become resistant.

Hopkins (1980) discussed how the immune system might exert its influence on *H. diminuta*. Serum antibodies appear to be produced both in mice and in rats in response to *H. diminuta*; IgA, IgM, IgG and IgE are probably all involved. However, the part that they play in rejection, if any, is unknown. They are produced in rats that do not reject *H. diminuta*, at least at low infection levels, and transfer of serum does not protect against the parasite. IgA coats the surface of *H. diminuta* in mice by day 8 or 9, but it also coats the surface of *Vampirolepis* (originally *Hymenolepis*) *microstoma*, a mouse tapeworm that is not rejected. There is no evidence that this attached IgA has any deleterious effect on *H. diminuta*, and Hopkins suggested that its function might even be beneficial, enhancing the tapeworm's survival by masking its 'foreign' surface with host protein (see schistosomes, section 18.3.11). No host cells have been found in contact with immune-damaged tapeworms, but this does not of course preclude attack by substances released by unattached cells in the neighbourhood.

An interesting interaction between two species of tapeworm inhabiting the same host was identified by Heyneman (1962). He discovered that *Hymenolepis* (= *Vampirolepis*) *nana* in its normal mouse host limited the abundance of *H. diminuta*. This interaction occurred when both species were present simultaneously, as well as when the previous infection of *H. nana* had been lost. *H. nana* was more immunogenic against *H. diminuta* than it was against itself. Even in the rat, the normal host of *H. diminuta*, an infection with *H. nana* adversely affected a concurrent infection with *H. diminuta*. This phenomenon of non-reciprocal cross-immunity was regarded by Schad (1966) as an adaptation that limited the population of competing parasitic species (see also section 20.3).

9.3.5 Locomotion

Pioneering work by Chandler (1939) provided interesting information on behavioural patterns in the tapeworm. He observed that *Hymenolepis diminuta* does not simply hang on for the rest of its life attached to the region of the intestine where it happened to excyst, but migrates forwards between the 7th and 10th day after infection. This was confirmed by Bråten and Hopkins (1969), who found that the worms migrate from day 7 to day 14 from a position of attachment 30–40% down the length of the small intestine to the 10–20% region (Fig. 9.7).

Anyone who has seen the vigorous activity of a freshly excysted young worm cannot doubt that it is a dynamic organism. The detached worm alternately moves pairs of suckers forwards on the scolex (Fig. 9.8); when attached to the mucosa,

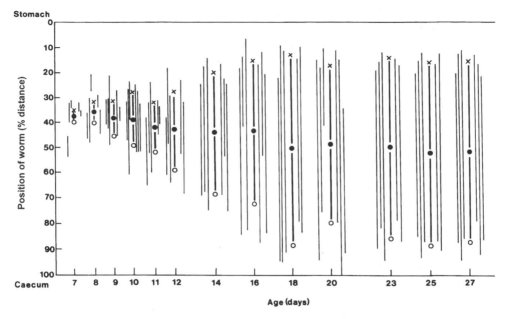

Figure 9.7 The positions in the intestine of the rat and lengths of single specimens of *Hymenolepis diminuta* administered experimentally as cysticercoids and aged from 7 to 27 days. The thin vertical lines indicate single parasites and the thicker vertical lines the average for each time interval. x = scolex; o = midpoint of body; o = position of 'tail'. Source: redrawn from Bråten and Hopkins, 1969.

effective and speedy locomotion could be achieved by this step-wise movement and this may be the way in which forward migration is effected from day 7.

By day 18 the worm has grown to occupy nearly the whole length of the small intestine (except for the anterior 10% and the posterior 5–10%), and this coincides with the achievement of maximum weight (350 mg dry weight) and sexual maturity with discharge of eggs (day 16–17).

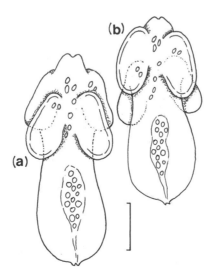

Figure 9.8 (*a*), (*b*) Two successive stages in the anterior movement of a freshly excysted scolex of *Hymenolepis diminuta* beneath a coverslip. Scale bar: 50 μm.

Bråten and Hopkins also transplanted 6.5- and 7.5-day-old worms from donor rats into the duodenum or posterior ileum, i.e. into sites anterior and posterior to their original habitat. Worms implanted both anteriorly and posteriorly returned within 24 h to the region of the intestine from which they had been removed.

This work reinforced the concept of a dynamic rather than a degenerate tapeworm and also revealed a surprising and hitherto unsuspected sophistication: the worm is capable of recognizing a specific region of the intestine, is able to detect when it is not in this region, has the sensory capability to select the right direction and has a well coordinated motor system able to overcome peristalsis. This ability is reflected in the well developed nervous system (Fig. 1.9*b*, *c*), with a concentration of nervous tissue ('brain') in the scolex giving rise to several nerve cords running the full length of the strobila.

The sensory capabilities of the tapeworm are reflected in the presence of sensilla projecting from the tegument (Fig. 9.5*a*). In the related *Vampirolepis nana*, Fairweather and Threadgold (1983) found sensilla within the cavities of the suckers, on the sucker margins and elsewhere on the scolex and neck, and these sensilla were of three kinds, one of which lacked a projecting cilium. However, integration of sensory input from the whole body of the tapeworm is indicated by the work of Hopkins and Allen (1979). They claimed that the position of a young worm with little or no strobila is dictated by the site preference of the scolex but as growth proceeds the end of the strobila is pushed back into a region unfavourable to it. 'Adverse' signals from the 'tail' override signals from the scolex and the whole worm moves forwards. This idea was supported by the observation that tapeworms from which the strobila had been experimentally removed migrated backwards to a position where a young worm of the same size would be found.

In addition to the anterior migration of *H. diminuta* during development, it has been established that adult worms, including the scolex, have a daily pattern of backward and forward movement (Read and Kilejian, 1969). In rats fed 1 g of glucose, worms migrate in an anterior direction, and injection of intestinal washes from glucose-fed rats stimulate a similar anterior migration (Sukhdeo and Kerr, 1992). In the natural world rats feed at night (see Arme, 1993), so it is at this time that forward migration will occur in natural infections, with the parasite in a more posterior location during the day. The most likely explanation for this phenomenon is that it keeps the greater part of the strobila within the region of the intestine where its nutrient uptake is optimal (Sukhdeo and Mettrick, 1987).

It is hard to believe that the tiny scolex is capable of dragging the whole adult worm forward in the intestine against gut peristalsis, and Sukhdeo and Kerr (1992) suggested that antero-posterior peristaltic contractions of the strobila serve to propel the worm forwards (Fig. 9.9*a*). The projecting posterior margins of the proglottides seem ideal for enhancing the effectiveness of forward motion by pushing backwards on the surrounding gut contents (see also section 10.5.5). Sukhdeo and Kerr regarded locomotory activity of this kind as routinely important to counter the expulsive effects of peristalsis. They considered that the worm's anterior migration after feeding was the result of an increase in this activity and that a decrease in activity would lead to posterior displacement of the worm.

There are many potential chemical cues, such as glucose, intestinal secretions and neurotransmitters, that might serve to stimulate increased worm peristalsis, but none of these appears to have any significant effect on the tapeworm *in vitro*. It

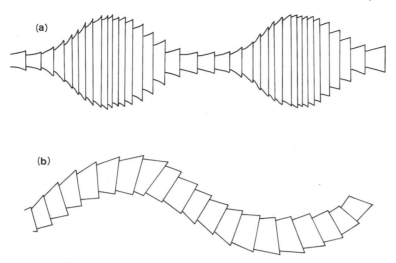

Figure 9.9   (*a*) Antero-posterior peristaltic movements of the strobila of *Hymenolepis diminuta*, generated mainly by the circular muscles. (*b*) Sinusoidal motility of the strobila, generated mainly by the long-itudinal muscles. Source: reproduced from Sukhdeo *et al.*, 1984, with permission from the American Society of Parasitologists.

seems that the most likely stimulant is mechanical pressure exerted on the strobila by the increased peristaltic activity of the gut itself induced by feeding. Sukhdeo (1992) showed that tapeworms responded to pressure, exerted on the strobila by a ligature tied around the intestine, by moving forwards. Sukhdeo and Kerr (1992) found that removal of the scolex did not significantly affect strobila behaviour, and that when the worm is cut into pieces each region continues to generate waves of contraction at the usual frequency. Thus, the locomotory behaviour appears to be a fixed action pattern controlled by local neural circuits and not by the 'brain'. Such a decentralized control system allows different regions of the strobila to respond to local changes in gut motility.

The view has been expressed that locomotion is not a universal feature of tape-worms and that *H. diminuta* has adopted peristaltic movement to compensate for the weakness of its scolex and to maintain its position in the gut (see, for example, Arme, 1993). This is thought to have led to a nutritional advantage in that the parasite can maintain itself in a region of the gut containing optimal concentrations of nutrients. However, it is hard to appreciate why the scolex of *H. diminuta* should have lost this power and even harder to understand why the parasite should compensate by adopting energetically costly peristalsis. It is indisputable that the scolex is small but arguable that it is weak and there are other tapeworms with relatively small scoleces. Tapeworm peristalsis as a means of reducing the load on the scolex may turn out to be more widespread, and the rigorous studies required to identify migration have been conducted on relatively few species (see also section 10.5.5).

9.3.6   Growth and reproduction

It was pointed out by Roberts (1980) that growth of *Hymenolepis diminuta* after excystment is best described as explosive and must rival or surpass that of any other metazoan tissue. Some 15 days after infection of the rat, the worm has produced 2200 proglottides, increased its length by 3400 times and its weight by up to 1.8 million times. Nevertheless, growth slows during the second half of this 2-week period to a rate just sufficient to replace spent proglottides (Insler and Roberts, 1980). All this poses interesting questions about how growth is regulated.

New proglottides are continually produced by a germinative zone just posterior to the scolex and each new proglottis then proceeds to grow in three dimensions, producing its first eggs about 9 days after its formation. Mature proglottides are narrow compared with those of *Taenia* (see Fig. 8.3), usually with only three testes arranged transversely (Fig. 9.10). There is an eversible cirrus adjacent to the vaginal opening on the lateral border of the proglottis. Wilson and Schiller (1969) observed that two adult worms maintained *in vitro* align their bodies such that the sexually mature proglottides of each worm correspond, and then undergo presumably multiple copulations. This again illustrates the remarkable sensory and motor capabilities of these worms.

In addition to the activity of the scolex and the peristaltic contractions of the strobila mentioned above, sinusoidal movements of the strobila also occur (Fig. 9.9*b*), mainly as a result of differential contraction of the longitudinal muscles on each side of the body. All of these movements could play a part in the alignment of individuals.

Nollen (1975), using isotopically labelled worms, showed that isolated tapeworms self-inseminate and that in multiple infections both self-insemination and cross-insemination occur.

The uterus is a blind-ending labyrinthine sac (Fig. 9.10), which first extends laterally and then anteriorly as it fills up with eggs (see Lumsden and Specian, 1980). The male system is said to degenerate first, but the female system must continue to function, at least until the uterus reaches its full egg-storage capacity. Gravid proglottides seem to be shed from the posterior end of the worm at a rate about equal to the growth rate of the worm, and this process is maintained, if not

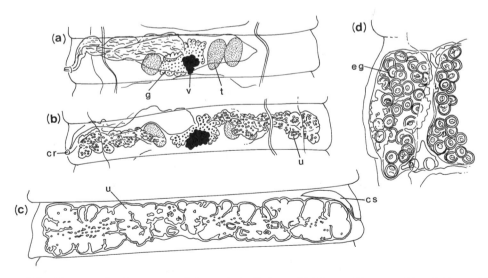

Figure 9.10   The reproductive system of *Hymenolepis diminuta*. (*a*) Mature, (*b*) pregravid and (*c*) gravid proglottis, from life. (*d*) Part of a gravid proglottis showing the labyrinthine structure. cr = cirrus; cs = cirrus sac; eg = egg; g = germarium; t = testis; u = uterus; v = vitellarium. Source: reproduced from Czaplinski and Vaucher, 1994, with permission from CAB International.

indefinitely, for a very long period of time (section 9.3.4). The detached proglottides liberate some of their eggs before the faeces are expelled.

9.4   THE EGG

In the egg of *H. diminuta* the layers surrounding the central oncosphere or hexacanth larva are bewildering in their complexity, but the work of Rybicka (1972) and Ubelaker (1980) has done much to establish the origin and relationship of these layers. The fertilized oocyte from the germarium is joined by only one vitelline cell from the relatively small vitellarium and the two cells pass through the ootype into the uterus. The zygote and vitelline cell become enclosed in a membrane or capsule, which seems to be homologous with the eggshell of a monogenean or a pseudo-phyllidean cestode, since it appears to originate from vitelline granules released in the ootype. The rest of the vitelline cell is incorporated into the developing embryo, which separates into three layers: the central preoncosphere, the inner envelope around the preoncosphere and a peripheral outer envelope (Fig. 9.11a).

A thin new outer covering develops at this stage (Fig. 9.11b). This covering eventually becomes relatively rigid and splits under pressure (Fig. 9.12) – it has all the attributes of an eggshell and is usually referred to as such. However, it is not homologous with the eggshell of a monogenean or pseudophyllidean cestode since it is probably secreted by the syncytial, cytoplasmic outer envelope and appears after the disintegration of the vitelline cell. This 'shell' replaces or perhaps incorporates the delicate capsule. According to Ubelaker (1980), the outer envelope eventually disintegrates, but some of it may persist as a 'subshell' membrane lying immediately beneath the 'shell' (see Fig. 9.12 and below).

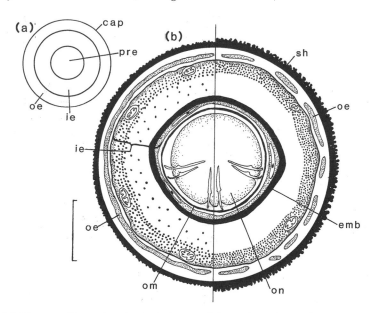

Figure 9.11   (*a*) Diagram illustrating the basic layers of the early embryo of *Hymenolepis diminuta*. (*b*) Diagram of early (left) and fully developed (right) egg of *H. diminuta*. cap = capsule; emb = embryophore; gl = gland cell; ie = inner envelope; oe = outer envelope; om = oncospheral membrane; on = oncosphere; pre = preoncosphere; sh = 'shell'. Scale bar: (*b*) 25 μm. Source: (*b*) reproduced from Rybicka, 1972, with permission from the American Society of Parasitologists.

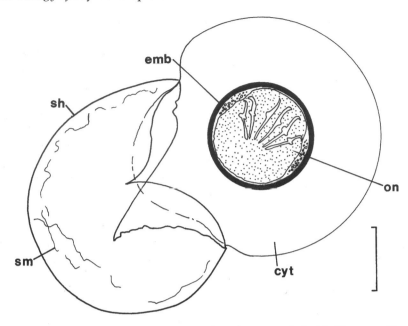

Figure 9.12   A fully developed living egg of *Hymenolepis diminuta* with the 'shell' ruptured by mechanical pressure. cyt = jelly-like cytoplasmic layer; emb = embryophore; on = oncosphere; sh = 'shell'; sm = 'sub-shell' membrane. Scale bar: 25 μm.

The inner envelope is also a syncytial embryonic layer. It delaminates from its inner surface a thin layer of cytoplasm that subsequently deteriorates. The new inner membrane of the inner envelope then becomes a thickened oncospheral membrane (Fig. 9.11*b*), which in turn is delaminated with a little cytoplasm and separates around the preoncosphere. The inner region of the inner envelope then differentiates as an embryophore layer, distinct from the rest of the inner envelope or cytoplasmic layer. Glycogen and mitochondria then shift towards the peripheral zone of the cytoplasmic layer.

9.5   INFECTION OF THE BEETLE INTERMEDIATE HOST

A variety of grain-infesting insects serve as intermediate hosts for *H. diminuta* (Fig. 9.1) and the hosts that have been used extensively in laboratory work are flour beetles of the genera *Tribolium* and *Tenebrio*. Eggs of *H. diminuta* within the rat faecal pellet at 10°C have an expected period of infectivity to the flour beetle *Tribolium confusum* of 11 days, although some may retain infectivity for up to 60 days (Keymer, 1982). Evans *et al.* (1992) found evidence to suggest that *T. confusum* preferentially ingests rat faeces contaminated with eggs of *H. diminuta* and, in a more rigorous series of experiments, Pappas *et al.* (1995) reached a similar conclusion using *Tenebrio molitor*.

The 'shell' and the 'subshell' membrane must be broken by the jaws of the beetle if ingested eggs are to hatch successfully (references in Ubelaker, 1980). Eggs that are swallowed with the 'shell' intact pass through the gut unharmed and retain their viability. The 'shell' consists of protein, which is resistant to attack by proteolytic enzymes and has other properties, such as dissolution in hypochlorite, that are shared with tanned protein eggshells (Lethbridge, 1971a). The 'subshell' membrane is resistant to hypochlorite, so Lethbridge was able to remove the 'shell' chemically, leaving the 'subshell' membrane immediately beneath intact. He went on to show that this thin layer is also remarkably resistant to proteolytic enzymes and impermeable to lipid-soluble dyes and preservatives.

Splitting of the 'shell' and 'subshell' membrane by the beetle's jaws exposes the surface of the cytoplasmic layer, which expands rapidly (Fig. 9.12). This often results in the swollen cytoplasmic layer, with its enclosed embryophore and oncosphere (= hexacanth), being expelled through the tear in the 'shell' into the midgut. It is at this stage that the oncosphere usually becomes active, and proteolytic enzymes from the beetle's midgut dissolve first the cytoplasmic layer then the embryophore, providing perhaps a small nutritional 'reward' for the beetle and freeing the active hexacanth. According to Lethbridge (1971b), dissolution of the cytoplasmic layer and embryophore takes less than 12 min in the larger beetle *Tenebrio molitor*.

The vigorous action of the hooklets of a newly hatched oncosphere, resembling the breast-stroke of a swimmer (Fig. 9.13), are always a surprise to the observer who has previously only seen the totally inactive oncosphere in the unhatched egg. Inactivity in the unhatched egg may be essential for survival, since it may prolong the life of the embryo and, thereby, increase its chance of being eaten. Caley (1975a), with reference to the similar egg of another tapeworm, *Moniezia expansa*, made the interesting suggestion that excretory products produced by the oncosphere within

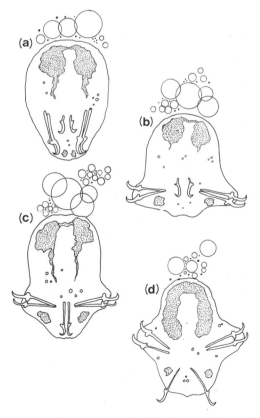

Figure 9.13(a)–(d) Successive stages in movement of hooklets in an active, newly hatched oncosphere of *Hymenolepis diminuta*.

the egg may accumulate and inactivate the larva, and that activation is brought about by escape of these inhibitors after rupture of the 'shell' and 'subshell' membrane (see also section 9.7).

The effectiveness of the movements of the three pairs of hooklets for penetration of the wall of the midgut of the beetle has been emphasized by Lethbridge (1971b), who observed living oncospheres penetrating excised pieces of midgut from *T. molitor*. The two lateral pairs of hooklets are the first to move, pushing forwards so that their blades emerge from the body on each side of the median pair of hooklets (Fig. 9.13a). The lateral pairs are then swept sideways and backwards (Fig. 9.13b), wedging the larva between adjacent midgut epithelial cells. Next, the two median hooklets push forwards until they project from the body (Fig. 9.13c). They then rotate downwards and backwards in the midline (Fig. 9.13d), pulling the larva forwards between the epithelial cells. The cycle is completed by the simultaneous withdrawal into the body of all three pairs of hooklets. In freshly hatched larvae the cycle takes 2–3 s, and hexacanths burst through the midgut wall into the haemocoel of *T. molitor* adults 45–75 min after the eggs are ingested by the beetle. Damage inflicted on the midgut is likely to be rapidly repaired (see Crompton, 1970).

The observation of Lethbridge that glands in the hexacanth were substantially reduced after penetration, and the finding of Moczon (1977) that beetle gut muscle exposed to the gland secretion was swollen and broken down, suggest that these glands have a histolytic role in penetration. A serine protease has been detected in the secretion by Moczon (1996a).

9.6   THE CYSTICERCOID

Over a period of about 10 days at 28°C the oncosphere of *H. diminuta* inside the haemocoel of *T. confusum* grows and develops into a cysticercoid, which is infective when ingested by a rat. This involves an increase in size from a roughly spherical oncosphere about 45 μm in diameter to an ovoid cysticercoid measuring about 400 μm by 300 μm. An unusual feature of beetle haemolymph is the high concentration of amino acids, of which two-thirds is proline, and the membrane transport systems of the cyst wall have special adaptations to avoid overloading with this amino acid (references in Arme, 1993).

The cysticercoid is essentially an encysted scolex with a complex, multilayered cyst wall (Fig. 9.14a), which betrays a substantial, highly-fibrous component when viewed with polarized light (Fig. 9.14b). Cysticercoids also have a fleshy 'tail' or cercomer and the hexacanth origin of the cysticercoid is usually revealed by the persistence of one or two redundant hooklets on the cercomer (see Fig. 9.17).

The development of the cysticercoid has been described by Caley (1974) in the related mouse tapeworm *Vampirolepis microstoma*, the scolex of which is attached within the bile duct (Litchford, 1963). Early growth involves differentiation of the growing cysticercoid into an anteriorly-situated developing scolex and a posterior midbody and tail, the last two regions containing a cavity (Fig. 9.15). During the seventh day at 25°C an important event takes place – within a period of 30–60 s the scolex is withdrawn into this cavity. This is brought about by special circular and longitudinal muscles beneath the surface of the scolex and midbody.

During the next few days the scolex completes its development and important changes occur in the cyst wall. The muscles involved in scolex retraction degenerate

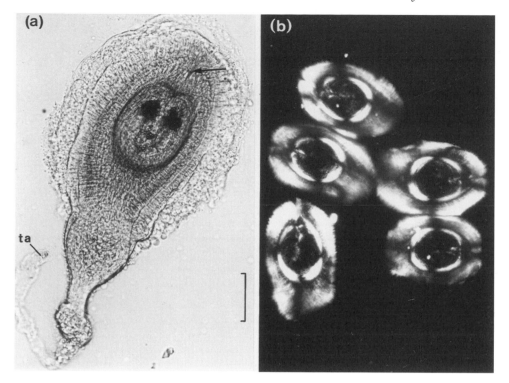

Figure 9.14 (*a*) Photograph of a fully developed cysticercoid of *Hymenolepis diminuta*. (*b*) Cysticercoids of *H. diminuta* viewed in polarized light. Arrow = channel through which the tapeworm emerges; ta = tail. Scale bar: (*a*) 100 μm.

and three layers of collagen fibres, an inner and an outer circular layer and a longitudinal layer between them, come to occupy a significant proportion of the bulk of the cyst wall (Fig. 9.16). The inner scolex becomes enclosed in a multilayered complex comprising thin sheets of parenchymal cell cytoplasm. These layers become compacted and thinner and eventually almost all of the cytoplasmic contents and extracellular material disappears, leaving essentially a stack of plasma membranes joined by tight junctions. Caley pointed out the similarity between this material and myelin.

The cysticercoids of *V. microstoma* are infective to rats on day 11. The functional significance of scolex retraction and the roles of the fibrous material and myelin-like layer will be considered shortly.

No density-dependent constraints on the establishment of cysticercoids of *H. diminuta* in individual specimens of *T. confusum* were found by Keymer (1980) and, with repeated exposures to infection, cysticercoid numbers per host increase linearly to 60 or more. The larger *T. molitor* may harbour more than 200 (Hurd and Fogo, 1991). However, the size of established cysticercoids declines as parasite burden increases, indicating competition for nutrients or space. A surprising feature of cysticercoids developing in the haemocoels of *Tribolium confusum* and *Tenebrio molitor*, and even in an unlikely host, the locust *Schistocerca gregaria*, is that they are not recognized as foreign and not encapsulated by host haemocytes (Lackie, 1976). In contrast, the larvae are normally encapsulated in cockroaches such as *Periplaneta americana*. Latex beads injected into *S. gregaria* at the same time as cysticercoids

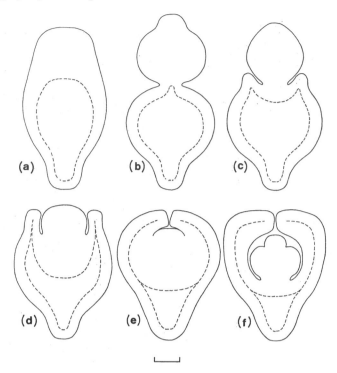

Figure 9.15 The retraction of the scolex of *Vampirolepis microstoma* shown as a sequence of six stages, (*a*)–(*f*). The broken line marks the extent of the midbody and tail cavity. The approximate times for each of these changes are given below for a parasite at 25°C: (*a*) to (*b*), 1 min; (*b*), 15–30 min; (*b*) to (*e*), 1 min; (*e*) to (*f*), 30–60 min. Scale bar: 20 μm. Source: reproduced from Caley, 1974, with permission from Cambridge University Press.

became thickly encapsulated. Lackie's experimental work suggested that the cysticercoid surface may resemble host tissue surfaces, permitting the parasite to escape recognition as 'not-self' by the host's haemocytes.

This specialization on the part of the cysticercoid for circumventing host defensive responses, and the fact that beetles with large numbers of cysticercoids are active, creates an image of a well balanced, mutually tolerant, parasite–host relationship. However, this impression is misleading because Keymer (1980) demonstrated that cysticercoids have deleterious effects on both survival and fecundity of *T. confusum*. She found that host mortality is linearly related to parasite burden, whereas the relationship between parasite burden and host fecundity is non-linear. Keymer suggested that host survival is related to the extent of damage done to the midgut wall and hence to the number of penetrating oncospheres. She related fecundity to the total biomass of cysticercoids, which increases non-linearly to a plateau as parasite burden increases (see above). Curiously, in *Tenebrio molitor* a similar reduction of egg output by parasitized beetles is not directly related to consumption by the cysticercoids of nutrients destined for the beetle's eggs (see review by Hurd, 1993, Webb and Hurd, 1996). In the unparasitized insect the eggs are provisioned with an egg yolk protein (vitellin) that is synthesized as a precursor (vitellogenin) in the fat body of the mature female; vitellogenin is then released into the haemolymph, from which it is taken up by the oocytes. In the presence of cysticercoids, the synthesis of vitellogenin is reduced and, although it is plentiful in the haemolymph,

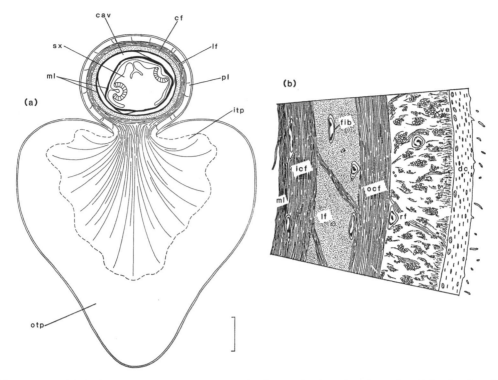

Figure 9.16 (*a*) A diagrammatic representation of a longitudinal section through a fully developed cysticercoid of *Vampirolepis microstoma*. (*b*) A diagrammatic representation of a section through a portion of the cyst wall of a fully developed cysticercoid, as seen with the transmission electron microscope. cav = cyst cavity; cf = circular fibres; dc = distal cytoplasmic layer of tegument; fib = fibroblast cell; icf = inner circular fibres; itp = inner tail parenchyma; lf = longitudinal fibres; ml = myelin-like layer; ocf = outer circular fibres; otp = outer tail parenchyma; pl = peripheral layer of cyst; rf = radial fibres; sx = scolex; ve = vesicles. Scale bar: (*a*) 50 μm. Source: reproduced from Caley, 1974, with permission from Cambridge University Press.

its uptake by the oocytes is also reduced. The cysticercoids do not themselves take up vitellogenin or limit the amount of leucine available to the fat body for vitellogenin synthesis, so the observed effects on the host seem likely to be produced by a parasite-derived effector molecule or molecules.

Keymer (1980) pointed out that, in many cestodes, the main reproductive potential is invested in a long-lived adult worm, which tends to compensate in terms of evolutionary fitness for the low probability of successful transmission between hosts combined with the high pathogenicity of the larval stages.

9.7 INFECTION OF THE RAT

Cysticercoids of a dilepidid cestode were found to alter the behaviour and appearance of their ant hosts (*Leptothorax nylanderi*) to such an extent that these infected ants were regarded at first as a new species by Plateaux (1972). There is evidence that cysticercoids of *H. diminuta* may influence *Tenebrio molitor* in less obvious but equally interesting ways. According to Hurd and Fogo (1991), beetles with 11- or 12-day-old infections exhibited a significant decrease in activity and an increased tendency to seek illuminated rather than dark areas. Hurd and Fogo also found that both male and female infected beetles failed to respond to aggregation pheromone and Blankespoor, Pappas and Eisner (1997) found that the chemical defences of the beetle are impaired in infected individuals. The fact that these changes are only demonstrable in beetles harbouring mature, infective cysticercoids and not in beetles with immature parasites that would not develop into adult tapeworms even if eaten is consistent with the hypothesis that these are adaptive manipulations of beetle behaviour promoting successful infection of the definitive host. As was suggested by Arme (1993), such behavioural effects in *T. molitor* may be mediated by changes induced by the parasite in the hormonal balance of the beetle rather than by mechanical damage produced by the cysticercoids.

Yan, Stevens and Schall (1994) reported changes in behaviour in both *Tribolium castaneum* and *T. confusum* infected with *H. diminuta*. However, since behavioural changes related to beetle activity were opposite and since the two beetle species share similar natural habitats and the same role in the life cycle, they rejected the possibility that the changes represented adaptive manipulation of the beetle's behaviour. Nevertheless, they emphasized that some behaviour changes, for example in the rate of cannibalism of beetle eggs, could have a profound influence on population and community ecology of the beetle hosts.

There remain two further hurdles that must be successfully negotiated before the parasite reaches its definitive site in the small intestine of the rat. These were referred to in the first paragraph of this chapter and are the threat of mechanical destruction of the cysticercoid by the molar teeth of the rodent and the dangers posed by the high acidity of the stomach. Secretion of hydrochloric acid by the parietal cells of the stomach wall creates a luminal pH as low as 1.2, which is the optimum for the activity of the proteolytic stomach enzyme pepsin. Caley (1974), working with *V. microstoma*, found that intact beetles (*T. confusum*) were resistant to digestion for at least 2 h in acid pepsin and that the cysticercoids remained enclosed within the insect's tissues. Thus, cracking of the insect exoskeleton is essential for liberation of the parasite, but carries with it a serious risk of damage by the teeth or by fragments of exoskeleton.

Caley demonstrated that the collagenous layers of the cysticercoid (Fig. 9.16) provide the mechanical protection needed; cysts with fully-developed collagen layers (11 days old or more) resisted vigorous grinding with glass beads for 5 min while 6-, 7- and 9-day-old cysts, which lacked fully developed fibre layers, were totally destroyed. Caley also found that the encysted scolex survived for 30 min at pH 2 and 37°C, long enough to survive passage through the stomach of the mouse, and that this protection was maintained even when pepsin was present, although the fibre layers were then digested. This suggests that the myelin-like layer (Fig. 9.16) is responsible for this protection. This is supported by the fact that experimentally excysted worms, no longer enclosed by this layer, were killed within a few minutes when exposed to pH 3 or lower and that intact cysts in which the myelin-like layer had been damaged by treatment with bile were no longer protected from acid. Caley suggested that the tight junctions in the myelin-like layer block the passage to the scolex of H^+ ions, and that bile may damage the layer by acting on its lipid component and by disrupting the tight junctions. Caley found that the scolex of *V. microstoma* is activated by treatment with bile salts but requires the presence of pancreatic enzymes for complete excystment.

Caley (1975b) extended her observations to the cysticercoids of *V. nana*, which develop in *T. confusum* and are structurally similar to those of *V. microstoma*. (*V. nana* is also able to omit the beetle host and develop directly in the mouse definitive host – see Chapter 11.) In an elegant study of activation of the scolex of *V. nana* she showed that bile is not always necessary – newly formed cysticercoids can be activated by trypsin. She hypothesized that inactivation of the scolex is established and maintained by one or more inhibitory molecules, a suggestion which she also made in relation to the hexacanth within the egg (section 9.5). Since the mature cyst is impermeable, substances excreted by the scolex would accumulate within the cyst and may serve as inhibitors. This would explain the gradual loss of scolex mobility with age, the negative feedback acting to reduce respiration rate and keep energy wastage to a minimum. She regarded the latter as a further necessity imposed on the fully developed scolex by its impermeable cyst wall, since the scolex is unlikely to be

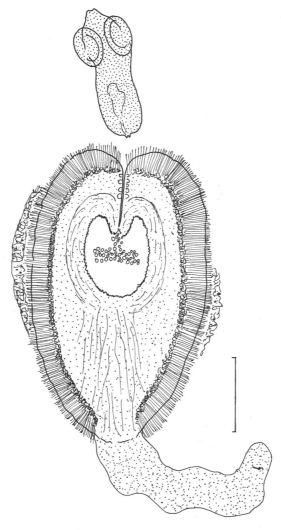

Figure 9.17 Drawing of a cysticercoid of *Hymenolepis diminuta* with a newly emerged tapeworm. Note persistent hooklet on tail of cyst. Excystation was promoted by immersion of the cysticercoid in saline containing 1% pancreatin and 1% bile salts at 40°C. Scale bar: 100 μm.

able to draw on the host for nutrients and therefore must rely on what it has stored for its larval life of perhaps several months and for the burst of activity required during excystment. Of the two common excretory acids produced by tapeworms, succinic, but not lactic, acid was shown to have a strong inhibitory effect and may be the natural inhibitor. Thus, bile salts may have an indirect role in activation by increasing the permeability of the cyst and permitting the inhibitory substance(s) to diffuse out.

Thus, scolex retraction in cysticercoid development is an essential event for survival of the tapeworm, since it is vulnerable to the teeth and stomach acid of the rodent definitive host. In *V. microstoma* and *V. nana*, and probably by analogy in *H. diminuta*, the wrappings around the scolex provide a truly remarkable food parcel. The collagenous layers protect the scolex from mastication and then, after their role is accomplished, are attacked by acid pepsin in the stomach. It is then the turn of the myelin-like layer to protect the scolex from acid pepsin and finally, in the duodenum, the myelin-like layer is itself damaged by bile, which serves directly or indirectly as a trigger for scolex activity and its emergence through a narrow channel at one end of the cyst (Fig. 9.17).

Barrett and Precious (1994) showed that absence of K^+, Mg^{2+} or Ca^{2+} has no significant effect on excystment, but Na^+ is essential, cysticercoids requiring at least 70 mmol/l sodium. Further experiments suggested that excystment may involve activation of a tegumental proton pump, leading to a change in internal pH. Whether this activation is a direct effect of bile salts or some other intestinal factor or is mediated by release of an inhibiting substance remains to be investigated.

9.8   THE WAY FORWARD

It is not easy to convince persons unfamiliar with flatworms that monogenean parasites of fishes are the nearest relatives of ribbon-like, gutless tapeworms like *Taenia* and *Hymenolepis*, with their multiple sets of reproductive organs, detachable proglottides and two-host life cycles. Nevertheless, the evidence is good that this is the case (see Chapter 3) and it is a remarkable example of natural selection that such dissimilar creatures have been derived from the same ancestor. However, *Taenia* and *Hymenolepis* are highly specialized cyclophyllidean cestodes of mammals and it is among the tapeworms of fishes, hosts of their monogenean and gyrocotylidean relatives, that we must look for clues to help us understand how such drastic changes might have come about. This fish tapeworms will be explored in the next chapter.

CHAPTER 10

The fish tapeworms

10.1 ACQUISITION OF AN INTERMEDIATE HOST

It is likely that modern monogeneans parasitic in the gut of fishes have single-host life cycles and spread to new fish hosts by way of free-swimming oncomiracidia hatching from tanned eggs passed out in host faeces (section 7.2.2). Their ancient gut-parasitic relatives that gave rise to the cestodes would probably have been disseminated in the same way. The gyrocotylideans are 'missing links', modern survivors of ancient gut parasites that already have at least one major tapeworm feature (section 7.3). Their gut has disappeared, but the adults attach themselves by the persistent haptoral end of the body, while at the same time developing a supplementary attachment organ at the anterior end. They lay tanned eggs that liberate a ciliated, free-swimming larva with hooklets. If there is an intermediate host, and there is still no sound evidence of this, then the parasite must undergo virtually no development in it, since the smallest parasites recovered from the gut of the fish host are little bigger than the free-swimming larvae (section 7.2.2).

In gut parasites with a single fish host, great demands are placed on the free-swimming larvae of the parasites, since they not only have to locate another fish but also have to find their way to the intestine. It could be advantageous for the parasite to recruit an intermediate host – an organism that is part of the diet of the definitive fish host in which the parasite reaches maturity. It would need to be easier to find and infect the intermediate host than the fish host.

This change in strategy and acquisition of a second host could have taken place in the following way. The tanned eggs of parasitic platyhelminths are likely to be eaten by small invertebrates. The embryos will be protected from digestion by the tanned shell and provided that the shell is not punctured by the jaws of the predator, they will probably still be alive when they are voided in the predator's faeces (section 1.4.5). Fishes will prey on these invertebrates and occasionally a definitive host will eat an invertebrate while parasite eggs are still in transit through its gut. However, this is unlikely to be a realistic alternative to free-swimming, fish-seeking larvae until there is a way of delaying transit of the parasite's eggs through the gut of the potential intermediate host. This could be achieved readily by hatching of eggs ingested by the invertebrate and by use of the hooklets to hang on to the gut lining.

This strategy has two advantages. First, the larva may live longer inside the intermediate host than it would as a free-living organism, increasing its chances of reaching its final destination. Secondly, by stowing away in a prey organism, the definitive host is left to do most of the work, not only by seeking out the infected prey organism but also by swallowing the prey and conveying the parasite passively to its definitive habitat, the intestine, where it is liberated by the host's digestive processes.

In view of the proto-monogenean ancestry of the cestodes we might expect the first tapeworms to be fish parasites with a single set of reproductive organs. There is one group of modern cestodes that has these features, as well as utilizing an invertebrate intermediate host in a manner similar to that outlined above. These are the caryophyllideans.

10.2   THE CARYOPHYLLIDEANS

Mud-feeding invertebrates such as oligochaete worms (annelids) are likely to ingest tapeworm eggs lying on the bottom, and it comes as no surprise that freshwater tubificid and naidid oligochaetes have been recruited as intermediate hosts by caryophyllidean tapeworms. The tanned eggs hatch in the gut of the oligochaete and the oncosphere uses its six hooklets not just to hang on but to penetrate the gut wall and enter the coelom (Fig. 10.1). Here the oncosphere grows and acquires a new anterior attachment organ or scolex, while retaining the hooklets on a posterior flap-like cercomer. In caryophyllideans such as *Biacetabulum macrocephalum*, development of the reproductive system begins but becomes suspended before maturity is reached, and the resting or procercoid stage will remain unchanged until the parasitized oligochaete is ingested by a benthic fish. In the fish definitive host, the cercomer with its hooklets is shed and at the site of this severance, there is a single median excretory pore. The single set of reproductive organs (Fig. 10.2) then resumes development and goes on to produce tanned operculate eggs that pass out in the faeces and sediment to the bottom.

Since there is no gut in cestodes, there is no need to retain freedom and mobility of the anterior end for feeding and specialization of the anterior scolex for attachment is possible. As we will see later, this shift in attachment from posterior to anterior paved the way for interesting innovations in more advanced tapeworms. Although there is some diversity among caryophyllideans in the morphology of their scoleces (Fig. 10.3), they appear unexpectedly feeble; they are rarely equipped with suckers and have no hooks like those that we will encounter later in tapeworms such as the cyclophyllideans. The function of caryophyllidean scoleces has attracted virtually no attention and we can only speculate that their apparently feeble structure may reflect relatively weak peristaltic movements in the gut of their fish hosts and/or the possible use of cement as a means of bonding the scolex to the host's gut lining.

Mackiewicz (1981) regarded the widely distributed, mud-dwelling oligochaetes as desirable hosts for cestodes but difficult to colonize because of the anoxic nature of their environment. In this context he highlighted an extraordinary feature of the vitelline cells of these tapeworms. In addition to cytoplasmic glycogen, the cells have an extra glycogen store inside the nucleus and, unlike tapeworms such as the pseudophyllideans, the caryophyllidean egg contains no lipid (Fig. 10.4). Mackiewicz suggested that this contributed to survival of the caryophyllideans in two ways. First, replacing light lipid with heavier glycogen increases the density of the egg, ensuring that it remains on the bottom where the oligochaete intermediate hosts are feeding. Secondly, their anoxic environment will increase the

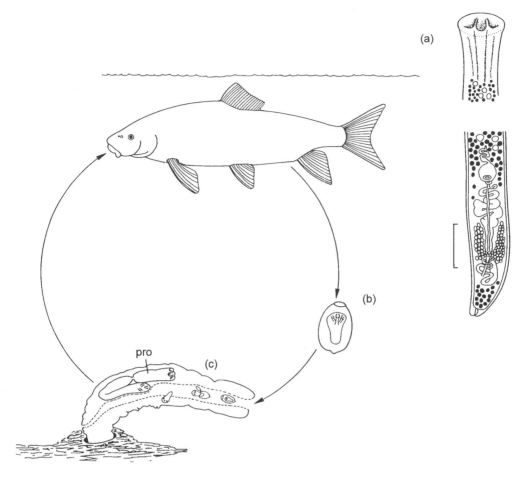

Figure 10.1 The life cycle of the caryophyllidean *Glaridacris catostomi*. Adult parasite (*a*) in the intestine of the freshwater fish, *Catostomus commersoni*, lays tanned eggs (*b*), which are voided in the faeces. Fully developed eggs are eaten by the oligochaete *Limnodrilus udekemianus* and hatch in the intestine. The oncosphere with six hooklets enters the coelom and develops into a procercoid (*c*). The fish is infected by eating oligochaetes containing procercoids. pro = procercoid. Scale bar for adult: 0.5 mm. Total length of adult about 2 cm. Source: based on Cooper, 1920 and Olsen, 1986.

demand for glycogen which, unlike lipid, can be metabolized in the absence of oxygen.

The eggs of some caryophyllideans have long lives. Calentine (1964) found that embryonated eggs of *Archigetes iowensis* survived for 80 days at room temperature. Kennedy (1965) observed that eggs of *A. limnodrili* in fresh water did not hatch and that the application of mechanical pressure dislodged the operculum in only 30% of the eggs. However, after incubation in amylases and hyaluronidase the operculum became dislodged in 80% of eggs subjected to pressure. Proteinases and lipases applied externally had no such effect but if the eggshells were broken before application of trypsin and pepsin the opercular cement was then weakened. He concluded that the opercular cement is weakened enzymatically when the eggs are eaten by the oligochaete host, but whether these enzymes are of host origin, secreted by the activated hexacanth or from both of these sources is unknown.

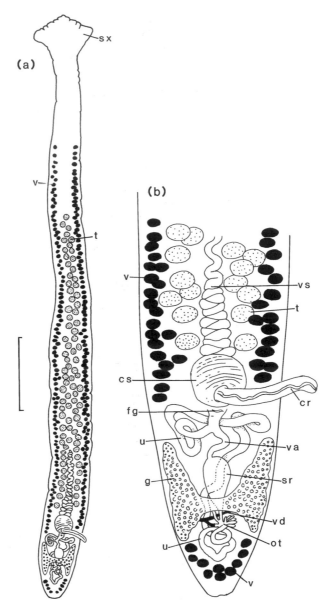

Figure 10.2   The reproductive system of a caryophyllidean cestode (*Caryophyllaeus laticeps*). (*a*) Whole adult parasite. (*b*) Enlarged posterior region of the body showing relationship of the ducts of the single set of reproductive organs. cr = cirrus; cs = cirrus sac; fg = female gonopore; g = germarium; ot = ootype and associated glands; sr = seminal receptacle; sx = scolex; t = testis; u = uterus; v = vitellarium; va = vagina; vd = vitelline duct; vs = vas deferens. Scale bar: (*a*) 0.5 cm. Source: reproduced from Fuhrmann, 1931.

Of great interest is a progressive tendency among caryophyllideans to undergo progenetic development in the oligochaete intermediate host, i.e. precocious development of the reproductive organs in the juvenile (procercoid) stage. This development may be partly favoured by the longevity of these annelids; Calentine (1964) recorded that tubificids infected with one or two procercoids of *A. iowensis* survived for 2 years. There are interesting parallels with the procercoids of spathebothridean

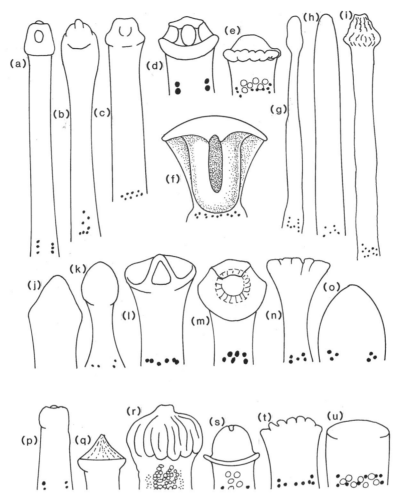

Figure 10.3 Scoleces of various caryophyllidean cestodes. (*a*) *Monobothrium*. (*b*) *Dieffluvium*. (*c*) *Promonobothrium*. (*d*) *Archigetes*. (*e*) *Balanotaenia*. (*f*) *Capingens*. (*g*) *Lytocestus*. (*h*) *Crescentovitus*. (*i*) *Monobothrioides*. (*j*) *Pseudolytocestus*. (*k*) *Atractolytocestus*. (*l*) *Glaridacris*. (*m*) *Biacetabulum*. (*n*) *Caryophyllaeus*. (*o*) *Hunterella*. (*p*) *Monobothrium*. (*q*) *Khawia* (=*Bothrioscolex*). (*r*) *Wenyonia*. (*s*) *Caryoaustralus*. (*t*) *Khawia*. (*u*) *Caryophyllaeides*. Not to scale. Source: reproduced from Mackiewicz, 1994, with permission from CAB International.

cestodes and the plerocercoids of some pseudophyllideans that will concern us later (sections 10.4, 10.5.4).

Progenesis in caryophyllideans ranges from parasites such as *Biacetabulum*, in which development of the genitalia in the procercoid is suspended early, to *Archigetes* spp., in which the procercoid is fully mature and able to assemble eggs (see references in Williams and Jones, 1994). A consequence of this trend is a progressive shortening of the period of time between ingestion by the fish and onset of egg laying, a development that would be advantageous if for some reason contact between fish host and oligochaetes became brief. A further step is for the progenetic procercoid to release its eggs directly into the environment by rupture of the body wall of the oligochaete (see review by Mackiewicz, 1981). This may follow the natural death of the host or may be promoted in some way by the procercoid. In most *Archigetes* spp. some procercoids remain immature until ingestion by a fish

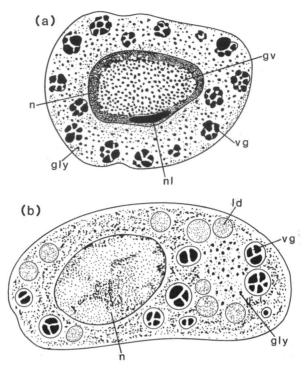

Figure 10.4 Sketches of transmission electron microscope sections through mature vitelline cells of (*a*) the caryophyllidean cestode *Glaridacris catostomi* and (*b*) the pseudophyllidean cestode *Bothriocephalus clavibothrium*. gly = glycogen particles in the cytoplasm; gv = intranuclear glycogen vacuole; ld = lipid droplet; n = nucleus; nl = nucleolus; vg = vitelline globules. Source: reproduced from Mackiewicz, 1981, with permission from Academic Press.

and others become ovigerous in the oligochaete intermediate host. There is a recognizable trend here. In *A. iowensis*, progenetic development is uncommon, while *A. limnodrili* and *A. sieboldi* are able to sustain a cycle in tubificids while retaining the ability to utilize fishes. *A. cryptobothrium* has been found only in tubificids and may have lost the fish host.

Another unique feature of caryophyllideans is the discovery by Mackiewicz and co-workers (see Mackiewicz, 1981) of triploid races that maintain their viability by reproducing parthenogenetically. Three of the four polyploid species recorded by Mackiewicz, which belong to the genera *Glaridacris* and *Isoglaridacris*, have diploid as well as triploid forms. The full significance of this phenomenon remains to be recognized.

10.3　BOOSTING EGG OUTPUT – REPLICATION OF GENITALIA

Many of the caryophyllideans have elongated bodies (Fig. 10.2*a*) and this is an expected development in a parasite that absorbs nutrients through the surface and inhabits a tubular environment in which such nutrients are available over considerable distances. The superabundance of food and the stability of the intestinal environment will remove constraints on egg output and this will be maximized within the limits imposed by the single reproductive unit (see Jennings and Calow,

1975). There will be a limit because of the bottleneck of the ootype (see Llewellyn, 1965). A minimum period of time is required for the assembly of a single egg in the ootype and once this limit is reached an increase in egg output by producing more egg components (oocytes, vitelline cells) is futile. By analogy with motor car manufacture, once the production line is working at full speed, the provision of more components like engines, bodies and wheels will not improve car output. The only way to do this is to have more production lines and this is precisely the evolutionary step taken by early cestodes. Their already elongated bodies provided space for the serial repetition of reproductive units (proglottization), each with its own ootype.

This is not necessarily an evolutionary step of any great complexity since replication of the genitalia could have come about by a relatively simple mutation; indeed similar developmental 'errors' have been recorded in tapeworms reared in the 'wrong' hosts (see Llewellyn, 1965). Modern survivors of this step forward in cestode evolution are the spathebothrideans and the pseudophyllideans, in which many sets of reproductive organs develop more or less simultaneously in an elongated body.

The effectiveness of multiplication of genitalia as a strategy for boosting egg output is dramatically illustrated by observations made on the pseudophyllidean cestode *Diphyllobothrium dendriticum* by Pronin, Timoshenko and Sanzhieva (1989). Sexually mature worms exceed 70 cm in length and contain over 600 reproductive units. Mean egg output per worm per day was estimated at about 10.4 million; in other words a single worm produces on average 12 eggs per second, compared with about 2 eggs per hour in the monogenean *Entobdella soleae*, which possesses only one ootype (section 4.8). However, the egg output of *D. dendriticum* is probably surpassed by pseudophyllideans belonging to the genus *Diplogonoporus*, in which each reproductive unit (proglottis) contains not one but two sets of reproductive organs (two ootypes) situated side by side. A similar development occurs in some cyclophyllideans (sections 11.4, 11.6).

10.4 THE SPATHEBOTHRIDEANS

Cyathocephalus truncatus is a representative of this small group of tapeworms. It is a parasite of whitefish (*Coregonus*) and trout (*Salmo*) in northern Europe and possesses 20–45 sets of reproductive organs in an undivided strobila (Fig. 10.5*a*). There are no internal partitions between the reproductive units (Fig. 10.5*b*) and no prominent external projections corresponding with them (a condition described as acraspedote, see section 10.5.5). The uteri have openings in the midline (Fig. 10.5*b*), and the operculate eggs are laid in the fish's intestine and pass out with the faeces. The eggs do not hatch until eaten by amphipod crustaceans, in the body cavity of which the procercoid develops. Progenetic development of the procercoid has been recorded (e.g. by Amin, 1978).

10.5 THE PSEUDOPHYLLIDEANS

10.5.1 Pelagic invertebrates as intermediate hosts

The pseudophyllideans have developed a different way of promoting infection of the intermediate host by the oncosphere. Unlike the eggs of caryophyllideans and

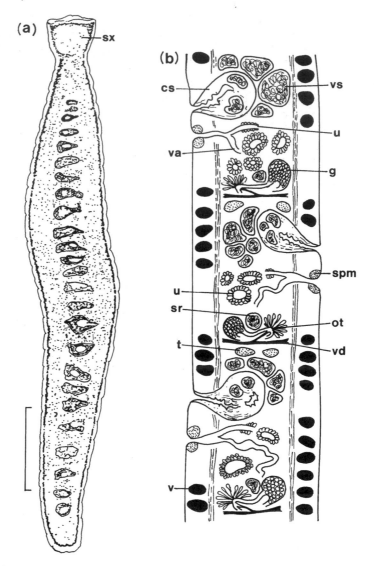

Figure 10.5 The spathebothridean cestode *Cyathocephalus truncatus*. (*a*) Whole adult parasite, showing numerous sets of reproductive organs in an undivided body. (*b*) Sagittal section through three sets of reproductive organs in part of the undivided body. Note that reproductive openings alternate from dorsal to ventral surface. cs = cirrus sac; g = germarium; ot = ootype and associated glands; spm = sphincter muscle surrounding female gonopore; sr = seminal receptacle; sx = scolex; t = testis; u = uterus; v = vitellarium; va = vagina; vd = vitelline duct; vs = vas deferens. Scale bar: (*a*) 2 mm. Source: (*a*) reproduced from Wisniewski, 1932; (*b*) based on Fuhrmann, 1931.

spathebothrideans, which fail to hatch until they are eaten by the intermediate host, those of most pseudophyllideans contain a ciliated larva or coracidium, which escapes into open water through an opercular opening in the tanned eggshell (Fig. 10.6). An active coracidium is more likely to attract the attention of a predatory invertebrate than an inert egg and this opens the way for exploitation of pelagic invertebrates, especially copepod crustaceans, as intermediate hosts. After the coracidium is eaten the ciliated epidermis is shed and the hexacanth within enters the body cavity of the pelagic crustacean and gives rise to a procercoid.

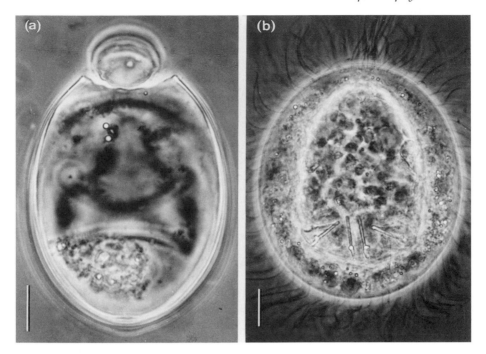

Figure 10.6 (*a*) An empty eggshell and (*b*) a freshly hatched living coracidium of the pseudophyllidean cestode *Triaenophorus nodulosus*. Phase contrast microscopy. Note the operculum in (*a*). Scale bars: 10 μm.

10.5.2 The three-host life cycle

There is another innovation in the pseudophyllideans and that is the recruitment of a second intermediate host. Caryophyllideans and spathebothrideans are parasites of what might be called microcarnivorous fishes, i.e. fishes that feed on small invertebrates. However, many fishes became larger and more powerful and shifted their diet to include the smaller microcarnivorous fishes. It might be supposed that these macrocarnivorous (piscivorous) fishes would escape the attention of cestodes, since the small invertebrate hosts of the juvenile tapeworms were no longer part of their diet, but young macrocarnivores are invariably microcarnivorous, preying on small invertebrates, and will be exposed to infection by tapeworms during this early phase of their lives. So, the infection window for newly evolved fish-eating fishes may at first have been restricted to their small microcarnivorous young, but the pseudophyllideans adapted to infect adult macrocarnivorous fishes directly, by ascending the ladder of the food chain.

The life cycles of *Bothriocephalus* spp. illustrate how this could have happened. *B. acheilognathi* has a two-host life cycle; it is a parasite of freshwater microcarnivorous (planktonivorous) fishes infected by eating copepods containing procercoids. The copepods, in turn, are infected by feeding on free-swimming coracidia. *B. gregarius* is a marine parasite of the turbot (*Psetta maxima* = *Scophthalmus maximus*), which is planktonivorous up to a length of about 11 cm and there after turns to a diet of fishes, especially gobies (Robert *et al.*, 1988). Young planktonivorous turbot ingest infected copepods and acquire adult *B. gregarius* (Fig. 10.7), but there is also a second option, an alternative route that leads to the piscivorous adult turbot. If a goby eats an infected copepod, the procercoid attaches itself within the

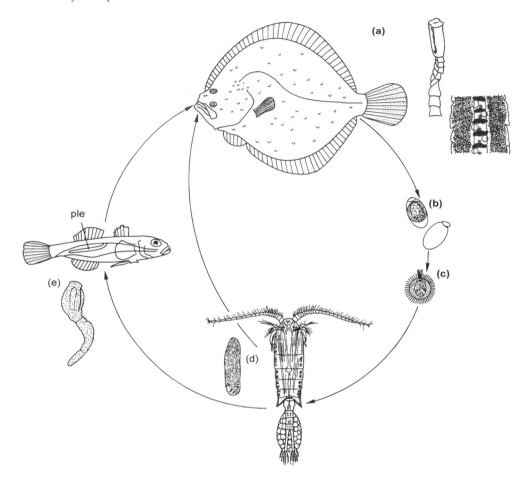

Figure 10.7   The life cycle of a marine pseudophyllidean cestode of the genus *Bothriocephalus*. The adult tapeworm (*a*) inhabits the intestine of the turbot. Tanned eggs (*b*), voided in the faeces, liberate a coracidium (*c*); this is eaten by a copepod and the oncosphere enters the body cavity and develops into a procercoid (*d*). Planktonivorous turbot are infected by eating infected copepods, but procercoids eaten by other plankton-feeding fishes such as gobies do not become adult. They remain in the intestine as immature plerocercoids (*e*). Piscivorous turbot are infected by eating infected gobies. ple = plerocercoid.

goby's gut, where it undergoes some development but fails to become adult. Development of this plerocercoid remains suspended until the goby is eaten by a large turbot, in the gut of which the plerocercoid resumes development and reaches maturity. These two options, a two-host and a three-host cycle, exist side by side, and the intermediate host, the goby, is a facultative one. The possibility that *B. gregarius* might achieve maturity in the goby, thereby acquiring a new definitive host, has been examined by mathematical modelling, but these simulations suggest that such a development would reduce the reproductive success of the parasite (Morand, Robert and Connors, 1995).

Triaenophorus nodulosus, a parasite of the freshwater pike (*Esox lucius*), has taken this one stage further (Fig. 10.8). When the procercoid is eaten by a small microcarnivorous fish such as a loach (*Noemacheilus barbatulus*), it no longer remains in the gut. Perhaps as a result of the persistence and proliferation of the glands used by the coracidium to breach the gut wall of the copepod intermediate host, the

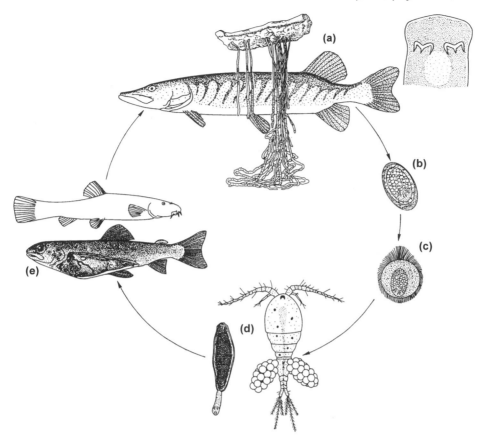

Figure 10.8 The life cycle of the pseudophyllidean cestode *Triaenophorus nodulosus*. (*a*) A piece of intestine of the definitive host (the pike, *Esox lucius*) with adult tapeworms attached; the scolex is also shown. Tanned eggs (*b*) pass out in the faeces and liberate a coracidium (*c*), which is eaten by a pelagic copepod. The hexacanth larva enters the haemocoel and develops into a procercoid (*d*). The infected copepod is eaten by a planktonivorous fish (e.g. trout or loach). The procercoid leaves the gut and develops into a plerocercoid (*e*), usually in the liver. Pike can acquire adult worms only by eating fishes with plerocercoids, not by eating infected copepods. Source: sketches of adult tapeworm and plerocercoids *in situ* redrawn from Petrushevski and Shulman, 1958.

procercoid passes through the fish's gut wall and enters the liver. Here it grows, giving rise to a plerocercoid with an immature body but a well developed scolex. Growth is continued and sexual maturity achieved when the parasitized loach is eaten by a pike. However, the invasive ability of the procercoid is not the only new feature. If an infected copepod is eaten by a young microcarnivorous pike, the procercoid does not reach maturity in the intestine but responds as it does in other microcarnivorous fishes like the loach – it enters the liver and becomes a plerocercoid. Thus, one of the two options open to *B. gregarius* is closed to *T. nodulosus*; a three-host life cycle is the only option and the second intermediate host is now obligatory.

Plerocercoids in young pike are not necessarily lost; adult pike have cannibalistic tendencies towards young of their own species.

The acquisition of an obligatory fish second intermediate host and the colonization of fish-eating fishes opened up new possibilities for the pseudophyllideans.

Many land vertebrates, especially mammals and birds, became fish eaters. The elevated body temperatures of these warm-blooded hosts would have provided an added boost to the reproductive output of tapeworms but, because such high temperatures permit more efficient and rapid digestion, this would create special difficulties for ingested plerocercoids, as would the teeth and gizzards of these vertebrates. Nevertheless, some pseudophyllideans have succeeded in switching final hosts and have established themselves in fish-eating animals as diverse as birds, bears, cats, otters, seals and man.

The plerocercoids that are eaten by warm-blooded vertebrates seem singularly ill-equipped to survive ingestion and digestion in their hosts when compared with cyclophyllidean cestodes (section 9.7), but survive they do and their ways of coping remain a challenge to be investigated. The plerocercoids of pseudophyllideans such as *Schistocephalus* (see below) have no physical protection in the form of a cyst wall (see Fig. 10.10), and their survival is reminiscent of the ability of 'naked' migrating juveniles of the monogenean *Pseudodiplorchis americanus* to survive passage through the gut of their toad host (section 7.1.4(b)). Other pseudophyllidean plerocercoids are enclosed in a cyst wall, possibly of host origin, but there must be some doubt about the protective capacity of this wrapping because Archer and Hopkins (1958) found that 44% of plerocercoids of *Diphyllobothrium* sp. stripped of their cyst wall became established in rats after administration by stomach tube.

10.5.3   Host specificity

Most adult tapeworms, like monogeneans, are closely adapted to life in their definitive host and display a narrow host specificity. However, it is not uncommon for relaxation of this host specificity to occur. For example, *Bothriocephalus acheilognathi* mentioned above is remarkably non-specific, both to its microcarnivorous fish final host (it has been reported from at least 40 different fishes) and to its copepod intermediate host. A similar relaxation and broadening of specificity occasionally occurs in other platyhelminth parasites, such as in the capsalid monogenean *Neobenedenia melleni* (section 5.3), but how some parasites are able to achieve this while other close relatives remain strictly host-specific is puzzling.

A similar broadening of specificity to the final host has been claimed for *Diphyllobothrium latum*, but, according to Roberts and Janovy (1996), many of these records are misidentifications. Nevertheless, some of the pseudophyllideans that have colonized birds display a genuine lack of host specificity, for which a possible explanation will be considered below.

10.5.4   Tapeworms of fish-eating birds

A life cycle involving more than one host is open to the danger that evolutionary changes in the behaviour of one of the hosts may weaken a link in the life cycle chain. This is particularly so when an obligatory second intermediate host is acquired, extending the life cycle to include three levels in the food chain. Life cycles involving fish-eating birds may be particularly vulnerable, because birds have great mobility and may spend relatively short periods of time in contact with the aquatic environment inhabited by the first and second intermediate hosts.

Meyer (1972) studied events in a lake in Maine (USA) where the gull *Larus argentatus* feeds on the smelt *Osmerus mordax*. Although the gulls and the smelt coexist for only part of the year (spring and summer), their life cycles are so interrelated that propagation of the gull tapeworm *Diphyllobothrium sebago* is ensured. The smelt spawn in late April/early May and then die. The arrival of the gulls in the area and their period of reproductive activity coincide with this die-off and the birds feed on the dead fish. The parasite does not appear to contribute significantly to the death of spent fish, since unparasitized fish also die, but the dead fishes are easy targets for the gulls, greatly enhancing transmission of the parasite, and the birds become heavily infected with worms. From early July to November, when ice covers the lake and the birds leave, parasite eggs will enter the lake, releasing coracidia to infect copepods.

In other diphyllobothriid pseudophyllideans there have been some interesting developments which Dubinina (1964) attributed to the acquisition of migratory behaviour by fish-eating birds. The fish-inhabiting plerocercoids undergo varying degrees of progenetic development. In *Ligula intestinalis* (Fig. 10.9), the plerocercoid is well differentiated with the primordia of the genitalia present. The parasite requires only 67–72 h to mature in the bird intestine and will mature *in vitro* in a

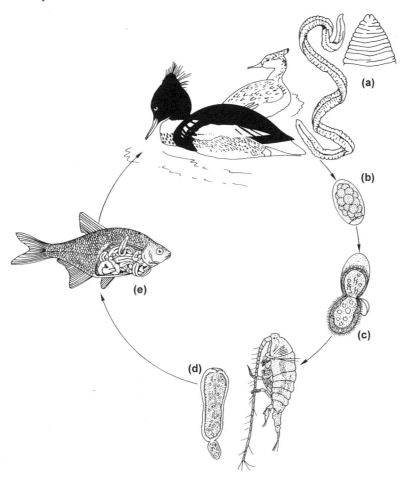

Figure 10.9 The life cycle of the pseudophyllidean cestode *Ligula intestinalis*. Adult (*a*) in fish-eating birds. Egg (*b*) enters water in faeces and liberates a coracidium (*c*), which is eaten by a copepod. Procercoid (*d*) in copepod eaten by a planktonivorous fish. Plerocercoids (*e*) in body cavity eaten by a bird. Source: based on Wardle and McLeod, 1952; Joyeux and Baer, 1961 and Petrushevski and Shulman, 1958.

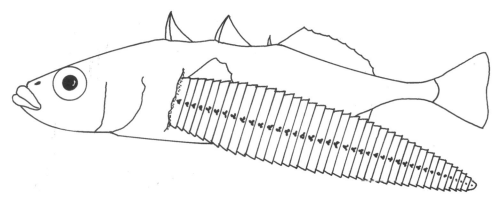

Figure 10.10   Stickleback with body cavity partly opened to permit exit of a plerocercoid of *Schistocephalus solidus*. Source: drawn from a photograph by Hopkins and Smyth, 1951.

non-nutrient medium at 40°C, although the percentage of fertile eggs is low (references in review by Smyth, 1962). In other words the parasite makes little nutritional demand on the bird and high levels of glycogen stored in the plerocercoid provide most of the reserves needed for maturation. Provided that the bird remains in the area for 3 days or so, the eggs of the parasite are likely to be returned to the aquatic environment inhabited by the intermediate hosts.

Schistocephalus solidus, with plerocercoids in the body cavity of the stickleback *Gasterosteus aculeatus* (Fig. 10.10), takes this further. Half of the body weight of the plerocercoid is glycogen and it matures in only 36 h in the bird, producing viable eggs *in vitro* at 40°C without any nutrient provision. In comparison, the plerocercoid of *Diphyllobothrium* sp. is small, undifferentiated except for the scolex and has little food reserves; it requires several days to mature in the definitive host and in non-nutrient media undergoes mitosis for a few hours and then degenerates.

Thus, in *Schistocephalus*, metabolic dependency has shifted from the bird, which merely acts as an incubator, to the fish intermediate host. Corresponding with this shift, host specificity to the definitive host is lost, with the parasite able to mature experimentally in a range of birds, including grain-eaters, and in mammals, while a new strict host specificity to the fish host has developed, with the plerocercoid unable to use even the closely related stickleback *Pungitius pungitius*. Such a link between metabolic dependency and host specificity is perhaps to be expected, but this makes it more difficult to understand how parasites that are metabolically dependent, like *Bothriocephalus acheilognathi* (see above), can exploit a range of hosts.

Hopkins, Law and Threadgold (1978) and Threadgold and Hopkins (1981) incubated plerocercoids of *Schistocephalus* and *Ligula* with electron-opaque molecules such as ruthenium red and claimed to have found this material in membrane-bound vesicles throughout the tegument and beneath the tegumental basal lamina (see also section 9.3.2(b)). They visualized two possible roles for this phenomenon: a nutritional role supporting the rapid and prolonged growth in the body cavities of their fish intermediate hosts and a defensive role against the host's immune responses by continual pinocytotic internalization of surface membrane and attached antibody. However, Conradt and Peters (1989) failed to confirm the work of Hopkins and co-workers, and, *in vitro*, found no evidence that pinocytosis occurs in *S. solidus*. In fact, Conradt and Schmidt (1992) have suggested that traffic may occur in the opposite direction. Using a fixative for electron microscopy containing osmium tetroxide and potassium ferrocyanide, they found that the apical surface of the tegument of the plerocercoid of *L. intestinalis* is covered by two closely apposed lipid bilayers. They raised the possibility that this surface membrane complex may be maintained by fusion with tegumentary vesicles, which also have two enclosing membranes, and that the double membrane may play a part in protecting the parasite from the host's immune response. Adult parasites from the gut of the definitive host display a single surface membrane. A double membrane covers the tegument of the blood-dwelling schistosome digeneans and is thought to be a significant feature in opposing host attack (section 18.3.9).

Polzer and Conradt (1994) compared proteases from the procercoid, plerocercoid and adult of *S. solidus* and found substantial differences: a chymotrypsin-like protease capable of digesting collagen in the procercoid and leucine aminopeptidases in the isolated syncytial teguments of the plerocercoids and adults. They suggested that the chymotrypsin-like protease may serve for penetration of the stickleback gut wall by the procercoid, while the aminopeptidases may contribute to worm nutrition by degrading oligopeptides at the tegumental surface.

Schistocephalus solidus may contribute almost a quarter of the total body weight of an infected stickleback and usually greatly distends the fish's abdomen (Fig. 10.10). It is to be expected that such a parasite load would have detrimental effects on the performance of the fish and that these effects might increase the fish's susceptibility to predation by birds. According to McPhail and Peacock (1983) many of the alleged detrimental effects have not been well documented, but they thought it likely that decreased swimming ability combined with increased conspicuousness was likely to make them easier prey for birds. Lester (1971) claimed that parasitized fishes inhabited shallower water and that this decreased their availability to piscivorous fishes, which do not enter shallow water and do not serve as definitive hosts, while, at the same time, increasing their availability to piscivorous birds.

Tierney, Huntingford and Crompton (1993) confirmed earlier indications that plerocercoids suppress the stickleback's antipredator responses, and found that fish with plerocercoids resume foraging or sculling with the pectoral fins (features of an undisturbed stickleback) more quickly after a simulated attack with a heron model. These effects on host behaviour appear to be elicited only by infective plerocercoids; small, uninfective plerocercoids, which cannot establish themselves and mature in a bird, are not associated with changes in stickleback behaviour. Thus, behavioural effects of the parasite that promote capture of the fish by the bird definitive host appear to have been delayed by natural selection until the parasite is mature enough to survive this predation.

LoBue and Bell (1993) discovered Alaskan populations of the stickleback *Gasterosteus aculeatus* containing conspicuous white individuals with depigmented skin and hyperpigmentation of the eyes. These white fishes were observed swimming at the surface and were infected with *S. solidus*. Evidence gathered by LoBue and Bell suggested that these features represent manipulation of the host by the tapeworm, rather than incidental consequences of infection.

In a Canadian population of infected sticklebacks with a 1-year life span, McPhail and Peacock (1983) found that the major adverse effects of *S. solidus* manifested themselves in the autumn, when stickleback breeding is over, and pointed out that this coincided with peak predator abundance (the autumnal migration).

Plerocercoids of *L. intestinalis* are reported to interfere with schooling behaviour of minnows. This may increase the susceptibility of the parasitized fishes to predation by birds, since schooling is a primary antipredator response of minnows (Barber and Huntingford, 1996). A similar phenomenon in the digenean *Crassiphiala bulboglossa*, which also infects fishes and their bird predators, may be another remarkable example of convergence (section 17.7.1).

10.5.5   Growth and maturation

In many pseudophyllideans all the reproductive units reach maturity at about the same time and the organism has a limited life. Adult *Schistocephalus solidus* seldom spend more than 3–4 days in the gut of their bird host (Hopkins and Smyth, 1951), but not all pseudophyllideans are short-lived and *Diphyllobothrium latum* in man is known to be capable of surviving for as long as 6 years (see Grove, 1990). The reproductive units appear to have a limited life, so that increased longevity can be achieved only by the development of a proliferative zone capable of generating fresh sets of reproductive organs. Like its short-lived relatives, *D. latum* first develops a primary strobila with repeated sets of reproductive organs of the same age, but the parasite then begins to produce secondary reproductive units. According to Freeman (1973), the 'neck' region immediately behind the scolex is the source of the secondary proglottides, a development that may have arisen independently in the apolytic cestodes (section 10.8.1). However, there are reports of tertiary reproductive units developing by some kind of transverse division of existing proglottides (see Freeman, 1973), illustrating the complexities of growth patterns in pseudophyllideans and how poorly we understand them.

For the duration of their limited reproductive lives, the units of *D. latum* shed their tanned eggs into the gut lumen *via* permanent uterine pores and then, according to Wardle and Green (1941), the exhausted proglottides shrink to one-half or one-third of their former weight, begin to break up and become detached from the strobila (= pseudoapolysis). According to Wardle (1935), strips of exhausted worm 2–3 feet (60–90 cm) in length may pass from humans about 5 weeks after initial infection and a similar expulsion of exhausted strips occurs roughly at monthly intervals.

A remarkable phenomenon was reported by Pronin *et al.* (1989) in the bird tapeworm *Diphyllobothrium dendriticum*, which has a relatively short reproductive life (7–10 days). They claimed that, in multiple infections, only one worm is mature at any one time and that as the egg-laying individual comes to the end of its short life another individual becomes mature. The process is repeated, with the consequence that there is considerable difference between individuals in their longevity, although long-lived worms are immature for all but the last week or so of their lives. It is not known how control of maturation and communication between the worms are effected – Pronin *et al.* suggested that mature worms secrete substances that inhibit maturation in other individuals.

In many pseudophyllideans we encounter for the first time an external 'segmentation' superimposed upon the repetitive sequence of reproductive units (Fig. 10.10). This manifests itself as the 'craspedote' condition, in which the posterior border of each proglottis extends beyond the anterior border of the next, like the eaves of a house. This is a puzzling development. In cyclophyllideans it is often associated with the habit of shedding egg-filled proglottides (apolysis), but pseudophyllideans either do not shed any proglottides, as in *Schistocephalus*, or jettison lengths of worm containing many proglottides, as in *D. latum*. However, the craspedote condition could have another function. In order to maintain their station in the gut, tapeworms must counteract intestinal peristalsis, which will exert a continual backward drag on the worm, especially if the strobila is long and if the host has enhanced intestinal mobility. We have already noted that the strobila of the cyclophyllidean *H. diminuta* relieves the strain on the scolex and 'neck' and maintains position by performing its own peristaltic movements, which counteract the drift due to host peristalsis (section 9.3.5). The repetitively arranged, backwardly projecting borders of the craspedote worm seem well suited for pushing against the gut contents and gut wall and enhancing like a ratchet the effectiveness of the locomotory movements. In fact, the forward locomotion of newly established specimens of *Diphyllobothrium* sp. exceeds the displacement by host peristalsis in the rat, and the young worm migrates in an anterior direction (Archer and Hopkins, 1958; see also section 9.3.5).

10.5.6 The scolex

The scolex of typical pseudophyllideans is a bilaterally symmetrical structure, usually with two diametrically opposite, shallow grooves or bothria (Fig. 10.11*a*, *b*). The effectiveness of this organ for attachment seems likely to be little better than that of the caryophyllidean scolex (section 10.2), but it is nevertheless capable of anchoring large tapeworms such as *D. latum* in the intestine of warm-blooded vertebrates, in which we might expect relatively powerful and frequent gut movements. That we may have underestimated the scoleces of these parasites is suggested by the work of Andersen (1975), who observed that each bothrium of *Diphyllobothrium* spp. clasps one or two intestinal villi (Fig. 10.11*c*) and that in *D. latum* and *D. ditremum* there is a layer of material (cement?) between the bothrial surfaces and the host tissue. Each shallow bothrium of *Triaenophorus* is supplemented by two three-pronged hooks (Fig. 10.8).

10.5.7 *Diphyllobothrium latum*

Diphyllobothrium latum, the broad tapeworm, is the third large tapeworm that we have encountered in man. Specimens may be significantly larger than *Taenia* spp., reaching 10 m in length (Fig. 10.12). It was not until the early part of the 17th century that it was realized that humans harboured more than one species of large tapeworm and recognizable descriptions appeared of the worms we now know as *Taenia* and *Diphyllobothrium* (references in Grove, 1990). Early distinctions were based on the shapes of the proglottides and the positions of the genital openings – lateral in *Taenia* and median in *Diphyllobothrium* – although, at the time, it was not appreciated that these were openings of the reproductive system. The first detailed description of the head of *D. latum* was given by Bonnet in 1777.

In spite of the fact that one of the first tapeworm life cycles to be partly elucidated was that of *Schistocephalus* by Abildgaard in 1790 (section 8.3) and that fish had been established experimentally as intermediate hosts, it was a long time before the connection between consumption of fish and human infection with *D. latum* was appreciated. Pointers to this connection were the observations that people dwelling near rivers and lakes harboured the tapeworm more commonly than those living elsewhere, and that orthodox Jews carried *D. latum* but never ate raw meat. It was not until 1881 that the fish connection was experimentally demonstrated. Braun in

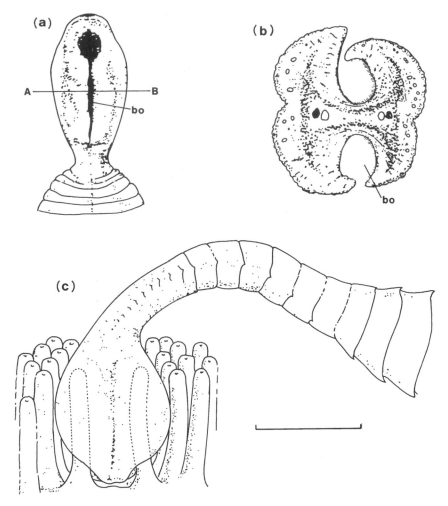

Figure 10.11   The scolex of *Diphyllobothrium* sp. (*a*) in lateral view; (*b*) in transverse section in the plane AB; (*c*) attached to the gut. bo = bothrium. Scale bar: (*c*) 1 mm. Source: reproduced from Andersen, 1975, with permission from the Australian Society for Parasitology.

Estonia succeeded in raising the parasite by feeding plerocercoids from pike to a dog and then in 1882 repeated the experiment successfully with three medical students as recipients of plerocercoids.

The recognition that copepods served as the first intermediate hosts was even more elusive. It was not until 1917 that Janicki found very tiny worms free in the stomach of perch that had been feeding on copepods and, in the same year, Rosen successfully infected the copepods *Cyclops strenuus* and *Diaptomus gracilis* by exposing them to coracidia.

Foci of human infection with *D. latum* occur throughout northern Europe and Asia, particularly in Scandinavia, the Baltic countries, Russia and Japan. Bauer (1958) gave a detailed account of the parasite in the former USSR, particularly in the regions of Leningrad (St Petersburg) and in Karelia, where human infection was common. The plerocercoids measured about 1 cm in length and were located in the muscle and viscera of several freshwater fishes, including pike (*Esox lucius*). In the

Figure 10.12   Dr S. Kamegai, Director of the Meguro Parasitological Museum, Tokyo, with a specimen of *Diphyllobothrium* sp. removed from a human patient. Source: photograph courtesy of Dr K. Ogawa.

Gulf of Finland, 88.8% of pike were infected, harbouring per fish an average of 152 (41–641) plerocercoids, about 9% of which were in the muscles and 91% in the viscera. Some 90% of the fish had worms in the gonads and this provided an important source of infection because of the common practice of eating uncooked, slightly salted roe.

Bauer described some interesting local recipes that enhance transmission of the parasite. Karelian fish pie is prepared by coating small, raw, ungutted fish with pastry, which prevents thorough cooking of the fish. Volga fishermen prepare *zharekhi* by impaling a whole fish on a vertical stick standing close to a fire; parts of the fish, especially the viscera, remain undercooked. Fishermen at Lake Ladoga split fish along the backbone into two fillets, joined by the abdominal muscles, add salt, place on the embers of a fire and cover with a wet tarpaulin. This extinguishes the fire, leaving the fish to be 'cooked' by the hot air.

Sporadic cases of human infection were first noted in North America, particularly in the Great Lakes region, in the latter half of the 19th century (see Grove, 1990). Over the years more cases were reported and it gradually became clear that *D. latum* was an immigrant to the New World, introduced probably from the Baltic region of Europe inside human émigrés. Suitable first and second intermediate hosts awaited the adult parasites and they became well established. It is interesting that in upper Michigan, female Jews became important hosts because of their habit of tasting raw fish as a test of their flavouring skills. Human emigration has also played an important part in the spread of digeneans of the genus *Schistosoma* (section 18.2).

With one exception, clinical manifestations of infection with adult *D. latum* are little more than discomforting. The exception is the tendency for a small proportion of infected patients to develop pernicious anaemia. An important factor, but not the only one, in the development of this condition, is the ability of the tapeworm to compete successfully with the host for dietary vitamin B_{12}, both in its free state in the gut lumen and bound to an intrinsic factor secreted by the gut (see Grove, 1990). In fact, Grove reported a transient fashion for treating anaemic patients with an extract of dried fish tapeworm, which contained some 50 times the level of vitamin B_{12} found in *T. saginata*. The reason for the accumulation of the vitamin by *D. latum* is unknown.

10.5.8 Spirometra

A discovery made by Mueller in 1933 (see Mueller, 1974) was to have important consequences. He found a pseudophyllidean tapeworm in the intestine of cats from the Syracuse area of New York State and, since at that time there was much concern about the recent appearance of the broad tapeworm of man in the western Great Lakes region, he at first regarded the cat parasite as *Diphyllobothrium latum* (see Mueller, 1966). He soon realized that this identification was incorrect and the parasite subsequently became known as *Spirometra mansonoides*. Nevertheless, Schmidt, as recently as 1986, placed the parasite in the genus *Diphyllobothrium*, in spite of the surprising findings about its biology (see below) made by Mueller in the decades following its discovery.

The plerocercoids of pseudophyllideans like *Diphyllobothrium*, *Ligula* and *Schistocephalus* are fish parasites, but fishes appear to be the only vertebrates that *Spirometra* cannot use as second intermediate hosts (Mueller, 1966). Except for fish and birds, plerocercoids (so-called spargana) of *S. mansonoides* have been found naturally in all groups of vertebrates and even birds can be infected experimentally. This freedom from the constraints imposed by the immune system of the host is achieved by only a handful of parasitic platyhelminths (section 10.5.3) and has particularly close parallels with digenean mesocercariae (section 17.7.3). However, broad specificity is not maintained by the adult parasites, which are restricted to cats and bobcats and occasionally dogs or raccoons.

There is currently considerable interest in how this broad specificity is achieved. Among the proteases secreted by the spargana of *S. mansonoides* that facilitate its migration through host tissues, one has been identified that is capable of splitting IgG *in vitro* (Kong *et al.*, 1994), and this may be part of the immune evasion mechanism.

In the Syracuse area the spargana of *S. mansonoides* are found most commonly in water snakes. The finding of a sparganum in the snake *Coluber constrictor* in Louisiana led Mueller to suggest that fish may not have been entirely eliminated from the life cycle – he proposed that these snakes become infected by eating small fishes with stomachs packed with plankton, including copepods infected with procercoids of *S. mansonoides*.

If the sparganum is eaten by a host in which it is unable to mature, it penetrates the wall of the intestine of the paratenic (transport) host and re-establishes itself as a tissue parasite. Spargana continue to grow in a suitable intermediate host, reaching at the end of a year a length of 400–500 mm (Mueller, 1974). A chronically infected laboratory mouse may be as much as 10% worms by weight. The scolex of the sparganum has an unlimited capacity for regeneration, if removed and transferred to a new host – Mueller maintained some heads in this way for 16 years. However, the really surprising feature is that the greatly elongated body of the sparganum is cast off by the scolex in the definitive host and makes no contribution to the adult worm – the adult strobila is a new development. In fact the plerocercoid is infective in the mouse after only 4.5 days, when the worm is a mere 1 mm in length, so it is a mystery as to why the plerocercoid produces such an enormous body destined only for rejection.

Humans claim their place in the life cycle of *Spirometra* with a creepy tale more typical of the pen of Edgar Allen Poe than the exploits of the eminent parasitologist Patrick Manson. According to Grove (1990), Manson was in China in 1881, looking for evidence to support his view that elephantiasis is caused by adult filarial worms (the nematode *Wuchereria bancrofti*). Unfortunately, the local populace were strongly opposed to post-mortem examinations and Manson was forced to carry out an autopsy in secret on the body of a 34-year-old man suffering from the disease. It is claimed that the autopsy was held at the dead of night in the Chinese cemetery, with the aid of a flickering candle and a trusted servant. What Manson did discover were ribbon-like worms (spargana), some 30 cm long and 5 mm wide, in the adipose tissue and elsewhere in the body. Thus, spargana are capable of using humans as intermediate hosts and cases have also been reported from North America. Drinking water containing infected copepods is the most likely source of human infection, but ingestion of spargana in the raw flesh of chickens has been implicated in the Far East. There is also a practice in this part of the world of applying a dismembered frog as a poultice to the eyes or other parts of the body, and the penetrative abilities of the spargana are such that they are capable of entering the human body from the poultice.

Sparganum proliferum is a rare aberrant form known only as a tissue parasite of humans. What is ordinarily a benign parasite becomes a malignant and invasive agent, continuously branching and budding and producing an elephantiasis-like enlargement of the affected trunk region or limb (Mueller, 1966). The abnormality of the organism is confirmed by its disturbed organization and symmetry and lack of a true scolex, and it fails to mature when transferred to presumptive definitive hosts such as dogs or cats.

The involvement of humans leads us to an intriguing feature of sparganum infections that has important implications for future research. Rodents infected with spargana of *S. mansonoides* undergo accelerated growth which is due to the production by the parasite of a plerocercoid growth factor (PGF; see review by Phares, 1987). Even more remarkable, PGF resembles human growth hormone (hGH) more closely than other growth hormones. Although there are differences between PGF and hGH, the pituitary recognizes PGF and decreases its own production of growth hormone. These similarities led Phares to suggest that the parasite gene for PGF is derived from a human gene (for hGH) that has been sequestered by the tapeworm during its evolution (see also section 20.3). Phares pointed out that such exchanges of genetic material are known to occur and that viruses are the most likely intermediaries in such a transfer.

More intriguing is the finding by Phares and Kubik (1996) that PGF not only has growth promoting properties but also is a cysteine protease. Thus, PGF is a single protein that stimulates growth while facilitating tissue penetration by the plerocercoid. Phares and Kubik suggested that the proteolytic activity of PGF may also contribute to the ability of plerocercoids to evade immune surveillance in their great range of potential intermediate hosts. The significance for the parasite of enhanced host growth is less obvious.

10.6   THE TRYPANORHYNCHS (TETRARHYNCHS)

Most adult trypanorhynchs inhabit the spiral intestine of elasmobranch fishes and knowledge of their life cycles and biology is limited. Nevertheless, an affinity with the pseudophyllideans is indicated by some of the life cycles that have been unravelled. The eggs of *Lacistorhynchus* spp. have tanned shells and are operculate, liberating a ciliated coracidium that is eaten by a copepod (Sakanari and Moser, 1985a, 1989). The procercoid of *L. tenuis* in the copepod develops into a plerocercoid (plerocercus of some authors) when fed experimentally to mosquito fish, *Gambusia affinis*, and plerocercoids were found naturally in white croaker, *Genyonemus lineatus* (see Sakanari and Moser, 1985b). Adult tapeworms were obtained from the leopard shark, *Triakis semifasciatus*, collected from Monterey Bay, California, and sharks were infected in the laboratory by feeding them with infected *Gambusia* and *Genyonemus*.

It is perhaps not well known that Dollfus (1966) reported immature trypanorhynchs (essentially a scolex and attached proliferative region, sometimes with one or two immature proglottides; Fig. 10.13) free in plankton hauls taken off the coast of Senegal. This was no accidental occurrence, since more than 50 individuals belonging to five species were found in $1\,cm^3$. Dollfus surmised that these were derived from plerocerci in unknown second intermediate hosts, from which they escaped into the water. He envisaged a route to the definitive elasmobranch host *via* a plankton-feeding organism, in which the parasite might be re-encapsulated, but we may be witnessing here an adaptation for directly infecting plankton-feeding elasmobranchs.

The most remarkable feature of trypanorhynchs is the scolex (Fig. 10.14). Two or four, often ear-like, muscular organs called bothridia are supplemented by four spiny tentacles or proboscides, which can be protruded and withdrawn in a most sophisticated way. Each tentacle operates independently of the others and, when withdrawn, is turned outside in with the spines on the inside (Fig. 10.15). The tentacle is enclosed in a tentacle sheath and the space between the sheath and the tentacle is filled with fluid, which communicates proximally with a large fluid-filled muscular bulb. Each tentacle is protruded hydraulically by contraction of the muscular bulb – the increased fluid pressure generated between the tentacle and its sheath leads to the eversion of the tentacle. As the tentacle turns inside out, the hooks will be pushed rapidly forwards, sideways and backwards, i.e. they will sweep through 180° or more. If the scolex is in contact with the host's intestinal lining, the hooks will engage the mucosal cells and as they hook into the mucosa and sweep outwards they will open a path for eversion of the next section of the tentacle. As the hooks in the next and subsequent sections drive forwards and sweep sideways and backwards in their turn, the tentacle will force its way into the mucosa.

Figure 10.13   Free-living, immature trypanorhynch cestode from the plankton. Scale bar: 0.5 mm. Source: redrawn from Dollfus, 1966.

Borucinska and Caira (1993) observed that the tentacles of two trypanorhynchs from the nurse shark, *Ginglymostoma cirratum*, were sometimes superficially embedded in the mucosal epithelium but sometimes penetrated deeper – in the larger of the two species as far as the mucosal muscle layers.

A more effective means of anchoring an animal to a tissue substrate by combining anchorage and penetration can hardly be imagined. Moreover, the tentacle can be readily disengaged by reversing the process – reducing the fluid pressure in the hydraulic system and withdrawing the tentacle by contraction of a retractor muscle that runs from the distal tip of the everted tentacle to the wall of the bulb (Fig. 10.15).

It is interesting that the same principle has been adopted for attachment by the unrelated acanthocephalans and by the copulatory organs known as cirruses, which are common in tapeworms and have evolved independently in some monogeneans (section 4.7). In some tapeworms, such as the tetraphyllidean *Dioecotaenia cancellata*, which has separate male and female individuals (gonochorism or dioecism), penetration of the intact wall of a female proglottis by a well-armed cirrus has been described (Fig. 10.16), while in others self- and cross-impregnation *via* a vagina are known (see Williams and McVicar, 1968). One wonders whether the trypanorhynchan tentacle and the cestode cirrus have evolved independently in response to separate and different needs for attachment or whether the tentacles are modified cirruses that have lost their connection with the reproductive system.

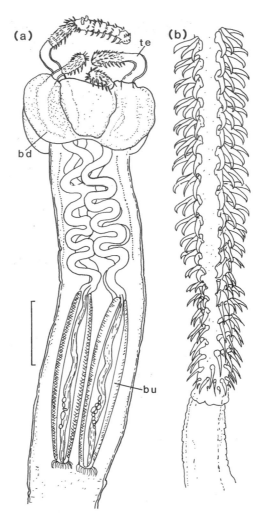

Figure 10.14 (*a*) The scolex of a trypanorhynch cestode (*Cetorhinicola acanthocapax*) with the basal region of one of the tentacles enlarged (*b*). bd = bothridium; bu = muscular tentacular bulb; te = tentacle. Scale bar for whole scolex: 0.5 mm. Source: modified from Campbell and Beveridge, 1994.

The wall of the tentacular bulb of the trypanorhynch *Grillotia erinaceus* consists of 25–35 distinct muscle layers, with fibres in adjacent layers at right angles to each other (Fig. 10.17). It is particularly interesting that these fibres have distinct transverse or oblique striations, a feature that occurs spasmodically in the parasitic platyhelminths (in the haptors of some monogeneans, section 6.4.1, and in the tails of digenean cercariae, sections 13.5, 15.3). The phenomenon appears to be associated with rapid, often repetitive, activity and it is interesting that, according to Ward, McKerr and Allen (1986), eversion of the tentacles in *Grillotia* is rapid and associated with an obvious twitch of the bulb muscle. Tentacle retraction is slower and the retractor muscle consists of a long cord of smooth (unstriated) muscle cells joined end to end.

The plerocercoids of *Gilquinia squali* occupy an interesting location. According to MacKenzie (1965, 1975), they occur in the eyes of whiting, *Merlangius merlangus*, usually lying free in the aqueous or vitreous humour but occasionally being

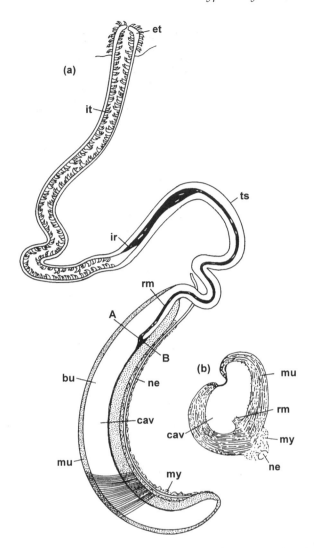

Figure 10.15 (*a*) A single scolex tentacle of the plerocercoid larva of the trypanorhynch *Grillotia heptanchi*. (*b*) Transverse section of the tentacular bulb in the plane AB. bu = tentacular bulb; cav = fluid-filled cavity of tentacular bulb; et = evaginated part of tentacle; ir = insertion of retractor muscle on wall of tentacle; it = invaginated part of tentacle; mu = muscular wall of bulb; my = muscle cell bodies; ne = nerve of tentacle; rm = retractor muscle; ts = tentacle sheath. Source: redrawn from Rees, 1950.

attached to the lens. Each plerocercoid consists mainly of a fully developed scolex enclosed in a fluid-filled space within a fleshy cyst (Fig. 10.18).

The eye is also a site that has been colonized by digenean metacercariae of the genus *Diplostomum* (section 17.7.2). The effector arms of the vertebrate immune system seem to be unable to penetrate into the eye (or indeed into the brain – section 17.7.2). Therefore, the eye, as an immunologically privileged site, may be an advantageous residence for parasites. There may be an added advantage for eye parasites in that, by influencing the host's vision, the host may be rendered more susceptible to predation by the final host. There is evidence for this in *Diplostomum* (section 17.7.2), but no information on whether infected whiting are more likely to be captured by the dogfish *Squalus acanthias*, the definitive host of *G. squali*.

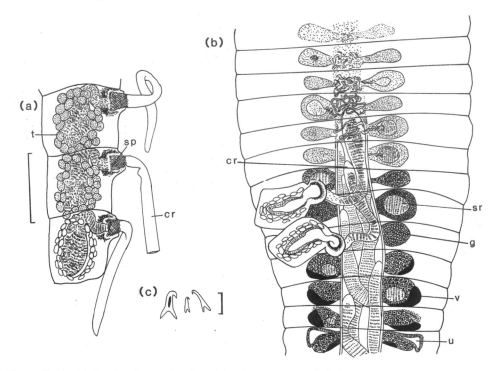

Figure 10.16   (*a*) Distal male proglottides of the dioecious tetraphyllidean tapeworm *Dioecotaenia cancellata*, with cirruses protruding and seminal vesicles filled with spermatozoa. (*b*) Part of the strobila of a female worm with two male proglottides attached by hypodermic impregnation. The spermatozoa have been forced into the cirrus. (*c*) Hooks from the spiny base of the cirrus. cr = cirrus; g = germarium; sp = spiny base of cirrus; sr = seminal receptacle; t = testis; u = uterus; v = vitellarium. Scale bars: (*a*), (*b*), 0.5 mm; (*c*) 10 μm. Source: reproduced from Schmidt, 1969, with permission from the American Society of Parasitologists.

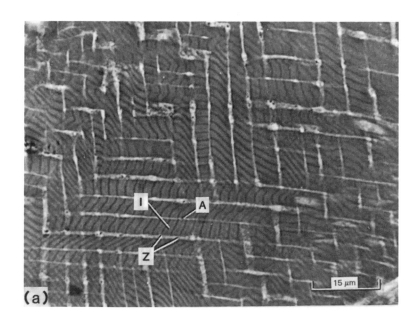

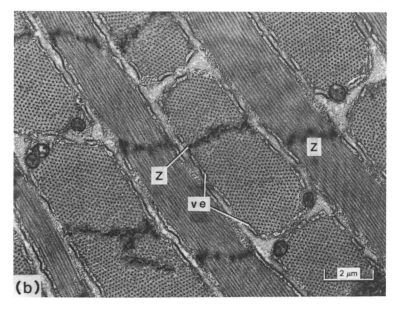

Figure 10.17 Tentacular bulb musculature of the trypanorhynch cestode *Grillotia erinaceus*. (*a*) Photomicrograph showing that muscle fibres in adjacent layers cross each other at 90°. Striations appear both transverse and oblique. (*b*) Transmission electron micrograph of muscle of tentacular bulb. A = A band; I = I band; ve = vesicle; Z = Z line. Source: reproduced from Ward, McKerr and Allen, 1986, with permission from Cambridge University Press.

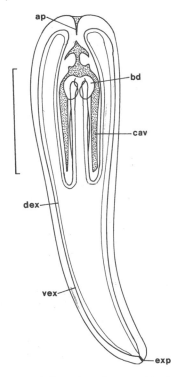

Figure 10.18 Diagrammatic representation of a plerocercoid of the trypanorhynch cestode *Gilquinia squali*. ap = anterior pore; bd = bothridium; cav = cavity surrounding scolex; dex = dorsal excretory vessel; exp = excretory pore; vex = ventral excretory vessel. Scale bar: 1 mm. Source: modified from MacKenzie, 1965.

MacKenzie failed to hatch the eggs of *G. squali* and failed to infect various potential first intermediate hosts by feeding eggs to them. However, MacKenzie suspected that the first host might be a benthic invertebrate. Whatever this first host may be, after ingestion by the whiting the parasite must find its way from the intestine through the body of the whiting to the eye, a small target for a migrating procercoid.

10.7   THE AMPHILINIDEANS

The special features of amphilinideans have led to two very different interpretations of their origins and relationships. One school sees them as distinct from the rest of the cestodes and relates them to the gyrocotylideans, a view that seems at first sight to be consistent with their single set of reproductive organs (Fig. 10.19) and apparent complement of 10 hooklets (Fig. 10.20). Their utilization of crustaceans as intermediate hosts is cestode-like, but the fact that the adults inhabit the body cavities of their definitive hosts (fishes, turtles) also sets them apart from the rest of the cestodes. There is, however, an alternative argument that offers an explanation for the habitat of the parasite and its single set of reproductive organs and places the amphilinideans among the cestodes – this is the view, proposed first by Pintner (1903, in Rohde and Georgi, 1983) and developed by Llewellyn (1965), that amphilinideans are progenetic plerocercoids that have lost their final host.

The hooklets are no stumbling block for this argument, since Llewellyn pointed out that only six of them are 'typical' in shape (Fig. 10.20). He quoted Dubinina's discovery that the other four sclerites develop in the body, not in the posterior cercomer where the hooklets arise, and appear later than the hooklets. Thus, the six typical hooklets were regarded as homologous with the six oncospheral hooklets of cestodes, with the four additional sclerites appearing at a different time, like the hamuli of monogeneans, in response to some unknown selection pressure.

Progenesis is an occasional trend in cestodes and has produced a sexually mature procercoid in *Archigetes* (section 10.2) and sexually advanced, coelom-inhabiting plerocercoids in the ligulids (section 10.5.4). The presence of an abdominal pore in sturgeons would have provided a ready-made exit for eggs laid by precocious plerocercoids inhabiting the body cavity. Eggs of *Dictyocotyle coeliaca*, a coelom-inhabiting monogenean parasite of rays (see Table 7.1), leave the host by the same route (Hunter and Kille, 1950). Popova and Davydov (1988) observed that most individuals of *Amphilina foliacea* initially invade organs such as the liver of the sturgeon; some of these may be encapsulated by the host but this does not prevent them returning to the body cavity, where eggs are released. The absence of abdominal pores did not stop the colonization by amphilinideans of the coelom of hosts such as freshwater turtles – perhaps maintenance of the tissue-penetrating capability of the early plerocercoid, as observed in *A. foliacea*, provided a means of escape from such sealed habitats, but Rohde and Georgi (1983) were unable to identify the escape route of *Austramphilina elongata* from the turtle *Chelodina longicollis*.

Even if amphilinideans are cestodes with life cycles curtailed by progenesis, it is not easy to relate them to any of the modern tapeworms. Their additional sclerites and the discovery of Rohde and Georgi (1983) that the ciliated larvae of *A. elongata*

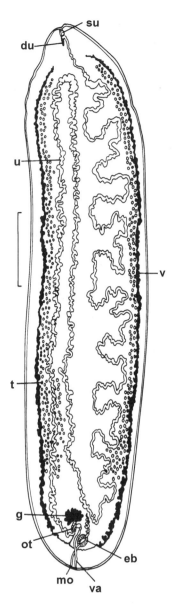

Figure 10.19 Adult specimen of *Amphilina japonica* in ventral view. du = ducts of frontal glands; eb = ejaculatory bulb; g = germarium; mo = aperture of male reproductive system; ot = ootype; su = anterior sucker; t = testis; u = uterus; v = vitellarium; va = vagina. Scale bar: 5 mm. Source: redrawn from Dubinina, 1985.

enter their crustacean first intermediate host by penetrating thin areas of cuticle are features not shared with any of the other tapeworms and the amphilinideans may be modern survivors of a group of cestodes of very ancient origin. That some of these parasites have changed very little over a very long period of time is supported by the close similarity between species of the genus *Nesolecithus* found in freshwater fishes on opposite sides of the south Atlantic; these parasites have probably been separated since South America and Africa drifted apart over 130 million years ago (Dönges, 1980, in Rohde and Georgi, 1983).

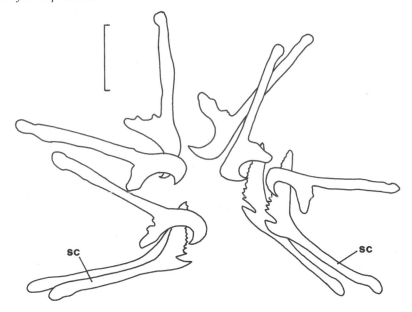

Figure 10.20   Larval hooklets and additional sclerites (sc) of *Amphilina foliacea*. Scale bar: 20 μm. Source: redrawn from Dubinina, 1985.

10.8   THE TETRAPHYLLIDEANS, PROTEOCEPHALIDEANS AND NIPPOTAENIDEANS

10.8.1   Loss of tanning and apolysis

The development of the pseudophyllidean egg has been studied by Rybicka (1966). As in the monogeneans, the vitellarium is well developed and this is related to the provision of many vitelline cells for each egg. Vitelline droplets released by these cells coalesce to form the operculate eggshell, which then undergoes tanning; the growing embryo separates into three layers: a central sphere of cells destined to become the oncosphere and inner and outer envelopes (Fig. 10.21*a*). The inner envelope becomes the ciliated epidermis of the coracidium larva. The remnants of the vitelline cells become incorporated into the outer envelope and together provide resources for the growth of the coracidium.

A significant evolutionary step now took place. Some tapeworms adopted a completely new way of ensuring that the hexacanth emerges in the intestine of the first intermediate host. They achieved this by abandoning the tanned shell and relying on the intermediate host to free the larva by digesting away the embryonic envelopes enclosing the hexacanth. This change is already foreshadowed in pseudophyllideans like *Eubothrium*, in which the embryo develops in a non-operculate shell that enlarges as development proceeds *in utero* (Rybicka, 1966). Fewer vitelline cells are associated with the oocyte and the indications are that the shell is not a stable, quinone-tanned protein.

Three other groups of fish tapeworms have adopted this strategy: the tetraphyllideans parasitizing the spiral 'valve' (= intestine) of elasmobranch definitive hosts; the proteocephalideans, which are essentially parasites of teleosts but have colonized some amphibians and reptiles; and the nippotaenideans found in the intestine

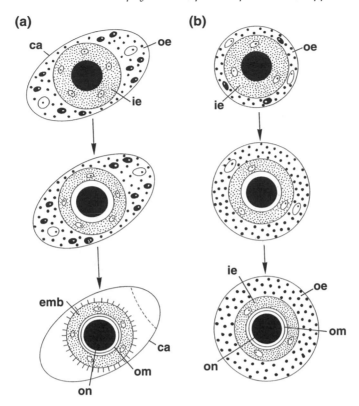

Figure 10.21 Schematic diagram showing the formation of the embryonic envelopes in (*a*) a pseudo-phyllidean and (*b*) a proteocephalidean. ca = capsule (shell); emb = embryophore; ie = inner envelope; oe = outer envelope (includes remnants of vitelline cells); om = oncospheral membrane; on = oncosphere. (For explanation see text.) Source: redrawn from Rybicka, 1966.

of freshwater teleosts. We know little about the embryology of these tapeworms, but the shell appears to be little more than a delicate membrane and there are no longer any cilia on the inner envelope (Figs 10.21*b*, 10.22). In fact, the inner and outer envelopes have become elaborate protective layers for the hexacanth. Some tetra-phyllidean eggs have elongated appendages and others clump together (Fig. 10.22). In some cyclophyllidean eggs there are similar adaptations that are thought to enhance their chances of consumption by specific invertebrate intermediate hosts (section 11.3.1).

The problem with the loss of tanning is that there is a serious risk of digestion of the egg envelopes by the digestive enzymes of the definitive host, especially if the eggs are laid in the anterior region of the intestine, where host digestive activity is at its peak. A tapeworm faces a dilemma: it needs to be in the anterior region of the intestine in order to compete successfully with the host for the absorption of nutrients newly released by host digestive processes but, if it loses the tanned eggshell, the envelopes of freshly laid eggs will be digested. Some means of protect-ing the eggs is necessary and a straightforward solution is to abandon egg-laying, close the uterine pore and store freshly assembled eggs in the uterus.

The eggs, protected inside the body of the worm, can reach the relative safety of the outside world by expulsion of the whole egg-filled (gravid) parasite or by the detachment and expulsion of each separate, gravid, but still living, reproductive

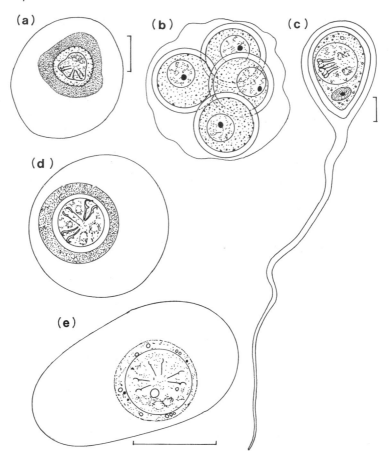

Figure 10.22 Eggs of some tetraphyllidean, proteocephalidean and nippotaenidean cestodes. The tetraphyllideans (*a*) *Acanthobothrium coronatum*, (*b*) *Echeneibothrium* sp., (*c*) *Echeneibothrium elongatum*, (*d*) *Proteocephalus* (=*Ichthyotaenia*) *filicollis* and (*e*) *Nippotaenia fragilis*. Scale bars: (*a*) 25 μm; (*c*) 10 μm; (*e*) 20 μm. Source: redrawn from: (*a*) Rees and Williams, 1965; (*b*), (*c*) Williams, H.H. 1966; (*d*) Meggitt, 1914; (*e*) Hine, 1977.

unit (the phenomenon of apolysis). The little known observations of Meggitt (1914) point to another possibility. He recorded the not uncommon sight of the proteocephalidean tapeworm *Ichthyotaenia* (= *Proteocephalus*) *filicollis* hanging from the anus of its stickleback host (*Gasterosteus aculeatus*). The worm apparently moves backwards until the gravid segments protrude from the anus, when the eggs are expelled by the splitting of the uterus and the body wall.

Logic demands that egg storage evolved in tapeworms before the worms abandoned eggshell tanning. In tapeworms with tanned eggs some limited storage is probably essential to permit the tanning process to run to completion before the eggs are laid and exposed to host digestive enzymes, but longer-term storage of tanned eggs is practised by trypanorhynch cestodes (see, for example, Sakanari and Moser, 1989 and personal observations on *Gilquinia squali*). We know so little about the general biology and life cycles of these worms that it is difficult to evaluate the advantages of this strategy, but, in general terms, it will permit the embryos to undergo some or all of their development while *in utero* and offers the opportunity to concentrate infective larvae in space and in time.

Many tapeworms that store untanned eggs have an expandable uterus, and this is a great asset because it prolongs the reproductive life of the egg assembly unit. An important limitation on this expansion and accumulation of eggs is likely to be the increasing load on the scolex, and before this load becomes critical the parasite must cast off the gravid proglottis. This will then be expelled from the intestine by host peristalsis, effectively putting an end to the life of the reproductive unit. How the gravid proglottides are set free has not been adequately explained because, contrary to popular belief, there is no septum between proglottides or any obvious specialization of the tissues at the junction (Mehlhorn *et al.*, 1981).

Apolytic tapeworms compensate for loss of proglottides by producing new proglottides from a zone of particularly active cell division in the neck region just behind the scolex. This development, which is already foreshadowed in the large pseudophyllideans (section 10.5.5), creates a very different kind of tapeworm, in which all proglottides are not at the same stage but progressively increase in size from the neck region backwards, becoming sexually mature and gradually expanding as the uterus fills with eggs. These worms have the potential for perennial survival rather than an annual or even shorter-term existence, and the continual replacement of old by new tissue may have immunological implications.

Excessive strain on the scolex, imposed by the accumulation of eggs in expanding uteri, can be circumvented by permitting the terminal proglottides to detach at an early stage of development, so that maturation, egg assembly and egg accumulation can take place during a period of free existence, independent of the parent strobila. However, this strategy will only be effective if the free proglottides are able to maintain their station and avoid expulsion from the host. There is no doubt that they can do this, because the detached proglottides of the nippotaenidean *Amurotaenia decidua* (Fig. 10.23) remain free in the intestine for up to 2 months, during which period they increase in length from 1.5 mm to about 5 mm and develop an enormously enlarged and convoluted egg-filled uterus (Hine, 1977). The anterior region of the free proglottis of *A. decidua* is specialized to form a holdfast organ (Fig. 10.23*b*) – essentially a tapering, spiny protrusion, which provides anchorage by expanding between the host's intestinal villi. The whole proglottis also initiates posteriorly directed waves of contraction, which may compensate for displacement of the free proglottis by host peristalsis.

This phenomenon of hyperapolysis has probably evolved independently in other groups of tapeworms. The cyclophyllidean *Cylindrotaenia* (= *Baerietta*) *hickmani* adopts this strategy (Jones, 1985), as do many tetraphyllidean cestodes. The convergence between these organisms is striking. The free proglottis of the tetraphyllidean *Acanthobothrium coronatum*, for example, has a contractile, anterior, spiny protuberance that presumably acts as a holdfast. This holdfast is separated from the rest of the proglottis by a sheet of muscle fibres and contains no reproductive organs (Fig. 10.24).

In tetraphyllideans and proteocephalideans, the vitellarium is still a substantial organ (Fig. 10.24), in spite of the reduction in its contribution to the eggshell, and it may retain a significant role as a provider of nutriment for the developing embryos. In nippotaenideans, however, the vitellarium is reduced to two small lobes (Fig. 10.23*b*) and nutrients must be supplied to the embryo *in utero* from other sources. Similar reductions of the vitellarium have been noted already in the polystomatid monogenean *Pseudodiplorchis americanus*, which, like the nippotaenideans, has a reduced eggshell and undergoes substantial embryonic growth *in utero* (section 7.1.4(b)).

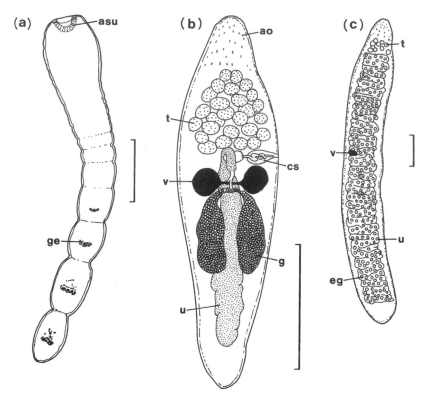

Figure 10.23   The nippotaenidean cestode *Amurotaenia decidua*. (*a*) Fully developed strobila. (*b*) Free mature proglottis. (*c*) Free gravid proglottis. ao = spiny 'adhesive organ'; asu, apical sucker (scolex); cs = cirrus sac; eg = eggs; g = germarium; ge = genital primordium; t = testis; u = uterus; v = vitellarium. Scale bars: 0.5 mm. Source: redrawn from Hine, 1977.

10.8.2   Scoleces

There are considerable regional differences in the microtopography of the mucosal surface in the spiral valves of elasmobranchs, as well as differences between species (see Williams, 1960). Surface features include folds, tubular crypts and projecting villi. Trypanorhynchs appear to be generalists, in that the scolex is probably capable of attaching itself to any surface provided that it is soft enough to permit penetration by the tentacles, but individual tetraphyllideans have become highly specialized for attachment to specific intestinal surfaces and consequently, as a group, show much greater diversity of scolex morphology (Figs 10.25, 10.26). They have also developed the tetraradiate pattern of scolex organization, already evident in the trypanor-hynchs. This pattern persists in the proteocephalideans and in the cyclophyllideans (section 11.2.1).

The scoleces of tetraphyllideans are dominated by four so-called bothridia, ear-like muscular organs that are generally regarded as suctorial. In some species they are stalked, permitting them to attach to widely separated gut folds (Fig. 10.26*a*) or to the tips of villi (Fig. 10.26*b*), while in other species they are sessile and mounted on a scolex small enough to fit inside a crypt or between villi (Fig. 10.26*c*). Bifid or trifid hooks are an accessory feature of crypt dwellers. In *Phyllobothrium* spp. the bothridia are foliaceous and provided with small, supplementary suckers (Fig. 10.25*e*). Williams (1968c) has described how each leaf-like bothridium of young specimens

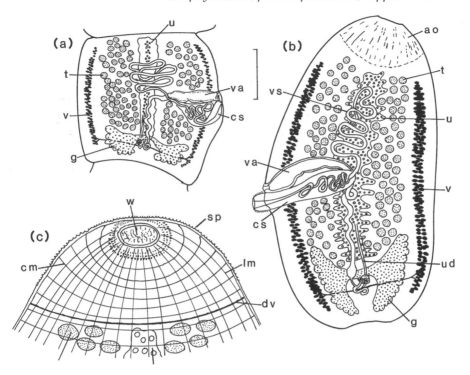

Figure 10.24 The tetraphyllidean cestode *Acanthobothrium coronatum*. (*a*) Mature proglottis. (*b*) Free mature proglottis. (*c*) Anterior region of free proglottis, showing the 'adhesive organ'. All semi-diagrammatic. ao = 'adhesive organ'; cm = circular muscle; cs = cirrus sac; dv = dorso-ventral muscles; g = germarium; lm = longitudinal muscle; sp = spine; t = testis; u = uterus; ud = uterine duct; v = vitellarium; va = vagina; vs = vas deferens; w = wound at apex of free proglottis. Scale bar: (*a*), (*b*) approximately 0.5 mm. Source: redrawn from Rees and Williams, 1965.

of *P. piriei* wraps around a single villus but as the parasite grows each bothridium becomes more foliaceous and eventually may embrace as many as 12 villi (Fig. 10.26*d*). The similarity with some gyrocotylideans (section 7.2.2) and calceostomatine monogeneans (section 6.4.1) is noticeable and the possibility that a cement is involved is worth investigating.

The terminal rostellar region is often a significant feature of the tapeworm scolex and in tetraphyllideans ranges from a rudimentary structure to a long flexible tentacle (= myzorhynchus; Figs 10.25*a*, 10.26*b*). It is assumed that the myzorhynchus has an attachment function but attention has also focused on its histolytic properties and there is no clear understanding of its contribution to the worm's biology.

The scolex of proteocephalideans has four compact muscular suckers, foreshadowing the cyclophyllidean scolex, and there may be a fifth apical (rostellar) sucker. In nippotaenideans, the single rostellar sucker performs the task of attachment (Fig. 10.23*a*) and radially directed suckers and accessory structures are absent.

10.8.3 Life cycles

The nippotaenideans and many proteocephalideans are tapeworms of microcarnivorous fishes, with copepods serving as first intermediate hosts (see Williams

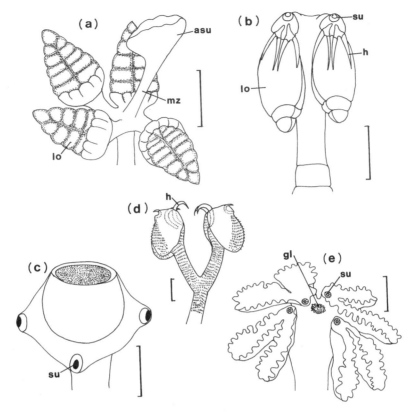

Figure 10.25   The range of scolex morphology in tetraphyllidean cestodes. (*a*) *Echeneibothrium variabile* (immature specimen). (*b*) *Acanthobothrium tripartitum*. (*c*) *Discobothrium fallax*. (*d*) *Yorkeria parva* (?). (*e*) *Phyllobothrium lactuca*. asu = apical sucker of myzorhynchus; gl = gland; h = hook; lo = loculus; mz = myzorhynchus; su = sucker. Scale bars: (*a*), (*b*), (*d*) 250 μm; (*c*), (*e*) 1 mm. Source: redrawn from: (*a*), (*c*) Williams, H.H. 1966; (*b*) Williams, 1969; (*d*) Williams, 1964; (*e*) Williams, 1968a.

and Jones, 1994). We have already encountered similar two-host life cycles, involving fish with an invertebrate diet, in primitive tapeworms such as the caryophyllideans and spathebothrideans and in some early pseudophyllideans (sections 10.2, 10.4, 10.5.2). The implication is that the reduction of the tanned eggshell and release of the hexacanth by digestion of the egg envelopes were developments that took place early in the evolutionary history of tapeworms, and indeed such developments may have occurred more than once.

The flexibility of the tapeworm life cycle is well illustrated by developments in the proteocephalideans. *Proteocephalus tumidocollus* parasitizes trout (*Oncorhynchus mykiss*) and procercoids occur in copepods but, curiously, copepods are rarely sought out and eaten by the trout. Cox and Hendrickson (1991) discovered that infected copepods are epiphytic on the alga *Enteromorpha* sp. and are consumed accidentally when the trout ingests the alga. During periods when the alga is absent, transmission of the tapeworm virtually ceases.

The trout *Salvelinus namaycush* is the definitive host of *P. parallacticus* and plankton-feeding individuals are infected by eating copepods containing procercoids (Freeman, 1964). However, there is a second option. When the infected copepod is eaten by a fish other than trout, the parasite maintains itself in the gut

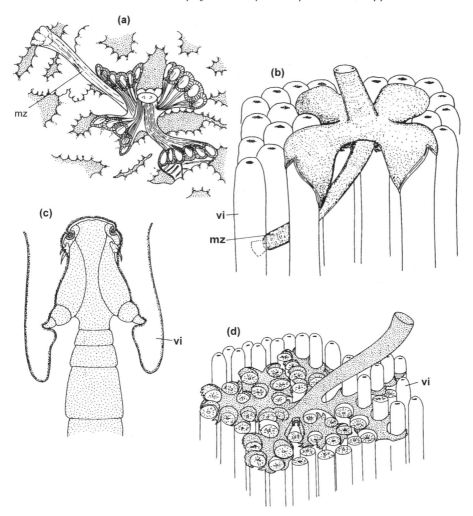

Figure 10.26   The adhesive attitudes of some tetraphyllidean cestodes. (*a*) *Echeneibothrium maculatum* in spiral valve (intestine) of *Raja montagui*. (*b*) *Pseudanthobothrium hanseni* in *R. radiata*. (*c*) *Acanthobothrium quadripartitum* in *R. naevus*. (*d*) *Phyllobothrium pirei* in *R. naevus*. mz = myzorhynchus; vi = villus. Source: (*a*), (*b*), (*d*) redrawn from Williams, 1961, Williams, H.H. 1966 and Williams, 1968c respectively; (*c*) based on Williams, 1968b.

where some development occurs without achieving sexual maturity. Maturity is reached when the fish is eaten by a larger macrocarnivorous trout. This is strikingly similar to the life cycle of the pseudophyllidean *Bothriocephalus gregarius* (section 10.5.2) and is perhaps the first step in the recruitment of a second intermediate host.

A surprising twist in the life cycle has evolved in *P. ambloplitis*. The smallmouth bass, *Micropterus dolomieui*, harbours adult tapeworms, while tissue-dwelling (parenteral) plerocercoids occur in planktonivorous fishes. It was originally supposed that these parenteral plerocercoids reached adulthood in the gut of predatory bass, but Fischer and Freeman (1969, 1973) showed that parenteral plerocercoids from the prey use a large apical proteolytic gland on the scolex to leave the gut of the predatory bass and resume essential development in the viscera. When lake temperatures increase in May and early June, these parenteral plerocercoids then travel in the reverse direction, return to the gut lumen of the bass, become sexually mature and lose their apical gland.

In keeping with the macrocarnivorous habits of most sharks and rays, tetraphyllidean plerocercoids are encountered in a variety of prey organisms, including squid, gastropod molluscs and teleost fishes (see Williams and Jones, 1994). Copepods have been identified as first intermediate hosts of tetraphyllideans (see, for example, Mudry and Dailey, 1971; Jarecka and Burt, 1984) but early events in the life cycles of most tetraphyllideans are entirely unknown.

I personally encountered tetraphyllidean plerocercoids in an unusual location in November 1986. A freshly dead sperm whale was stranded on the coast of Norfolk (UK) and blubber lying just below the skin was found to be seeded with large numbers of plerocercoids. These parasites may have been *Phyllobothrium delphini*, which has been widely reported as a plerocercoid from dolphins, whales and seals (Dailey, 1985). The idea of a whale as an intermediate host at first seems ludicrous, but there is evidence that the adult tapeworm may parasitize sharks (see Williams, 1968a) and it is conceivable that these fishes could become infected by biting pieces of flesh from the body of a living or dead whale (see Clarke and Merrett, 1972). In fact, the cookie cutter shark, *Isistius brasiliensis*, is known to feed in precisely this way on tuna and porpoise (Pough, Heiser and McFarland, 1990) and there are reports of the Greenland shark, *Laemargus glacialis*, taking flesh from living whales and other aquatic mammals (see Williams, 1968a). How the beached whale acquired the plerocercoids is unknown.

10.9   THE WAY FORWARD

All the tapeworms that we have studied so far are essentially tied to water for the propagation of their larvae and almost all of those that have succeeded in exploiting birds and mammals as definitive hosts are restricted to those with fish-eating habits. However, the changes in the egg envelopes and suppression of hatching that took place in some fish tapeworms permitted their descendants to make a complete break with water and opened the way for the evolution of the wholly terrestrial life cycles of the cyclophyllideans. This significant development will be considered in the next chapter.

CHAPTER 11

The cyclophyllideans

11.1 INTRODUCTION

The following features were inherited by cyclophyllideans from their fish-tapeworm ancestors: loss of the tanned shell and of conventional hatching *via* an opercular opening; replacement of the shell by multilayered envelopes derived from the embryo; abandonment of a free-swimming, ciliated coracidium; employment of the intermediate host to release the unciliated hexacanth. These developments created the potential for colonizing wholly terrestrial, vertebrate, definitive hosts. Some cyclophyllideans, such as the nematotaeniids, are parasites of amphibians and reptiles, but most cyclophyllideans have diversified as parasites of the warm-blooded, land-dwelling birds and mammals.

The terrestrial environment is characterized by the absence of a trophic level equivalent to plankton, and the shorter food chain is reflected in a reduction of most cyclophyllidean life cycles from three to two hosts (see Figs 8.9, 9.1). The three-host cycle of the mesocestoidids (section 11.7) may be primitive; the identity of the first intermediate host is not known but there is an insectivorous second intermediate host (mammal, bird or reptile) and a carnivorous avian or mammalian definitive host.

Many aquatic birds harbour cyclophyllideans and utilize aquatic intermediate hosts, but, with one exception (see below), these tapeworms preserve the typical two-host cycle of their wholly terrestrial relatives, suggesting that the definitive hosts of these parasites have developed a secondary association with aquatic habitats.

11.2 SPECIAL FEATURES

11.2.1 The scolex

The scoleces of many cyclophyllideans give an impression of power (Figs 8.2, 11.1), most of the volume of the scolex being occupied by four highly muscular suckers, often with a supplementary rostellum armed with strong hooks. This emphasis on secure anchorage may be a response to more powerful and more frequent gut movements in the intestines of their warm-blooded hosts. The tetrabothriideans, which are regarded by some (e.g. Schmidt, 1986) as cyclophyllideans and by others (e.g. Hoberg, 1994) as closer to the tetraphyllideans, inhabit the intestines of marine mammals and birds and also have four massive suckers (Fig. 11.2).

An intriguing study by Mount (1970) has revealed that each rostellar hook of the cyclophyllidean *Taenia crassiceps* originates from a single tegumental microthrix

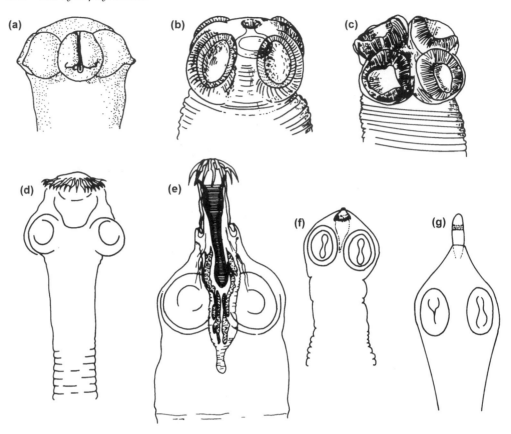

Figure 11.1   Scoleces of some cyclophyllidean cestodes. (*a*) *Ascometra longicirrosum*. (*b*) *Raillietina echino-bothrida*. (*c*) *Thysanosoma actinioides*. (*d*) *Taenia macrocystis*. (*e*) *Paricterotaenia porosa*. (*f*), (*g*) *Dipylidium caninum*, with rostellum contracted and expanded, respectively. Source: (*a*)–(*c*) reproduced from Joyeux and Baer, 1961, with permission from Masson S. A.; (*d*), (*e*) reproduced from Rausch, 1994 and Bona, 1994 respectively, with permission from CAB International; (*f*), (*g*) reproduced from Wardle and McLeod, 1952, with permission from the University of Minnesota Press.

(section 9.3.2(a)). The dense cap of the microthrix provides the core of the developing blade. New core material is added to the cap *via* a gap around the edge of the microthrix base plate, greatly increasing its length and girth. The developing blade projects initially into a space, but because of invagination of the developing scolex the tegument is not far away (Fig. 11.3). This adjacent tegument hypertrophies, gradually eliminating the space around the blade, and eventually makes contact with the blade surface. Secondary hook material is then produced by ribosomes in this tegument (ribosomes are not observed in the general body tegument of adult worms) and is added externally to the blade core, which serves as a temporary template for this deposition – later the primary hook core vacuolates and breaks down. Where the original base of the blade makes contact with the tegument, additional secondary thickening reinforces the root and numerous hemidesmosomes provide further anchorage by linking this region of the tegument to underlying fibrous material.

It is surprising that microtriches similar to those on the outer surface of adult worms and known to be involved in nutritional and perhaps immunological interactions with the host can undergo such drastic modification and development to

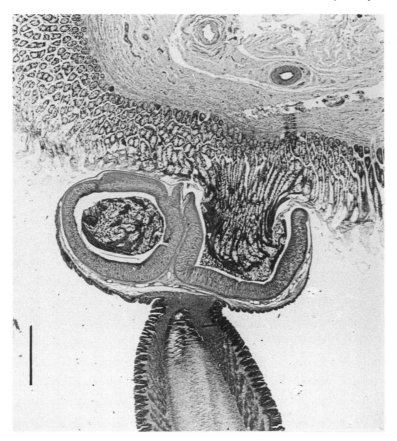

Figure 11.2 Longitudinal section through the scolex of *Tetrabothrius affinis*, attached to the mucosa of the blue whale, *Balaenoptera musculus*. Scale bar: 1 mm. Source: reproduced from Rees, 1956, with permission from Cambridge University Press.

produce hooks. It should also be noted that the hooklets of the hexacanth are not homologous with scolex hooks, since the former are not tegumentary derivatives but are secreted by special parenchymatous cells called oncoblasts (see review by Ubelaker, 1983).

11.2.2   Adaptations for the terrestrial environment

(a)   Survival and dissemination of eggs

Because of the changes in the hexacanth and its envelopes, water is no longer required as a medium for the dispersal of the larvae but desiccation of eggs deposited on land is undoubtedly a serious threat to survival. Extremes of temperature may also be a threat. Our knowledge of the survival and dispersal of cyclophyllidean eggs and the ecological interactions between their intermediate and definitive hosts is mostly limited to tapeworms with economic and/or public health significance, such as species of the genera *Taenia* and *Echinococcus*.

There is evidence that the eggs of the anoplocephalid *Moniezia expansa* are resistant to drying (see below), but the eggs of other cyclophyllideans seem to depend on moist conditions for prolonged survival. In a moist environment, taeniid eggs seem

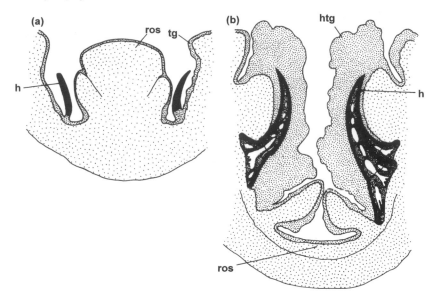

Figure 11.3   (*a*), (*b*)   Two stages in the histogenesis of the rostellar hooks of *Taenia crassiceps*, in bladder worms developing in the mouse. h = hook; htg = hypertrophied tegument; ros = rostellum; tg = tegument. Source: drawn from photographs by Mount, 1970.

able to tolerate temperatures within the range 38°C to −30°C and, under favourable conditions, some taeniid eggs are potentially able to maintain their infectivity for long periods (references in Lawson and Gemmell, 1983). At low temperatures (3–5°C) and high humidity, the eggs of *T. pisiformis*, for example, retain their infectivity to rabbits for more than 300 days, but this survival period is reduced at higher temperatures and drastically curtailed by low humidity at all temperatures. Eggs of *T. saginata* stored in dry conditions lost their infectivity to calves in as little as 1 day. In these circumstances it would seem advantageous to delay apolysis until all the eggs in the segment are infective but, according to Lawson and Gemmell, some taeniid eggs are not mature at deposition and their chances of completing development and becoming infective will depend largely on the availability of moisture.

Faeces from the definitive host may retain moisture for considerable periods of time and coprophagous animals are particularly suitable as intermediate hosts since there is no necessity for the eggs to leave the moist environment of the faeces. Other invertebrates may be attracted to faeces by the need for moisture or to feed on coprophagous animals such as insect larvae. Dung beetles serve as intermediate hosts for a number of tapeworms (references in Miller, 1961) and the stable-fly *Stomoxys calcitrans* and the house-fly *Musca domestica* serve as intermediate hosts for poultry tapeworms (see Lawson and Gemmell, 1983). Reference has already been made (section 9.5) to the coprophagous habits of the flour beetle hosts of *Hymenolepis diminuta* and the evidence to suggest that rat faeces containing eggs and egg-filled proglottides are more attractive to the beetles than faeces from uninfected rats. As we will see below, dung-associated arthropods may have an additional, unsuspected role in the dispersal of tapeworm eggs.

Coprophagous mammals like the pig are equally effective intermediate hosts. Viljoen (1937) described the efficiency and speed with which domestic pigs remove human faeces deposited in the open by South African farm workers, ensuring effective transmission of *Taenia solium* to the intermediate host. He also quoted a story from Bali of an old headman who, after defecation, simply whistled to summon his pig to clear up his excreta – this example of the definitive host summoning the intermediate host must surely be unique.

There are many intermediate hosts of cyclophyllidean tapeworms that are not normally associated with faecal matter. Cattle and sheep, for example, usually avoid grazing around faecal deposits (see Lawson and Gemmell, 1983) and yet are effective intermediate hosts for *Taenia* and *Echinococcus*. However, in drought conditions cattle have been seen to ingest human faeces and Viljoen (1937) associated this behaviour with increased prevalence of *T. saginata* in South African cattle after periods of drought.

In many cyclophyllideans, the detached proglottides retain their activity after expulsion and may migrate away from faeces. Gravid proglottides of *E. granulosus* can travel over 25 cm and those of *T. hydatigena* may move up to 90 cm away from faeces (references in Lawson and Gemmell). Webster (1949) observed that gravid proglottides of *Mesocestoides latus* migrate actively out of the faecal mass. The proglottides of the lizard tapeworm *Oochoristica vacuolata* may also leave the faeces and their locomotion as described by Hickman (1963) is leech-like.

Some taeniid eggs are disseminated over much greater distances than can be accounted for by the limited localized outward migration of proglottides. In a study in New Zealand, 'sentinel' sheep were grazed on paddocks at varying distances from a source of *T. ovis* in dogs kept in kennels (see Lawson and Gemmell, 1983). It was found that sheep in the paddock nearest the kennels had the highest cyst counts and infection decreased with increasing distance from the dogs. However, assuming that the kennel dogs were the source, eggs were dispersed over a distance of at least 175 m. Using a grazing circle subdivided into nine sectors radiating outwards from a central ungrazed area containing dogs infected with *T. hydatigena*, it was found that the cyst pattern in infected sheep showed no significant difference between the plots, indicating that eggs were dispersed equally in all directions. According to Lawson and Gemmell, the radial pattern of egg dispersal and the distances covered can only be adequately explained by the intervention of other animals, and among the invertebrates, flies, beetles and ants are the most likely dispersal agents. Flies in particular use faeces for feeding and breeding and taeniid eggs ingested by them can survive passage through their gut. Moreover, their mobility and behaviour are compatible with the generation of the observed egg-dispersal pattern. Where the intermediate hosts are grazing animals that cover a wide area as they feed, or, like mites, are ubiquitous inhabitants of ground vegetation, the random dissemination of eggs by dung-associated arthropods may be particularly effective, although the success of this tactic has to be balanced against the increased chance of desiccation as eggs are separated from moist faeces.

New Zealand studies on the distribution of infected sheep around foci of infections of *T. hydatigena* and *T. ovis* in dogs suggested that some eggs may travel several kilometres and observations by Torgerson *et al.* (1995) made in Scotland point to egg transport over even greater distances. The island of St Kilda off the west coast of Scotland has no definitive hosts of *T. hydatigena*, but resident sheep harbour the cestode. Occasional clandestine visits by infected dogs were rejected as sources of infection, but the distribution of the parasite was deemed consistent with transport of eggs against the prevailing wind by wildlife. The nearest source of infection was

60 km away and birds were thought to be the most likely carriers. Silverman and Griffiths (1955) suggested that herring gulls might be instrumental in transporting eggs of *T. saginata* from sewerage treatment plants to cattle pastures, and they claimed to have successfully infected cattle experimentally with faeces from herring gulls that had ingested eggs. However, it is not at all clear how infections were established *via* this route, since they also claimed that only immature eggs survived with their egg envelopes intact after passage through the bird.

Hall (1934) made the interesting observation that proglottides of the ruminant tapeworm *Thysanosoma actinioides* in freshly passed faeces usually had eggs in them but in older faeces rarely contained eggs. He also found that when a fresh proglottis was placed in a dry bottle and re-examined later, it was devoid of eggs, which were distributed in a ring at some distance from the proglottis. The inference is that the eggs are ejected in some way, and may leave the faeces if the proglottis has a superficial position in the faecal mass. Such an event is better suited to infection of a non-coprophagous host. In fact, similar expulsion of eggs has been recorded in other tapeworms of herbivores, such as *Moniezia* from sheep and *Cittotaenia* from rabbits (Stunkard, 1938), and appears to be associated with their exploitation of mites as intermediate hosts (section 11.4). Mites are not primarily coprophagous and are widely distributed in meadows.

(b)   Modifications of the uterus

In some cyclophyllideans, profound changes take place in the uterus as it fills with eggs, or new structures appear into which eggs are passed from the uterus. Most observers regard these modifications as responses to the demands of the terrestrial life cycle and, in particular, to the need to retard desiccation. In the dilepidid *Dipylidium caninum* and the anoplocephalid *Oochoristica anolis* the eggs become enclosed in capsules, which appear to be derived by reorganization of the uterine wall (Pence, 1967 and Conn and Etges, 1984 respectively). In *D. caninum*, these uterine capsules (see Fig. 11.19) usually contain 15–25 eggs (see Wardle and McLeod, 1952), while those of *O. anolis* contain only one egg (Conn and Etges, 1984). In genera such as *Paruterina* and *Mesocestoides*, eggs are transferred to one or more paruterine organs (Fig. 11.4). In *M. lineatus* this organ has a thick wall comprising several layers of cellular origin, the outermost layer consisting of well developed muscle bands (Conn, Etges and Sidner, 1984). The eggs are embedded in a cellular matrix, which can be resolved mainly into concentric cytoplasmic layers around each egg.

In the nematotaeniids, there is a more elaborate arrangement, the eggs being passed from the uterus into two paruterine organs and then from each of these into a so-called paruterine capsule (Fig. 11.5a, b). In the paruterine organs the eggs are surrounded by membranous components and lipid droplets, all of which accompany the eggs into the capsules (Jones, 1988). The wall of the capsule consists of two layers of contraorientated fibres and, near the junction of the capsule with the paruterine organ, muscles occur among the fibres. These muscles may control movement of eggs and associated material into the capsule from the organ and prevent movement in the reverse direction.

The hymenolepidid cestode *Ditestolepis* (*Soricinia*) *diaphana* goes one stage further. The posterior gravid proglottides (18–30 in number) unite together to form a 'syncapsule' containing a single elongated egg packet produced by the fusion of the uteri (Fig. 11.5c, d).

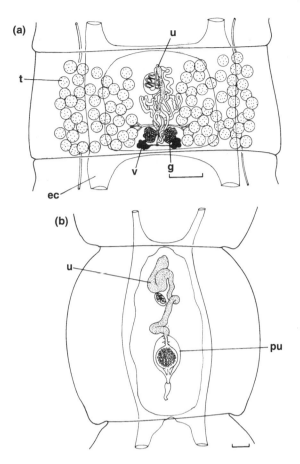

Figure 11.4 (*a*) A mature proglottis and (*b*) a gravid proglottis of *Mesocestoides kirbyi*. ec = excretory canals; g = germarium; pu = paruterine organ containing eggs; t = testis; u = uterus; v = vitellarium. Scale bars: 250 μm. Source: redrawn from Rausch, 1994.

With regard to the possible role of uterine and paruterine capsules in reducing water loss, it may be significant that many have lipid associated with them. A waterproofing function for this material has been suggested for the eggs of *Moniezia expansa* (section 11.4). Lipid droplets occur in cells around the uterine capsules of *Oochoristica anolis*, in the wall of the uterine and paruterine organ in *Dipylidium caninum* and *Mesocestoides lineatus* respectively and inside the paruterine capsule of the lizard parasite *Cylindrotaenia hickmani*. Webster (1949) found that dry and apparently desiccated proglottides of *M. latus*, 27 h after passage in host faeces, still contained eggs with oncospheres capable of movement, although this is no guarantee that they retained their infectivity. However, the complexity of some of these capsules sits rather uncomfortably alongside the idea of a simple waterproofing function, and we are bound to wonder whether this elaborate packaging has other practical advantages. For example, it would seem worthwhile to explore the possibility that some of these capsules may have attributes such as size, shape and provision of high-energy nutritional rewards (lipid reserves) that make them highly attractive food parcels for suitable intermediate hosts (see also section 15.4), analogous perhaps to the adaptations found in seeds and fruits that encourage predation.

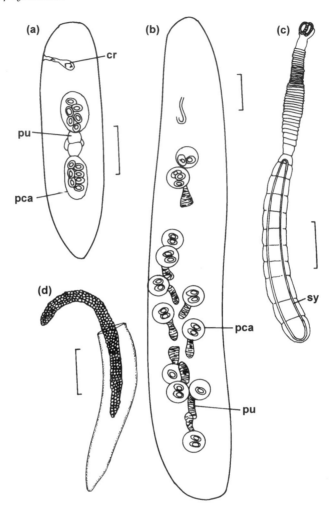

Figure 11.5 Special modifications of the gravid uterus in cyclophyllidean cestodes. (*a*), (*b*) Gravid proglottides of the nematotaeniids *Cylindrotaenia* (= *Baerietta*) *hickmani* and *Nematotaenia chantalae* respectively. (*c*) The hymenolepidid *Ditestolepis* (= *Soricinia*) *diaphana*, showing syncapsule (sy) derived from several gravid uteri. (*d*) Ruptured syncapsule of *D. diaphana* with emerging fused egg packet. cr = remains of cirrus; pca = paruterine capsule; pu = remains of paruterine organ. Scale bars: (*a*), (*b*) 100 μm; (*c*), (*d*) 0.5 mm (approximately). Source: (*a*), (*b*) modified from Jones, 1985 and Jones, 1987 respectively; (*c*) reproduced from Vaucher, 1971, with permission from Muséum d'Histoire Naturelle, Geneva; (*d*) reproduced from Kisielewska, 1961, with permission from the W. Stefanski Institute of Parasitology.

11.3   THE HYMENOLEPIDIDS AND DILEPIDIDS

11.3.1   Tapeworms of aquatic birds

The two-host life cycle, so typical of terrestrial cyclophyllideans, is also a feature of the many tapeworms that parasitize aquatic birds (Fig. 11.6). As already noted, this suggests that these life cycles are secondarily aquatic, the tapeworms having succeeded in exchanging a terrestrial for an aquatic intermediate host when their bird definitive hosts switched from a terrestrial to an aquatic environment.

Jarecka (1961) studied the rich hymenolepidid fauna of birds frequenting lakes in northern Poland and discovered considerable diversity in the eggs of these parasites. Moreover, she was able to relate features of the eggs, such as shape, size, and density, to the habitat, size and behaviour of their relatively specific intermediate hosts. She found that the eggs of *Cloacotaenia* (*Hymenolepis*) *megalops* are heavy (Fig. 11.7*a*) and are eaten by the benthic ostracod *Cypris pubera*, while other eggs, e.g. those of *Variolepis* (*Hymenolepis*) *furcifera* (Fig. 11.7*b*), are light, readily kept in suspension by slight water movements and fall prey to pelagic copepods and cladocerans. Eggs with long filamentous extensions are also common. In *Diorchis* spp. (Fig. 11.7*c–e*) these extensions favour the settlement of eggs on leaves of *Ceratophyllum*, where they are ingested by ostracods feeding on the leaves, and Belopolskaya (1958) described

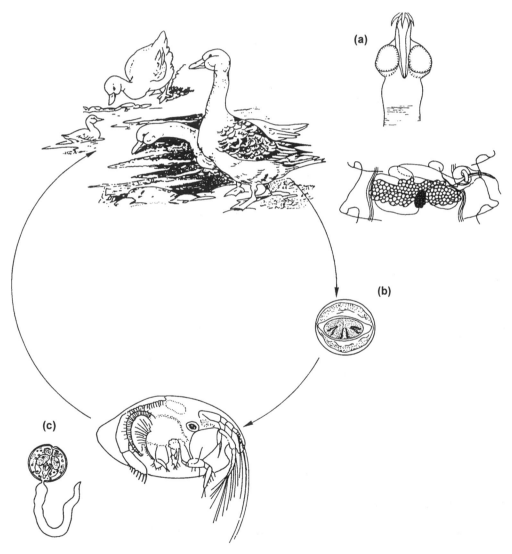

Figure 11.6   Life cycle of *Sobolevicanthus* (= *Hymenolepis*) *gracilis*. (*a*) Adult tapeworm in an aquatic bird (duck, goose). (*b*) Eggs enter water in bird's faeces and are eaten by an ostracod. (*c*) Cysticercoid develops in the body cavity and birds are infected by swallowing infected ostracods in drinking water. Source: based on Joyeux and Baer, 1961 and Czaplinski and Vaucher, 1994.

in some species of *Anomotaenia* even longer filaments (Fig. 11.7*f*, *g*), which retard sinking (like the appendages of some monogenean eggs, section 5.4) and promote ingestion by pelagic copepods. The intermediate hosts of the flamingo tapeworm, *Cladogynia* (*Flamingolepis*) *liguloides*, are brine shrimps of the genus *Artemia*; Robert and Gabrion (1991) observed that dense hair-like processes on the surface of the egg help to keep it in suspension.

Jarecka described how the eggs of *Aploparaksis furcigera* and *Microsomacanthus* (*Hymenolepis*) *abortiva* form relatively heavy packets (Fig. 11.7*h*, *i*), which provide suitably large prey items for bottom-dwelling oligochaetes and gammarids respectively. Jarecka also pointed out the striking similarity in shape and appearance between the ribbed embryophore of *Diorchis stefanskii* (Fig. 11.7*j*) and diatoms found in the gut of the ostracod intermediate host, *Dolerocypris fasciata*, the implication being that the eggs mimic the food items of their hosts.

Some female brine shrimps parasitized by *C. liguloides* undergo changes in pigmentation and tend to swim near the surface, features that may increase their vulnerability to predation by their flamingo hosts (Robert and Gabrion, 1991). The brine shrimps may also harbour cysticercoids of three other flamingo tapeworms that do not induce these changes, but these species may benefit from the influence of *C. liguloides*.

A three-host aquatic life cycle has been described by Jarecka (1970) for the dilepidid *Valipora campylancristrota*. This tapeworm uses copepods as first intermediate hosts but its definitive host, the heron, is infected by eating cyprinid fishes harbouring in their gall bladders what Jarecka called a plerocercus (essentially a scolex and a neck region). Thus, this secondarily aquatic cyclophyllidean has

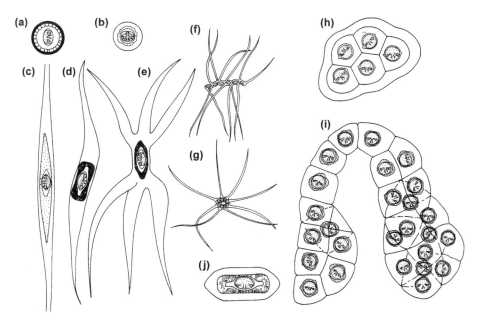

Figure 11.7 Eggs of some hymenolepidid and dilepidid cestodes parasitizing aquatic birds. (*a*) *Cloacotaenia* (= *Hymenolepis*) *megalops*. (*b*) *Variolepis* (= *Hymenolepis*) *furcifera*. (*c*) *Schillerius* (= *Diorchis*) *ransomi*. (*d*) *Diorchis nyrocae*. (*e*) *Diorchis* sp. (*f*), (*g*) *Anomotaenia* spp. (*h*) *Aploparaksis furcigera*. (*i*) *Microsomacanthus* (= *Hymenolepis*) *abortiva*. (*j*) *Diorchis stefanskii*. Not drawn to the same scale. Source: (*a*)–(*e*); (*h*)–(*j*) reproduced from Jarecka, 1961, with permission from the W. Stefanski Institute of Parasitology; (*f*), (*g*) redrawn from Belopolskaya, 1958.

already expanded its life cycle to accommodate the extended food chains in the aquatic environment, in much the same way as the pseudophyllideans and proteo-cephalideans described in the last chapter.

11.3.2 Tapeworms of shrews and host specificity

Kisielewska (1961), working in the Bialowieza National Park in Poland, conducted a detailed study of the population dynamics of nine tapeworm species (seven hyme-nolepidids and two dilepidids) that use the shrew *Sorex araneus araneus* as the definitive host. This work gives an interesting insight into the way tapeworms interact with their hosts and with environmental factors.

Kisielewska stressed that the intermediate host must form a natural, not acci-dental, food for the final host and that its habits and behaviour must create a permanent, or at least periodical, not accidental, chance of making contact with the parasite's eggs. She found that the tapeworms display some physiologically based host specificity to the intermediate hosts and that the habits of these hosts greatly influence the seasonal pattern of occurrence of the worms in the shrews. For exam-ple, the hymenolepidids *Ditestolepis* (= *Soricinia*) *diaphana* and *Staphylocystis furcata* are most abundant in shrews in the late summer and autumn, probably as a result of feeding on the coprophagous beetle *Geotrupes stercorosus*, which serves as the inter-mediate host and is active only during the warmer months. In contrast, another hymenolepidid, *Pseudodiorchis prolifer*, increases in numbers in shrews during the winter and early spring and, although its intermediate host, the myriapod *Glomeris connexa* is available all year round, it inhabits decaying tree trunks and is probably taken by shrews mainly in the winter, when the availability of insect food is restricted.

Kisielewska also found evidence of host specificity in the adult worms. Some tapeworms are restricted to *Sorex* spp., some to *Neomys* spp., and others parasitize both genera. *S. furcata* is a tapeworm found in *S. araneus araneus* but also occasionally in the water shrew, *N. fodiens*. Kisielewska believed that this reflects an enlargement of the living space of *N. fodiens* in the summer, permitting contact with the terrestrial intermediate hosts of *S. furcata*. Thus, the prevalence of the parasite in these two hosts may have an ecological rather than a physiological basis.

11.3.3 Rodent tapeworms

Rodents present a special challenge to cyclophyllideans because of their efficient grinding teeth, but the problem has been solved in tapeworms like *Hymenolepis diminuta* and *Vampirolepis microstoma* by the development in the cysticercoid of thick collagenous walls which enclose the vulnerable scolex and withstand mastication (see Chapter 9).

A feature of the eggs of these tapeworms is the presence of a substantial 'shell', but there is evidence (reviewed in section 9.4) that this is not homologous with the tanned shell of fish tapeworms like the pseudophyllideans; in other words it seems to be a secondary structure. The selection pressures that have led to this develop-ment are obscure. It has not yet been demonstrated that these 'shelled' eggs are any better at surviving desiccation than taeniid eggs that lack a shell but it might be related to the habit of coprophagy – consumption by the rodent of its own faeces.

Without protection a substantial proportion of the eggs of the tapeworm are likely to be destroyed as a result of this habit. However, other cyclophyllideans, e.g. some anoplocephalids, that parasitize hosts that do not habitually eat their own faeces have substantial shells.

Perhaps the significant common factor in relation to cyclophyllidean eggs with secondary 'shells' is their utilization of arthropod intermediate hosts. The 'shell' must be removed by the arthropod to permit host digestive enzymes to free the hexacanth. Lethbridge (1971b) observed that eggs of *H. diminuta* that passed with the 'shell' intact through the gut of the beetle *Tenebrio molitor* retained their infectivity to other beetles. This provides a mechanism for infecting a limited range of arthropods – those with jaws suitable for cracking the 'shell' – while, at the same time, promoting safe passage without hatching through the guts of animals that are unsuitable as hosts but useful as dispersal agents.

The presence of a protective 'shell' may also account for a limited premature discharge of eggs from proglottides still attached to the worm, reported in *V. nana* by Kumazawa (1992).

11.3.4   A single host life cycle – *Vampirolepis nana*

Rats and mice are known to eat some of their own faeces (Ebino *et al.*, 1989). This habit may well have led to an interesting life-cycle option in the mouse tapeworm, *V. nana*. Eggs eaten by a flour beetle give rise to cysticercoids in the haemocoel, similar to those of *V. microstoma* and *H. diminuta* (see Fig. 9.16a). However, eggs eaten by the mouse hatch in the mouse's intestine and the hexacanth penetrates an intestinal villus and also transforms into a cysticercoid. In 5–6 days the scolex emerges and establishes itself in the intestine. Thus, the intermediate host is no longer essential for the propagation of this tapeworm.

Caley (1975b) demonstrated that the mouse cysticercoid (Fig. 11.8) is strikingly different from the insect cysticercoid. The former has a thin, flexible and delicate cyst wall, with no myelin-like lining. Although there are collagen fibres, they are evenly distributed and sparse, being absent from large areas of the cyst wall. Caley showed that the mouse cysticercoid is vulnerable to low pH and unable to infect mice when experimentally introduced into the stomach. There is of course no need for protection from mastication or from stomach acid. The mouse cysticercoids were activated just as well in a salt solution as in solutions containing bile and trypsin; rupture of the cyst appeared to be effected by application of the suckers and the spiny rostellum (cf. the emergence of *V. microstoma* from an insect-derived cysticercoid, section 9.7). Trypsin accelerated the escape of the scolex by breaking down the surrounding villus tissue.

Because of their habit of eating some of their own faeces, the likelihood of mice becoming overloaded with *V. nana* by autoinfection seems great. However, invasion of the intestinal villi of a naive mouse by hexacanths from ingested eggs leads to an immunity against a challenge (secondary) egg infection that is discernible at 9 h and absolute at 24 h after initial infection (Smyth, 1969). This immunity is effective against oncospheres as they penetrate but also has a second stage operative against any oncospheres that slip through the first net and establish themselves in the villi. An initial egg-immunizing dose also has an influence on a secondary dose of insect-derived cysticercoids, only about one-fifth as many adults developing from this secondary dose as in controls with no egg immunization. Conversely, an initial dose

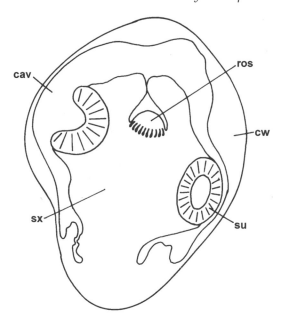

Figure 11.8 Diagrammatic representation of a longitudinal section through a cysticercoid of *Vampirolepis* (= *Hymenolepis*) *nana* from a mouse villus. cav = cavity of cyst; cw = cyst wall; ros = rostellum; su = sucker; sx = scolex. Source: reproduced from Caley, 1975a, copyright © by Springer-Verlag 1975, with permission from Springer-Verlag.

of insect-derived cysticercoids limits the number of adults that establish from a secondary dose of eggs but no immunity develops against adults from a challenge of insect-derived cysticercoids following initial administration of a dose of insect-derived cysticercoids.

A tapeworm morphologically identical with the mouse tapeworm and capable of direct transmission without the intervention of an intermediate host occurs in humans. This is a small tapeworm, rarely exceeding 40 mm in length and 1 mm in breadth, and has earned the name of dwarf tapeworm of man. It is particularly common in children, spreading most probably by dissemination and accidental ingestion of eggs. Infections of mice with tapeworm eggs from human infections and *vice versa* have been successful (Grove, 1990) and tapeworms from the two hosts are regarded as the same species, *V. nana*. However, higher rates of infection result from eggs obtained from the same host species than from eggs obtained from the other host, raising the possibility that speciation is under way (Roberts and Janovy, 1996).

11.3.5 Asexual multiplication in the intermediate host

A few hymenolepidids and dilepidids have acquired the ability to multiply asexually in the intermediate host. For example, the hexacanth of *Staphylepis* (= *Hymenolepis*) *cantaniana*, after entering the haemocoel of the dung beetle *Ataenius cognatus*, develops into a mycelium-like structure that buds off cysticercoids (Fig. 11.9). These may later become detached from the branching stem and as many as 2217 apparently mature cysticercoids were counted in a single beetle (Jones and Alicata, 1935).

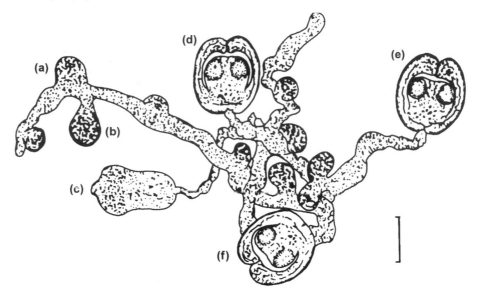

Figure 11.9   Asexually reproducing juvenile of *Staphylepis* (= *Hymenolepis*) *cantaniana* from the haemocoel of the beetle *Ataenius cognatus*. (*a*)–(*c*) Developing cysticercoids. (*d*)–(*f*) Fully developed cysticercoids still attached to branching juvenile tissue. Scale bar: 50 μm. Source: reproduced from Jones and Alicata, 1935.

According to Whitfield and Evans (1983), relatively few cestode families have members that multiply asexually in the intermediate host and almost all of these have terrestrial life cycles. Whitfield and Evans suggested that the phenomenon may have arisen independently in the Mesocestoididae (section 11.7), Dilepididae, Hymenolepididae and Taeniidae (see below) in response to selection pressures associated with transmission in the terrestrial environment. However, the proposed hypothetical advantages of this asexual multiplication do not have strictly terrestrial connotations: Moore (1981) suggested that proliferation in the intermediate host might be a transmission strategy, damage inflicted by the asexually expanding parasite rendering the intermediate host more vulnerable to predation (considered in more detail below); Whitfield and Evans (1983) proposed that asexual multiplication would promote the conservation of adaptive (successful) combinations of genes, since the population of sexually reproducing parasites in the final host are likely to be genetically identical. With regard to the latter suggestion, Whitfield and Evans saw a similar advantage for genetically identical cercariae (see Chapter 14), spatially and temporally clumped after emission from an aquatic snail.

11.4   THE ANOPLOCEPHALIDS AND HERBIVORES

Tapeworm life cycles depend on predator–prey relationships, so we might expect herbivorous animals to escape the attentions of these parasites. Some digeneans, like *Fasciola hepatica*, gain access to herbivores by encysting on the vegetation on which they feed (see Fig. 12.1), but tapeworms have come up with a surprisingly different solution to this problem. The eggs are not inhaled by the herbivore, as suggested for *Moniezia expansa* by Sinitsin (1931), but exploit the great abundance and ubiquitous

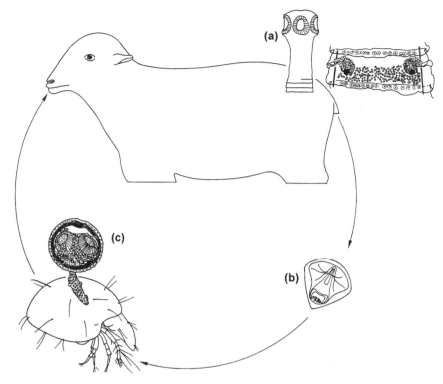

Figure 11.10 Life cycle of *Moniezia expansa*. Adult tapeworm (*a*) in sheep. Eggs (*b*) eaten by terrestrial oribatid mites. Cysticercoid (*c*) develops in haemocoel of mite; definitive hosts infected by ingesting infected mites accidentally while grazing.

nature of mites on grassland. Although mites are tiny, they are large enough to harbour cysticercoids and large numbers of these arthropods must be inadvertently ingested by grazing animals (Fig. 11.10). Some anoplocephalids such as *M. expansa* from the sheep have two sets of reproductive organs per segment (Fig. 11.10). This development, which has arisen independently in some pseudophyllideans (section 10.3) and in the dilepidid *Dipylidium caninum* (see below), undoubtedly greatly increases egg output.

Calcy (1975a) demonstrated some general similarities in egg structure and hatching between *M. expansa* and hymenolepidids such as *H. diminuta*. The egg of *M. expansa* has an external 'shell' with an underlying 'subshell' membrane (Fig. 11.11), the latter providing a barrier to the passage of water and solutes. These two layers are resistant to a variety of proteolytic, amylolytic and lipolytic enzymes. Their mechanical removal is essential to activate the hexacanth and to permit the digestive enzymes of the mite to remove the outer envelope and embryophore and liberate the hexacanth.

The ability of free proglottides in the faeces to eject eggs seems to be a common feature of herbivore tapeworms utilizing mites as intermediate hosts. This may have the advantage of seeding the surrounding vegetation with eggs where they are more likely to be eaten by the largely non-coprophagous mites (section 11.2.2(a)). We might predict that such a habit would demand some resistance to desiccation. Sinitsin (1931) allowed eggs of *Moniezia expansa* to dry out on a slide several times during a period of 45 days and claimed that the larvae inside

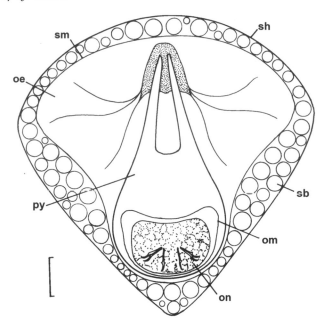

Figure 11.11 Diagrammatic representation of an egg of *Moniezia expansa*. oe = outer envelope; om = oncospheral membrane; on = oncosphere; py = pyriform apparatus (embryophore); sb = 'subshell' droplets; sh = 'shell'; sm = 'subshell' membrane. Scale bar: 10 μm. Source: modified from Caley, 1975b.

retained the ability to move. Caley suggested that the 'subshell' membrane may play an important part in water retention, but it is also possible, as suggested by Sinitsin, that a layer of lipid droplets between the 'shell' and the 'subshell' membrane and not found in *H. diminuta*, may provide an oily water-proofing coat.

Caley (1975a) infected mites with eggs of *M. expansa* and observed them manipulating the eggs with their mouthparts, sometimes for several minutes. Although some of her mites became infected, she did not observe intake of an entire egg and noted that the anterior part of the gut was probably too narrow to permit this. Oribatid mites take in food by the sucking action of a pharyngeal pump and Krull (1939) stated that mites feed on anoplocephalid eggs by piercing the shell and removing the contents. The embryophore of *M. expansa* is unusually pear-shaped (= pyriform apparatus; Fig. 11.11) and Caley suggested that this may be an adaptation for gaining entry into the narrow gut by suctorial means.

Freeman (1952) found evidence to suggest that the porcupine *Erethizon dorsatum* does not become infected with tapeworms of the genus *Monoecocestus* unless the exoskeleton of the mite intermediate host is cracked by the porcupine's molar teeth.

11.5   THE TAENIIDS AND CARNIVORES

11.5.1   General

By adopting mammalian intermediate hosts, especially rodents and herbivores, the taeniids have successfully exploited the large carnivorous mammals (cats and dogs; Fig. 11.12). Their fluid-filled bladder worms or cysticerci differ in location

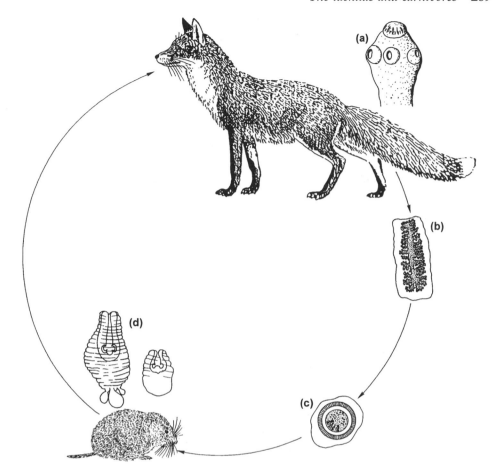

Figure 11.12 Life cycle of *Taenia crassiceps*. Adult tapeworms (*a*) in foxes and dogs. Gravid proglottides (*b*) voided in faeces. Eggs (*c*) eaten by rodents. Cysticercus (*d*) develops beneath the skin and buds developing at the posterior extremity become additional cysticerci. Carnivore becomes infected by preying on infected rodents. Source: based on Joyeux and Baer, 1961.

and construction from the cysticercoids of their relatives. Many cysticerci develop in the muscles and viscera of their mammalian intermediate hosts and lack the multilayered fibrous walls of the cysticercoids. These features reflect the fact that the large carnivorous mammals rarely chew their food; they swallow the prey whole or shear off large chunks of flesh, which are swallowed without mastication.

Bryant and Behm (1989) have highlighted an interesting difference between the metabolic pathways of tapeworms of carnivores and herbivores. In the sheep tapeworm *M. expansa*, fructose-1,6-biphosphate (FBP) is a feed-forward activator of the enzyme pyruvate kinase (Fig. 11.13). If glucose availability rises, the intracellular concentration of FBP also rises. This activates pyruvate kinase, permitting a greater carbon flux through the metabolic pathway. In contrast, in the dog tapeworm *Taenia serialis*, pyruvate kinase is not activated by FBP. Bryant and Behm suggested that this difference is related to the diets of their respective hosts. The carnivorous diet contains an abundance of fat and glycerol, while these substances are virtually absent from the diet of a herbivore. Consequently, in the glycerol-metabolizing

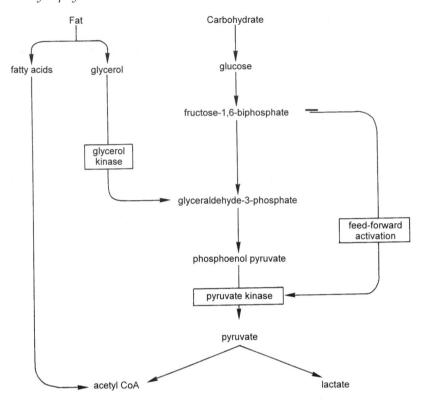

Figure 11.13   The feed-forward activation of pyruvate kinase by fructose-1,6-bisphosphate during catabolism. Source: reproduced from Bryant and Behm, 1989, with permission from Chapman & Hall.

tapeworm of a carnivore, FBP, which is derived from glucose intake, is a poor indicator of levels of carbon intake and therefore not a good controller of carbon flux.

We have already referred briefly to differences between the egg layers of cyclophyllideans like *Hymenolepis* and taeniids. As in *Hymenolepis*, the taeniid shell that is homologous with that of a pseudophyllidean cestode is greatly reduced, but in *Taenia* there is no secondary 'shell' like that of *Hymenolepis* (see Fig. 9.11). The *Taenia* egg has a thick outer envelope and the inner envelope gives rise to a substantial embryophore, consisting of blocks of keratin-like material cemented together, and an oncospheral membrane adjacent to the central oncosphere (see Fig. 8.4). The shell and the outer envelope may be lost in the faeces before the eggs are eaten by the intermediate host but, according to Gönnert (1970), the eggshell is not as fragile as previous literature would suggest.

Since the intermediate hosts of taeniids are mammals, their eggs face a particularly hazardous journey, having to cope with mastication and the highly acidic and proteolytic stomach contents. However, they are not the only tapeworms to have solved these problems, since the eggs of *Vampirolepis nana* habitually survive these hazards (see above). When the eggs of *T. saginata* are treated *in vitro* with artificial stomach juice (acid pepsin at 37°C) any remaining shell and outer envelope are removed and the cement between the embryophore blocks is digested away so that the embryophore disintegrates (Silverman, 1954b; Gönnert, 1970). Most of the oncospheres, still enclosed within their oncospheral membranes, remain quiescent if stored in an artificial intestinal juice containing pancreatin but no bile at 37°C. This state of suspended animation is terminated by the addition of bile, which puts an end to the impermeability of the oncospheral membrane. Nile blue sulphate, which fails to penetrate the oncospheral membrane prior to bile treatment, rapidly enters after treatment and stains the larva. The change in permeability is followed by activation and liberation of the larva. Thus, the oncospheral membrane appears to be the functional equivalent of the myelin-like layers of the cysticercoid of *V. microstoma*, which act as a protective barrier against the destructive H$^+$ ions of the stomach.

There may be differences between *Taenia* species in the way their egg layers respond to host enzymes. *T. saginata* eggs rarely undergo embryophore disintegration in pancreatin, even after prolonged treatment, unless they have been pre-treated with pepsin. In fact, after treatment in acid pepsin for 2–4 h the embryophores can usually be disrupted by vigorous shaking. On the other hand, *T. pisiformis* eggs are unaffected, even after 30 h in acid pepsin with constant mechanical shaking, whereas a 30 min treatment in intestinal juice only leads to disintegration of the embryophore (Silverman, 1954b). Presumably, differences in the molecular composition of the cement between the embryophore blocks are reflected in their vulnerability to different proteolytic enzymes.

11.5.2 Cysticercosis

Since both the definitive and intermediate hosts of taeniids are mammals, it is theoretically possible for the definitive host to take on the role of intermediate host, i.e. eggs eaten by the definitive host are potentially capable of hatching and invading the tissues (cf. *Vampirolepis nana*, section 11.3.4). This role reversal creates a sinister twist to the life cycles of taeniids involving humans, and the dangers were dramatically illustrated in an account by Gajdusek (1978), which will be summarized here.

In 1971, in the Wissel Lakes area of western New Guinea (now Indonesian New Guinea), physicians reported a sudden 'epidemic' of severe burns among members of a tribe known as the Ekari. These injuries, some so serious they required amputation of limbs, were all sustained when sleeping, patients suffering a seizure and falling unconscious into the household fire. The Ekari live at elevations of about 1500 m, where the cold at night necessitates a fire in the centre of each hut, in order to keep their virtually naked bodies warm. Individuals occupy bunks close to the fire. In the search for what appeared to be a new infectious agent, a survey of intestinal parasites was undertaken and about 8% of the faecal samples were found to contain eggs of *Taenia solium*, a tapeworm previously unknown in the area. Since the absence of eggs in the faecal samples does not necessarily mean that the individual is uninfected, a much higher level of infection with the worm was likely.

The question we are bound to ask is what have fits resembling epilepsy and severe burns to do with *T. solium*? The answer concerns an unpleasant feature of the eggs. If the eggs of *Taenia saginata* are eaten by a human, no infection ensues, but larvae of *T. solium* hatching from ingested eggs behave as they do in the pig intermediate host, i.e. they enter the wall of the human intestine and are carried in the circulating blood to tissue sites, where they develop into bladder worms or cysticerci (see Fig. 8.9). Thus, humans take on the role of intermediate host.

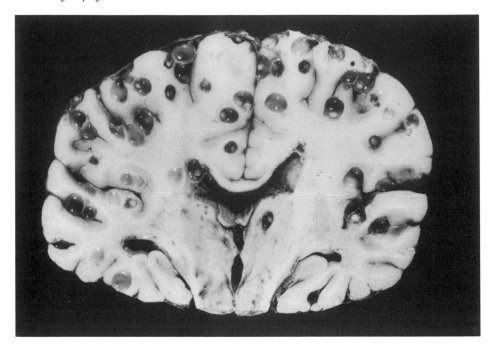

Figure 11.14 Section through the brain of a 9-year-old girl with extensive cysticercosis. The only symptom was mild chronic headache. A CSF puncture to determine the cause of symptoms led to ventricular collapse and death. Source: photograph by Dr Juan Olvera-Rabiela, reproduced from Flisser and Larralde, 1986, with permission from Dr A. Flisser and Academic Press.

Adult *T. solium* are rarely pathogenic. People harbouring a large tapeworm may be quite unaware of its existence – this was brought home to me by the experience of a Brazilian student who was horrified to find in his stool one day moving fleshy pieces which he thought were detached bits of his intestine. In contrast, the cysticerci may have a devastating effect. Some cysts may lodge in the brain (Fig. 11.14), where an inflammatory reaction may develop around them, leading to neurological disorders such as convulsions and bizarre personality changes. These manifestations may not appear until 2–5 years after contracting the infection. Occasionally cysticerci may develop in the eye (Fig. 11.15).

The Ekari patients themselves noted the presence of subcutaneous nodules, a phenomenon totally new to them, and, after surgical removal, these nodules were identified as cysts of *T. solium*. The presence of cysts in the brain of a young girl who died following progressive neurological complications confirmed that these were the cause of the fits and indirectly the burns. According to Desowitz (1981), serological tests in 1978 showed that at least 25% of the people had cysticercosis.

The rather remote region inhabited by the Ekari had been free of the parasite until 1971, even though pigs were highly prized and intimate members of their community. According to Desowitz (1981), the appearance of this disastrous disease was linked with politics. In 1969 the people of western New Guinea were directed by the United Nations to decide whether or not they wished to join the Republic of Indonesia. The Ekari were undecided and to help them make up their minds Indonesia sent troops to the area. An attempt was made to soften this overtly military

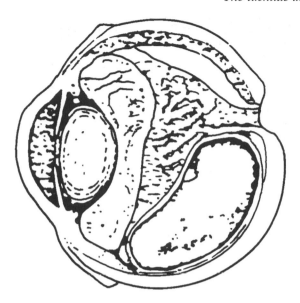

Figure 11.15 Subretinal cysticercus of *Taenia solium* in the human eye. Source: reproduced from Leuckart, 1886.

action by sending a gift of pigs, but, unfortunately, these animals came from the Indonesian island of Bali. The Balinese, being mainly Hindu, raise pigs for food and *T. solium* is endemic. The adult worm is common in humans but, because the Balinese have clean personal habits, human cysticercosis is almost unknown. In contrast, the Ekari have primitive toilet habits and, following the establishment of adult tapeworms in the Ekari from the Balinese pigs, cysticerci were widely contracted from eggs in the faeces.

A similar phenomenon in *Echinococcus* has even more serious health implications for man (see below).

11.5.3 Asexual reproduction

Asexual reproduction (cloning) in the intermediate host occurs more extensively in the taeniids than in the hymenolepidids and dilepidids (section 11.3.5). The oncosphere may give rise to many tapeworm heads by external (exogenous) or internal (endogenous) budding. In exogenous development, the scoleces are orientated towards the exterior (tegumentary) surface of the cyst-like organism, although this relationship may be obscured by invagination of each scolex so that it projects into the cyst cavity, as in *Taenia crassiceps* (fox–rodent life cycle; Fig. 11.12) and the coenurus cyst of *T. serialis* (dog–rabbit cycle; Fig. 11.16).

In *Echinococcus* spp. budding is endogenous. Grove (1990) gave a vivid summary of the experiences of Johann Goeze when he discovered in 1782 the tapeworm nature of what we now call a unilocular hydatid of *Echinococcus granulosus* (dog–sheep cycle; Fig. 11.17). Goeze was given a sheep liver full of watery vesicles of various sizes. When some of these were pierced liquid rushed out like a fountain. Within some of the hard and rather leathery vesicles he found smaller, soft, bluish vesicles and when these were opened he observed a greyish-white granular material connected to a very delicate membrane. When part of this membrane was placed in

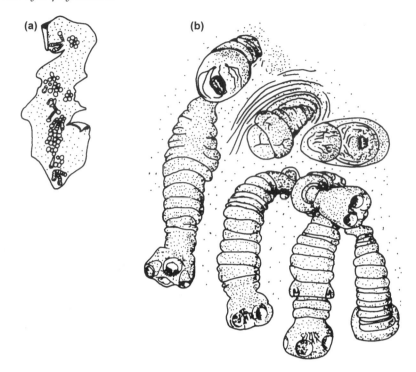

Figure 11.16 (*a*) View of the inner surface of part of the wall of an opened coenurus cyst of *Taenia serialis* from the body cavity of a rabbit. Note grouping of scoleces. (*b*) A group of scoleces from (*a*) at a higher magnification. Source: reproduced from Joyeux and Baer, 1961, with permission from Masson S. A.

water, the granules became detached and appeared to swim around individually. In one of the vesicles, the size of a pigeon's egg, he found many thousands of these granules which were barely visible to the naked eye and when he examined these with a magnifying glass he recognized on each of them the four suckers and hooks of a tapeworm.

We now know that the leathery wall of the primary cyst has an external fibrous coat derived from the host. The parasite cyst wall proper comprises an external laminated layer and an inner germinal layer, the latter being essentially similar to the adult tegument with surface microtriches projecting into the overlying laminated layer (see review by Thompson, 1995). This germinal layer buds off the bluish vesicles or brood capsules, which produce from their inner surfaces the tiny, greyish white granules, which are the tapeworm heads or protoscoleces. There are also reports of individual protoscoleces developing from the germinal layer (Fig. 11.18). Protoscoleces may be so numerous that when they sediment out they form a 'hydatid sand'. The fluid-filled cavity of the primary cyst may sometimes contain daughter cysts, which are identical in construction to the parent cyst and able to produce internally their own brood capsules from their germinal lining.

The alveolar or multilocular hydatid of *E. multilocularis* (fox–rodent cycle) contains several cavities producing protoscoleces. The alveolar cyst is invasive, growing and penetrating adjacent host tissue like a cancer and capable of liberating exogenous propagules that metastasize and establish themselves elsewhere in the host (see Thompson, 1995).

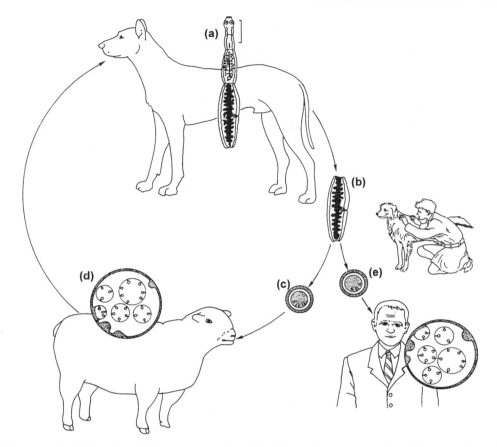

Figure 11.17 Life cycle of *Echinococcus granulosus*. The tiny adult tapeworms (*a*) (whole worm shown) occur in the dog intestine (scale bar: 1 mm). Gravid proglottides (*b*) are voided in faeces and eggs (*c*) are eaten by grazing sheep. Hydatid cysts (*d*) develop in sheep. Dogs are infected by eating protoscoleces from hydatid cyst in sheep carcass. Humans become infected with hydatid cysts by accidental ingestion of tapeworm eggs (*e*) picked up from fur of dog or from other contaminated surfaces.

Whitfield and Evans (1983) have summarized the most important features of the multiplication process in taeniid cysts. Exogenous and endogenous budding are usually regarded as fundamentally different proliferative processes because the outer surface of a hydatid is undoubtedly the tegumental outer surface and yet the developing protoscoleces are directed inwards. However, as Whitfield and Evans pointed out, most new protoscoleces are budded from the inner surfaces of the brood capsules and these capsules have their tegumental surface on the inside. Thus the polarity of scolex development in the brood capsules is similar to that in exogenous cysts.

An important distinction differentiates proliferative cysts of some taeniids from those of others. *T. crassiceps* is distinguished from related genera by the ability to produce scolex-bearing buds from the end of the cysticercus opposite to the scolex (see Fig. 11.12). These buds separate from the parent cyst and are themselves capable of budding off new cysts in what appears to be an unlimited process. Hydatids also seem to have a potentially unlimited capacity for asexual multiplication of proto-scoleces. An infective protoscolex, which would differentiate into a sexually repro-ducing adult worm in the gut of the definitive host, is capable of transforming to a

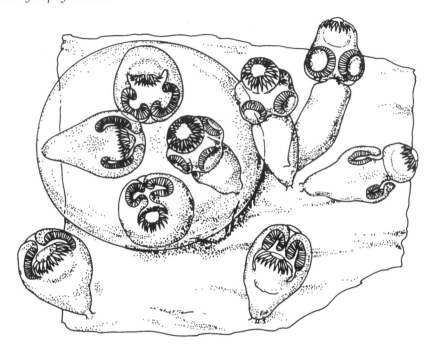

Figure 11.18 View of the inner surface of a small portion of the wall of a hydatid cyst of *Echinococcus granulosus*. Note separate protoscoleces, some evaginated, some not, growing from germinal membrane lining the cyst wall. Note also the brood sac, which has also grown from the cyst wall and contains four protoscoleces. Source: reproduced from Joyeux and Baer, 1961, with permission from Masson S. A.

hydatid cyst, itself able to produce new protoscoleces, if transplanted to the body cavity of a new intermediate host. This process has been repeated for at least seven generations with no apparent diminution of reproductive capacity. Presumably, infective protoscoleces of *Echinococcus* retain a supply of stem cells, which, like the germ cells of digenean germinal sacs (section 14.3.2), are totipotent and endow the protoscolex with the flexibility, when appropriately stimulated, to differentiate either into an adult worm or into a hydatid. Other taeniids capable of proliferation seem to be limited to a single round of histogenesis, producing more or less synchronously several new scoleces.

Whitfield and Evans (1983) also pointed out that extensive powers of regeneration after damage have only been recorded in tapeworm cysts that are capable of asexual multiplication (*T. crassiceps* and *Mesocestoides corti*). They suggested that the cells involved in repair and regeneration may be derived from the totipotent cells. Apart from limited regeneration after experimental damage in some digeneans, most other parasitic platyhelminths are unable to match the remarkable regenerative abilities of some free-living turbellarians.

In *Echinococcus*, asexual proliferation leads to large infrapopulations of small adults, each with few developing proglottides. Adult *E. granulosus* are only 4–6 mm long and usually have only three proglottides (Smyth, 1964; Fig. 11.17), but their large numbers compensate for their small size – 6000 or more in a dog, according to Mackiewicz (1988). This contrasts sharply with worms such as *Taenia solium* that have large adults and non-proliferative cysticerci. According to Moore (1981), the adults of proliferative species are short-lived and produce relatively resistant eggs at a high rate, while adults of non-proliferative species have long-lived adults producing less resistant eggs at a lower rate. Moore considered possible advantages of the proliferative strategy, some of which have already been mentioned (section 11.3.5), but it is apparent that we need a better understanding of the reproductive patterns and parasite–host interactions of small and large tapeworms before we can fully appreciate the significance of these contrasting strategies.

With regard to *Echinococcus*, Moore (1981) emphasized that a single hydatid is potentially capable of infecting more than one carnivore or scavenger. Since the cyst of *E. granulosus* is pressurized, the protoscoleces are likely to be scattered over the carcass when the cyst is punctured. However, this may have little real advantage since many potential intermediate hosts have multiple infections with small cysts (see Rausch, 1993).

Large cysts may affect the behaviour of the intermediate hosts in such a way as to increase their chances of falling victim to hunting carnivores. Coenurus cysts often lodge in the brain of sheep, where they produce a loss of coordination ('gid' or 'the staggers'), and there is evidence that the presence of hydatid cysts in some animals may increase their vulnerability to predators. Rau and Caron (1979) showed that moose, *Alces alces*, heavily infected with hydatid cysts are more likely to be killed by human hunters and Mech (in Rau and Caron) has shown that heavily infected moose are more vulnerable to wolves (*Canis lupus*), their natural predators. Rau and Caron found that the golf-ball-sized cysts are located in the lungs and that the parasite is overdispersed, i.e. relatively few animals have large numbers of cysts (up to 100). The suggestion has been made that the cysts interfere with breathing, not when the moose is resting but when it is forced to exert itself to escape from predators. Rau and Caron suggested that the role of parasites in the predator–prey equation deserves further attention, since the size of the predator population is determined by the number of available prey organisms, which may be related to the abundance and distribution of parasites; a reduction in the number of parasite-debilitated hosts could result in a reduction of the predator population or possibly a shift to alternative prey animals. Thus, parasites may determine not only which animals are taken by predators, but also, in part, how many are taken.

11.5.4 *Echinococcus granulosus* and man

Two major types of *E. granulosus* have been distinguished (see review by Rausch, 1993). These have been designated the Northern and European biotypes. The Northern biotype is indigenous to the tundra and boreal forest throughout the Holarctic and cycles between the wolf (*Canis lupus*) and deer (especially reindeer, *Rangifer tarandus*, and *Alces alces*). This biotype is considered to be ancestral to the European biotype, which has adapted relatively recently to domestic hosts associated with man (synanthropic hosts), having appeared since domestication of the wolf (= dog) and various ungulates. A parallel dichotomy occurs in *Taenia* spp., with biotypes of *T. ovis* and *T. hydatigena* being perpetuated in the north by the predator–prey relationship between wolf and deer and biotypes at lower latitudes cycling between synanthropic hosts.

As in *T. solium*, humans are acceptable intermediate hosts for *E. granulosus*. Regarding the Northern biotype of *E. granulosus*, most cases occur in indigenous peoples who hunt or herd reindeer. Infections are typically benign, with zero mortality, and the cysts are small (average diameter 4 cm) and asymptomatic, with low productivity of protoscoleces. In contrast, the close association between sheep, dog and man favours human infection with the more pathogenic hydatid cyst of the European biotype (Fig. 11.17). Surfaces contaminated by dog faeces, including the fur of the dog, provide a source of eggs, which may be transferred accidentally to the mouth, especially by children, and the feeding of offal from domestic animals such as sheep to farm and house dogs encourages transmission. Other domestic animals serve as intermediate hosts for the European biotype, and strains using horses, cattle, pigs and camels may each be genetically distinct (Schantz *et al.*, 1995).

It is generally supposed that hydatid cysts in modern man are unable to contribute further to the life cycle, but this may not be so in the Turkana tribesmen of north-western Kenya. According to Macpherson (1983), many of the dead are not buried but left on the surface where they are consumed by dogs or wild carnivores. Respected old men and married women with children may be buried, but not deep enough to prevent disinterment by carnivores. Since primary infections in the Turkana are fertile, they are likely to be an important source of infection for canids such as dogs or jackals.

In *E. granulosus* there is virtually no role reversal with regard to dogs, which rarely contract hydatid cysts. How this natural resistance is achieved is uncertain, although Berberian (1957, in Smyth, 1964) showed that few eggs hatched in dog bile and that the oncospheres were immobile.

More is demanded of the oncospheres of *E. granulosus* in terms of penetrative ability than those of *H. diminuta*, since they must bore their way into the thick and complicated wall of the mammalian intestine as far as the lamina propria (see Fig. 9.3) and enter a portal venule. In the pig, larvae appear in the portal vein within 8 h of eating eggs (Dew, 1925 in Grove, 1990). The liver capillary network traps many of the larvae arriving *via* the hepatic portal system, but some pass through and are trapped in their turn in the lung capillaries. Occasionally larvae manage to pass through this second filter, enter the systemic circulation and finish their journey in sites such as the brain, heart or bone marrow. Davaine (1860 in Grove, 1990) found that nearly 50% of cysts occur in the liver of man and 10% in the lungs, with the rest elsewhere in the body. More recent surveys recorded a somewhat larger proportion of liver hydatids at the expense of non-pulmonary cysts and revealed that one-third or more of infected humans had multiple cysts (references in Grove, 1990). According to Heath (1971), unilocular hydatids in ruminants commonly develop in the lungs (see above); he relates this to the relatively large villus lacteal of ruminants providing an opportunity for transportation *via* the lymphatic circulation.

The rate of growth of unilocular hydatids in humans is slow and variable, this variability being related to the site of the cyst and the resistance to expansion provided by the surrounding tissues. In fact, most clinical manifestations of echinococcosis are due to the pressure effects on the adjacent tissues. Because of slow growth, infections may be asymptomatic for many years, although cysts in the nervous system may betray their presence at an early stage. Cysts in the bone marrow have little space to expand and the pressure leads to necrosis, thinning of the bone and ultimate spontaneous fracture. Cysts in the heart may precipitate sudden death. Anaphylactic shock, often fatal, may ensue from spontaneous or traumatic rupture of a cyst, and complications may arise from secondary bacterial infections (see Grove, 1990; Roberts and Janovy, 1996).

High frequencies of echinococcosis occur in areas of the world where intensive sheep rearing is conducted, such as Iceland, Australia, New Zealand, Argentina, North Africa and the Middle East. Goats and camels make a contribution to transmission in the two last-named areas. *E. granulosus* was undoubtedly brought to Australia with sheep from Europe in the late 18th and early 19th centuries, but the existence of a dingo–macropod cycle raises the possibility that the parasite may have made a separate earlier entry into Australia in dingoes introduced by aboriginal immigrants (Thompson, 1979).

Iceland has many unique features. One of them, the absence of dogs in the capital, Reykjavik, is directly related to tapeworms and the following parasitological saga is taken from Grove (1990).

In the middle of the 19th century echinococcosis had become an epidemic in Iceland, with one in seven of the human inhabitants infected and an exceptionally high population of dogs – one for every five inhabitants. In 1862, an English physician, Arthur Leared, went to Iceland and suggested that the problem could be dealt with by treating all the dogs in the country simultaneously with an anthelminthic. He advised the use of kamala, a resin derived from glands and hairs of an Asian fruit and containing substances that paralyse tapeworms. His plan was translated into Icelandic, published in two newspapers and then submitted for approval to the College of Health in Copenhagen, Denmark, which held sovereignty over Iceland at that time. This move was a necessary preliminary to gaining approval from the Icelandic Legislative Assembly. The head of the College expressed his disapproval of foreign interference in the affairs of Denmark and Leared's plan was rejected, in spite of the established success of kamala against the worm and the fact that the plan had the support of the chief Icelandic physician.

Denmark sent an adviser, whose solution was to recommend the killing of large numbers of dogs, a proposal that was rejected by the Icelanders. The turning point came in 1890 when treatment of dogs with anthelminthic and burial of carcasses likely to be infected with hydatid cysts were enforced by Icelandic law, and an educational programme was set in motion to increase the awareness of the populace. Taxation was introduced to control dogs; the inhabitants of Reykjavik (a quarter of the Icelandic population) were forbidden to keep these animals. Perhaps the most important development was the instigation of a change in farming practice such that male lambs were killed at a younger age (4–5 months), before cysts had had time to mature. So effective were these measures that by 1927 there was only one dog per 15 inhabitants and only one hydatid infection per 1500 people.

Local attitudes, customs and practices may facilitate transmission of *E. granulosus* to humans. Higher levels of hydatidosis in Christian Lebanese compared with the Muslim population were revealed in a study by Schwabe and Daoud (1961). This may reflect differences in attitudes to the owning and handling of dogs, partly stemming, perhaps, from religious teaching. In the same study, the prevalence of hydatidosis among shoemakers was striking. The handling of street shoes by cobblers and the habits of placing tacks in the mouth and moistening thread with the lips are likely to increase the risk. In addition, when preparing leather, hides are often softened by immersion in concoctions of dog or bird faeces, a practice fraught with obvious dangers.

The Turkana people of Kenya have the highest rate of infection in the world with cysts of *E. granulosus*. This has been attributed to an alleged partiality for consuming dog intestine, but there is no evidence that they indulge in this practice (see Nelson and Rausch, 1963). It seems more likely that intimate contact between dogs and children is a significant factor in the high prevalence of the parasite in humans.

Surgery provides the most useful treatment for hydatidosis, but chemotherapy is advancing and a relatively new technique involving the temporary injection of ethanol into the cyst is effective and gives immediate relief (see Eckert *et al.*, 1995).

11.5.5   *Echinococcus multilocularis* and man

Echinococcus multilocularis is a parasite with a discontinuous distribution in the northern hemisphere, with endemic areas in North America, Europe, Russia, China and Japan. The parasite circulates between wild foxes and rodents (voles) – the so-called sylvatic cycle – but some of these wild rodents may be taken by domestic carnivores, especially dogs, providing a route for human infection with the alveolar cysts. Incidence of alveolar hydatid disease (AHD) in Europe is low – an annual incidence of 0.78 per 100 000 people was recorded in the worst affected canton in Switzerland. Much higher rates of between 65–170 per 100 000 have been reported among Eskimo people in parts of Alaska and in indigenous groups in northern Siberia, but some of the highest prevalence rates with AHD are reported from China, possibly as high as 410 cases per 100 000 in one area of study (Craig *et al.*, 1992).

Risk factors for infection in Europe are poorly understood (Schantz *et al.*, 1991), but an attempt was made by Stehr-Green *et al.* (1988) to define these factors in an Alaskan Eskimo population, in an area where the Arctic fox and voles are involved in the sylvatic cycle and domestic dogs provide the link with man. Case patients were matched with uninfected controls of the same age and sex living in the same area. The results suggested that lifelong ownership of domestic dogs was more significant as a risk factor than engagement in the trapping and skinning of foxes. The risk from dogs appeared to be linked more closely with tethering the animals near the house than with feeding or watering dogs, harnessing them for sledding or permitting them to enter the house. Patients were significantly more likely than controls to have always lived in dwellings built directly on the tundra rather than on a permanent or gravel foundation. Although vehicular transport has largely replaced sled dogs, some animals remain as pets and the risk of infection with AHD persists.

There is evidence that AHD is spreading in some areas (Schantz *et al.*, 1991). Factors that may have contributed to this are as follows: in Japan, the increasing tendency for foxes to adopt a commensal way of life, feeding on animal carcasses and wastes in farming areas; in the United States, the transportation of foxes and coyotes by hunting clubs from one area to another; in Europe, an increase in fox populations as a result of successful vaccination programmes against fox rabies.

AHD is the most lethal of the helminth infections of man and difficult to cure. Some of the factors related to treatment and control have been summarized by Schantz *et al.* (1991) and by Eckert *et al.* (1995). Surgery is still the only curative therapy, but its success depends on early diagnosis. Progress has been made in diagnostic screening using parasite antigens and enzyme-linked immunosorbent assays (ELISAs). In Japan, where serological screening of populations at risk has been practised for more than 20 years, the success rate of surgical treatment is higher than the success rates of 25–57% in other areas. Historically, 90% of patients with inoperable hydatids die within 10 years. Patient survival may be prolonged in such cases by treatment with benzimidazole derivatives, which inhibit larval growth but have limited parasitocidal effects. Some patients have been treated by liver transplantation.

A different approach, which has met with some success, is the treatment of dogs with the drug praziquantel on St Lawrence Island, Alaska and the use of praziquantel-loaded baits for red foxes in Germany.

11.6   *DIPYLIDIUM* AND THE CARNIVORES

As van Beneden discovered in 1858 (section 8.5), there is another way in which cyclophyllidean tapeworms can gain access to carnivorous mammals. Cysticercoids of the tapeworm *Dipylidium caninum* occur in the dog flea, *Ctenocephalides canis* (Fig. 11.19). They depend on the irritation inflicted on the dog host by the bite of the flea for their survival, since this irritation leads to grooming by the dog and

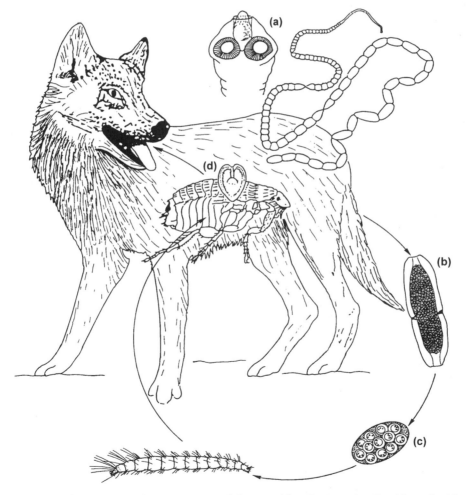

Figure 11.19 Life cycle of *Dipylidium caninum*. Adult worm (*a*) in dog intestine. Gravid proglottides (*b*), containing uterine capsules enclosing several eggs, pass out in faeces. Uterine capsules (*c*) eaten by flea larva and oncosphere enters haemocoel. At metamorphosis adult flea associates with dog and encysted parasites (*d*) complete their development. Irritation of flea bite leads to ingestion of infected flea by dog.

accidental ingestion of the infected flea. This seems a rather precarious means of propagation for the tapeworm, as we might expect natural selection to favour a reduction in the irritation caused by the feeding flea and a corresponding reduction in the proportion of fleas ingested by the dog.

The parasites develop very slowly until the adult flea begins to feed on a mammal. Pugh (1987) showed that this developmental spurt is brought about by the increased temperature provided by the warm body of the mammalian host and is not a consequence of the intake of blood by the adult flea.

Each proglottis of *D. caninum* contains two sets of reproductive organs; this is a feature that we have met previously and undoubtedly boosts egg output (sections 10.3, 11.4).

11.7   MESOCESTOIDES CORTI

Adult cestodes of the genus *Mesocestoides* live in the intestines of carnivorous birds and mammals, which acquire their worms by preying upon vertebrates (amphibians, reptiles, birds and small mammals) containing a so-called tetrathyridium stage in the liver or body cavity. It is assumed, but not known, that there is an arthropod first intermediate host, which is infected by eating the eggs or uterine capsules (section 11.2.2(b)) of the worms.

The tetrathyridium of *M. corti*, originally isolated from lizards (see Specht and Voge, 1965), has an exceptional feature, being able to multiply asexually not by budding as in other tapeworms but by longitudinal fission. In fact, the parasite has been maintained for experimental research for many years simply by intraperitoneal passage in rodents. This reproductive feature does not seem to be shared by other species of the genus. For example, tetrathyridia of *M. leptothylacus* from the body cavities of voles, *Microtus arvalis*, often survive but do not multiply if transferred experimentally to the body cavities of other rodents (Loos-Frank, 1980).

The tetrathyridium of *M. corti* is a flattened, maggot-like organism, measuring no more than 3 mm in length, with four apical suckers and a parenchymatous body containing muscles and a pool of germinative cells (Hess, 1980). The tetrathyridium is freely mobile and is not encumbered by a cyst wall of parasite or host origin. A median groove separates the suckers into two pairs and it is the deepening of this groove that initiates the longitudinal division (Fig. 11.20). Hess (1972) showed that division is only partial, the two daughter tetrathyridia, each inheriting two suckers from the mother, separating from the undivided posterior region which is lost. Each daughter regenerates a new pair of suckers to raise its complement to four. Further divisions continue unabated in experimentally infected mice until after 1.5–2 years the abdominal cavity is filled with a creamy mass of tetrathyridia and death ensues (Mueller, 1972).

The proliferative ability of this parasite extends in another direction. Eckert, von Brand and Voge (1969) fed dogs each with approximately 1500 tetrathyridia and found over 40 000 worms in each intestine several weeks after infection. In other words, the tetrathyridium is able to undergo asexual multiplication in the intestinal lumen, a phenomenon apparently unique in the cestodes. Even more surprising was their finding of many worms in the intestine of a dog injected peritoneally with tetrathyridia (cf. *Proteocephalus ambloplitis*, section 10.8.3), demonstrating another dimension in the versatility of these organisms, namely the ability to penetrate

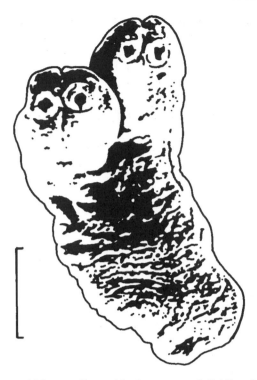

Figure 11.20   Tetrathyridium of *Mesocestoides corti* in the process of dividing. Scale bar: 50 μm. Source: drawn from a scanning electron microscope photograph by Hess, 1980.

host tissues. It was already known that tetrathyridia could move in the opposite direction from the gut to the body cavity in dogs and in other hosts such as mice. This invasive ability of tetrathyridia opens up another unique opportunity for transmission. Conn and Etges (1983) showed that tetrathyridia found their way to the mammary glands of mice and that nursing mothers passed on the parasite to their offspring in the milk. No evidence for *in utero* transmission was found, and it is not yet known whether the remarkable properties of the strain of *M. corti* maintained for so long in laboratory isolation are shared by natural populations of the parasite.

CHAPTER 12

Unravelling the digenean life cycle

As with the cestodes, the digeneans will be introduced by an historical account of the unravelling of their life cycle, with special emphasis on that of *Fasciola hepatica*. This task was made especially difficult because of extra complexity in the cycle.

Because of its association with liver rot in sheep, *Fasciola hepatica* was probably a familiar organism for a very long time before the first written reference to the animal in 1379 (see below). The flattened, leaf-like adults are large, reaching 3 cm in length and 13 mm in width, and cannot fail to attract attention as they emerge from the severed bile ducts of an infected liver. In an account of the disease in a book on animal husbandry published in 1523 and ascribed to an English lawyer, Anthony Fitzherbert, the parasites were referred to as 'flokes' (= flukes), a name thought to reflect their similarity in shape to a flat-fish (the Anglo-Saxon word *floc* means a flounder; see Grove, 1990).

The unravelling of the life cycle of *Fasciola hepatica* is a monumental detective story involving many great biological minds and spanning several centuries, from the first recorded reference to the adult parasite in 1379 to the discovery of how the parasites reach the biliary system of the liver in 1914. An excellent account of this story is given by Grove (1990) and will be outlined here (references as in Grove, unless consulted separately). The main features of the life cycle are summarized in Fig. 12.1.

Jean de Brie, commissioned by the French king Charles V to write a treatise on sheep management, first recorded the adult parasite in 1379, and he was close to the truth when he associated sheep liver rot with the consumption of plant material. However, he appeared to regard this food material as having a direct corrupting influence on the liver, with the flukes arising as a secondary consequence of this damage. During the next three centuries the adult worms spasmodically attracted the attention of researchers, including Francesco Redi, but it was not until 1698 that a significant advance was made. In that year, Bidloo, in a letter to the Dutch microscopist Antony van Leeuwenhoek, described his finding of eggs in the adult parasite and this led him to reject earlier suggestions that the worms were generated by factors such as decay, excessive wetness and heat. Bidloo believed that the worms bred in moist earth and were ingested by herbivorous animals, and this stimulated van Leeuwenhoek to search without success for microscopical creatures resembling the adults on pastures inhabited by infected sheep.

At this point stagnation set in, and further progress had to wait until 1773, when Müller discovered some free-living, freshwater organisms which he believed were independent adult animals. He gave them the generic name *Cercaria*. However, when Nitzsch took up their study he recognized their resemblance to flukes, but regarded their mobile tails as separate organisms (*Vibrio*) combined with the

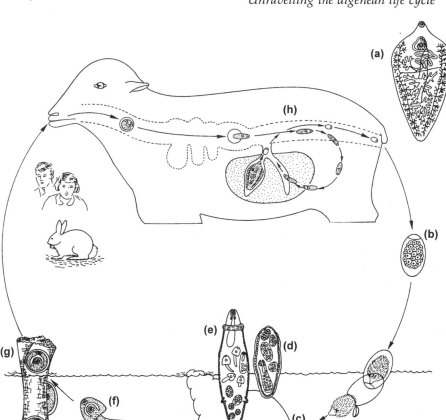

Figure 12.1 The life cycle of the digenean *Fasciola hepatica*. Adults (*a*) live in bile ducts in sheep liver. Eggs (*b*) in sheep faeces enter water, develop and liberate a ciliated miracidium (*c*). Miracidium penetrates amphibious snail *Lymnaea truncatula* and transforms in haemocoel into a sporocyst (*d*). Sporocyst liberates many rediae (*e*), each of which produces many cercariae (or another redial generation). Cercaria (*f*) escapes from snail and, propelled by a muscular tail, attaches to vegetation. Encystment follows loss of tail. Encysted cercaria (metacercaria) (*g*) is eaten by grazing sheep. Excystation occurs in intestine and juvenile fluke (*h*) passes through intestine wall, migrates to liver and penetrates to reach bile duct lumen, where sexual maturity is reached and egg laying begins. Rabbits are potential hosts and cysts on watercress are an occasional source of human infection. Source: based on Olsen, 1986.

Distomum body. Since the organisms had no reproductive organs, their origin was a puzzle and many believed that they owed their existence to spontaneous generation. Then, in 1818, Bojanus found cercariae issuing forth from large, worm-like, yellow sacs embedded in the viscera of freshwater snails. It was astonishing that cercariae were derived not from parents like themselves but from these peculiar sacs or 'royal yellow worms' as Bojanus called them, and this led Oken, the publisher of Bojanus's work, to remark that 'observations of this kind make one dizzy'. However, he was not too dizzy to make the perceptive remark that 'one might lay a wager that these cercariae are the embryos of distomes'. The royal yellow worms were given the generic name *Redia* in honour of Francesco Redi by de Filippi in 1837, but he later considered them to be a larval stage and recommended suppression of '*Redia*' as a generic name.

In 1831 Mehlis turned his attention to the eggs of digeneans that were called at that time *Monostomum flavum* and *Distomum hians*. He observed a ciliated larva escaping from them *via* an opening in the shell previously closed with a lid or operculum, and in 1837 Creplin showed that the eggs of *F. hepatica* likewise produced a ciliated larva. Braun later called the larva a miracidium (= 'youthful person'). This was another surprising discovery and at odds with the popular idea that infection of sheep was acquired by ingestion of eggs, but this hypothesis was not fully laid to rest until Simonds (1852) demonstrated that no flukes were found in sheep fed with fluke eggs.

It then became von Siebold's turn to make a contribution. He became interested in *Monostomum mutabile*, a fluke with an unusual habitat, the orbital cavities of geese, and a somewhat telescoped life cycle that fortuitously provided a link between miracidia and rediae. He observed the hatching of the ciliated miracidium, which after a while disintegrated and released a redia identical with cercaria-containing rediae that he had seen in snails. He postulated that free-swimming miracidia emerging from eggs laid by the adults penetrated the tissues of a snail and then died, releasing cercaria-sacs (rediae). Support for this theory was provided by Wagner (1857), who observed the metamorphosis of the miracidium of the frog parasite *Distoma cygnoides* into a redia. The fate of the cercaria remained unknown, although von Siebold reported that some cercariae encysted in snails.

In spite of these important advances, most authorities of the time were unable to make any sense out of these irregular and apparently disconnected observations. The situation was saved by the timely publication of a book entitled, in English translation, *On the Alternation of Generations: A Peculiar Form of Fostering the Young in the Lower Classes of Animals* by the Dane Johannes Steenstrup in 1842 (English translation, 1845). This expounded the discovery that some adult animals did not give rise directly to a new generation of similar adults but interposed a second generation of organisms that might differ substantially from the parents. Steenstrup quoted as one of his examples the coelenterates, in which the eggs produced by individuals of the free-swimming, medusoid generation produce an alternate generation of sedentary 'nurses' or polyps which produce a new generation of medusae. In 'trematodes' (= digeneans) he recognized the same phenomenon, eggs laid by the adult flukes producing one or more generations of 'nurses' (= rediae). He pointed out that 'several forms which have been considered as different species and genera are seen to be stages in the development of one and the same animal'. Steenstrup observed cercariae swimming freely using their tails and described the loss of the tails and encystation in other snail hosts. He appreciated that these encysted organisms (he called them pupae) were miniature adult flukes, but erroneously supposed that they emerged and became mature inside the snail. Nevertheless, Steenstrup's concept of the alternation of generations gave added impetus to the search for the missing pieces in the puzzle of the digenean life cycle and provided the kind of background needed for the correct assembly of all the pieces. It provided similar impetus to students of the cestode life cycle (Chapter 8).

Von Siebold then proposed that encysted cercariae had to be transmitted to another host before they could mature. Support for this was provided by La Valette de St George (1855), who showed that cysts (= metacercariae) of *Cercaria echinifera* collected from snails excysted in the intestine of warm-blooded animals and gave rise to *Distoma echinifera*. At this time the life cycle of *F. hepatica* was still unknown and Küchenmeister, stimulated by the work of his contemporaries on other digeneans, suggested that herbivorous or omnivorous mammals (including humans) become infected with *F. hepatica* by consuming snails on herbage or vegetables or in drinking water. However, Weinland (1875) found rediae in a small amphibious snail, *Limnea* (now *Lymnaea*) *truncatula* and observed that cercariae from these snails tended to leave the water by climbing upwards on foreign objects. He suggested that they might be *F. hepatica* and speculated that the cercariae might encyst on grass and be eaten by sheep.

At this point, two workers set out independently to investigate the elusive life cycle of *F. hepatica*. Two more different researchers could hardly be imagined. One of them, Leuckart in Germany, was arguably the most eminent parasitologist of the time, while the other, Thomas in England, was a 23-year-old demonstrator in biology at Oxford University's Museum. Leuckart won the race by a few days in terms of priority of publication, but Thomas's achievement is remarkable in view of his lack of previous experience. His account of the life cycle (Thomas, 1883) is lucid, accurate and informative.

Thomas was given his task by Rolleston, a professor of anatomy and physiology at Oxford. A major outbreak of sheep liver rot in Britain in 1879–1880 had killed more than three million sheep and the Royal Agricultural Society funded the investigation. Thomas began with the eggs (see Chapter 13 for detailed consideration of the biology of *F. hepatica*). He showed that at 23–26°C the miracidia began to hatch after 2–3 weeks, although hatching continued for weeks or even months. He observed that the miracidia survived for only 8 h in water. He assumed that a mollusc acted as intermediate host and he exposed a range of slugs and snails, including *L. truncatula*, but failed to infect any of these animals. He then changed his tactics and collected invertebrates, including snails, from farms near Oxford where there had been severe outbreaks of sheep liver rot. Among these animals was a single specimen of *L. truncatula* that released very active, tadpole-like cercariae with the ability to encyst on surrounding surfaces and, although Thomas suspected that this was *F. hepatica*, it was more than a year before he could collect *L. truncatula* for exposure to miracidia. He was perhaps encouraged by the high incidence of fascioliasis in the Faeroe Islands and by the fact that *L. truncatula* was one of only eight molluscs to be found there.

When this experiment finally became possible, he found that the exposed snails became heavily infected, the miracidium actively penetrating into the snail's body and transforming into a sac-like sporocyst which grew to a length of 0.5–0.7 mm. Within each sporocyst several rediae developed, each with its own pharynx and intestine – organs that were absent in the sporocyst. Thomas observed that the rediae escaped from the sporocyst by rupturing the body wall, the wound sealing up after their exit, but the offspring that developed inside rediae (see below) escaped through a birth pore. The rediae were very active, spreading through the body of the snail and demonstrating a preference for the digestive gland. The yellow gut contents of the rediae indicated the ingestion of snail digestive gland cells.

Thomas recognized two kinds of rediae, one kind producing daughter rediae and the other producing cercariae. He claimed that the former were smaller, with a larger pharynx and gut, and that they usually contained one to three daughter rediae (maximum ten), while cercaria-producing rediae were said to contain up to 23 offspring. He believed that redia-producing rediae occurred during warm weather, while only cercariae were produced by rediae in the cold months. This is discussed further in section 13.4.2.

The body of the fully developed cercaria was up to 0.3 mm long with a tail more than twice the body length. Thomas noted the presence of granular cells, which rendered the body opaque in transmitted light and white and conspicuous by reflected light. Although vigorously propelled by the tail on leaving the snail, Thomas found that their free-swimming life was short; they soon came to rest when they contacted the glass wall of the aquarium or water plants. Here the cercaria assumed a rounded form and poured forth a mucous substance together with granules from the body glands (cystogenous glands). This hardened to form the cyst wall. The tail was lost just before or during encystation. The cysts were snowy-white, but the contained metacercaria was transparent.

Thomas demonstrated the relatively strict host specificity of *F. hepatica* by exposing many other species of snails and slugs to miracidia. All failed to become infected, with the exception of young specimens of *L. pereger*, in which development did not proceed beyond an early stage. Thomas emphasized that the behaviour of *L. truncatula* was well suited for the infection of terrestrial herbivores, since this snail, unlike its relatives, spends more time out of water than in it and may be found not only on the banks of ditches but further away in the fields if the grass is damp. However, he indicated the evolutionary adaptability of the parasite by pointing out that it occurred in Australia where *L. truncatula* did not occur. (The most important snail host in Australia is the endemic lymnaeid *Austropeplea tomentosa*).

It was a Brazilian, Lutz, working in Hawaii with a parasite that was probably *F. gigantica*, who demonstrated experimentally in 1892 that livers of guinea pigs became infected with flukes after ingesting cysts. The final surprise was the discovery by a Russian, Sinitzin, in 1914 that, although young *F. hepatica* emerged from their cysts in the rabbit intestine, they did not take the obvious route to the liver *via* the bile duct opening but penetrated right through the intestine wall into the peritoneal cavity, where they remained for up to 2 weeks. They then entered the liver and migrated deeper and deeper until they reached the larger bile ducts, where they achieved maturity and began to lay eggs about 2 months after infection.

The biology of *Fasciola hepatica* will be considered in more detail in the next chapter.

CHAPTER 13

The biology of Fasciola hepatica

13.1   THE ADULT

Although a relatively unspecialized digenean and therefore a suitable text-book 'type', *Fasciola hepatica* has one unusual feature and that is its adult location in the bile ducts of the liver rather than in the intestine, the habitat of most adult digeneans. However, in spite of the potentially damaging effect of bile on cells and especially on their membranes, the biliary system has become the habitat for a number of platyhelminths; it has been colonized independently by, among others, the unrelated digenean *Dicrocoelium dendriticum* (section 16.3.4(f)), the aspidogastrean *Stichocotyle nephropis* (section 19.2) and the tapeworm *Vampirolepis microstoma* (section 9.6).

13.1.1   General features

Like monogeneans and pseudophyllidean cestodes, digeneans have all the equipment for making tanned operculate eggs but in spite of the intestinal location of most of them there has been no tendency to increase egg output by replicating the genitalia and digeneans have a single set of reproductive organs including a single ootype (Fig. 13.1). Digeneans also retain their alimentary canal. *F. hepatica* has an anterior mouth leading to a muscular pharynx and then to a branched intestine (Fig. 13.1a–c). The extensive dendritic gut of *Fasciola* (Fig. 13.1b) is a reflection of its large size and flattened shape (section 1.4.3) but smaller cylindrical digeneans have unbranched gut caeca.

The failure of digenean adults to replicate their genitalia and lose their gut may reflect a more recent colonization of the vertebrate intestine compared with that of the cestodes. Digeneans have an intimate and undoubtedly long-established relationship with molluscs, and vertebrate hosts may well have been acquired at a relatively late stage in their evolutionary history (section 3.3).

Digeneans attach themselves by muscular suckers, which are highly effective for attachment to soft and pliable substrates by drawing in a plug of host tissue and retaining it in the cavity of the sucker. This in itself is not a unique feature, since suckers have evolved independently in both the monogeneans and the tapeworms. What is unique in digeneans is the number and arrangement of these suckers, with one of them (the acetabulum) typically situated on the ventral surface and the other (the oral sucker) perforated by the buccal tube and incorporating the mouth as its aperture (Fig. 13.1c). So consistent is this pattern of two tandem suckers that the generic name *Distoma*, introduced by Retzius in 1786 (in Grove, 1990; later changed to *Distomum*), incorporated most of

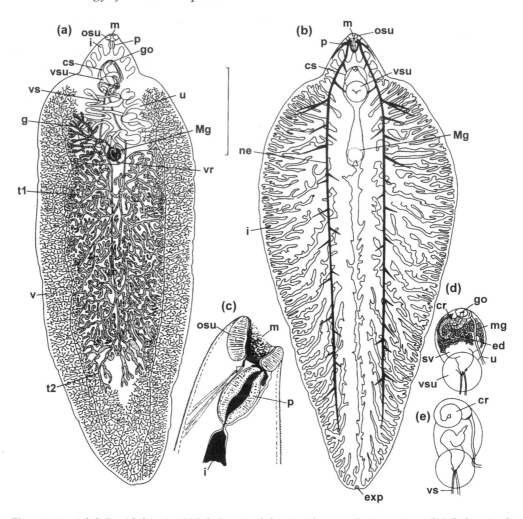

Figure 13.1 Adult *Fasciola hepatica*. (*a*) Whole animal showing the reproductive system. (*b*) Whole animal showing the gut and main nervous system. (*c*) Sagittal section through the oral sucker and pharynx. (*d*) Terminal genitalia with cirrus withdrawn. (*e*) Terminal genitalia with cirrus everted. cr = cirrus; cs = cirrus sac; ed = ejaculatory duct; exp = excretory pore; g = germarium; go = genital opening; i = intestine; m = mouth; Mg = Mehlis's gland; mg = male accessory glands; ne = longitudinal nerve; osu = oral sucker; otg = ootype glands; p = pharynx; sv = seminal vesicle; t1, t2 = testis 1 and 2; u = uterus; v = vitellarium; vr = vitelline reservoir; vs = vas deferens; vsu = ventral sucker. Scale bar: (*a*), (*b*) 5 mm. Source: redrawn from Sommer, 1880.

the digeneans known at the time and would include most of the digeneans recognized today. Stepwise locomotion can be achieved by using these two suckers alternately and exploiting the extensibility and contractility of the body region between the suckers, the bulky reproductive organs being housed behind the suckers. The oral sucker has an additional role in feeding, which will be considered below.

The tegumental architecture of *F. hepatica* is basically similar to that of a monogenean or cestode, with a syncytial cytoplasmic layer at the surface linked by internuncial processes to nucleated cytons beneath the tegumentary muscle layers. The structure and development of the tegument is considered more fully below. The exposed surface is not microvillous but contains backwardly projecting spines. The scolex hooks of cyclophyllidean tapeworms also have a tegumental origin but appear to be derived in a unique way from microtriches (section 11.2.1). Like the hooks of the tapeworm scolex, digenean spines are proteinaceous but they lack the amino acid cystine, which is abundantly present in the keratin-like hooks of cestodes (Pearson *et al.*, 1985). It was suggested by Cohen *et al.* (1982) that digenean spines were composed of actin, but this is not supported by the finding of Pearson *et al.*, who failed to detect the amino acid 3-methylhistidine, a characteristic constituent of actin. Hooklets of subtegumentary origin equivalent to the marginal hooklets typical of monogeneans and cestodes are always absent in digeneans.

In addition to an extensive system of flame cells (protonephridial system), with collecting ducts draining into a bladder opening posteriorly by a median pore, *F. hepatica* also possesses a so-called paranephridial system (= reserve excretory system) of uncertain function. This consists of branching and anastomosing tubules without flame cells and connects with the bladder (Fig. 13.2). According to Pearson (1986), this system is so widespread in digeneans that it is undoubtedly a major feature of the group; he considers that the so-called lymphatic system of paramphistomids (section 15.6.3) is a paranephridial system that loses its connection with the

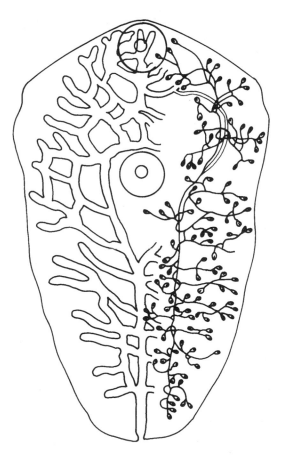

Figure 13.2 Protonephridial system (shown on the right side only) and paranephridial system (shown only on the left side) of a 2-week-old adult *Fasciola hepatica*. Source: redrawn from Kawana, 1940.

excretory system during development. Thus, the paranephridial system may have been an ancestral feature that has been lost or modified in some digenean lineages.

13.1.2 The diet and feeding

In view of the attention focused on *F. hepatica* it is rather surprising to find that there used to be a difference of opinion regarding the food of the adult parasite. The two opposing views were represented by Jennings, Mulligan and Urquhart (1956), who claimed that the adult parasite feeds almost exclusively on blood, and Dawes (1963b), who proposed that they are tissue feeders. Dawes claimed to have evidence that migrating juveniles are tissue feeders and thought it unlikely that they would change this habit on entering the bile ducts. He also offered observations that provided more convincing support for his suggestions: his descriptions of the caecal contents as nondescript and a kind of homogenized cellular debris; his finding that the epithelium lining the bile duct began to hypertrophy before the migrating juveniles arrived, burying the few blood vessels present and rendering them inaccessible; his failure to find any evidence of haemorrhage into the bile duct. Dawes proposed the interesting idea that the inflammatory reaction generated by the flukes provided a 'pasture' of hyperplastic epithelium and connective tissue on which the parasite browsed. Sawma, Isseroff and Reino (1978) also showed that proline, an amino acid released by *F. hepatica*, induced remarkably similar hyperplasia when administered to infected rats.

The work of Todd and Ross (1966) was central in the settlement of this dispute. First, they pointed out that many flukes disgorged their caecal contents unless collected immediately after killing of the host. A high proportion of freshly removed flukes contained black gut contents, which matched blood, but not bile, in spectroscopic tests (Fig. 13.3). Although the caecal contents contained appreciable quantities of haemoglobin or its breakdown products, the parasite itself did not and homogenized flukes with and without gut contents were negative for bilirubin, indicating that little bile is ingested. Their estimates of the quantities of blood ingested were incompatible with consumption of appreciable quantities of bile duct epithelium. Determinations of copper and iron suggested that the caecal contents were largely blood and that bile, bile duct epithelium and liver tissue made little or no contribution to the diet. Moss (1970) found that adult *F. hepatica* excreted ten times as much ammonia per unit weight as *Hymenolepis diminuta* and related this to the difference in diet, *F. hepatica* feeding mainly on blood and *H. diminuta* mainly on carbohydrate (section 9.3.2(b)).

This work did not solve the problem of how the parasite gains access to blood vessels. However, Sukhdeo, Sangster and Mettrick (1988), using fast-freezing to preserve parasites in near-natural attitudes, found evidence to suggest that in the rabbit the adult parasite is relatively sedentary and feeds at permanent sites, where haemorrhagic ulcers develop.

An interesting feature of the caecal epithelial lining of adult *F. hepatica* is that, according to Robinson and Threadgold (1975), the cells are of a single type but exist in one of two phases, a secretory phase and an absorptive phase (Fig. 13.4). The implication is that digestion is at least partly extracellular and Yamasaki, Kominami and Aoki (1992) have localized a cysteine protease, known to hydrolyse haemoglobin, in granules within caecal cells in the secretory phase and also in the caecal lumen. The apical surfaces of the caecal cells give rise to triangular or rhomboidal

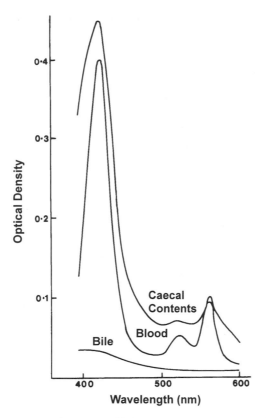

Figure 13.3 Absorption spectra of 'reduced alkali haematin' test on blood, bile and caecal contents of *Fasciola hepatica*. Source: reproduced from Todd and Ross, 1966, with permission from Academic Press.

lamellae which may divide, anastomose or cross from one cell to another (Thread-gold, 1978).

The adult parasite can obtain nutrients by another route. Isseroff and Read (1969, 1974) showed, by ligating the oral sucker to prevent uptake of nutrients *via* the gut, that *F. hepatica* can take in small molecules through the tegument. Absorption of glucose is a non-linear function of concentration and there are competitive inhibitory interactions between glucose and its other sugars. However, unlike *Hymenolepis diminuta* (section 9.3.2(b)), *Fasciola* cannot accumulate glucose to concentrations above those of ambient media and uptake is characterized as mediated transport (facilitated diffusion). The fluke also differs from *H. diminuta* in having a separate locus for the uptake of fructose, again by a mediated process. Uptake of xylose is a linear function of concentration, i.e. it is absorbed by diffusion, and amino acids are also taken up by diffusion, not by active uptake as in *H. diminuta*.

Thus, *Fasciola* is not restricted to ingested blood as a source of nutrients but is also potentially capable of extracting material from the bile bathing its tegument. The relative sizes of the contributions to the nutritional economy of the fluke made by the tegument and the gut remain to be determined.

Adult *F. hepatica* will ingest host hormones with its blood meals and this led Sukhdeo and Sukhdeo (1989) to test the effects of certain gastrointestinal hormones on behaviour. The significant result was that some of these hormones do elicit changes in the behaviour of the adult fluke *in vitro*, albeit at rather higher concentrations

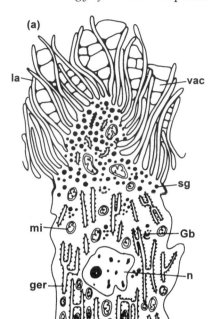
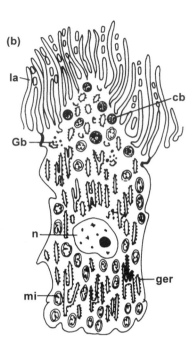

Figure 13.4 Schematic diagrams of gastrodermal cells of *Fasciola hepatica* adults in (*a*) predominantly secretory phase and (*b*) predominantly absorptive phase. cb = cytoplasmic body; Gb = Golgi body; ger = granular endoplasmic reticulum; la = lamella; mi = mitochondrion; n = nucleus; sg = secretory granule; vac = vacuole. Source: reproduced from Robinson and Threadgold, 1975, with permission from Academic Press.

than in the host. Some of the responses were difficult to interpret in terms of parasite biology, but changes in ventral sucker activity provided grounds for speculation on function. Cholecystokinin enhances bile flow by relaxing the sphincter of Oddi at the duodenal junction and inducing contraction of the gall bladder. This hormone stimulates ventral sucker activity, which may translate as strengthening of attachment, preventing the parasites from being expelled with the bile. In contrast, motilin stimulates closure of the sphincter, preventing bile flow, and decreases ventral sucker activity. These observations are interesting in relation to proposals that the behaviour of polystomatid and other blood-feeding monogeneans may be regulated by host hormones ingested with their blood meals (section 7.1.1).

13.2   THE EGG AND HATCHING

Given the similarity between monogeneans, digeneans and many cestodes in the assembly and chemical nature of the tanned eggshell, it is surprising to find that the hatching mechanisms of *Entobdella soleae* (section 4.9) and *Fasciola hepatica* have little in common. They share the first step, which seems to be the activation of the larva (oncomiracidium and miracidium respectively), but subsequent events are very different. In *E. soleae*, a hatching fluid produced by glands in the head is spread around the opercular seal, which eventually dissolves, releasing the operculum. In

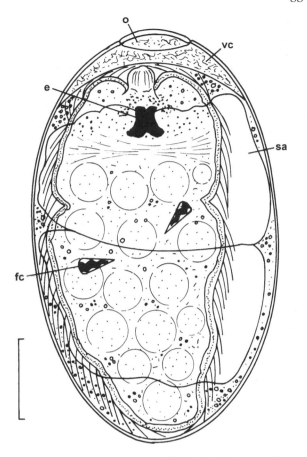

Figure 13.5 Egg of *Fasciola hepatica* containing a fully developed miracidium. e = eye; fc = flame cell; o = operculum; sa = sac; vc = viscous cushion. Scale bar: 25 μm.

F. hepatica, the miracidium is unable to enter the opercular corner of the ovoid egg because it is occupied by a so-called viscous cushion (Fig. 13.5). According to Wilson (1968a), this cushion consists of a fibrillar mucoprotein complex and is in a dehydrated or semi-dehydrated state, with the concave side separated from the larva by a membrane. About 5 min after activation of the miracidium, the cushion expands and the first sign of this is a change in refractive index on the concave surface of the cushion. Wilson suggested that the miracidium initiates this change by altering the permeability of the adjacent cushion membrane, permitting the entry of the fluid egg contents. The change spreads through the cushion as a 'wave of solution', interpreted by Wilson as hydration and hence expansion of the mucoprotein complex. This compresses the miracidium and also two separate thin-walled sacs, regarded by Wilson as remnants of the vitelline cells. The internal pressure gradually increases until the opercular seal ruptures violently about 30–60 s after expansion of the cushion. Thus, Wilson attributed dislodgement of the operculum to increased internal pressure and offered arguments against the intervention of a hatching enzyme in the weakening of the opercular seal.

The emergence of the miracidium appears to be facilitated by expansion of the sacs, which burst after escape of the larva, and Wilson described successful ejection of larvae facing the wrong way, against the efforts of their cilia to propel them in the abopercular direction.

Wilson found that light is a major factor in the activation and subsequent hatching of the miracidium, eggs kept in total darkness failing to hatch, but he also found that prolonged vibration and abrupt changes in temperature are effective activation stimuli.

According to Wilson (1967a), two closely apposed cell membranes (= vitelline membrane) line the eggshell and open out to enclose the viscous cushion. This double membrane configuration points to a cellular origin and cells derived from the embryo come to line the eggshells of the digeneans *Parorchis acanthus* (see Rees, 1940) and *Schistosoma mansoni* (section 18.3.2). Viscous cushions have not been reported in the tanned eggs of monogeneans and cestodes, and we are bound to ask whether digeneans differ fundamentally from these groups in their embryology and egg hatching mechanism. However, surprisingly, hatching has been studied in relatively few digeneans. A viscous cushion appears to be present in *Dicrocoelium lanceolatum* (see Ratcliffe, 1968), but was not reported in *P. acanthus* by Rees (1940). Until we know more about hatching in a wider range of digeneans, it is not possible to judge whether the viscous cushion mechanism is a fundamental feature of the group.

13.3   THE MIRACIDIUM

13.3.1   Anatomy

The miracidium of *F. hepatica* is roughly cone-shaped with the narrow end posterior (Fig. 13.6a). The body, with the exception of the anterior projecting papilla, is covered by plate-like ciliated epidermal cells arranged in five tiers of six, six, three, four and two cells respectively from anterior to posterior (Fig. 13.6b). These cells do not quite meet and are separated by intercellular ridges which are connected to subsurface vesiculated cell bodies (Fig. 13.7; Wilson, 1969a). Thin cytoplasmic extensions from these vesiculated bodies form a discontinuous cytoplasmic layer beneath the ciliated plates, separating them from two layers of muscle fibres, outer circular and inner longitudinal (Wilson, 1969b). Each ciliated plate has a single nucleus near its posterior border and a desmosomal junction running around the periphery of each plate joins it to the intercellular ridges. The inner membrane of each ciliated plate is fused to the outer membrane of the underlying cytoplasmic extensions at several points, creating 'tight junctions'.

There is no gut, although earlier workers were understandably misled by a large syncytial apical gland occupying a position where we would expect the gut to be situated (Fig. 13.6a; Wilson, 1971). This gland is flask-shaped with three or four nuclei and numerous tubular cytoplasmic connections (ducts) to the anterior surface of the apical papilla. The gland is flanked by a pair of smaller, uninucleated, accessory gland cells, the ducts from which also open on the apical papilla, dorsal to the apical gland. The apical papilla is furnished with sensilla and there are sensilla embedded in the intercellular ridge separating the first and second tiers of ciliated plates (Fig. 13.6c; Wilson, 1970). The miracidium has a single pair of pigment-shielded eyes. Each pigment cup encloses two rhabdomeres, but Isseroff and Cable (1968) showed that there is a fifth small rhabdomere occupying a median, posterior chamber in the left pigment cell and imparting an asymmetry to the eyes

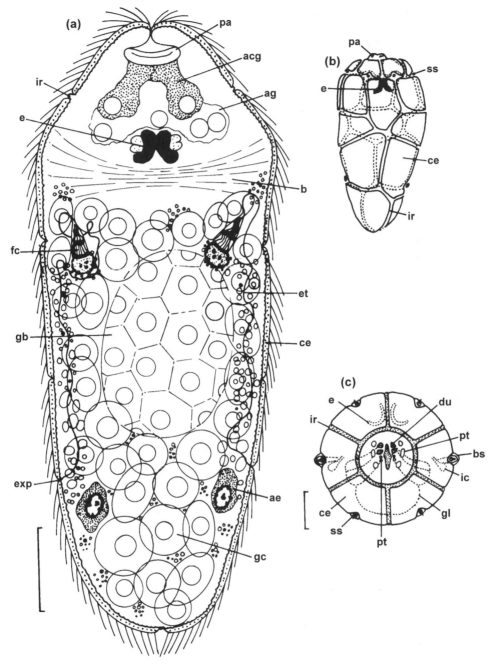

Figure 13.6   The miracidium of *Fasciola hepatica*. (*a*) Whole larva, drawn from life, slightly flattened with apical papilla withdrawn. (*b*) Sketch of whole larva showing outline of ciliated epidermal cells, as seen after impregnation with silver. (*c*) View of anterior (apical) region of the miracidium with apical papilla exposed, showing outlines of the first tier of ciliated epidermal cells and sensory endings. acg = accessory gland; ae = accessory excretory cell; ag = apical gland; b = 'brain'; bs = bulbous nerve ending; ce = ciliated epidermal cell; du = gland duct termination; e = eye; exp = excretory pore; ext = excretory tubule; fc = flame cell; gb = germ ball (?); gc = germ cell; gl = gland cell; ic = internal ciliary eye (?); ir = intercellular ridge; pa = apical papilla; pt = pit containing sensilla; ss = sheathed sensillum. Source: (*a*) original; (*b*) reproduced from Bayssade-Dufour *et al.*, 1980, with permission from Masson S. A.; (*c*) modified from Wilson, 1970.

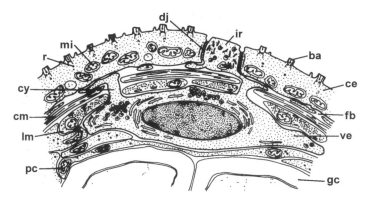

Figure 13.7   Diagrammatic reconstruction of part of a transverse TEM section through the body wall of the miracidium of *Fasciola hepatica*, in the region of the germ cell cavity. ba = base of cilium; ce = ciliated epidermal cell; cm = circular muscle; cy = cytoplasmic connection; dj = desmosomal junction; fb = fibrous material; gc = germ cell; ir = intercellular ridge; lm = longitudinal muscle; mi = mitochondrion; pc = parenchyma cell; r = ciliary rootlet; ve = vesiculated cell. Source: reproduced from Wilson, 1969a, with permission from the American Society of Parasitologists.

(Fig. 13.8). In addition, there is an internally located pair of structures similar to the ciliary 'eyes' (?) described in monogeneans (section 4.10.4).

The musculature of the apical papilla is arranged like that of the body but forms a separate muscle system (Fig. 13.9; Wilson, 1969b). There are approximately six circular muscle fibres and beneath them six longitudinal papillar retractor fibres.

The miracidium contains a single pair of flame cells, each giving rise to a convoluted and laterally situated duct that opens on the intercellular ridge between the fourth and fifth (posterior) tiers of ciliated plates (Fig. 13.6a; Wilson, 1967b). Wilson

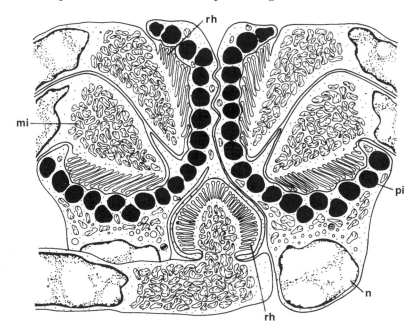

Figure 13.8   Diagrammatic representation of a TEM section through the eyes of the miracidium of *Fasciola hepatica*. mi = mitochondrion; n = nucleus; pi = pigment granule; rh = rhabdomere. Source: reproduced from Isseroff and Cable, 1968, with permission from Springer-Verlag.

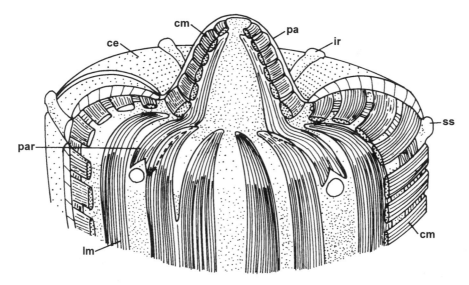

Figure 13.9 Schematic reconstruction of the anterior region of the miracidium of *Fasciola hepatica*, showing arrangement of muscle fibres. ce = ciliated epidermal cell; cm = circular muscle fibre; ir = intercellular ridge; lm = longitudinal muscle fibre; pa = apical papilla; par = papillar retractor fibre; ss = sheathed sensillum. Source: reproduced from Wilson, 1969b, with permission from the American Society of Parasitologists.

pointed out that the duct convolutions may do no more than permit the miracidium to extend its body without rupturing the ducts.

There is a central cavity containing germ cells and there are no hooklets like those in monogenean and cestode larvae.

13.3.2 Behaviour and invasion of the snail

Free-swimming miracidia normally follow a straight path and swim towards a light source at speeds of up to about 1.4 mm per second (Wilson and Denison, 1970a). Since *L. truncatula* is an amphibious snail a photopositive response may be advantageous since snails are likely to be near the surface and the parasite eggs on the bottom. In the presence of a snail or snail mucus, the rate of change of direction of the miracidium increases dramatically. This has been confirmed by Kalbe (in Haas *et al.*, 1995a), who also detected turnback swimming (reversal of direction in decreasing chemical gradients). Wilson and Denison (1970b) found that short-chain fatty acids (C_6–C_9) all caused a significant increase in the rate of turning, which could be correlated with chain length. Moreover, fatty acids are present in small amounts in the mucus produced by *L. truncatula* (see Wilson, 1968b). This led Wilson and Denison (1970b) to suggest that these acids might pass into solution from mucus on the surface of the snail and that they may serve as natural stimulants when miracidia come under their influence, inducing rapid change of direction and a greater chance of making contact with the snail. A sudden decrease in light intensity also induces an increase in the rate of turning, but the effect persists for only a few seconds compared with at least 3 min for snail-induced changes (Wilson and Denison, 1970a). The significance of this response to decreasing light, and of responses to gravity detected by Wilson and Denison (1970a), is not clear in terms of host location.

An account of snail penetration has been given by Wilson, Pullin and Denison (1971). The upper surface of the head–foot and the mantle of the snail are the most common sites penetrated by miracidia. Initial attachment involves the extension of the apical papilla and the formation of a sucker-like cup at the tip of the papilla by invagination of the anterior part. These changes are presumably brought about by appropriate contractions of the special circular and longitudinal muscles respectively of the apical papilla (Fig. 13.9). The miracidium is capable of attaching itself to glass, especially if snail mucus is present, indicating that mucus contains an attachment stimulant. Short-chain fatty acids (C_7–C_9) will stimulate an attachment response, but only at rather high concentrations that are eventually damaging to miracidia. Thus, these acids do not appear to be natural stimulants, although they may have similar properties. C_6–C_9 acids are effective stimulants of increased turning at much lower concentrations, so it seems that the same compounds stimulate two different behavioural responses depending on concentration.

Unless the glass surface is roughened, attached miracidia do not form a stable attachment, continuing to rotate about the longitudinal axis. Such rotation is momentary on snail skin and once stability of attachment is reached the miracidium begins to undergo slow peristaltic contraction and relaxation of the body (45 s after initial contact). These contractions of the circular and longitudinal body muscles probably serve to expel glandular secretions from the apical papilla. Body peristalsis and secretion are hard to induce *in vitro* (in glass vessels) probably because stable attachment is difficult to achieve, and this suggests that appropriate stimulation from host skin is an important factor. Wilson, Pullin and Denison (1971) highlighted an obvious size correlation between the width of pits on the surface of the apical papilla and microvilli projecting from the surface of snail epithelial cells and suggested that an interaction between these parasite–host features may be important in establishing a stable attachment.

Observations on miracidia plucked from snails at this stage indicate that the contents of the apical gland and the accessory glands are shed. Their functions are unknown, but Wilson, Pullin and Denison (1971) suggested that the apical gland is the source of the cytolytic enzymes that digest a path through the snail's columnar epithelial cells. This leaves the accessory glands with the possible role of stabilizing attachment.

About 60 s after stable attachment is established the first indications of shedding of the ciliated plates are detected. The order of shedding is from anterior to posterior and only a single tier of cells is shed at any one time, the plates peeling off as the miracidium penetrates, so that at no time is the surface underlying the plates exposed to fresh water. Wilson, Pullin and Denison (1971) observed that miracidia stimulated to shed their plates in water by adding *Lymnaea* serum very quickly became cytolysed whereas miracidia shedding plates in saline did not, suggesting that the ciliated plates provide osmotic protection for the miracidium. In snails preserved for electron microscopy no more than 10 min after initial attachment of a miracidium, the host's single layer of columnar epithelial cells had already been breached and both first and second tiers of plates had been shed. At this stage the apical papilla is already retracted into the body, an event that may be mediated by the sensilla embedded in the intercellular ridge between the first and second tiers of plates; these sensilla will make contact with the snail epithelium at about this time.

The thin cytoplasmic layer above the muscles and the interconnected intercellular ridges form the basis of the new surface of the sporocyst. This is filled out by material extruded from the vesiculated cells and this extrusion process spreads backwards synchronously with the forward penetration of the miracidium and peeling off of the ciliated plates. Indeed, the extrusion process may lead to break-down of the desmosomal attachments and tight junctions that hold the plates in place, and may be the cause of their detachment. Not more than 2.5 h after initial attachment to the host, the new body wall of the parasite covers the entire musculature and contacts between the new surface layer and underlying cells appear to be broken. Within the first 24 h the surface of the syncytial covering is thrown into folds and metamorphosis into the sporocyst can be considered complete.

13.4   INTRAMOLLUSCAN STAGES

13.4.1   Feeding

The sporocyst (Fig. 13.10*a*) has no gut and consequently must absorb all its nutrients through the tegument. There are reports of active uptake of glucose and absorption of organic acids by simple diffusion (e.g. in *Microphallus similis* by McManus and James, 1975), but sporocysts may also have a digestive influence on host tissue. Thomas and Pascoe (1973) claimed that the tegument of sporocysts from the marine gastropod *Gibbula umbilicalis* secretes hydrolytic enzymes capable of degrading exogenous glycogen to glucose and sucrose to glucose and fructose, *in vitro*. This is compatible with the apparent histolytic effects of sporocysts on nearby cells of the host digestive gland (see Erasmus, 1972, p. 71).

Each redia reaches a length of about 1 mm; it has a collar just behind the pharynx and two bluntly conical, backwardly projecting protrusions in the posterior region of the body (Fig. 13.10*b, c*). Thomas (1883) found that the diameter of the collar could be greatly increased, especially in rediae that were actively moving through the snail's tissues. He observed that the expanded collar anchored the redia, permitting the region anterior to it to extend into the snail's tissues, and he stressed the importance of the backwardly directed posterior 'feet' in preventing the animal from slipping backwards during its migrations.

The redia has a gut and there is no doubt that the parasite is able to ingest host cells (see Chapter 12 and below). Rees (1983a) described secretory bodies in the gastrodermal cells of the redia of *Parorchis acanthus* and suggested that these bodies contain enzymes involved in extracellular digestion, but there is a real possibility that host digestive enzymes from ruptured digestive gland cells may make a significant contribution to the release of nutrients for the parasite. We have already identified the importance of this source of nutrients in turbellarian symbiotes of molluscs (section 2.3.2). Moreover, Rees (1983b) has emphasized that feeding by mouth is more prevalent in young active rediae of *P. acanthus* than in larger immobile ones, and the suggestion has been made that tegumental uptake of nutrients may occur, perhaps more actively, in older rediae. This is supported by the presence of mitochondria in the redial tegument of *P. acanthus* and of surface microvilli (Rees, 1981), features which have also been reported in the redia of *F. hepatica* (see McDaniel and Dixon, 1967).

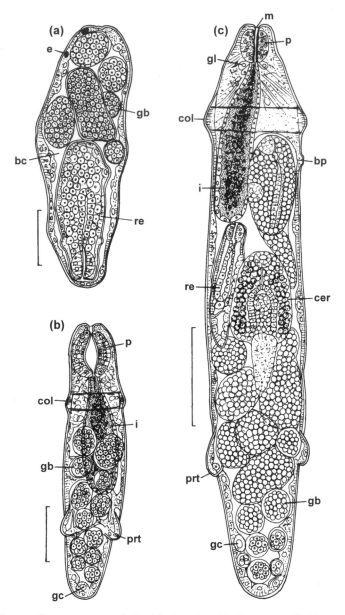

Figure 13.10 Intramolluscan stages of *Fasciola hepatica*. (*a*) Sporocyst. (*b*) Young redia. (*c*) Fully grown redia. bc = body cavity; bp = birth pore; cer = developing cercaria; col = collar; e = eye; gc = germinal cell; gb = germ ball; gl = gland cell; i = intestine; m = mouth; p = pharynx; prt = protuberance ('foot'); re = developing redia. Scale bars: (*a*), (*b*) 100 μm; (*c*) 250 μm. Source: reproduced from Thomas, 1883.

Rediae of *F. hepatica* are also capable of ingesting and destroying germinal sacs of other digeneans (see section 14.1 for the possible significance of this habit). In experimental infections of *Galba* (= *Lymnaea*) *truncatula* with *F. hepatica* and *Paramphistomum* spp., the former in most instances completely eliminated the latter (Samnaliev, Kanev and Vassilev, 1978).

13.4.2 Sequence of stages

According to Thomas (1883; see Chapter 12), rediae released by the sporocyst produce either daughter rediae or cercariae, depending on the ambient temperature. Recent studies have shown that the sequence of development of the intramolluscan stages is more complicated (Rondelaud and Barthe, 1981) and that this pattern of development is maintained irrespective of seasonal changes of temperature and other factors (Rondelaud and Barthe, 1978). Rondelaud and Barthe (1981) infected snails with a single miracidium and kept the snails at 20°C. They used serial histological sectioning to locate sporocysts and rediae, and morphological criteria to distinguish rediae of different generations ('generation' is used here to describe a stage in the succession of germinal sacs; section 14.3.3).

They identified three distinct generations of rediae during the first 49 days after exposure to the miracidium, but they also found a curious asymmetry in redial production (Fig. 13.11). The first redia (belonging to generation 1) produced by the sporocyst developed rapidly in the body of the mollusc and produced early second-generation rediae. The other rediae of generation 1 produced both rediae and cercariae (as in Fig. 13.10c), but these late daughter rediae of the second generation were fewer. The early rediae of generation 2 produced rediae of generation 3 at the same time as cercariae. The emergence of these rediae was synchronized with that of the late rediae of generation 2. The first-born redia of generation 1 was particularly productive, producing on average 17.91 rediae, and there was some evidence that, should this redia die, its place would be taken by a redia of generation 1 or 2, which would increase its redial output in partial compensation for loss of the first-born redia (Rondelaud and Barthe, 1982).

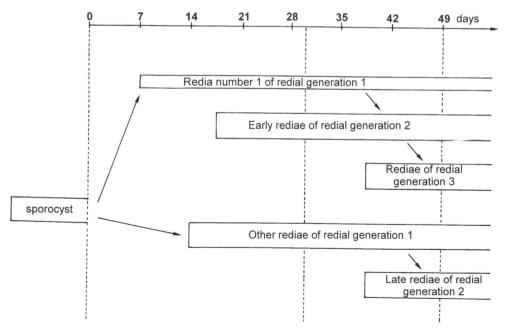

Figure 13.11 Sequence of development of rediae from a single sporocyst of *Fasciola hepatica*. The left-hand edge of each box indicates the date when the first rediae of the generation in question became independent. Source: redrawn from Rondelaud and Barthe, 1982.

13.4.3　Effects of germinal sacs on the snail

Wilson and Denison (1980) compared growth of snails each infected with five miracidia with growth of uninfected control snails. No difference in growth rates was detected for the first 21 days, but thereafter shell length and dry mass of infected snails continued to increase while those of the controls tended to an asymptote. At the end of the experiment the biomass of infected snails was twice that of the controls and the shell 1.34 times as long. This increased growth or gigantism is a feature of other digenean–mollusc associations (see also tapeworms, section 10.5.8).

Wilson and Denison recorded a significant effect of the parasite on reproduction, with egg production in infected snails virtually ceasing by 21 days post-infection, while controls continued to lay eggs steadily throughout the experiment. The first rediae reached the visceral mass on about day 17 and their gut contents indicated that they actively consumed the ovotestis during the next few days, destroying it completely in many snails. The paradoxical aspect of this relationship is that in spite of the active consumption of tissue by the parasite, the infected host grew larger than uninfected controls. Wilson and Denison suggested that this situation resulted from a reallocation of nutrients, not from an effect of the parasite on host hormonal control of reproduction (but see section 18.3.6). Parasitic castration removes the burden of reproduction, creating an excess of energy, which is sufficient not just for growth of the parasite but also for continual somatic growth of the host. Wilson and Denison pointed out that curtailed reproduction in other animals is often correlated with increased life span and they suggested that infected *L. truncatula* may be better fitted to survive than uninfected snails carrying the burden of reproduction. In fact, in their experiments, infected snails survived as long as uninfected controls.

13.5　THE CERCARIA

The process of multiplication in the mollusc produces, from a single miracidium, large numbers of cercariae (Fig. 13.12), which escape into the water. This mass exodus seems likely to be a traumatic process for the snail, but remarkably little is known about it. We do not know whether the cercariae have preferred exit sites on the snail's surface or how the cercariae breach the snail's skin (but see sections 16.3.4(e), 18.3.7). That this exodus is not necessarily a serious threat to the mollusc is underlined by Meyerhof and Rothschild (1940), who recorded cercarial production over a period of 5 years from a single marine snail, *Littorina littorea*, infected with *Cryptocotyle lingua* (section 17.5).

Once the cercaria is free it is propelled not by cilia but by a muscular tail, and in this respect is unique in the platyhelminths. A free-swimming cercaria is a delight to observe as it propels itself through the water with oscillations so rapid that the tail appears as a blurred 'figure of eight'. The tail muscles are unusual in being striated (Fig. 13.13), a rare feature in platyhelminths. We have already encountered striated muscle in the haptors of some monogeneans (section 6.4.1) and in the scoleces of some tapeworms (section 10.6), and striations seem to be a feature of muscles that contract rapidly and repeatedly. Paramyosin, a protein thought to provide structural support permitting rapid muscle contractions in striated insect

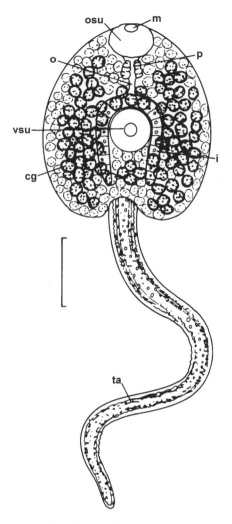

Figure 13.12 The cercaria of *Fasciola hepatica*. cg = cystogenous cells; i = intestine; m = mouth; o = oesophagus; osu = oral sucker; p = pharynx; ta = tail; vsu = ventral sucker. Scale bar: 100 μm. Source: reproduced from Thomas, 1883.

flight muscle, has been detected in cercarial tail muscle of *Schistosoma japonicum* by Gobert *et al.* (1997).

Most cercariae of *Fasciola hepatica* are emitted during the night (Audousset *et al.*, 1989). Dreyfuss and Rondelaud (1994) infected pre-adult specimens of *L. truncatula* (4 ± 0.2 mm in shell height) each with two miracidia and recorded the numbers of encysted metacercariae produced per day during the lifetimes of the snails. The patent period (period during which cercariae are emitted) began on day 44 ± 7.8 days and lasted 46 ± 27.6 days. A total of 102 snails produced 24 325 metacercariae, most of which were attached to the substrate but a small number (5%) were free-floating. A total of 102 snails shed parasites on the first day of the patent period and 56 on the second day, but thereafter the number of productive snails declined and on each day from day 76 to day 114 involved only one to four snails. Intermittent production persisted for a few more days and ceased on day 124. More

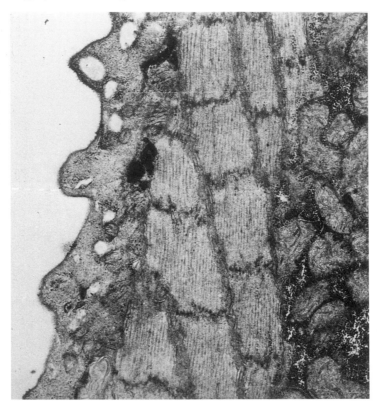

Figure 13.13 Transmission electron micrograph of an oblique section through the tail of an echinostomatid cercaria, showing striated muscle.

metacercariae were produced per snail during the first 30 days (mostly between 10 and 20 per snail per day) but, over the whole 124-day patent period, there was no obvious rhythmical shedding pattern other than the nocturnal emission noted above.

13.6   THE METACERCARIA

13.6.1   Encystation

According to Dixon (1968), the free-swimming cercaria of *F. hepatica* may attach itself briefly and then resume swimming; this ties in with experimental evidence indicating that cercariae may exercise some selectivity with regard to texture and colour of encystment substrate (Jimenez-Albarran and Guevara-Pozo, 1980a, b). Selection of a permanent attachment site by the cercaria is indicated by its behaviour – an apparent doubling of size occurs as the cercaria attaches itself by the ventral sucker and flattens its body against the substrate (Dixon, 1968).

The fully encysted metacercaria of *F. hepatica* is enclosed in a double capsule consisting of an outer and an inner cyst wall separated from each other by a gap (Fig. 13.14; Dixon, 1965). The outer and inner cyst walls each consist of two separate layers, but only the two layers of the inner cyst wall completely enclose the meta-

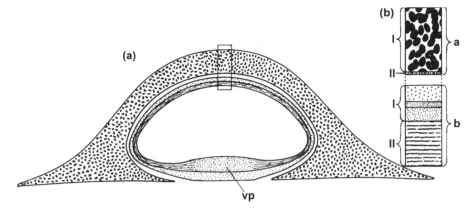

Figure 13.14 (*a*) Diagrammatic vertical section through the metacercarial cyst wall of *Fasciola hepatica*. (*b*) An enlargement of the section indicated in (*a*). a = outer cyst wall; b = inner cyst wall; vp = ventral plug. See text for details of main layers (indicated by Roman numerals). Source: redrawn from Dixon, 1965.

cercaria. The outermost layer of the cyst (layer aI; Fig. 13.14) is thick and at its base spreads out in both directions, i.e. outwards away from the inner cyst and inwards beneath its edge, forming a bond with a relatively large surface area where it contacts the substratum. Since there is no bonding material in the centre between the surface of the inner cyst and the substratum, the bonding zone is ring-shaped. This incomplete outermost layer is brown in colour and appears to consist of tanned protein (section 1.4.5). It is lined by a thin layer of mucoprotein and mucopolysaccharide (layer aII).

Dixon (1968) observed that each layer of the cyst wall is derived from a specific cystogenous cell type in the body of the cercaria. The four cystogenous cell types can be distinguished by cytological and histochemical features. The outer cyst wall is the first to be put in place, immediately after attachment of the cercaria to the substrate by the ventral sucker. The secretory bodies that produce the outer cyst wall appear to be already present in the tegument of the free-swimming cercaria, but Dixon's claim that the tegument is shed along with the secretion was not confirmed by Køie, Nansen and Christensen (1977). The tail, which remains outside the cyst wall, is shed at this stage and the cercarial body contracts away from the outer cyst, creating a gap that persists after secretion of the inner cyst wall and permits easy separation of the outer cyst from the inner cyst. There is a pause in activity and then secretion of the raw material of the inner cyst begins from the tegument.

The outer layer of the inner cyst wall (layer bI) consists of three sublayers composed of carbohydrate–protein complexes. The inner layer of the inner cyst wall (layer bII) consists of a keratin-like protein – closely packed proteinaceous lamellae stabilized by disulphide bonds and embedded in a matrix containing protein and lipid. Dixon (1968) observed that these lamellae are rolled up to form tight scrolls in the cytoplasm of the corresponding cystogenous cells. Rotatory movements of the metacercaria that persist for about 1h seem to be important in spreading the secretions of the inner cyst wall and particularly in ensuring that the keratinous scrolls unroll, since unrolling of secreted scrolls did not take place if the metacercaria was anaesthetized.

In the completed metacercarial cyst there are no keratinous lamellae in the central region of the surface in contact with the substrate. This is the so-called ventral plug region and Dixon attributed this lack of lamellae to the fact that during formation of the inner cyst the rotating metacercaria pivots about the ventral sucker, which remains more or less in the ventral plug region. Since the ventral sucker contains no cystogenous cells, no scrolls are secreted here.

13.6.2   Hazards faced by the cyst wall

The longer the cysts of *F. hepatica* survive the greater their chance of being eaten by the vertebrate host and, since the cysts are exposed to the outside world they will be targets for bacterial and fungal attack. The cysts have limited resistance to toxic agents, such as corrosive sublimate, alcohol, formalin and pepsin (references in Dixon, 1965, 1966), indicative of the general resistance of tanned proteins to chemicals (section 1.4.5), and it seems likely that these properties prevent the entry of bacteria and fungi. The fact that the outer layer is absent from the ventral plug region suggests that it has no role in protecting the metacercaria from chemical attack on its way to the small intestine of the vertebrate host. In fact, Dixon (1966) reported that it was relatively easy to separate the outer layer from the inner cyst by pressure and deduced from experimental infections of guinea pigs by Schumacher that most of the cysts lost their outer layer mechanically as a result of mastication and agitation of the stomach or rumen contents.

The metacercariae may be exposed to the osmotic stress of the freshwater environment and cysts are likely to be left high and dry on the emergent vegetation. They are known to survive drying for 24 h (reference in Dixon, 1965). Dixon has proposed that the keratinous inner layer of the inner cyst wall, with its laminated construction and lipid matrix, is well suited to resist desiccation and thus may preclude osmotic stress. Køie, Nansen and Christensen (1977) remarked on the porous nature of the outer layer of the cyst and suggested that it might protect against desiccation by absorbing water or dew. Dixon (1965) also noted the structural strength imparted to the cyst by the keratinous layer and emphasized the difficulty of piercing the inner cyst wall, features that are consistent with resistance to mechanical damage such as that inflicted by the host's teeth. In sheep and cattle, the secondary mastication of food regurgitated from the rumen ('chewing the cud') is another potential hazard for the cysts, but there appears to have been no study of the extent to which cysts are regurgitated or whether this habit reduces significantly the proportion of viable metacercariae reaching the small intestine. Another major hazard is gastric juice, with its low pH and strong proteolytic activity, and the two layers of the inner cyst wall, including the ventral plug, may protect the metacercaria while it is in transit through the stomach, while permitting its escape *via* the ventral plug when the small intestine is reached.

13.6.3   Excystation

Dixon (1964) observed that successful excystation of *F. hepatica* could be achieved *in vitro* by the application of treatment media lacking pepsin, trypsin or pancreatin. The significance of this discovery is that excystation appears to be an active process, the metacercaria itself being capable of perforating the cyst wall without any assistance from host enzymes, although there is evidence that pretreatments with host enzymes enhance both the percentage and rate of excystment (Dixon, 1966).

Two separate phases of excystation have been recognized: activation and emergence. According to Dixon (1966), the main activation stimulus is carbon dioxide at a temperature of 39°C. Reducing conditions, produced by the addition of sodium dithionite or cysteine, increased the rate of excystment but when applied alone at 39°C were not responsible for activation. Conditions suitable for activation prevail in the rumen of sheep and lead to vigorous rotation of the metacercaria for 5–20 min, followed by a gradual cessation of this movement and the appearance during the next 30 min of a gap between the metacercaria and the cyst wall. Sukhdeo and Mettrick (1986) suggested that this gap is created by the expulsion of the clear fluid contents of the parasite's gut caeca by peristaltic contractions.

Dixon (1966) maintained activated cysts with quiescent metacercariae for as long as 24 h without emergence, which requires the application *in vitro* of sheep bile. This stimulates renewed activity of a different kind, namely vigorous thrusting movements of the anterior end which are directed in the later stages at the ventral plug region of the inner cyst wall. Sukhdeo and Mettrick (1986) reported attachment of either the oral or the ventral sucker to the ventral plug and strong backward waves of contraction passing down the body; an important implication is that the metacercaria is able to recognize the ventral plug and concentrate its efforts on this specialized region. Sukhdeo and Mettrick regarded the evacuation of the gut contents as an important event, since they proposed that this fluid contained the enzymes that soften the ventral plug, which is then disrupted by the violent thrusting movements of the body and the activities of the suckers. Dixon reported that metacercariae required on average about 20 min to escape from the cyst wall after the receipt of the bile stimulus; 2–5 min of this was needed for the metacercaria to pull itself, always head first, through the hole in the plug. Most metacercariae emerged 1.25–1.75 h after the receipt of the activation stimulus and the shortest time recorded was 17 min.

It is significant that the rumen provides stimuli for activation but not emergence; the latter must await passage of the cyst, with its double wall and ventral plug intact, from the rumen through the highly acid abomasum to the intestine, where bile provides the stimulus for the completion of the excystation process. Bile is an extremely reliable cue for signalling arrival in the small intestine, so much so that it has been adopted as such by many gastrointestinal parasites (Lackie, 1975), including tapeworms (sections 9.7, 11.5.1).

High concentrations of carbon dioxide and strongly negative redox potentials of the rumen contents of sheep correspond with optimum *in vitro* conditions for excystment. Dixon's observation that cysts could be maintained in an activated state for 24 h and retained their capability to complete excystation within a further 2 h when stimulated with bile is consistent with rumen clearance times in sheep of about 24 h and transit times through the small intestine of 1–2 h. Dixon (1966) suggested that lower levels of carbon dioxide combined with the shorter small intestines of carnivores may be important in restricting the host range of *F. hepatica* to non-carnivorous animals. In addition, Sukhdeo and Mettrick (1986) reported that glycine-conjugated bile salts typical of herbivores were more effective in stimulating emergence than taurine-conjugated bile salts, which are typical of carnivores.

It is generally supposed that bile facilitates excystation by one of the following: a direct effect on permeability of the cyst wall; a potentiating effect on enzymes secreted by the parasite; non-specific stimulation of muscular activity (references in Sukhdeo and Mettrick, 1986). However, the work of Sukhdeo and Mettrick provided indirect evidence that recognition of bile occurs through sensory receptors. Not only was the glycine conjugate more effective than the taurine conjugate, indicating molecular specificity, but changing the molecular conformation of the aromatic part of the cholic acid molecule completely abrogated the emergence response. Also, the log dose–effect relationship to glycocholic acid was a sigmoid curve, which is characteristic of the binding of a ligand to a receptor where a significant fraction of the receptors must be occupied before a response is initiated.

13.7   THE JOURNEY TO THE LIVER

Newly emerged juveniles of *F. hepatica* penetrate the wall of the intestine and migrate through the body cavity to the liver. Sukhdeo, Sukhdeo and Mettrick (1987) found that juveniles *in vitro* often released their grip with the suckers on the substrate and underwent contraction and elongation of the body. These detachment episodes were significantly reduced when mouse duodenal extract was present. They considered this to be adaptive, since loss of adhesion by newly excysted juveniles in the intestine would increase the probability of parasites being swept away through the gut. A probing response with the oral region was also elicited in newly emerged juveniles by mouse duodenal extract but not by other treatments and this may be a first stage in penetration of the wall of the intestine.

Having passed through the gut wall, juveniles would presumably no longer be under the influence of duodenal substances and Sukhdeo, Sukhdeo and Mettrick (1987) suggested that loss of attachment by the suckers at this stage would enhance their passive transport between the viscera to the ventral body wall. Studies *in vitro* did not support suggestions that the juveniles locate the liver by chemotaxis. In fact, diffusion gradients of substances from sources like the liver are unlikely to be established and maintained in the body cavity, which is constantly stirred by the movements of the highly mobile gut. Moreover, studies *in vivo* by Sukhdeo, Sukhdeo and Mettrick (1987) showed that juveniles did not migrate directly to the liver of the mouse but travelled indirectly *via* the body wall. Sukhdeo *et al.* referred to an observation of Day and Hughes that most liver penetrations in cattle occur on the surface of the organ that lies against the diaphragm; they pointed out that the most likely way for parasites to reach the diaphragmatic surface is by crawling along the body wall. Parasites recovered from the abdominal cavity of mice 48 h after infection, responded to mouse liver extract by increasing their leech-like locomotion (section 13.1.1), but newly excysted juveniles showed no such response.

According to Sukhdeo, Sukhdeo and Mettrick (1987), adults of the related digenean *Fasciolopsis buski* live in the lumen of the small intestine and feed by browsing on the gut mucosa. These authors proposed that *Fasciola hepatica* has evolved from a similar ancestor, whose browsing behaviour led to accidental penetration of the gut wall. The work of Sukhdeo *et al.* makes it easier to understand how the wanderings of *F. hepatica* could have developed from parasites that strayed in this way, since they would require no special capability to locate the liver after entering the body cavity.

This evolutionary hypothesis is also compatible with the prevailing views on the way in which the migrating juveniles penetrate the gut wall and the liver on their way to the bile duct lumen. The detailed histological studies of Dawes (summarized in Dawes, 1963a) indicated that the juveniles literally eat their way through these

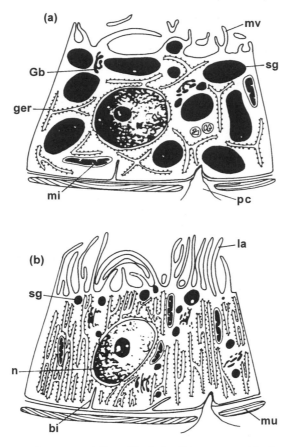

Figure 13.15 Diagrammatic representation of typical cell types present in the caecal epithelium of (*a*) a juvenile recovered from the abdominal cavity of the mouse, 3 days post-infection; (*b*) a juvenile from the liver capsule, 3 days post-infection. bi = basal invagination; Gb = Golgi body; ger = granular endoplasmic reticulum; la = lamella; mi = mitochondrion; mu = muscle; mv = microvillus; n = nucleus; pc = extension of parenchyma cell; sg = secretory granule. Source: reproduced from Bennett, 1975, with permission from Academic Press.

organs, and this is supported by the features of the caecal epithelium. Bennett (1975) showed that there is a profound change in this epithelium during migration to the liver of the mouse. Newly excysted juveniles have caecal cells packed with large secretory granules (1.3 μm in diameter) and lacking features associated with an absorptive function. Secretory granules are lost during penetration of the intestine wall and it is likely that this secretion disrupts host cells and facilitates penetration. Whether there is any contribution to the nutrition of the fluke at this stage from the products of digestion of host cells in the path of the juvenile is uncertain, but the caecal cells of juveniles newly arrived in the abdominal cavity still appear to be in a secretory mode (Fig. 13.15*a*). After a day or so in the body cavity morphological features of the caecal cells indicating limited absorptive function appear, such as the amplification of the surface membrane into short lamellae and the establishment of connections with parenchymal cells by junctional complexes; some digestion and uptake of nutrients seems likely at this stage since limited growth of the juvenile takes place. However, massive secretion by the caecal cells is maintained and seems likely to be involved in perforation of the liver capsule, since it is not until this is breached that the caecal cells switch production from the large secretory granules to small ones (0.5 μm in diameter) typical of the adult. In the liver parenchyma, transformation of the caecal cells to the adult type (Fig. 13.15*b*) is completed and those in the lateral caeca embark on a cycle of secretory and absorptive activity (see below).

13.8   IMMUNITY TO JUVENILES AND ADULTS

In the rat, an 'unnatural' host for *F. hepatica*, the parasite generates resistance to secondary (challenge) infections (Rickard and Howell, 1982), while the primary infection persists in the liver. This pattern is compatible with provocation of the host immune system by the tissue-invading juvenile stages of a primary infection, followed by escape from the tissues into the relatively safe immunological haven of the bile duct lumen (Hanna, 1980); the immune system is then primed to eliminate subsequent tissue invaders. We encountered concomitant immunity in *Hymenolepis diminuta* (section 9.3.4) and we will meet it again in *Schistosoma* (section 18.3.11). An advantage of concomitant immunity is that it may prevent the host being over-burdened with parasites, a possible life-threatening situation for host and parasite, and this may have been a factor in favour of the establishment of a tissue-invading phase in the life cycle of *Fasciola*.

Counter to this idea is the fact that in the sheep, regarded as a 'natural' host by Hanna (1980), there is no effective resistance to secondary and subsequent infections. In fact, flukes originating from a secondary infection migrate more rapidly to the bile ducts (Chauvin, Bouvet and Boulard, 1995). It might be inferred from this that *F. hepatica* is not immunogenic to sheep, but this is not the case because specific antibody appears in the blood stream and reaches a maximum during the first 6 weeks of infection, declining once the flukes are established in the bile ducts (references in Hanna, 1980). There is evidence that the immune response to secondary infection is reduced (Chauvin, Bouvet and Boulard, 1995). Thus *F. hepatica* generates ineffective immunity, i.e. the parasite is able to evade the sheep's immune response and, at least during primary infections, the immune responses of other animals. It has been suggested by Bryant and Behm (1989, p. 99) that selective breeding of sheep may have had the side effect of eliminating any effective immunity that the ancestors of the domesticated sheep may have had.

Hanna (1980) argued that immunological attack is most likely to be directed against the components of the tegumental surface of *F. hepatica*, since this is continually exposed to the host, but Hughes and Harness (1973a, b) found no evidence that protection is achieved by absorption of host antigens on to this surface (disguising the parasite as host tissue), a phenomenon that has been implicated in the evasion of the host response in blood-dwelling schistosomes (section 18.3.11). However, the glycocalyx of *F. hepatica*, which is continually secreted by the tegument, may have an important protective role; the duration of exposure of the tegument to antibody seems likely to be strictly limited because antibody (IgG) from sheep immune serum binds to the glycocalyx and is sloughed off with it (Hanna, 1980). Moreover, Hanna found some evidence to suggest that glycocalyx turnover is more rapid in the presence of specific antiserum than in non-immune serum.

According to Hanna (1980) glycocalyx turnover is likely to continue throughout the life of the fluke, although at a slower rate in the adult, but there is an interesting change as development proceeds. The glycocalyx of the migrating juveniles is derived from so-called type 0 secretory bodies produced by the tegumental cytons. Later these cytons switch to the production of a different kind of secretory body (type 1), which is characteristic of the adult tegument and contributes to the adult glycocalyx (Threadgold, 1963, 1967; Bennett and Threadgold, 1975). A second kind of secretory cyton contributes distinctive type 2 secretory bodies to the adult tegument (Threadgold, 1967). These changes in tegumental secretions presumably reflect the environmental differences between migrating juveniles and adults, especially the changing nature of the challenge from the host's immune system and the possible role of the adult tegument in uptake of nutrients (see above).

Another possible way in which the parasite counters the host immune system is revealed by observations made on the properties of the proteolytic enzymes referred to above, which serve to digest a route through the tissues of the host and provide nutrients for the fluke. Smith *et al.* (1993) found that a cysteine protease from the fluke cleaved host immunoglobulin and Carmona *et al.* (1993) demonstrated that a similar enzyme prevented the attachment of host eosinophils *in vitro* to newly excysted juveniles. Thus, as suggested by Dalton and Heffernan (1989), these proteinases may combine the functions of immune evasion, penetration and nutrient provision for migrating parasites. It might be supposed that flooding the host's tissues with proteolytic enzymes would be highly provocative to the host's immune system but according to Simpkin, Chapman and Coles (1980) a protease purified from gut exudate of adults does not seem to be antigenic.

A similar protease that may combine the functions of immune evasion and tissue penetration has been identified in the plerocercoid (sparganum) of the tapeworm *Spirometra mansonoides* (section 10.5.8).

13.9   CHANGES IN METABOLISM

Substantial changes occur in energy generation in *F. hepatica* reflecting corresponding changes in the habitat of the parasite during its life cycle (see review by Tielens, 1994). It is generally accepted that free-living stages such as eggs, miracidia and cercariae, with oxygen readily available, are mainly dependent on the citric acid cycle (see Fig. 9.6) and the mitochondrial respiratory chain for energy production. This is also the situation in the juvenile fluke freshly emerged from the cyst, but there is a gradual change to an aerobic pathway producing acetate (cf. tapeworms, section 9.3.3). This requires oxygen for the reoxidation of NADH, which is produced during the process, and is the most important source of energy during migration through the liver. Growth during this migration means that the inner tissues of the parasite receive progressively less oxygen by diffusion and this seems to initiate yet another change to an anaerobic fermentation with acetate and propionate as end products (see Fig. 9.6). The bile duct lumen contains little oxygen and the whole fluke is forced to adopt this process. What little oxygen there is in bile may be used up by the parasite in egg tanning (Moss, 1970) and relevant to this may be the discovery of a haemoglobin in liver flukes which seems to be of parasite origin (McGonigle and Dalton, 1995).

According to Burren, Ehrlich and Johnson (1967) an iridescent film appears on the surface of media containing adult *F. hepatica* and this reflects the fact that the parasite excretes appreciable quantities of lipid. The presence of lipid in the excretory system has been demonstrated histochemically and the quantity excreted amounts to as much as 2% of the organism's wet weight per day. It has been suggested that this lipid may represent the end product of respiratory metabolism (Moss, 1970), but Barrett (1981) pointed out that the range of lipids excreted is more consistent with a general loss of lipid than excretion of a specialized end product. According to Barrett, adult helminths are unable to catabolize lipids and the disposal of excess dietary intake of fats and the turnover of tissue lipids may depend on excretion. Lipid has been demonstrated in miracidia, sporocysts, rediae and some cercariae, but evidence that this is catabolized is indirect (Popiel and James, 1976). There is similar indirect evidence that some monogenean larvae utilize lipid (section 5.2). Lipid metabolism in platyhelminths is considered further in Chapter 20.

13.10   HUMAN FASCIOLIASIS

According to Grove (1990), one of the earliest detailed descriptions of fascioliasis in man dates from the severe epidemic of sheep rot that took place in Britain in 1879–1880. A 52-year-old labourer from Corfe Castle in Dorset developed a range of unpleasant symptoms that led to his death 4 months later. At post-mortem, 26 adult *F. hepatica* were recovered from his bile ducts. Fortunately, infections in humans are not common and many are asymptomatic or mild. Occasional outbreaks of human fascioliasis arise from the habit of eating watercress (*Nasturtium officinale*). Six patients infected in Hampshire, England in 1958–1959 and 44 patients infected in Monmouthshire in 1968–1969 had all consumed water cress, the latter group having taken the plant from a bed located close to infected sheep and cattle (references in Grove, 1990).

F. hepatica has also been blamed for the human affliction known as *halzoun*, an oedematous condition of the lining of the throat not uncommon in the Lebanon. This disease is clearly linked with the consumption of raw liver and sometimes lymph nodes from sheep and goats and less commonly from cattle and camels. The idea has developed that flukes in the liver attach themselves temporarily to the bucco-pharyngeal lining, creating the irritation and other symptoms associated with the condition. However, Schacher, Khalil and Salman (1965) came to the conclusion that a more likely causative agent is the nymph of the pentastomid *Linguatula serrata*, which encysts in liver and lymph nodes, and they pointed out that all attempts to induce *halzoun* by feeding fluke-infected liver to dogs and humans had failed.

13.11   THE WAY FORWARD

The process whereby digeneans multiply inside the bodies of their molluscan first intermediate hosts has no counterpart in any other group of parasitic platyhelminths and there are other features of the relationship between digeneans and molluscs that are so unusual that these intramolluscan events will be considered separately in the next chapter.

CHAPTER 14

Digenean germinal sacs, germinal lineage and genetic events

14.1   GERMINAL SACS AND MOLLUSC DEFENCES

In *Fasciola hepatica*, rediae, produced by the mother sporocyst into which the miracidium transforms, are the germinal sacs responsible for producing cercariae, which, propelled by their muscular tails, provide the link in the life cycle between the molluscan and vertebrate hosts. Many digeneans have cercaria-producing rediae, but this is not a universal feature – there is a dichotomy in the Digenea, with members of several families lacking rediae and relying on tubular, sometimes branched, gutless sporocysts to produce their cercariae (Figs 14.1, 14.2).

Most students who crack open an infected snail for the first time are surprised not just by the volume of parasitic material in the form of rediae and sporocysts that spills out of the body but also by the impression prior to dissection of an apparently healthy, actively moving, feeding snail and the absence of any obvious physical barrier such as a cyst wall between the parasite and host tissue. A natural reaction is to suppose that molluscs have no internal defence system that they can direct against intruders, but this is not so. Molluscs are capable of mounting highly efficient humoral and cellular defences against parasites (see Loker, 1994). The inference is that digenean germinal sacs have ways of circumventing these defences.

It is emerging that there are two fundamentally different ways in which digeneans protect themselves in their molluscan hosts. Moreover, the evidence, such as it is at present, supports a link between these two methods and whether or not rediae feature in the intramolluscan multiplication process (see review by Loker, 1994). Soon after invading their molluscan hosts, echinostomes, which are redia-producing relatives of *Fasciola hepatica*, embark on a strategy of rapid and strong interference with the functioning of the haemocytes that provide the cellular arm of the host's internal defence system. The effects on haemocytes include loss of their inherent stickiness and their ability to recognize non-self, accumulation in regions of the snail that are irrelevant to the infection and reduced ability to engage in phagocytosis. For example, excretory–secretory products from cultured sporocysts of *Echinostoma paraensei* inhibit haemocyte spreading and phagocytic activity. In contrast, schistosomes (see Chapter 18), which lack rediae and produce cercariae in daughter sporocysts produced in turn by the miracidium–mother sporocyst, rely less on overt immunosuppression in the early stages of mollusc infection. Instead, they depend more on molecular camouflage, either by producing molecules that mimic those of

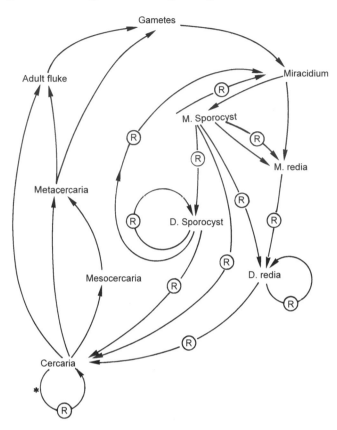

Figure 14.1 Diagram summarizing the possible sequences in the life histories known in the Digenea. 'R' marks the point at which non-sexual multiplication may occur. M =mother; D =daughter. The complications arising from the incorporation of different hosts have not been considered. The arrow marked with an asterisk indicates events in the gymnophallid *Parvatrema homoeotecnum*. Source: modified from Clark, 1974.

the host or by acquiring a disguise of borrowed host molecules (section 18.3.6). Later in the course of infection they may also interfere with the mollusc's defence system (section 18.3.6).

Digeneans such as echinostomes, which aggressively suppress the molluscan defence system, lay themselves open to a serious risk, especially if the parasite needs a long period of development within the host. Other opportunistic parasites with short generation times might invade the host unopposed and overrun and kill it. In this respect, it is interesting that echinostome interference with haemocytes seems to be selective, because some defensive activity such as wound repair and encapsulation of nematode larvae is retained (see Loker, 1994). However, closely related parasites (other digeneans) may present a greater threat. It has been demonstrated that strains of the snail *Biomphalaria glabrata*, normally resistant to *Schistosoma mansoni*, become susceptible if first infected with *E. paraensei*; haemocytes from echinostome-infected snails are no longer able to kill sporocysts of *S. mansoni*. Sousa (1992) reported from field-collected snails several examples of positive associations between two different digeneans, suggesting that interference by one digenean with the mollusc defence system followed by colonization by a second opportunistic digenean may be common.

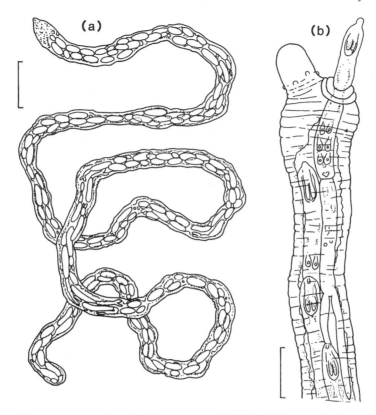

(a)

(b)

Figure 14.2 Daughter sporocyst of *Posthodiplostomum cuticola*. (*a*) Whole germinal sac. (*b*) Cercaria emerging from birth pore. Scale bars: (*a*) 200 μm; (*b*) 100 μm. Source: reproduced from Dönges, 1964, with permission from Springer-Verlag.

In this situation rediae need to be looked at in a new light. They are active, have a suctorial pharynx and a gut and, surprisingly, they are capable of ingesting and destroying germinal sacs of other digeneans (see section 13.4.1, and, for example, Lie, 1973, Sousa, 1993). Thus, we have the curious scenario in which one endoparasite is an active predator on a second endoparasite inhabiting the same host individual. This component of the redial diet may have more than nutritional significance. Some/all rediae may have a policing role, roving around the bodies of their molluscan hosts and destroying any opportunistic secondary infections, especially of other digeneans, before they pose a serious threat. Digeneans relying on molecular camouflage rather than on immunosuppression to evade the host's responses would leave the host's internal defence system relatively intact, have no need for predatory rediae and may have eliminated them from their life cycles.

The remarkable bird schistosome *Austrobilharzia terrigalensis* has succeeded in turning the tables on redia-producing digeneans and exploits their propensity to impair the host's defence system (section 18.3.6).

As Esch and Fernandez (1993, p. 288) have pointed out, although rediae of one digenean species may prey on germinal sacs of another inhabiting the same snail host, cannibalism has never been reported and this raises the question of how self-recognition operates among predatory germinal sacs.

14.2 EFFECTS OF GERMINAL SACS ON MOLLUSCS

As mentioned above, morbidity is not a word that comes to mind when snails infected with germinal sacs are observed. However, such impressions can be misleading. Lafferty (1993a), for example, found evidence to suggest that infection with germinal sacs reduces population density in the salt marsh snail *Cerithidia californica*, but increased mortality of parasitized snails is only one possible contributory factor. Interference with host reproduction is also common. This manifests itself as sterilization and, although it may affect both male and female gonads, is usually referred to as parasitic castration (since this term is in common use its employment will continue here). This not only curtails the reproductive output of the snails but may also reduce the fitness of unparasitized snails because castrated snails may remain alive and compete with unparasitized snails for the limited supply of epibenthic microalgae. According to Lafferty (1993a, b), parasitic castration is common not only among gastropod hosts but also in other marine invertebrates and its influence on host populations may have been underestimated.

Low prevalence in the molluscan host is a feature of digenean life cycles and this has been attributed to high rates of mortality in infected snails (references in Wilson and Denison, 1980). This explanation may well account for the situation in many digeneans, but not in *F. hepatica*, in which infection does not curtail the life of the snail host. On the contrary, infected molluscs undergo increased growth or gigantism, which, according to Wilson and Denison (1980), occurs because parasitic castration permits the snail to divert resources from reproduction to increased somatic growth (section 13.4.3).

Some snails may have adapted to the risk of parasitic castration. In the snail *Biomphalaria glabrata*, infection with the digenean *Schistosoma mansoni* leads to a sudden burst of reproductive effort before castration occurs (Thornhill, Jones and Kusel, 1986). There are indications that *C. californica* responds to the risk of castration by maturing earlier, since Lafferty (1993b) found a negative association between size of snails at maturity and the prevalence of parasitic castration by germinal sacs. Two advantages of earlier maturation are the reduction in generation time and the increase in probability that the individual will survive to maturity. Lafferty considered that such an adaptation could come about by natural selection against snails that delay maturation in populations that are at a high risk of parasitic castration. Alternatively, the snails may be phenotypically plastic, i.e. a snail may have the potential to develop in more than one way. Some kind of environmental cue reflecting the risk of parasitism and hence the risk of castration may act as a developmental switch leading to earlier maturation. Some evidence for genetic influence was revealed by experimentally transplanting snails between sites, but it was not possible to rule out phenotypic plasticity. According to Lafferty (1993b), factors such as penetration by miracidia, presence of free cercariae and level of interaction between snails (parasitized snails show reduced copulatory behaviour) could be used by young snails to 'assess' the risk of parasitic castration and could serve as cues for developmental switching. That such a mechanism is realistic is supported by the observation of Kuris (1980) that when young snails of the genus *Biomphalaria* are exposed to echinostome miracidia they grow slower than unexposed snails, irrespective of whether or not the exposed snails become infected.

Of special interest is the growing body of evidence from studies of interactions between snails and germinal sacs providing correlational support for the maintenance of sex by parasites (so-called Red Queen hypothesis). For example, the results of field studies of the snails *Potamopyrgus antipodarum* and *Bulinus truncatus* were both consistent with the idea that castrating digeneans select for at least partial cross-fertilization (Johnson *et al.*, 1995; see also section 17.7.1).

14.3 REPRODUCTION OF GERMINAL SACS

There is evidence from experiments involving repeated surgical transplantation of single rediae or sporocysts from snail to snail that the multiplication of germinal sacs may be potentially unlimited (Dönges and Götzelmann, 1988). A similar potential has been detected in the tapeworm *Hymenolepis diminuta* (section 9.3.4).

It is generally assumed that most hermaphroditic adult digeneans inseminate each other or self-inseminate and that the miracidial embryos in the eggs are the product of a typical sexual process. In contrast with the situation in the adult, the reproductive events involved in the process of multiplication in the molluscan host have been hotly debated and the problem is still a long way from satisfactory resolution. Essentially, the question is whether the sequential production of sporocysts, rediae and cercariae involves an asexual process (polyembryony, budding) or is a modified sexual process (meiotic or ameiotic parthenogenesis). It is also widely accepted that digenean reproduction involves the phenomenon of germinal segregation (germinal lineage; see below), a process that has no parallel in other parasitic platyhelminths as far as we know.

14.3.1 Early genetic events in the egg

The now classical account of sexual reproduction in the adult digenean *Parorchis acanthus* by Rees (1939) established that a sperm penetrates the primary oocyte in the first part of the uterus and that this event is followed, after the assembly of the eggshell, by prophase of the first meiotic division. Eleven chromosome bivalents ($2n = 22$) separate, 11 chromosomes going to each pole of the spindle. One chromosome group is extruded in the first polar body and the other is incorporated in the secondary oocyte, which then undergoes the second meiotic division. This produces a second polar body and a mature haploid ovum containing the female pronucleus, which subsequently fuses with the haploid male pronucleus from the spermatozoon.

14.3.2 Germinal lineage

This phenomenon is represented diagrammatically in Fig. 14.3 and is well illustrated by the pioneering studies of Rees (1939, 1940) on *P. acanthus*.

After fertilization, the zygote undergoes unequal cleavage, producing a somatic cell and a smaller propagatory cell. The former gives rise to the body of the miracidium and the latter contributes a few cells to the body but persists as a germ cell that enters a small cavity near the posterior end of the miracidium. *P. acanthus* differs from *F. hepatica* in that the first redia develops inside the miracidium and this redia is produced by multiplication of the germ cell. However, as in the development of the miracidium, the germ cell divides into a somatic cell that produces the body of the redia and a propagatory cell that gives rise to germ balls destined to become second generation rediae. The germ plasm is segregated in

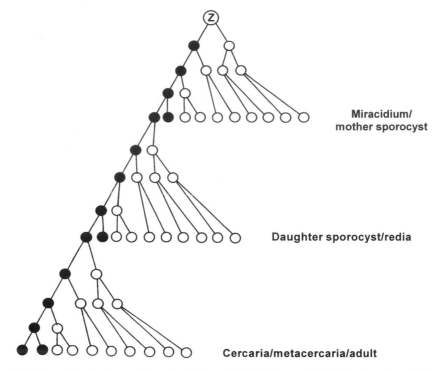

Figure 14.3 Generalized scheme of germinal lineage in the Digenea. Multiplication of germinal cells in the cercaria–metacercaria–adult generation gives rise to the germarium and testes. See text for further explanation. Z =zygote; filled circle =germinal cell; empty circle =somatic cell. Source: redrawn from Whitfield and Evans, 1983.

exactly the same way in second generation rediae and in their offspring, cercariae. The germ cells inherited by the cercaria produce the genital primordium, which will differentiate into the genital organs and germ cells of the adult. Thus, the germ cells inside the miracidium, redia, cercaria and adult can be traced back to one of the cells produced by the first cleavage of the egg. In other words a lineage of cells is established from the original zygote and this lineage is conserved and maintained separately from the somatic cells throughout the life-history phases.

This phenomenon of germinal segregation seems to be unique in the parasitic platyhelminths and the variety and flexibility of life cycle patterns in digeneans (Fig. 14.1) has been attributed to the ability of the totipotent germ cells in the germinal sacs to initiate development along different pathways (Clark, 1974; Whitfield and Evans, 1983). This flexibility extends to the ability of some germinal sacs to produce concurrently more than one type of offspring (e.g. redia and cercaria) and to switch production from one type of offspring to another (see Whitfield and Evans, 1983). However, it must be said that Haight *et al.* (1977a) failed to find support for germinal segregation in their cytological studies of cercarial development in the schistosome *Trichobilharzia ocellata*.

14.3.3 Genetic events in the germinal sacs

Whether reproduction in the germinal sacs is an asexual or modified sexual process hinges on whether or not the germinal cells that give rise to the various germinal sacs can be considered to be eggs. None of the sacs possesses a recognizable germarium, so the only way that this question can be answered is to look for evidence of meiosis in the germinal cells. It was in *F. hepatica* that Bednarz (1962) claimed to have found support for this. In developing germinal sacs (germ balls) he reported the precocious appearance of small, darkly staining bodies, which he interpreted as polar bodies. These bodies were pushed to the surface of the germ ball and then absorbed.

A few years later, evidence for parthenogenesis in the germinal development of another digenean, *Philophthalmus megalurus*, was published by Khalil and Cable (1968). They described a typical germinal lineage (Fig. 14.4) like that of *P. acanthus*. They claimed that, during the development of the germinal sacs (mother, daughter and granddaughter rediae), the propagatory cell of *P. megalurus* divides to produce small, darkly staining, germinal cells and that these cells are oogonia, which transform to oocytes by an abbreviated meiotic process in the development of all stages except cercariae. According to Khalil and Cable, the nuclei of the oogonia advance to diakinesis with the appearance of 10 bivalent chromosomes ($2n = 20$), but do not undergo reduction division and, therefore, do not produce polar bodies, the nuclei returning to interphase. Each oocyte enlarges and when it undergoes the first mitotic division to produce the somatic and propagatory cells the diploid number of chromosomes appears and is passed into each daughter cell.

Khalil and Cable cast doubt on Bednarz's claim that meiosis advances to polar body formation in *F. hepatica*. They pointed out that their 'oogonia' in developing germinal sacs are similar in size and appearance to the small cells interpreted as polar bodies by Bednarz. They stressed the incongruity of Bednarz's identification of polar bodies in germ balls destined to become cercariae, since the germinal cells of cercariae undergo complete meiosis in the adult.

The assertion of Khalil and Cable that diploid parthenogenesis occurs in the germinal development of digeneans has received wide acclaim, together with the conclusion that follows from this that the germinal sacs reproducing in the mollusc are not larvae. However, the work of Khalil and Cable has been placed in perspective by Clark (1974). In a carefully reasoned account Clark pointed out that the evidence for sexual reproduction as a general phenomenon in digenean sporocysts and rediae rests on the isolated observation of a possible first meiotic prophase in *P. megalurus*. Khalil and Cable did not state that they detected the double nature of the bivalents, nor did they offer any evidence that chiasmata formation or crossing over occurred. Clark questioned whether this transitory chromosome behaviour constitutes meiosis at all and made the point that, in the absence of any further reports of even a rudimentary meiosis in any other digenean, the information on *P. megalurus* falls far short of establishing that sexual reproduction occurs in the germinal sacs of even the philophthalmids, let alone the Digenea as a whole. Clark concluded that in the absence of maturation phenomena, or of recognizable ovaries, there is no evidence that germinal cells are ova or oogonia and thus there is no evidence of parthenogenesis, meiotic or ameiotic.

Important cytological studies by Haight *et al.* (1977b) have also cast doubt on the idea that the small intensely staining cells in germ balls represent a proliferating oogonial class of cells. Working with germ balls destined to become cercariae in the schistosome *Trichobilharzia ocellata*, they found no evidence for DNA synthesis and obtained DNA content values for these cells of 2C only, features that are inconsistent with cells undergoing meiosis.

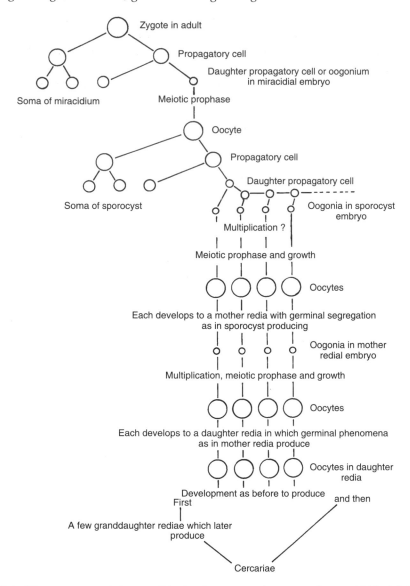

Figure 14.4   Diagram summarizing possible development in the germinal sacs of *Philophthalmus megalurus*. Source: reproduced from Khalil and Cable, 1968, with permission from Springer-Verlag.

Clark pointed out that to invoke polyembryony in the process of multiplication in the mollusc would require an extension of the concept of this term to include delayed disruption of the polyembryonic mass, a move he believed would render the term irrelevant and unhelpful. Clark found no cogent argument against regarding the usual propagation of daughter sporocysts, rediae and cercariae as anything other than budding.

Clark also challenged the concept of alternation of generations in the digenean life cycle and preferred to regard the germinal sacs as a sequence of morphs within a generation, with multiplication by budding. This interpretation depends very much on the definition of 'generation', which Clark chose to regard as 'the interval between the maturation divisions of the egg which mark the initiation of the development of a new individual, and the meiotic divisions which mark the beginning of embryogenesis in its offspring'. However, the term 'generation' is widely used in a different sense to describe each stage in the succession of germinal sacs (section 13.4.2) and there are no hard and fast rules to determine which of these interpretations is correct.

Clark argued that sporocysts and rediae fulfil the criteria of larval stages and he recognized functional equivalents of digenean larval stages in other groups such as the cnidarians. He emphasized that multiplication of larval and post-larval stages is not uncommon in a variety of animal groups after a hazardous stage, whether this be pelagic dispersal to a suitable substrate as, for example, in the planula larvae of cnidarians or transfer between hosts as in digeneans.

Clark's essay is important because it keeps the whole question of the nature of germinal events in digeneans very much alive and cautions against generalizations based on isolated observations. Clark's interpretation of the digenean life history as a sequence of polymorphic larval forms, multiplying asexually rather than parthenogenetically and linked by metamorphosis to the sexually reproducing ovigerous adult, is supported by the cytological work of Haight *et al.* (1977b) and endorsed by Whitfield and Evans (1983). Further extension of our understanding of intramolluscan events in digeneans, including germinal lineage, requires more of the cytological approach pioneered by Haight *et al.* (1977a, b).

CHAPTER 15

Digeneans with two hosts

15.1 DIGENEAN FAMILIES AND THEIR RELATIONSHIPS

At the family level, digeneans have undergone diversification that is unparalleled by the monogeneans and the cestodes. Schmidt and Roberts (1989), for example, listed 137 surviving digenean families. However, relationships between them are unclear. Yamaguti (1958) created a very asymmetrical division into two taxa, based on the position of the mouth in the adult. A more popular and more symmetrical subdivision is based on whether or not the excretory bladder has a lining of mesodermal cells (Epitheliocystidia and Anepitheliocystidia, respectively), but ultrastructural studies have cast doubt on the reliability of this criterion (see Gibson, 1987). Other features are useful, such as presence or absence of rediae and whether the cercarial tail is forked or single, and on the basis of these criteria four subdivisions can be recognized (Appendix). These names will be used in the account that follows, although there is doubt as to whether they reflect real phylogenetic relationships (see Gibson and Bray, 1994).

15.2 ADOPTION OF A VERTEBRATE HOST

In modern aquatic and terrestrial habitats, molluscs are ubiquitous and pervasive organisms in terms of numbers of species and individuals, and there are few vertebrates that do not come into contact with molluscs at some time during their lives. There is no reason to believe that the situation would have been different during the period in the distant past before digeneans acquired a vertebrate host (see Chapter 3). These early digeneans can be envisaged as endoparasites of molluscs, perhaps already multiplying asexually in germinal sacs like their modern descendants, and seeding the aquatic environment with free-living, actively swimming, tailed adults. Frequent contacts with contemporary aquatic vertebrates (fishes) would have occurred and some fishes may have included free-swimming adult digeneans in their diet. Two developments may have followed: some digeneans, the forerunners of the transversotrematids, may have become fish ectoparasites, while others survived being eaten and adopted an endoparasitic life style in the alimentary canal of predatory fishes. Some descendants of these early gut parasites survived with little change and gave rise to the modern azygiids and bivesiculids, but other descendants had much greater potential and eventually came to exploit vertebrates of all kinds.

Throughout the course of this evolutionary expansion, digeneans maintained the molluscan phase of their life cycle in parallel with the vertebrate phase. It may well have been the preservation of this molluscan reservoir of multiplication and infection and the frequent ecological contacts between molluscs and vertebrates mentioned above that encouraged the diversification of digeneans and their exploitation of such a great range of vertebrates. It is the link between mollusc and vertebrate that has been the focus for much evolutionary change in the digeneans and will provide the central theme in this and in the next two chapters.

The transversotrematids, the azygiids and the bivesiculids all possess a forked tail for swimming after leaving the molluscan host. The implication is that the forked tail is primitive and may have been a feature of free-swimming adult digeneans before the vertebrate host was acquired. It follows that the single, unforked tail is a secondary condition, created by reduction of the tail furcae or possibly by fusion of the tail forks. We have evidence of such tail reductions in brachylaimids (section 16.2.2) and in fellodistomatids (sections 15.5, 16.3.2), and similar changes may have occurred independently in other digenean lineages.

15.3   THE TRANSVERSOTREMATIDS

Transversotrematids occur on marine, brackish and freshwater fishes and a life cycle is shown in Fig. 15.1. The free-swimming stage emerging from the gastropod host is virtually an adult, with the reproductive organs well developed and spermatozoa present (Fig. 15.2). However, egg assembly does not begin until they are established on the fish host. The maximum life span of the free-swimming stage of *Transversotrema patialensis* is about 44 h at 24°C, but infectivity persists for only half of that time (Anderson and Whitfield, 1975).

Whitfield, Anderson and Moloney (1975) have shown that the free-swimming stage of *T. patialensis* is superbly adapted for taking advantage of any chance contact with the surface of a suitable fish host. The parasites are somewhat bizarre in that the body is transversely elongated, with two arms projecting laterally from the dorsal surface of the anterior region of the tail (Fig. 15.2). However, the configuration shown in Fig. 15.2 is not normal. Whitfield *et al.* showed that the free-swimming organism adopts a curious flexed attitude in which the body is folded backwards in a ventral direction and clasps the tail (Fig. 15.3). The ventral sucker makes contact with, and probably attaches to, the tail stem, locking the flexed body in position. At the extremity of each arm is an adhesive pad, facing posteriorly, and, facing anteriorly, an array of nine mammiform receptors (Fig. 15.3), each of which terminates in a sensillum ensheathed, apart from its tip, by a cylindrical collar.

The animal indulges in bursts of active swimming lasting for 1–3 s. Progression is by rapid beats of the tail with the tail leading, overall movement being in an upward direction (eyes are present). Between swimming bursts, the animal sinks passively, with the body downwards; the broad tail forks are extended at the top and possibly retard the descent. During the sinking phase the two arms lead, with their arrays of mammiform receptors in advance. If these receptive surfaces should touch a fish during the descent (or during the active swimming phase) a complicated and extremely rapid sequence of behavioural events occurs that firmly establishes the animal on the fish's skin (Fig. 15.4). If the receptors 'recognize' the surface as suitable (no response occurs on making contact with glass), the ends of the arms undergo

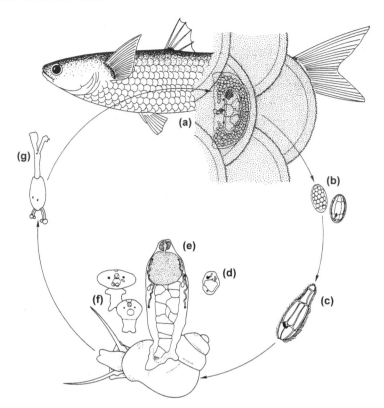

Figure 15.1 Life cycle of *Prototransversotrema steeri*. Adults (*a*) live beneath scales of fish such as *Mugil cephalus* inhabiting freshwater. Eggs (*b*), released into water, develop and hatch in 2 weeks. Miracidia (*c*) invade snail *Posticobia brazieri*. Mother sporocyst (*d*) produces rediae (*e*). Cercariae (*f*) undergo much development in snail after early release by rediae. After escape from snail, fully developed cercaria (*g*) makes contact with fish skin, jettisons tail and creeps beneath a scale. Source: based on photographs and drawings provided by Drs T. H. Cribb and S. Pichelin; stages redrawn from Cribb, 1988.

rapid rotation to bring the adhesive pads on the arms into contact with the fish. Whitfield *et al.* pointed out that the arms contain striated muscle, which is often associated with rapid or repetitive movements (see also sections 6.4.1, 10.6, 13.5), and suggested that it is this muscle that is responsible for arm rotation. Once the adhesive pads on both arms are attached, the body of the organism releases its grip on the tail and swings forward so that the dorsal surface of the body is adjacent to the fish's skin. The body then rolls over so that the ventral sucker is in contact with the fish and able to establish strong attachment, at which point all attachment by the arms is relinquished. This sequence, from initial pad adhesion to pad detachment, is achieved in 2–4 s.

The body of the parasite is able to glide over the surface of the fish until its anterior border becomes lodged in a recess beneath a fish scale. Powerful body undulations combined with tail movements force the body completely into the recess and the tail, including the arms, is discarded.

The transversely elongated shape of the parasite is well suited to fit neatly into the space beneath a scale (Fig. 15.1*a*). Colonization of this niche may have contributed to survival on the fish host by offering shelter from water currents passing over the host and protection from cleaner organisms that prey on ectoparasites (section 5.3).

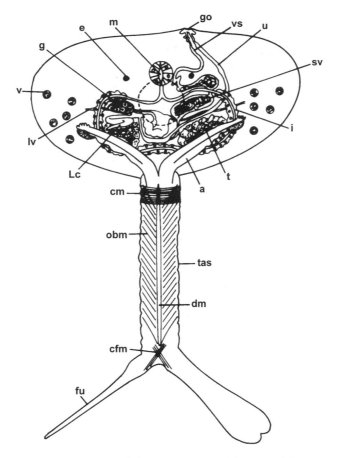

Figure 15.2 Diagrammatic representation of the main anatomical features of the cercaria of *Transverso-trema patialensis* in dorsal view. The body is shown extended forwards (see Fig. 15.3 for folded natural attitude) and the right tail furca is rotated through 90° to reveal its facial aspect. The broken circle represents the position of the ventral sucker. The musculature of the tail is highly schematic and the circular muscle extends down the entire tail stem. a = arm; cfm = cruciform muscles of furcae; cm = circular muscle; dm = dorsal muscle band; e = eye; fu = furca; g = germarium; go = genital opening; i = intestine; Lc = Laurer's canal (a duct of uncertain function opening on the surface); lv = lateral vitelline duct; m = mouth; obm = oblique muscles; sv = seminal vesicle with spermatozoa; t = testis; tas = tail stem; u = uterus; v = developing vitellarium; vs = vas deferens. Source: reproduced from Whitfield, Anderson and Moloney, 1975, with permission from Cambridge University Press.

At the same time, commitment to such a restrictive environment may have stifled further evolutionary expansion of the transversotrematids. In contrast, monogeneans have not colonized the subscale recesses and have diversified greatly (Chapter 5).

The occupation of subscale recesses also places constraints on the population size of transversotrematids. Mills, Anderson and Whitfield (1979) showed that survival and reproduction of *Transversotrema patialensis* on the fish *Brachydanio rerio* are reduced at high population densities. They found no evidence that this phenomenon reflects the mounting of an immunological response by the host and they suggested that it is competition for suitable scale recesses that limits their numbers. Few scales are large enough to offer full concealment and numbers of suitable recesses are further reduced by excessive disturbance in regions of the fish such as

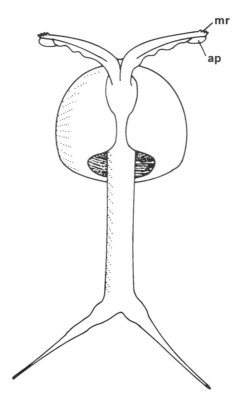

Figure 15.3   The normal flexed attitude of the cercaria of *Transversotrema patialensis* in dorsal view; cf. Fig. 15.2. ap = adhesive pad; mr = mammiform receptor. Source: reproduced from Whitfield, Anderson and Moloney, 1975, with permission from Cambridge University Press.

the tail flanks, which undergo a high degree of flexure during swimming. Food seems less likely to limit population size because the parasites are epidermis feeders, like skin-parasitic monogeneans (Mills, 1979a) and regeneration of subscale epidermis is probably rapid (section 4.4).

Brachydanio rerio, the fish host of *T. patialensis*, occurs in brackish and freshwater environments, and it is not surprising that the cercariae of *T. patialensis* show tolerance to a similar range of salinities and infect hosts equally well in fresh and brackish water (Mills, 1979b). Mills found that adult parasites survive less well *in vitro* in media of low ionic concentration. This suggests that parasites in subscale recesses on hosts living in fresh water must experience some degree of physiological isolation from the external environment.

Many monogenean eggs have a rhythmical daily pattern of hatching and these daily hatching periods may coincide with times when the fish host is more vulnerable to invasion by oncomiracidia (sections 4.10.2, 4.10.7). Hatching rhythms appear to be much rarer in digeneans (see sections 13.2 and 18.3.4 for hatching in *Fasciola hepatica* and *Schistosoma mansoni* respectively), but this is possibly because few people have looked for them. However, Bundy (1981) discovered that the miracidia of *T. patialensis* emerge rhythmically. The eggs are undeveloped when released into the water and take about 3 weeks to embryonate at 25°C. When exposed to LD 12:12 at either 30°C or 25°C, with the light turned on and off abruptly at 'dawn' and 'dusk', hatching occurred over a period of a few days and was restricted on each

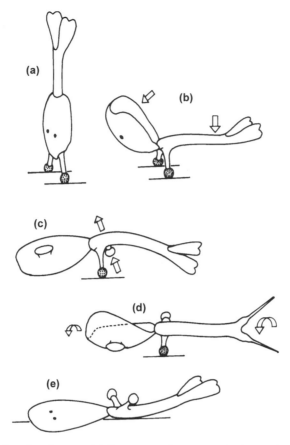

Figure 15.4 (*a*)–(*e*) A series of sequential diagrams to show events in the attachment of a falling cercaria of *Transversotrema patialensis* to the surface of a fish. See text for explanation. The adhesive pads at the tips of the arms are represented schematically as spheres; when stippled they are attached to the surface of the fish (represented by a horizontal line), when unstippled they are unattached. Arrows indicate cercarial movements. Source: reproduced from Whitfield, Anderson and Moloney, 1975, with permission from Cambridge University Press.

day to the last 9 h of the light period, with the hatching peak in the 'afternoons' (Fig. 15.5*a*). The hatching pattern persisted at 30°C in continuous light (Fig. 15.5*b*), suggesting that there may be an endogenous component to the rhythm as in the monogenean *Entobdella soleae* (section 4.10.2), but, unlike *E. soleae*, little hatching occurred in eggs kept in continuous darkness, and development itself may be retarded in the absence of light. Eggs of *T. patialensis* kept in continuous darkness and then exposed to light did not hatch, showing that light does not act as a hatching trigger as it does in *F. hepatica*. Bundy also found that exposure to LD 12:12 *in utero* for as little as 24 h was sufficient to entrain hatching rhythmicity in continuous light.

The snail intermediate host, *Melanoides tuberculata*, is crepuscular in the laboratory and probably also in the wild, ascending and feeding on the vertical walls of an aquarium or on aquatic vegetation at dawn and at dusk. Since miracidia have an average life span of only 2.59 h at 25°C, it seems advantageous for eggs on the bottom to hatch when snails are on the bottom. Such is the case during the hours of full daylight and the hours of full darkness, but miracidial emergence only occurs

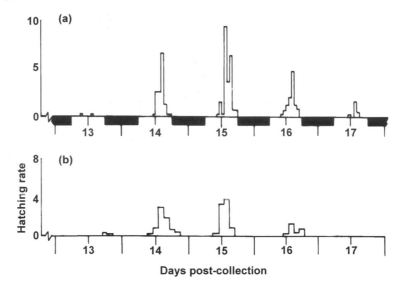

Figure 15.5 Hatching patterns of *Transversotrema patialensis*. (*a*) 30°C, LD 12:12, hourly observations. (*b*) 30°C, continuous light, 2-hourly observations. The hatching rate is the number of miracidia emerging during the observation period as a percentage of all eggs present. Bars on the abscissae (18.00h–06.00h) indicate the dark periods. Source: reproduced from Bundy, 1981, with permission from Cambridge University Press.

during the afternoon. Bundy suggested that nocturnal hatching does not occur because the miracidium has eyes and may require light for orientation and host finding. The absence of hatching in the morning is less understandable but may be related to some unknown feature of snail behaviour during this period.

The presence of a redial generation in transversotrematids (Fig. 15.1), in other digeneans regarded as primitive such as the azygiids and bivesiculids (see below), supports the notion that rediae appeared early in the evolution of the digeneans. According to Pearson (1972), their appearance most probably occurred after establishment of the two-host (mollusc–vertebrate) life cycle, the new redial generation being derived from cercariae in which germ balls proliferated and developed. If rediae did evolve in this way, then the cercariae that gave rise to them may have had only a single, rhabdocoel-like gut caecum, since all modern rediae have this feature.

Another feature of the transversotrematids regarded as primitive by Pearson is the release by the rediae of cercarial embryos at a very early stage, development being completed in the host's haemocoel (Fig. 15.1).

15.4 THE AZYGIIDS AND BIVESICULIDS

Predation is a potent selective force and we are all familiar with spectacular examples of its handiwork such as camouflage, colour change and mimicry in free-living animals. The intriguing feature of the azygiid and bivesiculid digeneans is that they inhabit a topsy-turvy world where their survival depends on soliciting predation rather than avoiding it (Fig. 15.6). Consequently, they have a range of characteristics that set them apart from most other cercariae; names such as *Cercaria mirabilis* and *C. splendens*, applied to some of the first azygiids to

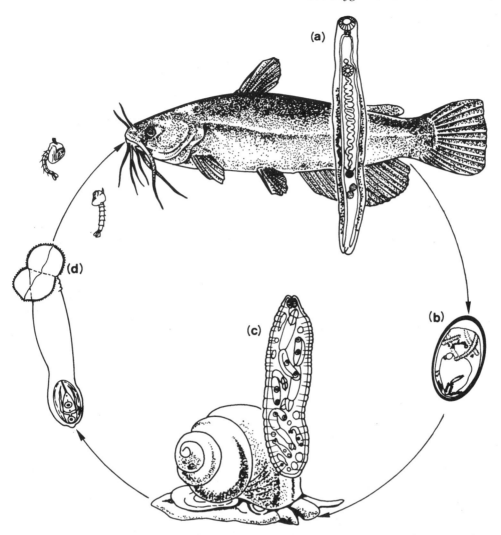

Figure 15.6 Life cycle of an azygiid digenean. Adult parasites (*a*) inhabit mainly the stomach of freshwater fishes and eggs (*b*) containing fully developed miracidia leave in faeces. Eggs are eaten by snail and cercariae produced in rediae (*c*). Fishes infected when they eat free-swimming cercariae (*d*). Resemblance of cercariae to aquatic pupae of midges and mosquitoes may improve likelihood of predation by fishes. Source: based on the life cycle of *Azygia acuminata* (in Yamaguti, 1975).

be recognized in Europe, eloquently reflect the striking appearance of these organisms (see Fig. 15.8*a*, *b*).

Throughout the literature these spectacular, free-swimming azygiids and bivesiculids are described as cercariae, and I will perpetuate this by employing the same convenient term. However, many of these organisms are free-living adults, the body possessing functional gonads with a uterus accommodating a variable number of eggs. Free-living individuals of *Proterometra macrostoma* have fewer than 50 undeveloped eggs, but *P. autraini* contains between 200 and more than 300, many of which enclose active miracidia (LaBeau and Peters, 1995).

P. dickermani is equally prolific, but its life cycle is atypical. According to Anderson and Anderson (1963), the most likely sequence of events is as follows. Secondary

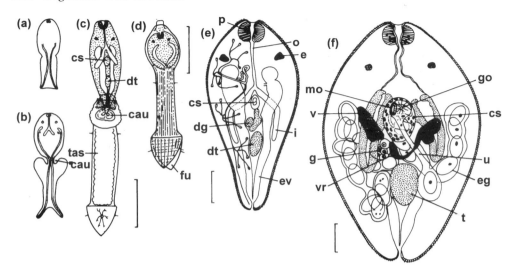

Figure 15.7   Development of the bivesiculid *Paucivitellosus fragilis*. (*a*), (*b*) Developing cercariae from within a redia. (*c*) Mature cercaria before retraction of the body into the tail. (*d*) Cercaria after retraction. (*e*) Cercarial body (ventral view), with complete excretory system shown only on left side. (*f*) Adult (ventral view). cau = caudal chamber; cs = cirrus sac; dg = developing germarium; dt = developing testis; e = eye; eg = egg; ev = excretory vesicle; fu = tail furca; g = germarium; go = genital opening; i = intestine; mo = male opening; o = oesophagus; p = pharynx; t = testis; tas = tail stem; u = uterus; v = vitelline follicle; vr = vitelline reservoir. Scale bars: (*c*), (*d*) 200 μm; (*e*), (*f*) 50 μm. Source: modified from Pearson, 1968.

germinal sacs in the gonoducts of the snail *Goniobasis* (*Elimia*) *livescens* produce cercariae, which mature without leaving the sacs. Eggs are deposited within the sacs and miracidia hatching from them leave the sacs and transform to a new generation of primary sacs. Entire secondary sacs are extruded from the snail's genital aperture from time to time and liberate their contents. Other snails are infected by ingesting eggs or miracidia, completing their life cycle without the intervention of a fish host. No vertebrates naturally infected with *P. dickermani* were found, but Anderson (1962), after numerous feeding attempts, obtained a few adults experimentally from fishes, indicating that natural infection of a fish is possible.

The forked tails of azygiids and bivesiculids are unusual in that only the posterior region, consisting of the tail stem and two terminal, paddle-like furcae, is concerned with propulsion, the anterior part being a flask-shaped vesicle. Before the cercaria leaves the snail or shortly afterwards, the body of the cercaria becomes enclosed within this tail vesicle (Fig. 15.7). Anderson and Anderson (1967) observed the process in azygiids of the genus *Proterometra* and likened it to embryonic involution, with the cercarial body resembling a passive yolk plug – the body remained quiescent while the margin of the tail cavity advanced over its surface, and at the final moment the opening contracted forcefully so that the body appeared to 'pop' into the cavity. In many species the caudal vesicle inflates soon after emission from the snail.

Some of these free-swimming cercariae reach a relatively enormous size: *Azygia lucii* (Fig. 15.8*a*) may be 7 mm in length (Szidat, 1932) and *Bivesicula caribbensis* (Fig. 15.8*d*) is almost 3 mm long and more than 1 mm wide, while the enclosed cercarial body is less than a quarter of a millimetre in diameter.

Cercariae of most other digeneans are relatively transparent, unless their bodies contain dense cystogenous cells. The large cercariae of azygiids and bivesiculids contain no cystogenous cells (they do not encyst) and yet are often opaque, coloured or banded. Yellow, brown or orange pigmentation is common – Cable and Nahhas (1962a) described the caudal vesicle of *B. caribbensis* as brownish orange and *Cercaria mirabilis* (=*Azygia lucii*) has a yellow-brown tail stem and furcae (Horsfall, 1934). According to Le Zotte (1954), *C. caribbea XLIV* (Fig. 15.8*e*) has two bands of reticulated purplish-black pigment, one at the posterior end of the tail vesicle and the other in the tail stem, *C. caribbea XLVI* (Fig. 15.8*f*) has a dark red tail vesicle and an unpigmented tail and *C. caribbea XLVII* (Fig. 15.8*g*) has a pink-red tail vesicle and a lighter pink tail. *C. caribbea XLVIII* (Fig. 15.8*h*) lacks noticeable pigmentation.

The tails of azygiid cercariae lead during locomotion, pulling the organism through the water as the tail moves from side to side with extended furcae (Horsfall,

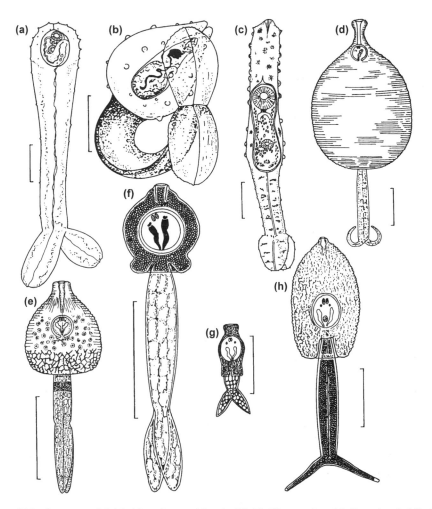

Figure 15.8 Some azygiid (*a*)–(*c*) and some bivesiculid (*d*)–(*h*) cercariae. (*a*) *Cercaria mirabilis* (= *Azygia lucii*). (*b*) *Cercaria splendens*. (*c*) *Proterometra edneyi*. (*d*) *Bivesicula caribbensis*. (*e*) *Cercaria caribbea XLIV*. (*f*) *Cercaria caribbea XLVI*. (*g*) *Cercaria caribbea XLVII*. (*h*) *Cercaria caribbea XLVIII*. Scale bars: (*a*) 1 mm; (*b*)–(*f*) 0.5 mm; (*g*), (*h*) 0.25 mm. Source: (*a*), (*b*) reproduced from Szidat, 1932; (*c*) modified from Uglem and Aliff, 1984; (*d*) reproduced from Cable and Nahhas, 1962 and (*f*)–(*h*) from Le Zotte, 1954, with permission from the American Society of Parasitologists; (*e*) reproduced from Cable, 1963, with permission from Springer-Verlag.

1934). The cercariae move vertically to the surface then sink slowly to the bottom, where renewed locomotion is initiated. Cable and Nahhas (1962a) reported that *B. caribbensis* swam clumsily by side-to-side lashing of the comparatively small tail stem and furcae. There can be little doubt that these movements, combined with the large size of the cercariae and their conspicuous coloration, attract predatory fishes. In fact, Horsfall added *Cercaria* (=*Proterometra*) *macrostoma* one at a time to vessels containing single centrarchid fishes and watched each of the cercariae being eaten, and Cable and Nahhas reported that seven marine squirrel fish, *Myripristis jacobus*, quickly ate all of several hundred cercariae of *B. caribbensis* placed in their aquarium. In both of these studies parasites were later recovered alive from the digestive tracts of the fishes.

Szidat (1932, in Ginetsinskaya, 1958) drew attention to the likeness between the shape and movements of the cercariae of *A. lucii* and those of aquatic larvae of mosquitoes (culicids; Fig. 15.6) and Dogiel (1962) went so far as to promote this as an example of mimicry. This is an intriguing concept although rather difficult to prove, based as it is on human perceptions.

The tail vesicle, as well as its role in enhancing the attractiveness of the morsel to the predatory fish, probably serves to protect the adult from mechanical and chemical damage during swallowing. (Significantly, there is no protective tail vesicle around the cercarial body of *P. dickermani*; see Anderson, 1962.) After liberation of the adult in the oesophagus, stomach or intestine of the fish, the huge discarded tail is likely to be digested, providing perhaps a nutritional 'reward' for the host. In fact, a strikingly similar biological parallel is provided by the strategies adopted by the fruits of many plants to ensure consumption and dissemination of their seeds by birds.

The behavioural and morphological features of azygiid and bivesiculid 'cercariae' as described above seem well suited to the recruitment of pelagic, planktonivorous fish hosts. However, many fishes are neither pelagic nor planktonivorous and the exploitation of these fishes has led to some interesting adaptations in the parasites. The cercaria of *Proterometra edneyi* (Fig. 15.8c) rarely swims, spending most of its time on the bottom performing irregular worm-like movements. This behaviour exposes the cercaria to predation by darters (Percidae), which are bottom feeders (Uglem and Aliff, 1984). Similarly, the host of the bivesiculid *Paucivitellosus fragilis* (Fig. 15.7) is a blenny, which, judging by its teeth, is a grazer, and Pearson (1968) noted that the cercaria attached itself to the rocky surfaces on which the host was likely to feed.

Young planktonivorous pike (*Esox lucius*) become infected with *Azygia lucii* by ingesting the free-swimming cercariae (Fig. 15.8a), but, although the cercariae are consumed by other plankton-eating fish species, they remain in a state of arrested development in these hosts. This may provide a different route of infection for pike; piscivorous adult pike may acquire the parasite by eating infected carrier (paratenic) hosts (Ginetsinskaya, 1958). A parallel trend is displayed by the tapeworm *Bothriocephalus gregarius* (section 10.5.2).

Evidence for a different kind of carrier host was unearthed by Stunkard (1956). He found undeveloped cercarial bodies (*Azygia* sp.) in the pharyngeal pockets of the planarian *Dugesia tigrina*, noted that they survived there for weeks and suggested that they may be a source of infection for fishes feeding on infected planarians.

An interesting polymorphism has been studied in *P. macrostoma* by Uglem (1980, 1987). The parasite is capable of living in the stomach as an endoparasite or on the gills as an ectoparasite. Naturally infected longear sunfish (*Lepomis megalotis*) may have stomach and gill parasites, but, in experimental infections, the parasite is found exclusively in the stomach of largemouth bass (*Micropterus salmoides*) and primarily in the gills and pharynx of the green sunfish (*L. cyanellus*). Gill parasites are darker and larger than stomach parasites and only the latter possess a functional tegumental glucose transport system (Na^+-independent facilitated diffusion) as in the cercaria. Thus an ingested cercaria retains sugar transport function unless it adopts an ectoparasitic location. Cercariae incubated in 50 mmol/l Na^+ retain glucose transport capacity but lose it in 0.5 mmol/l Na^+ (fresh water). Glucose transport capacity can be restored by returning parasites incubated in fresh water to 50 mmol/l Na^+, but incubation of endoparasitic adults of *P. macrostoma* and cercariae of *P. edneyi* in 0.5 mmol/l Na^+ had no effect on their glucose transport. The absence of a regulating effect of low sodium on *P. edneyi* cercariae may reflect the fact that they are always endoparasitic in their fish hosts.

P. macrostoma has a broader spectrum of fish hosts than other species of *Proterometra* and variability in the cercariae may reflect intraspecific adaptations to particular hosts. Riley and Uglem (1995) used morphological features to distinguish eight strains, each of which had a characteristic host preference, with distinctive swimming behaviour and time of emergence from the snail. These strains were so distinctive that it was possible to identify snails that had been naturally infected with more than one miracidium. Two-thirds of infected snails produced only one strain of cercaria (consistent with infection by ingesting a single egg), but some single snails were found shedding two, three or four strains simultaneously.

Some azygiid cercariae, like those of other digeneans, have strong daily emergence patterns. Lewis, Welsford and Uglem (1989) established that peak emergence in *P. edneyi* occurs early in the daylight period, while in *P. macrostoma* it occurs during the night (Fig. 15.9). Moreover, since the snail host for both species is the same, *Goniobasis* (*Elimia*) *semicarinata*, these contrasting emergence patterns are unlikely to be related to features such as activity patterns of the snail. They demonstrated that the patterns were altered when laboratory LD cycles were reversed, indicating that LD cycling modulates emergence of both species, but they also found evidence that emergence of *P. macrostoma* may be influenced by other factors such as daily temperature fluctuation and endogenous rhythms.

The authors believe that the emission patterns enhance transmission to the final host. *P. edneyi* emerges during the day when its visually feeding hosts (darters) are active (see above). *P. macrostoma* is eaten by sunfishes (Centrarchidae), whose daily foraging patterns are variable; Lewis *et al.* suggested that nocturnal emergence by *P. macrostoma* will decrease its susceptibility to predation by diurnally feeding non-host fishes. The possible importance of daily rhythms in hatching and host-finding in monogenean parasites has already been emphasized (section 4.10.2).

The strategy employed by azygiids and bivesiculids to link the molluscan phase of their life cycle with the vertebrate phase is costly in terms of energy. Since the cercaria cannot feed during its free-swimming life, it must be fully provisioned while inside the mollusc; the accumulated resources are required both for the maturation of the reproductive system and to provide the energy for the locomotor activity needed to attract the vertebrate host. The importance of this activity for survival was underlined by the observation of Uglem and Aliff (1984) that immobile cercariae of *P. edneyi* are ignored by potential fish hosts. Clearly, such activity cannot

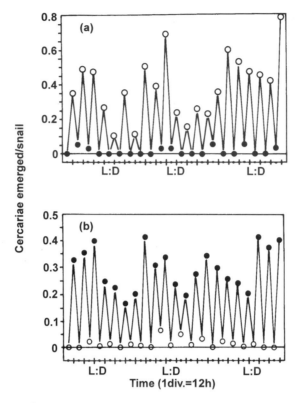

Figure 15.9   Patterns of emergence of cercariae of the azygiids (*a*) *Proterometra edneyi* and (*b*) *P. macrostoma*, exposed to LD 12:12 at 20°C over a period of 3 weeks. Data expressed as cercariae emerged per infected snail per 12 h (*n* = 5). Open circles = emergence during light; closed circles = emergence during dark. Source: reproduced from Lewis, Welsford and Uglem, 1989, with permission from Cambridge University Press.

be maintained for long and, although cercariae of *Proterometra* can survive for more than 24 h, their swimming activity declines rapidly after about 12 h (Lewis, Welsford and Uglem, 1989).

These energy demands and the need to be large enough to attract the attention of the predator make the cercariae of azygiids and bivesiculids very expensive to produce and, according to Lewis *et al.*, the shedding rates of *Proterometra* spp. are 100- to 1000-fold less than those of most other digeneans. In other words azygiids and bivesiculids produce small numbers of relatively large, continuously active and short-lived cercariae.

Reference has been made above (section 15.3) to the occurrence of a redial generation in digeneans which are regarded as primitive, such as the transversotrematids, azygiids and bivesiculids, and the view of Pearson (1972) that the redial generation appeared early in digenean evolution. Sillman (1962) describes the regression of the intestine and rudimentary nature of the pharynx in older rediae of *Azygia longa*, and this is in line with Pearson's suggestion that daughter sporocysts have evolved from rediae, perhaps on more than one occasion, by promoting the tegumental absorption of food at the expense of the gut.

A picture has emerged of striking similarity between the azygiids and bivesiculids in their mode of transmission. In spite of this, there are many reasons for regarding these groups as only remotely related, including differences in their miracidia, in their excretory systems and in the absence in the bivesiculids of oral and ventral suckers (Pearson, 1968).

15.5   THE FELLODISTOMATIDS

If the active life of the cercaria can be prolonged, its chances of being eaten by the vertebrate host will be improved. There is a tendency in azygiids and bivesiculids to delay sexual maturity so that the bulk of egg production and in some cases maturation of the gonads takes place after the cercaria is eaten by the fish (see Fig. 15.7). Delayed maturity is also displayed by some fellodistomatids such as *Monascus filiformis* and *Steringophorus furciger*, the cercariae of which are eaten by marine fishes. The reproductive systems of these cercariae are at the primordial stage, and Køie (1979) reported that the cercaria of *M. filiformis* could survive for a week at 4°C, alternately swimming and resting on the bottom. Their bodies do not have a protective tail vesicle and how they survive the traumas of being eaten by the fish host is unknown. They also lack the colouring of azygiids and bivesiculids, although their excretory vesicles contain opaque material, which makes them conspicuous.

Although these two digeneans are related, *M. filiformis* has a cercaria with a forked tail while that of *S. furciger* has a single tail. Cable (1954a) found evidence to suggest that the single tail is derived from a forked tail by reduction of the tail forks (furcae).

Pearson (1972) regarded the two-host life cycles of fellodistomatids as primitive but the possibility exists that this situation is secondary, the life cycle being derived by reduction from fellodistomatids with three hosts (section 16.3.2).

15.6   ENCYSTING ON THE VERTEBRATE HOST'S FOOD

15.6.1   General

A much more effective way of increasing the longevity of the cercaria would be to encyst on the food of the vertebrate host. Provided that only a brief period of activity was required to reach a suitable substrate and that development of the reproductive system was postponed, then, by being totally quiescent after encystation, the resting metacercaria would require little in the way of maintenance energy and this would be translated into greatly increased longevity (see below). A longer life would mean a greater chance of the metacercaria being eaten by the vertebrate host. The protective role of the bulky tail vesicle of azygiid-like ancestors would be taken over by a multilayered cyst wall and the tail, having performed its brief locomotive function at the time of encystation, could be shed. Since there would no longer be a premium on large size for cercariae, they could be assembled and released in larger numbers.

This novel 'sit-and-wait' strategy of encysting on the food of the vertebrate is suitable for infecting herbivorous definitive hosts. Haploporid and probably mesometrid digeneans gain access to herbivorous fishes, and *Fasciola hepatica* and its relatives are able to infect land-dwelling vertebrates that feed on aquatic vegetation (see Fig. 12.1). The surfaces of some aquatic invertebrates also provide suitable sites for encystation, opening the way for recruitment of carnivorous vertebrate final hosts. It is likely that the habit of encysting externally on the food of the vertebrate final host has evolved on more than one occasion, and some of the modern digeneans that have adopted this strategy are listed in Table 15.1. The special features of some of these digeneans will be dealt with below.

The effect of these life cycle innovations on cercarial longevity is impressive. Dönges (1969) studied a variety of cercariae that encysted externally and found

Table 15.1   Some digeneans encysting in the open and their substrates

Parasite	Substrate	Final host	Authority
Echinostomatidae			
Aporchis massiliensis	Surfaces of snails, arthropods, vegetation	Birds	Prévot, 1971
Notocotylidae	Vegetation	Muskrat meadow mice	Herber, 1942
Quinqueserialis quinqueserialis			
Notocotylus spp.	Shells of marine gastropods	Birds	El-Mayas, 1991
Philophthalmidae			
Philophthalmus gralli	Exoskeleton of crayfish	Birds	West, 1961
Parorchis acanthus	Mollusc shells, crustacean exoskeletons	Birds	Rees, 1967
Haploporidae			
Saccocoelium spp.	On vegetation, in experimental container	Fish	Pearson, 1972 Fares and Maillard, 1974
Mesometridae			
Mesometra brachycoelia	Vegetation	Fish	Palombi (in Bartoli, 1987)
Fasciolidae			
Fasciola hepatica	Vegetation	Sheep, cattle, man	see Chapters 12, 13
Paramphistomidae			
Paramphistomum microbothrium	Vegetation	Cattle	Lengy, 1960
Diplodiscus subclavatus	Skin of adult frogs; free in water	Frogs	Grabda-Kazubska, 1980
Psilostomidae			
Psilotrema spiculigerum	Vegetation	Birds	Mathias, 1924
Sphaeridiotrema (= *Astacatrematula*) *macrocotyla*)	Exoskeleton of crayfish	Birds	Macy and Bell, 1968
Haplosplanchnidae			
Haplosplanchnus acutus (= *Schikhobalotrema acutum*)	In experimental container	Fish	Cable, 1954b

that most were infective immediately after encystation and lived for 3–7 months (1 year in favourable conditions). I have kept metacercariae of freshwater notocotylids (Table 15.1 and below) for 9 months and found them alive at the end of this period.

15.6.2   The haploporids and mesometrids

Haploporids have single-tailed cercariae (Fig. 15.10) and, according to Pearson (1972), encyst on vegetation, their vertebrate hosts being fishes such as mullets that feed on algae. However, there may be a previously unsuspected dimension to the biology of these parasites since Fares and Maillard (1974) found evidence to suggest that small mullets may selectively feed on the cysts rather than ingesting them accidentally. Fares and Maillard recorded a free-swimming life of only 5–7 min in the cercaria of the haploporid *Saccocoelium tensum*.

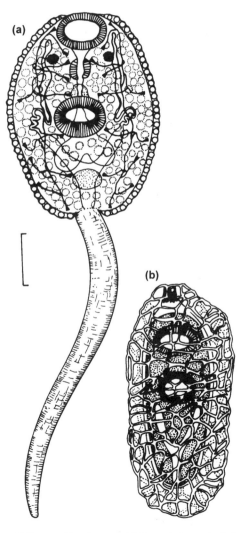

Figure 15.10   The haploporid *Saccocoelium tensum*. (*a*) Cercaria in ventral view. (*b*) Encysted metacercaria. Scale bar: 50 μm. Source: redrawn from Fares and Maillard, 1974.

The mesometrids are poorly known but have some remarkable features (see Bartoli, 1987). The cercariae of *Mesometra brachycoelia* encyst on vegetation eaten by the herbivorous marine fish *Sarpa salpa* (Palombi, in Bartoli, 1987). They have no ventral sucker and the oral sucker contains a denticulate filter, which may serve to remove plant fibres from a diet of host gut contents. The whole body, or the anterior part of it, seems to be modified for suctorial attachment to the smooth (virtually villus-free) gut lining of the host.

15.6.3   The paramphistomids

The rumen of herbivorous mammals is a voluminous chamber derived from the proximal region of the stomach. It receives ingested vegetable matter and subjects it to prolonged incubation and fermentation at $39^\circ C$ with the aid of special symbiotic bacteria and ciliates. It is not until this fermentation process is complete that the food passes through the rest of the stomach and is exposed to acid digestion in its distal chamber (the abomasum). Consequently, there is no other compartment of the gut with features comparable with those of the rumen: its liquid phase is populated by symbiotic organisms and rich in volatile fatty acids, but largely devoid of glucose and amino acids; its gas phase consists mainly of carbon dioxide and methane; it is a turbulent environment, food being mixed and regurgitated for extra mastication.

It is remarkable that some paramphistomid digeneans have established themselves in this challenging environment. The parasites are attached to the wall of the rumen by a posteriorly located ventral sucker (so-called amphistome condition), which is suitably large and powerful to resist detachment by turbulence. Little is known about the physiological and biochemical adaptations that enable these parasites to survive, but Sharma and Hanna (1988) found that, unlike other digeneans, there are no mitochondria in their tegumental syncytia. They interpreted this as indicative of the lack of active uptake of nutrients by the tegument, a situation that might be expected in an environment containing little or no glucose or amino acids. The diet of these parasites also appears to be unusual – Parshad and Guraya (1978) found that the gut contents of adult *Cotylophoron cotylophorum* contained plant material, rumen bacteria and protozoans in sufficient quantity to indicate that rumen contents of the sheep host form a significant part of their diet. Paramphistomids are reported to excyst in the duodenum and then migrate forwards to the abomasum within the mucosa; they then re-enter the lumen and complete the journey forward to the rumen (Roberts and Janovy, 1996).

Some paramphistomids inhabit the rectum of frogs and colonize these animals by exploiting a special behavioural feature of their hosts (Fig. 15.11). Frogs shed sheets of redundant epidermal cells which are then eaten. Grabda-Kazubska (1980) found that single-tailed cercariae of *Diplodiscus subclavatus* readily encysted on the skin surface of adult frogs (but not tadpoles) and that they gained access to the gut of the frog when the skin was sloughed off and eaten. In experimental infections, practically all the surface metacercariae reached the frog intestine within 24 h.

Evolutionary flexibility has not been lost in these frog paramphistomids. According to Bourgat and Kulo (1977), the cercariae of *D. fischthalicus* (referred to as *D. subclavatus* by Bourgat and Kulo) are drawn in with water currents through the nostrils of the tadpoles. On reaching the gut they encyst temporarily and then excyst and become adults. The cercariae are also able to encyst directly in water.

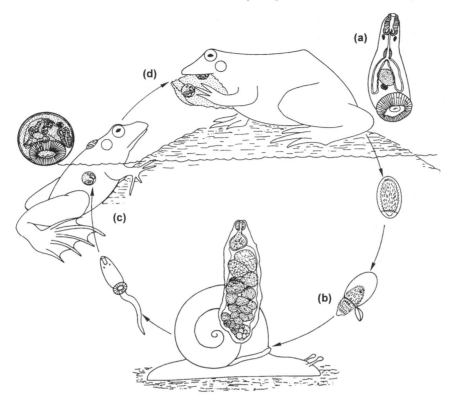

Figure 15.11 The life cycle of the paramphistomatid *Diplodiscus subclavatus*. (*a*) Adult parasites in the rectum of the frog lay eggs, which are voided in the faeces. (*b*) Eggs hatch and miracidia infect the snail *Planorbis planorbis*. (*c*) Cercariae encyst on the skin of immersed frogs. (*d*) Frogs are infected by eating pieces of shed skin with attached metacercarial cysts. Source: based on Grabda-Kazubska, 1980 and Olsen, 1986.

15.6.4   The philophthalmids

Species of *Philophthalmus* (Fig. 15.12*a*) are found in an unusual location, in the orbit of birds, and they have some remarkable features related to this habitat. West (1961) gave a graphic account of the life cycle of *P. gralli* from the eyes of kingfishers and herons. The cercaria (Fig. 15.12*b*) is a poor swimmer and is propelled as much by flexure of the body as by movements of the weakly muscular single tail. The tip of the tail contains adhesive glands and appears to be used for initial attachment to the exoskeleton of crayfish. The ventral sucker then takes over and the cercaria secretes a flask-shaped cyst (Fig. 15.12*c*–*e*). The metacercaria is immediately infective and excystation takes place before reaching the acid proventriculus (stomach), either in the mouth or in the crop. Cheng and Thakur (1967) showed that it is the rise in temperature experienced when the cyst enters the mouth of the warm-blooded bird that stimulates excystation *via* the neck of the flask-shaped cyst, and West (1961) found evidence that the acidity of the proventriculus repels any newly excysted parasites that descend as far as the stomach. The parasites migrate to the orbit *via* the nasolacrimal duct and some reach the eye within 3–5 h of ingestion. West suspected that lacrimal secretions might provide a chemical signpost for migrating parasites, but his attempts to investigate this experimentally were inconclusive.

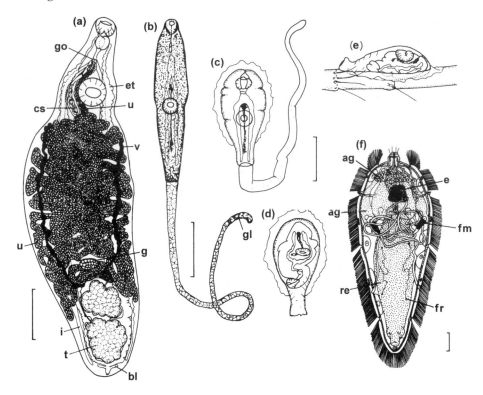

Figure 15.12 *Philophthalmus gralli*. (*a*) Adult from a kingfisher. (*b*) Cercaria. (*c*) Cercaria encysting. (*d*) Metacercarial cyst. (*e*) Cyst in side view on crayfish antenna. (*f*) Miracidium. ag = anterior glands; bl = bladder; cs = cirrus sac; e = eye; et = excretory trunk; fm = flame cell of miracidium; fr = flame cell of redia; g = germarium; gl = tail glands; go = genital opening; i = intestine; re = redia; t = testis; u = uterus; v = vitellarium. Scale bars: (*a*) 1 mm; (*b*) 200 μm; (*c*)–(*e*) 150 μm; (*f*) 10 μm. Source: reproduced from West, 1961, with permission from American Midland Naturalist.

According to West, the eggs of *P. gralli* are non-operculate and the shell stretches to accommodate the growing miracidium, which is fully grown when the eggs are laid. The eggs rupture suddenly in water and each miracidium already contains a redia (Fig. 15.12*f*). By exposing miracidia experimentally to magnetic fields, Stabrowski and Nollen (1985) made the exciting discovery that they exhibit a positive north-seeking magnetotaxis, which, because of the inclination of the earth's magnetic field in North America, would lead them in a downwards direction. This may be a factor in the strong downward movement (positive geotaxis), which will take the larvae to the bottom of ponds and streams where the snail host is found. No magnetite particles that might be components of a magnetic sensor have been found in miracidia. The miracidium only partially penetrates the snail (*Goniobasis* sp.) and liberates the redia into the snail's body.

15.6.5 The notocotylids

In notocotylids the ventral sucker is absent but is functionally replaced in the cercaria by a pair of pockets (Fig. 15.13*a*), each of which occupies a posterolateral position on the body near the origin of the single tail and is supplied with adhesive secretion from a unicellular gland (Southgate, 1971). In addition to its role in

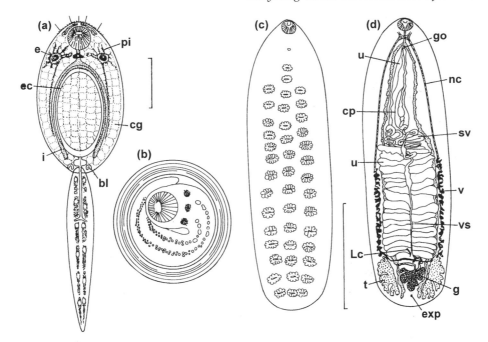

Figure 15.13 *Notocotylus triserialis.* (*a*) Cercaria. (*b*) Encysted metacercaria. (*c*) Adult, showing ventral glands. (*b*) Adult, showing general anatomy. bl = bladder; cg = cystogenous glands; cs = cirrus sac; e = eye; ec = excretory canal; exp = excretory pore; g = germarium; go = genital opening; i = intestine; Lc = Laurer's canal; lop = locomotory pocket; ne = nerve cord; pi = pigment granules; sv = seminal vesicle; t = testis; u = uterus; v = vitellarium; vs = vas deferens. Scale bars: (*a*), (*b*) 100 μm; (*c*), (*d*) 1 mm. Source: modified from Pike, 1969.

locomotion, the adhesive secretion may cement the cyst wall to the substrate (Fig. 15.13*b*).

Adult notocotylids (Fig. 15.13*c, d*) live in the caeca and hindgut of birds and in the intestine of mammals. The pockets are absent in the adult but new developments are mushroom-shaped eversible papillae arranged in longitudinal rows on the ventral surface (Fig. 15.13*c*). These unique papillae are pressed firmly against the mucosa of the host (MacKinnon, 1982a, b), but have no features that would indicate a digestive or absorptive function, such as the presence of glands, phosphatases and esterases, surface microvilli and pinocytotic activity. Evidence for a primary function as attachment organs is also lacking, but the presence of large numbers of mitochondria and haemoglobin (the papillae are red) led MacKinnon to suggest that they may be concerned with uptake of oxygen from the caecal mucosa.

The eggs of notocotylids deserve special attention. Not only do they have long opercular and abopercular appendages of uncertain function (Fig. 15.14), but they contain an inert sporocyst (Fig. 15.15), not a miracidium, and have a unique means of infecting the mollusc (Murrills, Reader and Southgate, 1985a, b). In *Notocotylus attenuatus*, the egg fails to hatch unless eaten by the snail host (*Lymnaea peregra*). Laboratory experiments indicated that the following events take place in the snail's gut: the operculum opens and a cord-like delivery tube shoots out rapidly from the egg and penetrates the gut wall of the snail; the sporocyst is then propelled along the tube and released into the snail's haemocoel (Fig. 15.16). In addition to the sporocyst

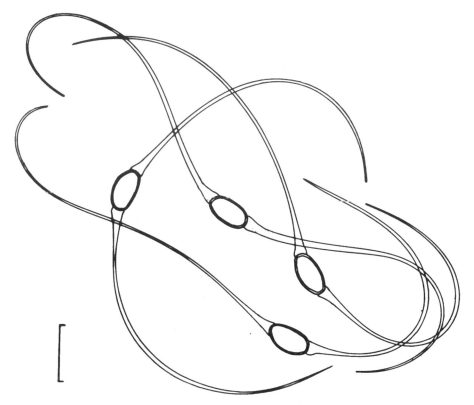

Figure 15.14   Eggs of *Notocotylus attenuatus*. Scale bar: 30 µm. Source: drawn from a photograph in Murrills, Reader and Southgate, 1985a.

and the delivery tube, the egg contains two cells, one of which is packed with glycogen (Fig. 15.15). This food reserve may enable the egg to survive for long periods in the absence of a suitable host – eggs kept in a refrigerator by Wright and Bennett (1964) were still infective after 105 days.

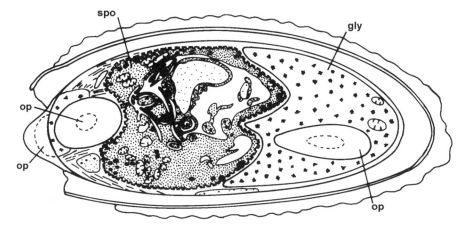

Figure 15.15   Diagram of a longitudinal section through an egg of *Notocotylus attenuatus*, showing opercular cord (op), sporocyst (spo; stippled) and glycogen-containing cell (gly). Source: modified from Murrills, Reader and Southgate, 1985a.

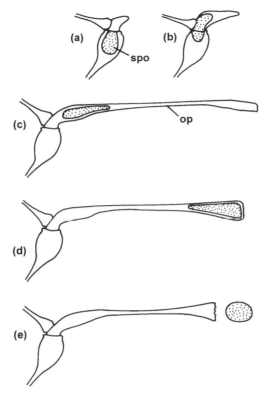

Figure 15.16   (*a*)–(*e*) Diagram illustrating sequential stages in the ejection of the sporocyst (spo; stippled) *via* the opercular cord (op), during *in vitro* hatching of the egg of *Notocotylus attenuatus*. Source: reproduced from Murrills, Reader and Southgate, 1985b, with permission from Cambridge University Press.

15.6.6   The echinostome *Aporchis massiliensis*

The first host of *Aporchis massiliensis* is a prosobranch mollusc, *Vermetus triqueter*, living in the infralittoral zone on exposed rocky shores (Prévot, 1971; Bartoli and Prévot, 1986). The cercaria is unable to swim and has a reduced forked tail containing adhesive glands (Fig. 15.17). It attaches itself to the substrate (algal fronds of *Cystoseira* spp.) by the everted tail forks. The cercaria is sensitive to touch and will attach itself to and encyst on any suitable animal (*Gammarus*, snail) with which it comes into contact. However, the unencysted cercaria also tends to move towards the tips of the algal fronds, on which encystation occurs. Herring gulls (*Larus argentatus*) feed on these apical fronds when insufficient food is available to them elsewhere. The loss of swimming ability and the adhesive specialization of the tail enable the parasite to maintain itself in its wave-swept environment, and it is possible that encystation on the external surface of the potential food of a vertebrate, an unusual feature for echinostomatids (see below), may be a secondary development related to its specialized way of life. An alternative possibility is that this habit of excysting in the open is a primitive feature and represents the ancestral way of life of echinostomatids. Whichever of these is the correct interpretation, there is no doubt that the Echinostomatidae is one of the most flexible and progressive groups of digeneans in terms of evolution of life cycles, a

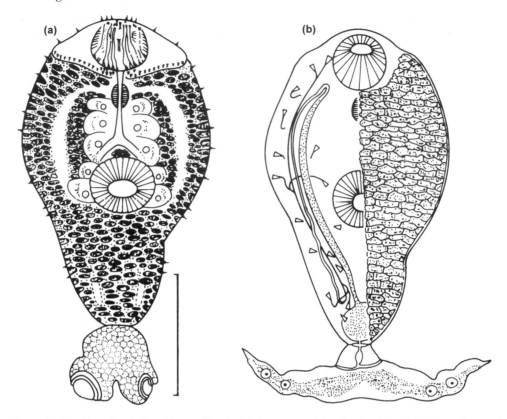

Figure 15.17 Cercaria of *Aporchis massiliensis*. (*a*) Anatomy with tail retracted. (*b*) Outline of cercaria showing extended tail. Scale bar: 100 μm. Source: modified from Prévot, 1971.

statement supported by the prominence of the echinostomatids in the next two chapters.

15.7   PREDATION ON MOLLUSCAN FIRST INTERMEDIATE HOSTS

15.7.1   General

Molluscs are ubiquitous animals and make a contribution of varying proportion to the diets of many vertebrates. We shall see in the next chapter that many digeneans have exploited this by recruiting molluscs as second intermediate hosts. Some of the molluscs consumed by vertebrates will support digenean germinal sacs, which will be digested and destroyed by the vertebrate. Selection will favour any development countering such a disastrous loss of investment by the parasite, and a tendency for cercariae to achieve infectivity to the vertebrate before they leave the mollusc will have obvious survival potential. This has led on more than one occasion to failure of cercariae to leave the mollusc first intermediate host, so that consumption of the mollusc by the vertebrate is now the only way that the parasite can be transmitted between its two hosts.

Heronimus chelydrae, the sole representative of the family Heronimidae, and the cyclocoelids have abbreviated life cycles of this kind and, before their ancestors suppressed the free-swimming phase of cercarial life, may have reached their vertebrate hosts by encysting on vegetation or on suitable external surfaces of prey organisms. The life cycles of these parasites are accelerated in other ways, allowing rapid establishment and rapid generation of infective stages in mollusc communities. This may be the consequence of selection pressures exerted by the short summer season in northern latitudes and the need to complete the life cycle within this period (see Gibson, 1987). Also, the sooner the intramolluscan parasites become infective the better will be their chance of survival when the mollusc is eaten.

Some microphallid digeneans that typically penetrate arthropods now encyst in the mollusc first intermediate host and have lost the arthropod second intermediate host (section 16.3.4(c)).

15.7.2　*Heronimus chelydrae*

A range of unusual characteristics has led Brooks, O'Grady and Glen (1985) to regard this unique parasite as the most primitive extant digenean, a viewpoint that has generated vigorous debate and found little support (see Gibson, 1987; Pearson, 1992). In fact, its remarkable features probably reflect the telescoped and specialized life cycle, and the original interpretation of the parasite as an ally of the paramphistomes (Echinostomata), made in 1956 by Cable and Crandall, has stood the test of time.

The ciliated miracidium penetrates the freshwater snail *Physa integra* and already contains well developed germ balls (Fig. 15.18*a*). These germ balls are enclosed in a thin sac, which Crandall (1960) believed may be the vestiges of a redia. The miracidium transforms into a mother sporocyst, identified even when fully developed by the miracidial eye spots (Fig. 15.18*b*), and this mother sporocyst produces cercariae directly – there are no daughter germinal sacs. Symmetrically arranged lateral pouches increase the space for developing cercariae, each of which has an oral and a ventral sucker (the latter posteriorly located on the body) and a single tail (Fig. 15.18*d*). The cercariae issue from a birth pore but do not emerge from the snail, which is eaten by a freshwater turtle such as *Chelydra serpentina*. Cercariae artificially released from the snail are able to swim, and Crandall suggested that this ability would permit cercariae released under water when the snail is crushed by the turtle's jaws to reach the lining of the host's mouth. The cercariae use their well separated suckers for leech-like locomotion and migrate to the lung of the turtle, probably *via* the trachea. In the lung they feed on blood, the ventral sucker degenerates and the parasite undergoes differential growth such that the excretory pore comes to occupy an unusually anterior and dorsal location.

Eggs accumulate and complete their development *in utero* but eggs or miracidia were not found in washings or scrapings from the lungs, respiratory passages or gut of infected turtles. Crandall believed that the eggs are not laid but are transported by the adult parasite to the outside world. This was based on his finding on several occasions of gravid parasites in the lumen of the respiratory tract of naturally infected turtles that had been killed and examined. Gravid parasites placed in water did not lay their eggs but the miracidia hatched *in utero* and emerged through the genital pore.

Shortening of the generation time of *H. chelydrae* is achieved by direct predation on the snail first intermediate host and by telescoping of events in the germinal sacs.

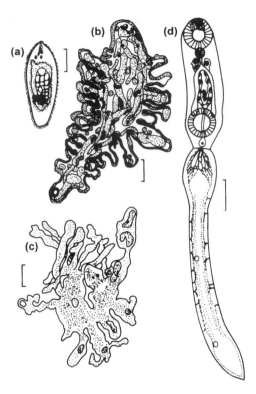

Figure 15.18   *Heronimus chelydrae.* (*a*) Miracidium. (*b*) 20-day-old sporocyst. (*c*) 40-day-old sporocyst with germinal elements almost exhausted. (*d*) Cercaria. Scale bars: (*a*), (*d*) 100 μm; (*b*), (*c*) 200 μm. Source: reproduced from Crandall, 1960, with permission from the American Society of Parasitologists.

Crandall (1960) found full-sized cercariae 17 days after infection with a miracidium, and he believed that cercariae were infective at 22 days. Infection of the mollusc is short-lived; many snails dissected 6 weeks after infection contained little or no trace of the parasite (Fig. 15.18*c*).

15.7.3   The cyclocoelids

Cyclocoelids use birds as definitive hosts. When their eggs are laid they are embryonated and hatch within hours of entering the water. Each miracidium contains a single small redia, which it injects into a gastropod mollusc before detaching and falling off. This redia produces tailless cercariae, which encyst without leaving the snail. The shortening of the generation time by direct predation on the mollusc first intermediate host and by telescoping reproduction in the germinal sacs is similar to the situation in *H. chelydrae. Cyclocoelum mutabile* produced metacercariae 2–3 weeks after infecting the snail in the laboratory (McLaughlin, 1976).

McKindsey and McLaughlin (1995a) found that specificity to the mollusc host in *C. mutabile* was surprisingly low; the parasite was found to be capable of infecting eight out of nine species of snail, six of which were highly susceptible. *C. mutabile* is a parasite of the air sacs of coots (*Fulica americana*) on the Canadian prairies and fails to overwinter in snails. Coots arriving in spring lose their infections and perpetuation of the parasite relies on its rapid re-establishment in the relatively few snails that have survived the winter (McKindsey and McLaughlin, 1995b). Reduced host specificity towards the snails not only increases the size of the available pool of intermediate hosts, but maintains a high degree of genetic variability by exposure to a range of selective pressures (several snail species) and protects against local declines of any single host species (McKindsey and McLaughlin, 1995a; see also section 16.2.1).

15.8   THE WAY FORWARD

Encysting on the external surfaces of prey animals has its limitations as a means of gaining access to the vertebrate intestine. Exposed cysts may be vulnerable to predation by organisms capable of perforating their cyst walls and, apart from molluscs, relatively few animals have external surfaces that provide a suitably hard, permanent encystation site. Cysts have been reported on the exoskeletons of crustaceans (section 15.6.4, Table 15.1), but the use of arthropods involves the risk of being shed with the exoskeleton during moulting. The surfaces of other aquatic animals provide poor sites for encystation. Annelids have a cuticle, but it is thin and perhaps too flexible to provide a secure substrate. The exposed cellular coverings of other aquatic animals, such as free-living cnidarians, platyhelminths and fishes, are not only soft and impermanent but may produce substances that deter settlement. In cnidarians, nematocysts may kill invaders and settlers on echinoderms may be mechanically removed or destroyed by pedicellariae.

Cercariae that found ways of entering the bodies of prey animals not only avoided cyst predators but, perhaps more importantly, greatly extended their range of potential second intermediate hosts; consequently they were able to exploit a much wider variety of vertebrate final hosts. The enormous evolutionary potential that was released by this development is reflected in the range of digeneans with three-host life cycles that will be considered in the next two chapters.

CHAPTER 16

Digeneans with invertebrate second intermediate hosts

16.1 ENTERING THE BODY OF THE SECOND INTERMEDIATE HOST

The least demanding way to enter the body of an intermediate host is *via* an external orifice such as the opening of the kidney, and this route has been extensively exploited, probably independently, by the echinostomes and brachylaimids. These parasites utilize the ubiquitous molluscs as second intermediate hosts. To gain access to other internal encystation sites the surface of the host must be breached and this requires some means of effecting penetration (glands, spines). Cercariae contain a variety of glands that could be functionally diverted to a penetration role. Strong candidates are those with histolytic properties, such as the glands used by cercariae to escape from the mollusc, glands used by some metacercariae to escape from the cyst and anterior glands thought to play a part in feeding in the adult.

The widespread availability and vulnerability of the soft-bodied molluscs gave them high priority as potential targets for tissue-penetrating cercariae and they have been recruited as second intermediate hosts on several occasions. Some of the echinostomes have acquired the ability to penetrate molluscs, as have some strigeids and most gymnophallids. Lepocreadiids and fellodistomatids are not closely related, but have independently developed pelagic cercariae that penetrate cnidarians and ctenophores and, as if this is not remarkable enough as an example of convergence, each of these groups has representatives that have turned to the penetration of molluscs. Some arthropod-penetrating specialists, namely the opecoelids and monorchiids, have also turned their attentions from arthropods to molluscs. Some of the microphallids have achieved the same end by abandoning the arthropod second intermediate host and failing to emerge from the mollusc first intermediate host. Similar life cycle abbreviations that exploit the tendency of many vertebrates to prey on molluscs have already been encountered in *Heronimus* and the cyclocoelids (Chapter 15).

Arthropods must have been a rich prize for the evolving digeneans. According to Barnes (1987), there are at least three quarters of a million species, more than three times the number of all other animal species combined. Moreover, they are perhaps the most successful of all the invaders of the terrestrial habitat. Their vast numbers ensured a place in many food chains and their utilization as intermediate hosts opened the door to a great range of predatory vertebrates. We have already noted that arthropods appear to be underutilized by digeneans that encyst externally, and the exoskeleton provides a barrier to penetration from the outside. However, some digeneans have acquired a stylet for the mechanical task of cutting through the cuticle, backed up by penetration glands. Arthropod diversity is reflected in the large number of digenean families with stylet-bearing cercariae or xiphidiocercariae, including plagiorchids, lecithodendriids, gorgoderids, dicrocoeliids, microphallids and troglotrematids.

Those digeneans that enter the bodies of their second intermediate hosts *via* external orifices seem to gain little in terms of increased longevity of metacercariae. Dönges (1969) reported that echinostomes encysted in molluscs survive for 10–14 months. This is a modest improvement on the longevity of their externally encysting relatives (section 15.6.1), and there is no evidence that these metacercariae obtain nutrients from the host – there is no detectable growth or development and they are quickly infective (see Pearson, 1972). However, adoption of an internal encystation site permitted a reduction in thickness and complexity of the cyst wall, releasing resources that could be used for other purposes.

The thinning of the cyst wall may have provided a new opportunity. If the thinning was accompanied by increased permeability, the parasite could absorb nutrients from the host and continue to grow and develop towards maturity. This would have the effect of shortening the time spent in the vertebrate definitive host, a possible advantage if the host is migratory or requires time to develop its antiparasite defences. Reduced metabolic dependency on the definitive host may also permit exploitation of a wider host range (see also section 10.5.4). A disadvantage, which is presumably outweighed by the possible advantages given above, is that a prolonged period of growth and development in the second intermediate host usually delays the advent of infectivity for weeks or even months (see Pearson, 1972).

Ultimately, the sequestration of resources from the second intermediate host could lead to the attainment of sexual maturity in the second intermediate host and loss of the definitive host, thereby reducing the hosts from three to two. This step has been taken independently on more than one occasion. Other ways in which the three-host life cycle has been shortened will be considered below.

A problem with a cyst wall is that it restricts growth in the intermediate host. Pearson (1972) pointed out that digeneans have solved this problem in three ways: by secretion of an inflated cyst with room for expansion of the metacercaria; by adopting a cyst wall capable of increasing in size as growth proceeds; by delaying cyst wall formation until growth is completed. A fourth solution would be to dispense with the cyst wall altogether, but this is a perilous step since the parasite faces defences mounted against it in the intermediate host and physical and chemical assault when the metacercaria is eaten by the definitive host. Nevertheless some metacercariae have adopted this solution and are able to survive without a cyst wall.

16.2 CERCARIAE WITHOUT PENETRATION GLANDS

16.2.1 The echinostomatids (echinostomes)

The echinostomatids exploit marine and freshwater molluscs extensively as second intermediate hosts and use them to gain access mainly to birds and mammals, including man (Fig. 16.1). The name 'echinostome' means 'spiny mouth' and refers to the presence of a collar of backwardly directed spines in the head region (Fig. 16.2), but these spines are not involved in penetration into the second intermediate host.

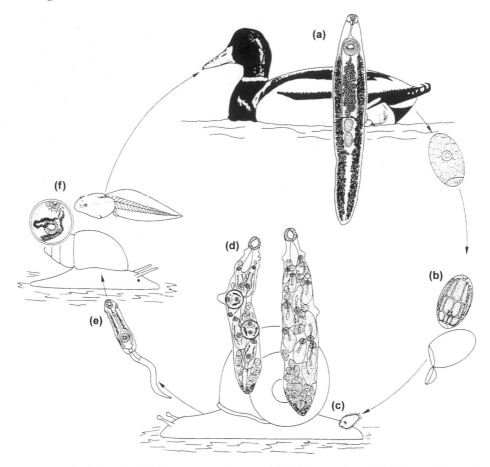

Figure 16.1 The life cycle of *Echinostoma revolutum*. Adult (*a*) in intestine of bird. Eggs (*b*) from faeces embryonate and hatch in water. Miracidium (*c*) penetrates snail; produces mother and daughter rediae (*d*). Some cercariae encyst within daughter rediae; most cercariae (*e*) enter water and swim to another snail or tadpole. Cercariae migrate across skin to renal openings and encyst (*f*) in kidney or pericardial sac (snails). Infected snail or tadpole ingested by definitive host. Source: based on Johnson, 1920, Beaver, 1937 and Olsen, 1986.

Indeed, in some echinostomatids the spines do not appear until after the cercarial stage (e.g. in *Mesorchis denticulatus*; see section 17.2), and Kim and Fried (1989) have shown that they are involved in attachment of *Echinostoma caproni* to the intestine of the definitive host. Fujino, Fried and Hosier (1994) found muscles in *E. trivolvis* that retract the tegument around the spines, thereby exposing their points.

The chemical gradients emanating from snails are likely to be little disturbed by the slow locomotion of these invertebrates and it is probably significant that chemicals have been implicated in snail location by echinostome cercariae (for example, Fried and King, 1989, McCarthy, 1990b). Kinesis, in which the cercariae turn back when small molecular components of snail origin (molecular weight < 500 Da; not peptides) decrease in concentration, and chemotaxis, in which cercariae swim up a concentration gradient of small peptides, have been identified (Haas, 1994; cf. cercarial location of fish intermediate hosts, section 17.5). It is particularly interesting that chemotactic orientation in *Hypoderaeum conoideum* is delayed until 1 h after

Figure 16.2   Scanning electron micrograph showing the anterior spined collar of an echinostome (15-day-old adult *Echinoparyphium recurvatum*). Source: reproduced from McCarthy, 1990a, with permission from Cambridge University Press.

emission; this may serve to reduce overloading of the mollusc first intermediate host with metacercariae (Haas *et al.*, 1995b).

The free-swimming cercariae of *Echinoparyphium recurvatum* from the freshwater pulmonate *Lymnaea peregra* alight first on any exposed part of the body of another *L. peregra* and move across its surface using the oral and ventral suckers. I have observed migrating cercariae arriving at the pneumostome, the opening of the lung-like mantle cavity, within a few minutes of exposure. The renal opening is just inside the mantle cavity and by carefully slitting open the pneumostome and mantle cavity I was able to see that almost all of the cercariae had passed through the renal opening within 10–20 min of making contact with the snail, and most of them had already discarded their tails. About 1 h post-exposure, many cercariae were at the distal end of the kidney, and after about 2 h some cercariae had passed through the reno-pericardial duct into the pericardial sac. Encystation took place in the kidney and in the pericardial sac.

How these cercariae find the kidney opening with such speed and accuracy is not known, but even more surprising is the broad spectrum of molluscan second intermediate hosts that this parasite can utilize, including bivalves and prosobranch gastropods as well as pulmonates (Evans, Whitfield and Dobson, 1981). Evans *et al.* suggested that such broad specificity provides protection from the deleterious effects of a local population decline in any single molluscan host species (see also section 15.7.3), a frustrating phenomenon that is well known to researchers on freshwater digeneans.

I found no evidence to indicate that any cercariae of *E. recurvatum* reached the kidney lumen or pericardial sac by penetrating snail tissue, but Adam and Lewis (1992) reported cysts in the lining of the mantle cavity of *L. peregra*, suggesting that some strains of *E. recurvatum* may have some ability to penetrate tissue (see also below). Metacercariae of *E. recurvatum* have been reported from the visceral mass, but only in snails naturally infected with rediae, and it is likely that these were derived from cercariae that failed to leave the snail (see Adam and Lewis, 1992; Fig. 16.1).

Work on echinostomatids from molluscs is likely to have major implications for our understanding of the evolutionary biology and speciation of echinostomatids in particular and of digeneans in general. McCarthy (1990a) studied apparently identical echinostome cercariae produced by the pulmonate *Lymnaea peregra* and by the prosobranch *Valvata piscinalis*, living in the same body of fresh water in the south of England. Cercariae from these two sources produced apparently identical metacercariae in *L. peregra* and apparently identical adults in ducklings. All of the parasitic stages fitted published descriptions of *Echinoparyphium recurvatum*, but attempts to infect *L. peregra* with miracidia derived from parasites originating from cercariae from *V. piscinalis*, and *vice versa*, failed. There seems little doubt that we have here two sympatric sibling species, which are morphologically indistinguishable but differ in the chosen first intermediate host and also in the position of the adult in the intestine of the final host. A record of the pulmonate snail *Planorbis planorbis* as first intermediate host for *E. recurvatum* in France (see McCarthy, 1990a) and differences in encystation site preferences in the second intermediate host (see above) suggest that '*Echinoparyphium recurvatum*' and indeed other echinostome 'species' named in the literature may mask even more complex species-groups.

The echinostomes have been used extensively in studies of the interactions between germinal sacs and the internal defence systems of their molluscan first intermediate hosts (section 14.1). The aggressiveness of their rediae towards the germinal sacs of other digeneans is well known.

16.2.2   The brachylaimids and leucochloridiids

Brachylaimids share with echinostomatids the ability to locate the excretory openings of snails but they achieve this in the terrestrial not the aquatic environment (Fig. 16.3). Their cercariae have rudimentary tails, but they have relatives with forked tails (e.g. *Leucochloridiomorpha constantiae*, see Allison, 1943), indicating that the brachylaimids are specialized strigeatans. The similarity between brachylaimid and echinostomatid life cycles is another example of convergence.

Ulmer (1951a) found that the first intermediate host of the brachylaimid *Postharmostomum helicis* is the terrestrial snail *Anguispira alternata*, which is abundant in wooded areas, beneath the bark of decaying logs and under leaves. The cercariae are produced in branched sporocysts with birth pores at the terminations of the branches (Fig. 16.3). As the snails move along, the emergent cercariae are left behind in their slime trails and are picked up when another snail of the same or a different species crosses the path of the first. The cercariae move over the body surface of the second snail, enter the renal aperture and ultimately reach the pericardial cavity *via* the kidney and reno-pericardial duct. The metacercariae undergo development in the pericardial cavity (rarely in the kidney), but do not encyst. Mice (*Peromyscus*) chew away the apical whorls of snail shells and eat the soft parts. Adult parasites localize in the mouse caecum, and eggs from mouse faeces do not hatch until ingested by *A. alternata*.

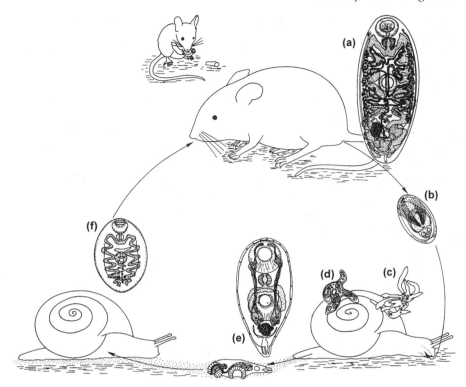

Figure 16.3 The life cycle of the brachylaimid *Postharmostomum helicis*. Adult (*a*) in caecum of rodent (e.g. *Peromyscus maniculatus*). Embryonated eggs (*b*) voided in faeces and ingested by land snail *Anguispira alternata*. Eggs hatch in gut; miracidium enters haemocoel and develops into branched mother sporocyst (*c*), in turn producing branched daughter sporocysts (*d*). Virtually tailless cercariae (*e*) move along snail slime trails and enter second snail *via* renal opening. Unencysted metacercariae (*f*) localize in pericardial cavity and infected snails are eaten by rodent definitive host. Source: based on Ulmer, 1951a, b and Olsen, 1986.

Ulmer (1951a) noted that 88% of *A. alternata* contained metacercariae but that only 1.1% of the collected snails contained sporocysts. This observation and the laboratory finding that sporocyst-infected snails did not acquire metacercariae convinced him that cercariae enter a different snail from the one that produced them. In contrast, Mas-Coma and Montoliu (1987) found that land snails (*Trochoidea caroli*) infected with sporocysts of *Dollfusinus frontalis* were more heavily infected with metacercariae (prevalence 91%; average intensity 32 metacercariae) than snails lacking sporocysts (prevalence 37%; average intensity three metacercariae). They regarded this as an adaptation to the hot and dry environment of the Mediterranean island of Formentera. There is rarely enough ground moisture to support exposed cercariae during transfer from snail to snail, so, by entering the kidney of the snail harbouring the sporocysts in which the cercariae developed, this dangerous transit is avoided and the three-host cycle is reduced to two. *D. frontalis* adults inhabit the frontal sinuses of hedgehogs and the nasal sinuses of dormice, sites reached by migration from the mouth.

The North American brachylaimid *Hasstilesia tricolor* has taken this a stage further, the metacercariae developing without leaving the sporocyst (Rowan, 1955). *H. tricolor* is also of interest because it has succeeded in colonizing herbivorous mammals, rabbits of the genera *Sylvilagus* and *Lepus*. This is achieved by utilizing the minute land snail *Vertigo ventricosa*, which is less than 2 mm in length and is consumed accidentally on vegetation. This is an interesting parallel with the infection of sheep by accidentally eating mites infected with cysticercoids of *Moniezia* (section 11.4) and ants harbouring dicrocoeliid metacercariae (section 16.3.4(g)).

Bargues and Mas-Coma (1991) made the fascinating discovery that germ cells from the tiny miracidium of *Brachylaima ruminae* in some unknown way take up an intracellular location within the acinar cells lining the ductules of the snail digestive gland. Later the germinal mass escapes into the haemocoel *via* the basal region of the host cell and develops directly into the branched, cercaria-producing sporocyst.

Retention of metacercariae in a branched sporocyst is also a feature of the leucochloridiids, which are regarded as relatives of the brachylaimids. However, their sporocysts are involved in one of the most spectacular displays in the parasitic platyhelminths. Distended terminal brood sacs containing mature metacercariae

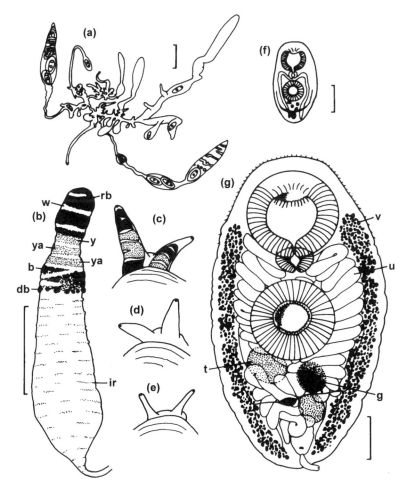

Figure 16.4 *Leucochloridium variae*. (*a*) Sporocyst with immature terminal brood sacs. (*b*) Mature brood sac. (*c*) Infected snail, with brood sacs occupying tentacles. (*d*) Infected snail, with brood sacs retracted. (*e*) Uninfected snail. (*f*) Metacercaria. (*g*) Adult. b = brown; db = dark brown; g = germarium; ir = incomplete light brown rings; rb = reddish brown; t = testis; u = uterus; v = vitellarium; w = white; y = yellow; ya = yellow amber. Scale bars: (*a*) 1 mm; (*b*) 5 mm; (*f*), (*g*) 0.25 mm. Source: reproduced from Lewis, 1974, with permission from the American Society of Parasitologists.

develop bands of colour and enter the tentacles of their terrestrial snail hosts (*Succinea* spp.; Fig. 16.4). The colours are visible through the stretched wall of the distended tentacle and are accentuated by pulsations of the brood sac. Lewis (1974) noted that the snail *S. ovalis*, with brood sacs of *Leucochloridium variae* extending into the tentacles, could sometimes be recognized at distances of 3 m by the human observer, and were likely to be equally conspicuous to passerine birds that serve as final hosts. Birds may eat the whole snail or peck the brood sac out of the tentacle. It is also thought that in some species of *Leucochloridium* the tentacle bursts, liberating the brood sac, which may be collected separately and possibly fed to nestlings. Adult parasites inhabit the cloaca.

16.3   CERCARIAE WITH PENETRATION GLANDS

16.3.1   More echinostomes

An echinostome cercaria identified tentatively as *Cercaria spinifera* readily penetrates planarians (Cleveland and Kearn, 1989). On encountering a planarian, *C. spinifera* attaches itself with its suckers and moves about across its surface. Then, while attached only by the ventral sucker, it adopts a curious attitude in which the anterior region of the body is curled ventrally, so that the dorsal surface of the anterior region is pressed against the planarian's body (Fig. 16.5*a*). The tegument covering this anterior region of the dorsal surface is perforated by the openings of so-called paraoesophageal gland cells (Fig. 16.5*b, c*), and it is these glands, absent in *E. recurvatum*, that seem to be responsible for penetration. The cercaria gradually sinks through the epidermis into the parenchyma of the planarian, leaving the tail behind at the surface, and penetration is completed in some cases within 9 min postexposure, with the paraoesophageal glands then virtually devoid of secretion. Encystment is at least partially completed about 30 min post-exposure. The scattered pattern of distribution of the paraoesophageal glands calls to mind the possibility that they may represent modified cystogenous glands.

There are many other echinostomatid cercariae that are able to penetrate the tissues of their intermediate hosts, although we know little about the glands that effect entry. *Himasthla militaris* from the marine snail *Hydrobia stagnorum* utilizes the polychaete *Nereis diversicolor* as a second intermediate host. According to Vanoverschelde and Vaes (1980) the cercariae enter the gut of the worm *via* the anus but then perforate the gut wall and migrate through the coelom to the anterior segments, where they encyst in the muscles. The cercariae emerge predominantly during darkness, when the polychaetes are most likely to leave their burrows and hence will be more vulnerable to infection. Vanoverschelde and Vaes suggested that the anterior concentration of metacercariae in *Nereis* might work to the advantage of the parasite, since only the anterior segments may emerge from the tube and this may be the only part of the worm taken by bird predators.

16.3.2   Penetrating pelagic cnidarians

The finding of one or two cercariae in plankton samples taken off Plymouth, England (Lebour, 1917) and frequent reports of unencysted metacercariae in planktonic animals (medusae, ctenophores, *Sagitta*) on both sides of the Atlantic

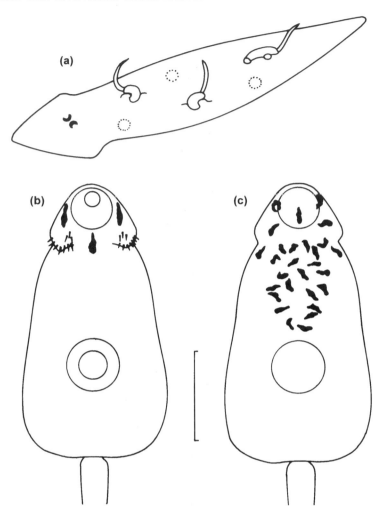

Figure 16.5 (a) The echinostome *Cercaria spinifera* penetrating a planarian. (*b*), (*c*) Ventral and dorsal views, respectively, of a cercaria, showing distribution of paraoesophageal glands. Scale bar: (*b*), (*c*) 100 μm. Source: (*b*), (*c*) redrawn from Cleveland and Kearn, 1989.

(Lebour, 1916, 1917; Stunkard, 1980) demonstrate that some digeneans are able to maintain a much extended link between bottom-living, molluscan, first intermediate hosts and pelagic organisms. This is reflected in unique features of their cercariae.

In the lepocreadiid *Opechona bacillaris*, produced in the gastropod *Nassarius pygmaeus*, the cercarial tail has 25 tufts of setae with five to seven setae per tuft and the tufts project from the lateral surfaces of the tail in opposite pairs (Fig. 16.6). These cercariae are large, extended individuals measuring 2 mm in length including the tail. Bartoli (1984) noted that *Cercaria setifera* from *Conus ventricosus* swims very rapidly, covering a distance of 1 cm in 2–8 s, and since the cercaria swims continuously it is likely to cover a distance of 13 m in 1 h. The cercaria moves with the tail leading and locomotion is achieved by rapid lateral undulations of the tail, in such a way that the tufts of setae sweep backwards, pushing against the water like the parapodia of a free-swimming polychaete worm (Fig. 16.7*a, b*).

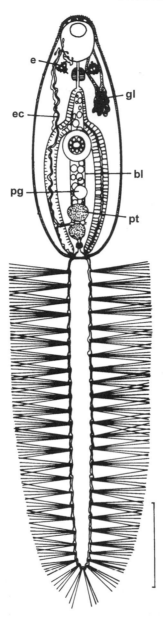

Figure 16.6   Cercaria of the lepocreadiid *Opechona bacillaris*. bl = bladder; e = eye; ec = excretory canal; gl = gland cells; pg = primordium of germarium; pt = primordium of testis. Excretory system shown on left only and anterior glands on right only. Scale bar: 200 μm. Source: redrawn from Køie, 1975.

Køie (1975) was the first to describe an unusual pattern of behaviour in *O. bacillaris*, since reported also in *C. setifera* by Bartoli (1984), during which the cercaria stops swimming, clasps the root of the tail with the body and draws the tail from root to tip through the folded body (Fig. 16.7c–g). This undoubtedly serves to remove from the tail detritus that would impair swimming, since Køie noted that isolated tails, which continued to swim for hours, were always covered with attached debris.

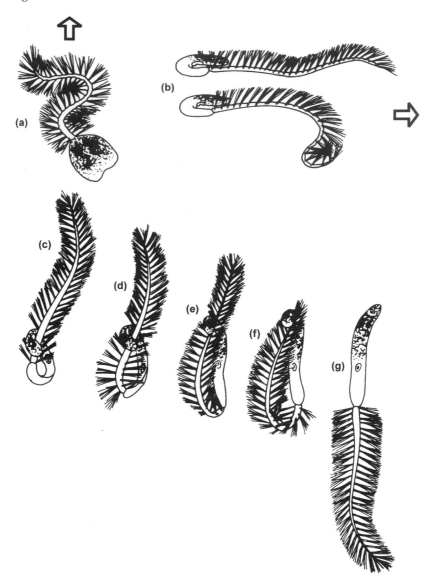

Figure 16.7   The lepocreadiid *Cercaria setifera*. Swimming attitude, viewed (*a*) from above and (*b*) from the side (tail in two positions). Arrows show direction of movement. (*c*)–(*g*) Successive stages in cleaning the tail setae. Source: reproduced from Bartoli, 1984, with permission from Masson S. A.

Køie noted that the eyed cercariae of *O. bacillaris* were slightly photopositive. They readily penetrated ctenophores, medusae and *Sagitta*; plankton-feeding fishes like the mackerel, *Scomber scombrus*, were common definitive hosts.

The investment of energy required to exploit planktonic hosts is considerable and it is perhaps not surprising that some lepocreadiids, such as *Lepocreadium pegorchis*, have reduced this investment, remaining in the benthic zone and penetrating bivalve molluscs inhabiting the sediment (Bartoli, 1983, 1984). Although freshly emerged cercariae swim vigorously, their overall size is reduced (total length including tail about 420 µm) and the setiferous tail is shorter than the body. Fishes serve as final hosts.

Lepocreadiid metacercariae in these cnidarian and molluscan hosts do not encyst, although encystment has been reported in other lepocreadiids (see Køie, 1985). Cnidarians are generally regarded as primitive multicellular animals in terms of their body organization, and they may be unable to mount a significant defensive response against tissue-invading cercariae, so that there is no requirement for encystation. However, molluscs do have cell-mediated defences (van der Knapp and Loker, 1990) and exposed metacercariae like those of lepocreadiids and brachylaimids (section 16.2.2) must have some means of circumventing these host responses. Moreover, cyst walls have a secondary protective role when the second intermediate host is eaten by the vertebrate final host, so how these apparently unprotected parasites survive their hazardous journey is unknown.

Some fellodistomatid digeneans using bivalves as first intermediate hosts have planktonic cercariae with a striking similarity to those of the lepocreadiids. For example, the cercaria of *Lintonium vibex* (described by Cable, 1954a, as *Cercaria laevicardii*) has a powerful tail with laterally projecting bundles of setae (Fig. 16.8); these setae are linked by a thin membrane to form paddles, resembling even more closely the parapodia of annelids. The undivided tail is derived from a forked tail by reduction of the furcae (section 15.2). The similarity with lepocreadiids is even more remarkable because, according to Stunkard (1978), the cercaria of *L. vibex* enters the body of a ctenophore, where it remains unencysted, and adults are found in the pharynx of a puffer fish.

16.3.3 Penetrating molluscs

(a) The gymnophallids
Gymnophallids have fork-tailed cercariae (Fig. 16.9) developing in sporocysts in bivalve molluscs and most species use bivalve molluscs as second intermediate hosts. Their metacercariae inhabit the extrapallial space between the mantle and the shell, where they remain unencysted. The periostracum, the horny outer layer of the shell, is secreted by the free edge of the mantle and prevents free access to the space from the outside world. Consequently, the cercariae have to pass through host tissue to reach the space. The cercariae are sucked into the mantle cavity *via* the inhalant siphon of the sand-dwelling bivalve and, at least in *Gymnophallus fossarum* in *Cardium glaucum*, most of them are carried by the ciliary currents to the gills and labial palps (Bartoli, 1973). After penetration, the cercariae curiously retain their forked tails and disperse through the tissues, many going astray. Those that reach the subarticular extrapallial space beneath the hinge lose their tails and survive as metacercariae. A few metacercariae find a suitable environment at the opposite end of the extrapallial space near the peripheral periostracal seal, but they reach this site by directly penetrating the mantle near its peripheral margin. How penetration is achieved is unknown, since Bartoli (1972) reported that the cercaria lacks cephalic glands.

The supposition that the extrapallial space is a haven where parasites can avoid host defensive responses is not supported by the observations of Campbell (1985) on *G. rebecqui* in the bivalve *Abra tenuis*. Metacercariae in the subarticular space evoked hyperplasia of the mantle epithelium and became enclosed, apart from a hole opposite the mantle, in a 'blister' of shell material. There is a similarity here with the bony, host-derived capsules enclosing the metacercariae of the heterophyid *Apophallus brevis* in the muscles of their fish second intermediate host (section

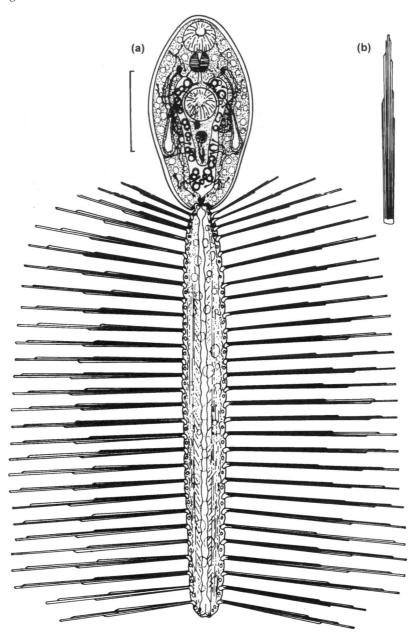

Figure 16.8   (*a*) *Cercaria laevicardii* of the fellodistomatid *Lintonium vibex*. (*b*) A tail 'fin' seen from above. Scale bar: 100 μm. From Cable, 1954a, with permission from the Marine Biological Laboratory, Wood's Hole, USA.

17.5). Campbell proposed that the blister in *G. rebecqui* may afford protection from mechanical damage in the gizzard of the bird final host.

The host reaction to *G. rebecqui* may have a secondary role, since the parasite appears to ingest the hyperplastic tissue. These metacercariae grow substantially and the reproductive system develops and stops short at egg production. *G. fossarum* took at least 4 months to become infective to an experimental host, the herring gull

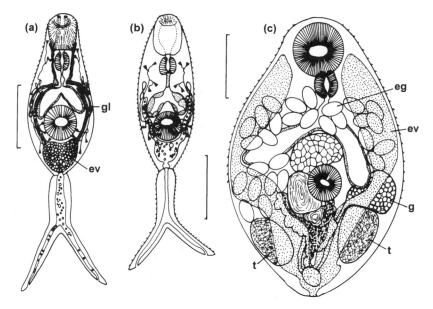

Figure 16.9 (*a*) Cercaria of *Gymnophallus nereicola*. (*b*) Cercaria of *G. fossarum*. (*c*) Adult of *G. fossarum*. eg = egg; ev = excretory vesicle; g = germarium; gl = gland cell; t = testis. Scale bars: 50 μm. Source: redrawn from Bartoli, 1972.

(*Larus argentatus*), and already contained eggs 3 days after ingestion by the gull (Bartoli, 1972). Thus, the lack of a cyst wall permits ingestion of host tissue and substantial growth in the second intermediate host.

Large numbers of metacercariae of *G. fossarum* in the bivalve *Venerupis aurea* seriously affect the growth and development of the mollusc, such that the shell fails to close fully and the position of the inhalant siphon is shifted (Bartoli, 1976–1978). The latter change induces the bivalve to adopt a different attitude in the sand (Fig. 16.10). Consequently, the host may be more susceptible to invasion by cercariae, more exposed to foraging oystercatchers and easier for them to open.

G. fossarum and its close relative *G. nereicola* provide an interesting example of host-switching. Bartoli (1981) has suggested that competition for relatively few bivalve second intermediate hosts led to the colonization by *G. nereicola* of an abundant alternative, the polychaete *Nereis diversicolor*. The worms live in burrows that are ventilated by body undulations, so cercariae responding to currents by cessation of swimming are passively drawn into the burrows as readily as they are 'inhaled' by filter-feeding bivalves. The parapodia of the worm serve for gaseous exchange and consequently have a thin cuticular covering, which the cercariae are able to penetrate using well developed cephalic glands (absent in *G. fossarum*, see above; Fig. 16.9). The unencysted metacercariae live inside the sleeve enclosing the large parapodial chaetae (acicula; Fig. 16.11). As in the echinostome *Himasthla militaris*, there are more metacercariae in the anterior region of the worm (section 16.3.1). According to Bartoli (1981), the presence of *G. nereicola* reduces the speed of retraction of the worm into its burrow and thereby facilitates its capture by bird predators.

The gymnophallid *Parvatrema homoeotecnum* has undergone remarkable evolutionary changes that are unparalleled in any other digenean. The cercariae of other species of the genus *Parvatrema* develop in sporocysts in bivalve molluscs and the

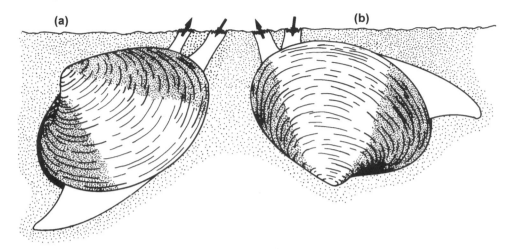

Figure 16.10   Attitude of the bivalve *Venerupis aurea* (*a*) when uninfected and (*b*) when heavily infected with metacercariae of *Gymnophallus fossarum*. Source: redrawn from Bartoli, 1976–1978.

unencysted metacercariae are found between the mantle and shell of bivalve or gastropod molluscs. In other words their life cycles are characteristically gymnophallid. *P. homoeotecnum* uses only one intermediate host, the littoral gastropod *Littorina saxatilis tenebrosa*, where it inhabits the haemocoel (James, 1964). Here the parasite undergoes multiplication usually associated with the first intermediate host, except that the primary germinal sac and the daughter germinal sacs produced by the primary have features of the cercaria–metacercaria phase of the life cycle, with oral

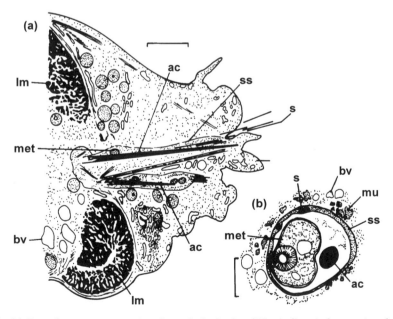

Figure 16.11   (*a*) Part of a transverse section through the body of *Nereis diversicolor*, passing through one parapodium and showing the location of metacercariae of *Gymnophallus nereicola* inside the setal sheath. (*b*) Section in a plane at right angles to (*a*), showing metacercaria inside the sheath. ac = aciculum; bv = blood vessel; lm = longitudinal muscle; met = metacercaria; mu = muscle; pa = parapodium; s = seta; ss = setal sheath. Scale bars: (*a*) 0.5 mm; (*b*) 100 μm. Source: modified from Bartoli, 1974.

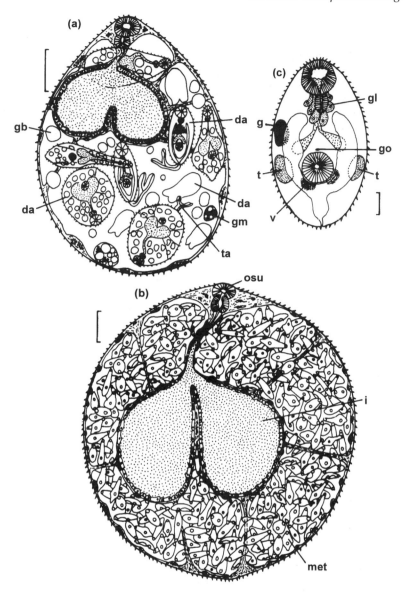

Figure 16.12 Intramolluscan stages of *Parvatrema homoeotecnum*. (*a*) Primary germinal sac containing developing daughter germinal sacs; note degenerating tails. (*b*) Fully developed daughter germinal sac after release from primary sac. (*c*) Metacercaria. da = daughter germinal sac; ev = excretory vesicle; g = germarium; gb = germ ball; gl = cephalic glands; gm = germinal mass; go = genital opening; i = intestine; met = metacercaria; osu = oral sucker; t = testis; ta = tail; v = vitellarium. Scale bars: (*a*) 50 μm; (*b*) 100 μm; (*c*) 10 μm. Source: reproduced from James, 1964, with permission from Cambridge University Press.

sucker, ventral sucker (this disappears later), pharynx and bilobed gut (Fig. 16.12). The primary germinal sac is enclosed temporarily in a membranous covering that may represent all that remains of the sporocyst phase. Inside, germ balls differentiate into five to 20 similar daughter germinal sacs, which start life with a forked tail that gradually withers (Fig. 16.12*a*). The primary sac eventually ruptures, liberating the daughters. Germ balls inside the daughters develop into fork-tailed cercariae, which do not emerge but ultimately lose their tails and become metacercariae, sometimes as many as 2000 in a single daughter sac (Fig. 16.12*b, c*). The final host, the oystercatcher, is infected by eating the snails.

As proposed by James, the most likely interpretation of this intriguing scenario is that *P. homoeotecnum* has evolved from a typical gymnophallid by shifting the characteristic molluscan multiplication process from the first (probably bivalve) to the second (gastropod) intermediate host and hence from the typical sporocyst phase of the life cycle to the cercaria–metacercaria phase. This would make the first intermediate host redundant and lead to its elimination.

(b)   The strigeids

Many fork-tailed strigeid cercariae penetrate and encyst in molluscs (Fig. 16.13), but the contrast between their development in the mollusc and that of the

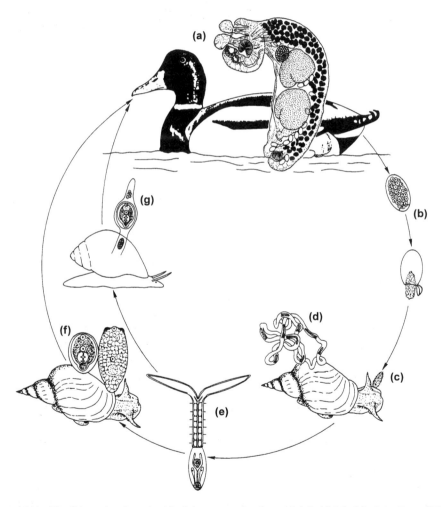

Figure 16.13   The life cycle of a strigeid of the genus *Cotylurus*. Adult (*a*) inhabits intestine of bird. Eggs (*b*) in faeces develop in water. Miracidium (*c*) penetrates freshwater snail and gives rise to sporocysts (*d*) and fork-tailed cercariae (*e*). Cercariae penetrate another snail and undergo growth and metamorphosis producing encysted 'tetracotyle' (*f*). *C. flabelliformis* can infect inhospitable host snails only if able to shelter and develop in germinal sacs of another digenean already present (*g*). Bird infected by eating snails containing tetracotyle cysts. Source: based on Olsen, 1986.

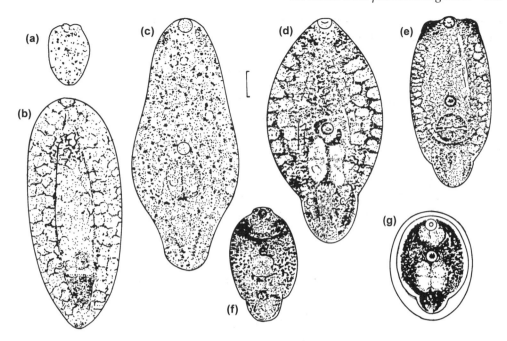

Figure 16.14   (*a*)–(*f*) Metamorphosis of the strigeid *Cotylurus flabelliformis*. (*g*) Encysted tetracotyle. Scale bar: 100 µm. Source: reproduced from Ulmer, 1957, with permission from the American Microscopical Society.

gymnophallids could not be greater. In the mollusc, the strigeids undergo what can only be described as a metamorphosis (Fig. 16.14). The unencysted cercarial body becomes flattened, with a corresponding dissolution of many organs and tissues. For example, in *Cotylurus brevis* in the freshwater snail *Lymnaea stagnalis* only the excretory system and the genital primordia seem to escape drastic reduction; the oral and ventral suckers, gut, tegumental spines and probably much of the musculature are dismantled (Nasir, 1960). Considerable growth of this precystic stage takes place, presumably by tegumental absorption of nutrients, and the oral and ventral suckers and the gut are reconstituted, together with special anatomical features of the adult. These drastic developmental changes culminate in the production of a cyst wall, which is often thick and complex with a corresponding decrease in size of the parasite. How the events of dissolution and rebuilding are orchestrated by these flatworms is a challenging question.

No explanation has been advanced hitherto to explain this unique developmental upheaval. However, strigeid adults, most of which inhabit the intestines of birds, are morphologically more elaborate than those of other digeneans and metamorphosis may be a more effective and faster way of converting a strigeid cercaria into a complex adult than modifying and adding to cercarial organs.

Strigeid adults of the genera *Strigea*, *Apatemon* and *Cotylurus* have a cup-shaped forebody (Fig. 16.13) with the oral sucker on its dorsal rim. The ventral sucker lies deep within the cup, which also encloses the projecting lobes of a so-called tribocytic organ. On each side of the oral sucker is a glandulo-muscular organ or lappet and these, together with the oral and ventral suckers, have given rise to the term 'tetracotyle' ('four suckers'), which is applied to the encysted metacercaria.

A great deal of attention has been focused on this complex forebody, but there is still no clear picture of the function of the component organs. Although a host villus is grasped by *Apatemon gracilis minor* between the lobes of the tribocytic organ, the host epithelium is eroded (Erasmus, 1969a), pointing to a nutritional rather than an adhesive role. The cup-shaped forebody may well provide an enclosed microenvironment where extracorporeal digestion of host tissue can occur and from which the products of digestion can be ingested. Erasmus has stressed the high degree of intimacy between the thin tegument of the lobes and the exposed host lamina propria. He has suggested that this is a 'placental' arrangement whereby nutrients can pass from the nearby blood capillaries of the host villus into the relatively large lacunae of the reserve excretory system (section 13.1.1), whence they can be distributed around the body. The ventral sucker is regarded as redundant as an attachment organ, and it is generally supposed that the lappets assume this role (Ohman, 1966; Erasmus, 1969b), although the reported secretion of enzymes by these organs suggests that they may be multifunctional.

The genital openings cannot be sited within the forebody cup and they have shifted to the posterior end of the hindbody (Fig. 16.13), permitting mating and the release of eggs without detachment of the forebody.

The complex cyst wall enclosing the metacercaria of *Apatemon minor*, which uses leeches as second intermediate hosts, has acquired a remarkable new property. When treated with acid pepsin, followed by a mixture of bile salts and trypsin, thereby simulating the chemical experiences of the cyst when swallowed by a bird, the cyst wall suddenly expands inwards, forcing the metacercaria out in an explosive manner through a canal at one end of the cyst (Fig. 16.15).

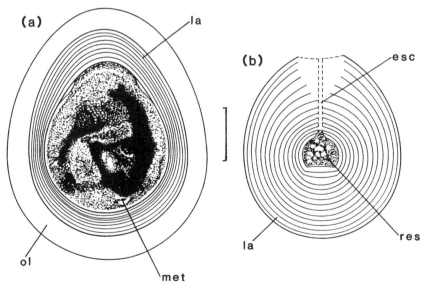

Figure 16.15 The metacercarial cyst of *Apatemon minor*. (*a*) Cyst freshly dissected from a freshwater leech. (*b*) Cyst after discharge of the metacercaria, induced by pretreatment with pepsin followed by treatment with bile salts. esc = escape canal; la = birefringent, lamellated cyst wall; met = metacercaria; ol = outer, non-birefringent layer of cyst (derived from host and vulnerable to pepsin digestion); res = residual material in cyst lumen. Scale bar: 100 μm. Source: reproduced from Kearn, Cleveland and Wilkins, 1989, with permission from CAB International.

A fascinating feature of some strigeids is that, after penetration of the snail, their cercariae are capable of entering and completing their development inside the germinal sacs of other digeneans already present. In *Cotylurus flabelliformis* these hyperparasitic individuals usually develop more quickly than in lymnaeid snails lacking germinal sacs (Cort *et al.*, 1945). Experimental infections revealed a surprising additional advantage. Cercariae of *C. flabelliformis* readily penetrate physid and planorbid snails but usually fail to develop. However, when these abnormal hosts are already infected with germinal sacs of other digeneans, the cercariae enter these sacs and develop normally (Fig. 16.13*g*). Thus, *C. flabelliformis* seems to shelter from host defences within the foreign germinal sacs, permitting the strigeid to greatly extend its range of second intermediate hosts. Cort *et al.* also found that cercariae of *C. flabelliformis* usually fail to penetrate normal lymnaeid hosts that already harbour sporocysts of *C. flabelliformis*.

16.3.4 Penetrating arthropods

(a) General

An anatomical feature of most cercariae that penetrate arthropods is the presence of a stylet or spike embedded in the midline dorsal to the oral sucker. The pointed tip of the stylet protrudes from a pit and projects just beyond the anterior extremity of the parasite (Fig. 16.16*a*). Stylet-equipped cercariae are known as xiphidiocercariae and they include some of the smallest cercariae, with the total length of the preserved specimens, including the tail, sometimes barely exceeding 150 μm. Their name is derived from the Greek *xiphos*, 'a sword', and this name is apposite, because the stylet combines cutting and puncturing operations, although in overall function it resembles a scalpel more closely than a sword. It is possible that the stylet may have evolved more than once (see Pearson, 1992).

The cutting edge of the blade of the cercarial stylet of the plagiorchiid *Plagiorchis megalorchis* is on the dorsal side of the sharply pointed tip (Fig. 16.16*b, c*) and its use in penetrating the cuticle of the larva of *Chironomus riparius* has been described by Rees (1952). With the cercaria attached by the oral and the ventral sucker, the dorso-ventrally orientated stylet is driven through the cuticle. Then, by rotating the tip of the stylet in a ventro-dorsal arc, the blade makes a straight cut through the cuticle. This is repeated several times, after a slight shift each time in the position of the oral sucker, so that a series of cuts is made along a line which, in length, is equal to or a little greater than the diameter of the oral sucker. The blade then works along the line a second time, until all the cuts are joined up to form a single slit. The cercaria then promptly passes through the slit into the insect, usually, but not always, leaving its tail on the outside, the whole penetration process taking as little as 2 min and sometimes as long as 13 min. Following an active period of 30 min to 1 h, the cercariae encyst. The stylet is eventually shed and can be found inside the cyst.

The exoskeleton of an arthropod varies in thickness from place to place. Where joints occur, the cuticle must by necessity be thinner and more flexible, and gaseous exchange in aquatic arthropods also demands that the gills have a thin covering. Many xiphidiocercariae are sufficiently discriminating to select these thin areas for penetration. *P. megalorchis*, for example, always enters *C. riparius* between two adjacent segments (Fig. 16.16*d*), while *P. neomidis* enters the bases of the gills of the aquatic larvae of the insect *Sialis lutaria* (see Theron, 1976).

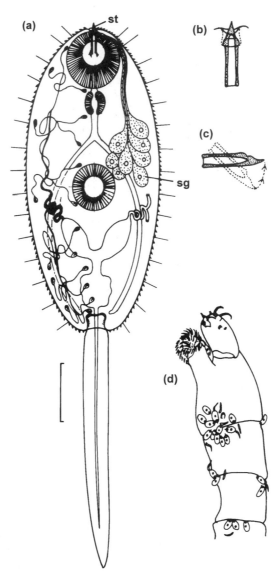

Figure 16.16   (*a*) Cercaria of *Plagiorchis megalorchis*; dorsal view, showing stylet glands on the right and excretory system on the left. (*b*) Stylet in dorsal view. (*c*) Stylet in lateral view, showing arc through which stylet moves during process of penetration. (*d*) Anterior region of a chironomid larva, showing accumulation of cercariae on the surface in the junctions between the segments. sg = stylet glands; st = stylet. Scale bar: (*a*) 50 µm. Source: redrawn from Rees, 1952.

Penetration into arthropods may require more than the mechanical action of the stylet because most xiphidiocercariae are also equipped with glands with ducts opening near the stylet (Fig. 16.16*a*). Moczon (1996b) has shown that in *Plagiorchis elegans* these glands secrete a serine protease to which a penetrative role has been attributed. In addition, the cercariae of lecithodendriids have a so-called virgula organ inside the oral sucker (Fig. 16.17). This is a single reservoir or paired reservoirs, which are filled during development of the cercaria with secretion produced by gland cells in the body (Kruidenier, 1951). The reservoirs open inside the oral

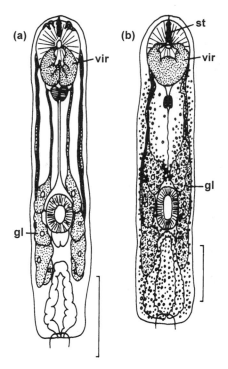

Figure 16.17 The bodies of two lecithodendriid cercariae with virgula organs (vir). (*a*) *Allassogonoporus vespertilionis.* (*b*) *Cercaria guttata.* gl = stylet glands; st = stylet. Scale bars: 50 μm. Source: redrawn from Burns, 1961b.

sucker near the mouth. Some authors have expressed the view that the virgula secretion plays a part in penetration, but there is better evidence that the virgula is the source of mucous threads produced while the cercaria is free-swimming (see Kruidenier, 1951). Burns (1961a,b) described such threads in *Allassogonoporus vespertilionis* and *Acanthatrium oregonense*, two lecithodendriids from bats that penetrate into caddis fly larvae and presumably infect bats when they consume adult flies. The threads drift like gossamer in the water and not only keep the larvae in suspension but also enhance their chances of infecting their insect hosts by becoming entangled in the gills.

A. vespertilionis and *A. oregonense* illustrate another facet of the penetration behaviour of some xiphidiocercariae. After making contact with the cuticle covering the gills of the caddis fly larva, these cercariae become enclosed in a cyst, which in *A. vespertilionis* consists of mucus but in *A. oregonense* is a thin but compact capsule (Fig. 16.18; Burns, 1961a). Most cercariae take less than 10 min to penetrate the cuticle from within the cyst and enter the body cavity. Both species then encyst for a second time, although this does not occur in *A. oregonense* until after emergence of the imago.

A similar phenomenon was described by Helluy (1982) in the microphallid *Microphallus papillorobustus* which enters *Gammarus* spp. through the gill surface. Helluy called the external cysts 'pseudocysts' or 'cysts of penetration' and she offered the following suggestions concerning their function: the cyst will protect the cercaria from dislodgement by the gill-ventilating current and by movements of the crustacean; enclosure of the cercaria will prevent dissipation and dilution of penetration gland secretions; the cyst wall will provide a platform permitting the

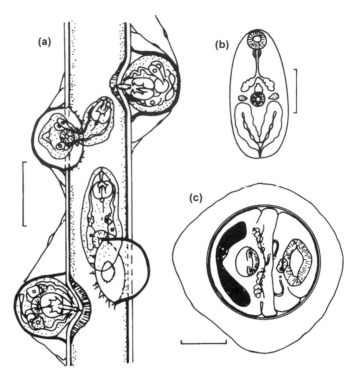

Figure 16.18   (*a*) Cercariae of the lecithodendriid *Acanthatrium oregonense* encysting on and penetrating the gill of a caddis fly larva. (*b*) Unencysted metacercaria. (*c*) Encysted metacercaria. Scale bars: (*a*), (*b*) 100 μm; (*c*) 50 μm. Source: reproduced from Burns, 1961a, with permission from the American Society of Parasitologists.

cercaria to exert a stronger force during penetration and a spring board for propelling itself into the host; the cyst wall will prevent pathogenic microorganisms from entering the host *via* the penetration wound. The occurrence of these temporary external cysts in more than one family raises the possibility that digeneans with stylet-bearing cercariae may have evolved from ancestors that made permanent external cysts on arthropods.

The gut wall of arthropods may be a less formidable barrier than the external body surface and more than one group of digeneans with xiphidiocercariae have adopted this as the route of entry. There may be a saving here in the expenditure of energy required for penetration and, if access to the gut lumen is assured by inducing the arthropod to eat the cercaria, then this combination could be a particularly favourable strategy.

That such an evolutionary change can occur is illustrated by the observations of Rees (1952) on *Plagiorchis megalorchis*. As described above, this parasite normally gains access to the haemocoel of chironomid larvae by penetrating the external cuticle but Rees also observed that some free-swimming cercariae were swallowed by the chironomids or ingested during grooming of the outer surface and that some of these larvae succeeded in penetrating the gut wall and entering the body cavity.

Cable (1985) described a lecithodendriid cercaria in which penetration of the gut is an established feature of its biology. He noticed sparkling at the surface of water in a dish containing the freshwater snail *Amnicola limosa* and discovered minute cercariae attached to the surface film by the tips of their tails (Fig. 16.19). Constant

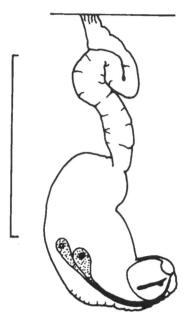

Figure 16.19 A lecithodendriid cercaria from the snail *Amnicola limosa*, attached to the surface film by the spiny tip of the tail. Scale bar: 50 µm. Source: redrawn from Cable, 1985.

wriggling distorted the surface film, causing it to sparkle in reflected light. These cercariae were readily ingested by mosquito larvae, encysted near the brain after pupation and persisted in the head of the adult mosquito. Cable was unable to identify the definitive host.

Many predatory arthropods are larger than mosquito larvae and respond to larger prey items. The cercariae of gorgoderids and hemiurids–halipegids have independently increased their size and attractiveness to predators by enlarging the tail (see below), like the azygiids and bivesiculids (section 15.6.1).

(b) Plagiorchis elegans *and intermediate host behaviour*
Most of the xiphidiocercariae of *Plagiorchis elegans* emerge nocturnally from the snail *Stagnicola elodes* and penetrate a wide range of aquatic insects, including mosquito larvae (Lowenberger and Rau, 1994a). Mammals and birds serve as definitive hosts. Lowenberger and Rau found that within 15 min of the start of cercarial emergence, the snail moved to the top of the water column and remained there for 2–3 h. Some 79% of all cercariae emerged in a dense cloud during this period and then dispersed passively. This mass emergence is unlikely to favour infection of a specific second intermediate host, since the parasite is a generalist, but the combination of snail vertical movement and mass release of cercariae may collectively enhance transmission by promoting dispersion of the cercariae. The infectivity of the cercariae increased from less than 20% upon emergence to greater than 75% 4–6 h later; Lowenberger and Rau suggested that this delay in maximum infectivity, giving time for dispersal, may reduce superinfection and subsequent parasite-associated mortality of second intermediate hosts.

Cercarial longevity was greater than 30 h and exceeded the period of infectivity (less than 5% after 24 h); this may be related to steadily falling glycogen levels.

Webber, Rau and Lewis (1987a) found that mosquito larvae infected with one or two metacercariae of *P. elegans* (identified as *P. noblei*) spent more time near the bottom and, since the cercariae generally settle to the bottom, may be more likely to acquire additional infections. Such superinfected larvae (with more than three metacercariae) spent more time suspended from the surface, hence escaping further superinfection at the bottom but possibly increasing their vulnerability to consumption by surface-feeding predators. Webber, Rau and Lewis (1987b) found that the metacercariae had no overall impact on the susceptibility of mosquito larvae to guppies (*Poecilia reticulata*) but did enhance the likelihood of predation by voles (*Microtus pennsylvanicus*) feeding at the water's edge.

Lowenberger and Rau (1994b) reported evidence to suggest that gravid mosquitoes are deterred from ovipositing in water containing mosquito larvae infected with *P. elegans*.

(c)   The microphallids

Microphallids have been particularly successful in the exploitation of marine and brackish water crustaceans as second intermediate hosts and their recruits include amphipods, isopods, barnacles and crabs. A vivid picture of how these parasites exploit amphibious amphipods inhabiting the particularly difficult habitat of the shoreline has been given by Bartoli and Combes (1986).

In the sheltered Lagune du Brusc on the Mediterranean coast of France, the amphipod *Orchestia gammarus* inhabits vegetation (*Cymodocea nodosa* or *Zostera nana*) cast up on the beach. Offshore, free-swimming xiphidiocercariae of *Maritrema misenensis* released from the snail *Cerithium mediterraneum* swim up to the surface and attach themselves to the underside of the surface film by the oral sucker. Many of these cercariae drift in passively and accumulate along the shoreline. The exposed stranded vegetation is a relatively thin layer and tends to dry up in hot or windy weather. This forces the amphipods to make frequent excursions into the water to rehydrate themselves, and as they cross the air–water interface they pick up cercariae on their bodies. An additional interesting feature is that females are more heavily infected than males of the same age, perhaps because they immerse themselves more frequently in order to keep their eggs damp. The definitive host is the shore bird *Charadrius alexandrinus*.

In the transition zone between the lagoon and the sea, environmental conditions are different, with moderate wave action and the accumulation of much thicker layers of *Posidonia oceanica* in the supralittoral zone. This vegetation rarely dries out, except at the surface, and the amphipod *Orchestia montagui* has no need to leave this habitat. It might be supposed that this would effectively disrupt the life cycle by severing the contact between the cercaria and the crustacean, but this is not the case because the cercariae, carried shorewards at the surface by the waves, are projected on to the stranded vegetation in droplets of spray produced by the breaking of the waves on the beach.

Another remarkable development that we have already met in the tapeworms is the finding by Helluy (1982, 1983, 1984) that the metacercaria of *Microphallus papillorobustus* can render brackish water amphipods of the genus *Gammarus* vulnerable to predation by their bird definitive hosts. She found that gammarids with a metacercaria lodged in the brain became strongly photopositive and switched from being geopositive to geonegative. They no longer responded to mechanical perturbations by cessation of movement or by seeking shelter, but retained a lively activity at the surface. This altered behaviour may attract birds. Helluy also found that mature cysts elsewhere in the body did not induce these changes and neither did immature cysts in the brain.

There are two significant major trends in microphallids. First, many of them are able to utilize resources provided by the arthropod intermediate host. For example, encystation is delayed in *Maritrema linguilla* after penetration of the marine isopod *Ligia oceanica* and the free metacercaria inside a pleopod trebles its size (Benjamin and James, 1987). The metacercaria then migrates to the dorsal haemocoel and growth ceases while three of the five cyst layers are laid down. Growth then resumes and, after formation of the last two layers of the cyst wall, a further 50% increase in body size occurs and the reproductive system undergoes considerable development. This shift in physiological and metabolic dependency from the final host to the second intermediate host finds parallels in other parasites such as the ligulid cestodes (section 10.5.4), and may be associated with a shift in specificity from the final to the second intermediate host. In *M. linguilla*, the second intermediate host even takes over some of the resource provision that normally falls to the first intermediate host, since development of the cystogenous cells is delayed until the cercaria penetrates the isopod.

This trend goes further in *Microphallus opacus*. The metacercariae in the digestive gland of the crayfish *Cambarus propinquis* grow and become sexually mature, usually encysting at this time, but they go on to produce eggs. Caveny and Etges (1971) suggested that eggs may be released by death and disintegration of the crayfish, or by digestion of the crayfish by a vertebrate predator. A vertebrate final host with resource provision may no longer be necessary.

The success of the echinostomes has taught us the importance of the ubiquitous molluscs as prey items for vertebrates, and the tendency in some echinostomatid species for a proportion of their cercariae to remain inside the first intermediate host and encyst has already been noted (section 16.2.1). This brings us to the second trend in microphallids. In many species they have gone further than the echinostomes, totally suppressing cercarial emission from the mollusc and losing the arthropod host, the life cycle changing direction and exploiting vertebrates that feed on molluscs rather than on arthropods. In *Levinseniella minuta*, recorded by Stunkard (1958) from the snails *Amnicola limosa* and *Hydrobia minuta*, the cercariae lack a tail and emerge from sporocysts when relatively immature. According to Stunkard, the cercariae encyst within the same snail and continue their growth. The consequences of this strategy are a relatively low prevalence (5% of *H. minuta* infected), and a high intensity (hundreds of metacercariae per snail, in spite of cercarial output per sporocyst rarely exceeding three). Nevertheless, these levels are sufficient to maintain the parasite in diving ducks that feed largely on molluscs.

In other microphallids, the cercariae do not leave the sporocyst, either remaining unencysted as in *Microphallus pygmaeus* in the marine snail *Littorina saxatilis* (see James, 1968), or encysting as in *M. breviatus* in the brackish water snail *Hydrobia ventrosa* (see Deblock and Maillard, 1975). In *M. breviatus*, the cercariae are poorly developed with a rudimentary stylet, ill-defined penetration glands and, although a tail is present, weak powers of locomotion when released experimentally from the sporocyst. According to Deblock (1980), the cercariae of *M. scolectroma* and *M. abortivus* encyst inside the sporocyst at such an early stage of development that they are little more than a ball of cells with a rudimentary oral sucker and no stylet, penetration glands, tail or excretory system

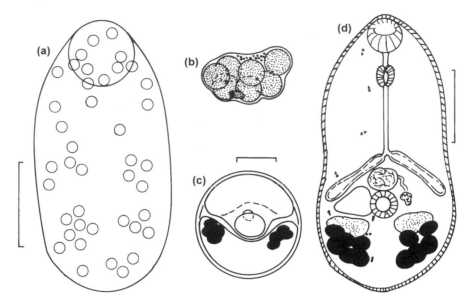

Figure 16.20   *Microphallus abortivus* in *Hydrobia ulvae*. (*a*) Appearance of cercaria at time of encystment in sporocyst. (*b*) Sporocyst containing encysted metacercariae. (*c*) Fully developed metacercarial cyst. (*d*) Excysted metacercaria. Flame cells shown on one side only. Scale bars: 50 μm. Source: redrawn from Deblock, 1980.

(Fig. 16.20). A similar abbreviation of the life cycle from three to two hosts has taken place in the monorchiid *Asymphylodora tincae* and in some gorgoderids of the genus *Phyllodistomum* (see below).

(d)   The gorgoderids

Like the azygiids and bivesiculids (section 15.4), the small cercarial body of gorgoderids becomes enclosed in a tubular extension of the tail, but this does not occur until after emergence of the cercaria into the water (Fig. 16.21).

Another interesting feature of gorgoderids is that they utilize bivalves as first intermediate hosts. Their sporocysts were thought to be ectoparasitic, being attached to the bivalve gill by a conical extension, but Mitchell and Mason (1980) showed that the sporocyst is enclosed by a thin host layer. Filter-feeding freshwater bivalves of the genus *Pisidium* live buried in the mud and the cercariae of *Gorgodera vitelliloba*, escaping from sporocysts on the gills, ascend to the surface of the mud *via* the ventilation shaft of the mollusc (Combes, 1968). They are occasionally eaten by insect larvae (*Sialis*) hunting in the mud and detecting their prey by touch, but more frequently they are ingested at the surface of the mud by tadpoles. In the gut of the predator, the cercarial body leaves the tail, penetrates the gut wall and encysts. Adult frogs are infected when they consume either newly metamorphosed frogs or adult *Sialis* flies containing cysts. The excysted parasites migrate from the frog intestine to the bladder with an intervening period spent in the kidney (Mitchell, 1973).

In the laboratory, Combes (1968) had no difficulty in infecting *Sialis* larvae and tadpoles with metacercariae of both *G. vitelliloba* and the closely related *G. euzeti*, but found that in the wild *G. vitelliloba* cysts occurred mainly in tadpoles while *G. euzeti* cysts were restricted to *Sialis* larvae. This pattern is a reflection of

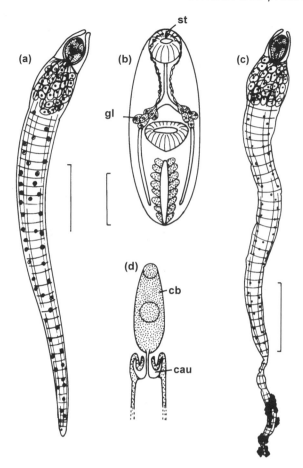

Figure 16.21 Gorgoderid cercariae. (*a*) *Gorgodera vitelliloba*. (*b*) Cercarial body of *G. vitelliloba*. (*c*) *G. euzeti* with adhesive tail tip. (*d*) Diagram showing attitude of cercaria just prior to release from the sporocyst and before enclosure of the body (cb) within the cup-shaped tail chamber (cau). gl = stylet glands; st = stylet. Scale bars: (*a*), (*c*) 0.5 mm; (*b*) 100 μm. Source: (*a*)–(*c*) redrawn from Combes, 1968; (*d*) redrawn from Mitchell and Mason, 1980.

the behaviour of the cercariae, those of *G. vitelliloba* ascending to the surface of the mud while those of *G. euzeti* attach themselves by the adhesive tail tip (Fig. 16.21*c*) to the shells of *Pisidium* buried in the mud. The anchored cercariae of *G. euzeti* are located and eaten by mud-dwelling *Sialis* larvae, but do not become available to tadpoles.

In parallel with some microphallids and monorchiids (sections 16.3.4(c) and 16.3.5 respectively), some gorgoderids of the genus *Phyllodistomum* encyst within the sporocyst, not only shortening the life cycle but taking advantage of aquatic vertebrates (fishes) that feed on molluscs (see, for example, Wanson and Larson, 1972).

(e) *The hemiurids and halipegids*
The cercariae of hemiurid digeneans do not have a stylet but share with gorgoderids the utilization of crustacean second intermediate hosts and the encapsulation of the cercarial body in a vesicle derived from the tail. These similarities, however, are misleading and hide an amazing and unique method of entering the haemocoel of

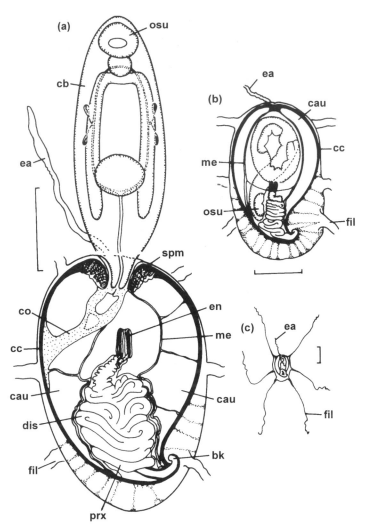

Figure 16.22 The cercaria (*Cercaria vaullegeardi*) of the hemiurid *Lecithochirium furcolabiatum*. (*a*) Cercaria removed from the anterior region of the daughter sporocyst before retraction of the cercarial body into the tail. (*b*) After retraction of the cercarial body. (*c*) Whole primed cercaria. bk = 'beak'; cau = caudal cyst cavity; cb = cercarial body; cc = caudal cyst; co = connection of body to caudal cyst; dis = distal delivery tube; ea = excretory appendage; en = end piece; fil = caudal filament; me = membranous capsule; osu = oral sucker; prx = proximal delivery tube; spm = sphincter muscle. Scale bars: (*a*), (*b*) 50 μm; (*c*) 100 μm. Source: reproduced from Matthews, 1981a, with permission from Dr B. F. Matthews and Cambridge University Press.

the arthropod. This is well illustrated by the work of Matthews (1981a, b) on the mode of infection of the harpacticoid copepod *Tigriopus brevicornis* with *Cercaria vaullegeardi* (the cercaria of *Lecithochirium furcolabiatum*) from the marine gastropod *Gibbula umbilicalis*.

The anterior region of the tail of the mature cercaria is a hollow vesicle with six external caudal filaments (Fig. 16.22). Still recognizable as tail-like is the vestigial posterior region of the tail or 'excretory appendage', although it occupies an anterior position because of differential growth of the rest of the tail (Fig. 16.22*a*). Retraction of the body of the cercaria into the vesicle usually occurs within seconds of the

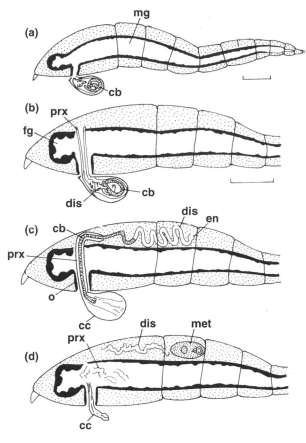

Figure 16.23 (*a*)–(*d*) Diagrams illustrating successive stages in the infection of the copepod *Tigriopus brevicornis* with the cercaria (*Cercaria vaullegeardi*) of the hemiurid *Lecithochirium furcolabiatum*. (Copepod appendages omitted; copepod gut wall shown in black, haemocoel lightly stippled.) See text for explanation. fg = foregut; met = metacercaria; mg = midgut; o = oesophagus. Other lettering as in Fig. 16.22. Scale bars: 100 μm. Source: modified from Matthews, 1981b.

cercaria entering water; the aperture is sealed by a sphincter muscle when retraction is completed and the connection between the body and the inner wall of the vesicle is severed. The vesicle also contains a coiled delivery tube, free at one end and connected at the other to a projecting beak-like structure at the posterior end of the vesicle (Fig. 16.22*a*, *b*).

The tail of *Cercaria vaullegeardi* is reduced and the cercaria is unable to swim. Other hemiurid cercariae are propelled by a motile forked appendage (tail?), possibly indicating a strigeatan origin for these parasites. In *C. vaullegeardi*, the caudal filaments probably serve to attach the inert cercariae to vegetation, where they must be sought out by foraging copepods. However, these bizarre cercariae are not swallowed. Indeed, if this happens infection fails to occur. Successful infection depends on the copepod seizing the inactive cercaria with its mouthparts, manipulating it so that the 'beak' projects into the mouth and severing the beak with the mandibles (Fig. 16.23*a*). This leads to the rapid eversion of the delivery tube, the tip of which forcibly penetrates the dorsal wall of the midgut, and simultaneous expulsion of the cercarial body through the tube into the haemocoel (Fig. 16.23*b–d*). The collapsed tail is crushed and eaten.

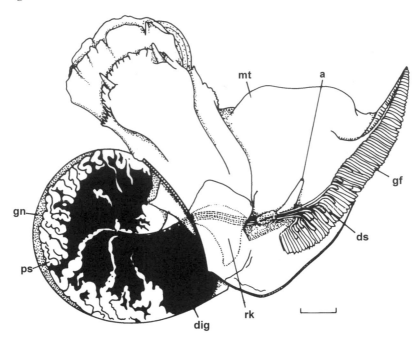

Figure 16.24   A specimen of the marine snail *Gibbula umbilicalis* showing the position of the sporocysts of *Lecithochirium furcolabiatum*. The shell has been removed, the mantle cavity opened along the left side and the mantle wall and gills deflected to the right. a = anus; dig = host digestive gland; ds = distal region of sporocyst with birth pore; gf = gill filament; gn = gonad; mt = mantle; ps = proximal germinal region of sporocyst; rk = right kidney. Scale bar: 2 mm. Source: reproduced from Matthews, 1980, with permission from Dr B. F. Matthews and Cambridge University Press.

In their gastropod first intermediate host, the site of emission of these cercariae is rather remote from their site of development. The cercariae escape into the water from the gills, but develop in sporocysts lying beneath the shell in the haemocoel of the digestive gland. The parasite has a novel way of transporting these inert and bulky cercariae from one site to the other. Matthews (1980) discovered that the sporocyst has a narrow tubular extension that runs from the digestive gland through the afferent renal vein, the right kidney haemocoel and the transverse pallial vessel into the blood channel within a gill filament (Fig. 16.24). Peristaltic movements of the tube assist the passage of the cercariae, which are thereby guided to the gill epithelium. This is punctured by the muscular and possibly histolytic action of the sporocyst tip, and the cercariae are released *via* a terminal birth canal (Fig. 16.25).

In another hemiurid, *Cercaria calliostomae*, from the prosobranch *Calliostoma ziziphinum*, the cercariae develop in rediae, which then actively transport the cercariae *via* the afferent renal vein from the haemocoel of the gonad to that of the right kidney (Matthews, 1982). Matthews suggested that each redia may then use its pharynx to disrupt the kidney tubule wall, permitting the bulky cercariae to escape directly to the exterior after emergence through the terminal redial birth pore. Since most hemiurid cercariae develop in rediae, Matthews (1980) suggested that the 'sporocyst' of *C. vaullegeardi* is a redia in which the gut has given rise to the birth canal.

Most adult hemiurids parasitize fishes, but the small size of their crustacean intermediate hosts (copepods, ostracods) does not restrict them to small microphagous fishes. There is a striking parallel between some hemiurids and tapeworms like

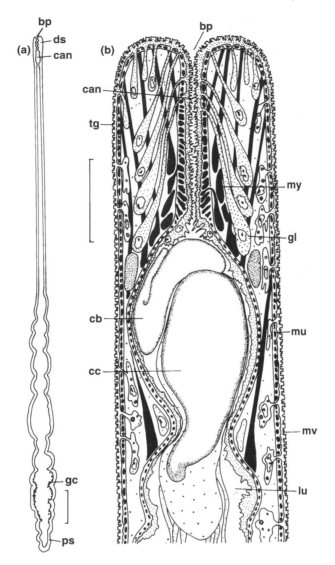

Figure 16.25 (*a*) Daughter sporocyst of the hemiurid *Lecithochirium furcolabiatum*. (*b*) Diagrammatic longitudinal section through the distal region of the sporocyst. bp = birth pore; can = birth canal; cb = cercarial body; cc = caudal cyst; ds = distal region of sporocyst; gc = germinal cells; gl = gland cell; lu = lumen of sporocyst; mu = muscle; mv = microvillus; my = myocyton; ps = proximal region of sporocyst; tg = tegument. Scale bars: (*a*) 400 μm; (*b*) 50 μm. Source: reproduced from Matthews, 1980, with permission from Dr B. F. Matthews and Cambridge University Press.

Bothriocephalus (section 10.5.2). *Genarchopsis goppo* infects ostracods and specimens of the freshwater fish *Channa punctata* of all sizes (Madhavi, 1978). Young *C. punctata* become infected by eating ostracods but larger fishes are piscivorous. However, the small microcarnivorous fish *Aplocheilus panchax* acts as a paratenic host, ingesting infected ostracods and harbouring immature *G. goppo* in the stomach. Large *C. punctata* are infected by eating *A. panchax* containing these immature parasites.

    Halipegus ovocaudatus, which is related to the hemiurids, has taken this further. It is a parasite of the buccal cavity of the frog *Rana ridibunda*. Cercariae from the snail *Planorbis planorbis* are injected into the haemocoel of copepods, where they exist without encysting, but tadpoles fed parasitized copepods by Kechemir (1978) failed to become infected and adult frogs feed on terrestrial invertebrates. This problem has been solved by recruitment of a dragonfly that transports the parasite from the water to the air. The parasites (mesocercariae of Kechemir) survive when infected copepods are eaten by dragonfly larvae and persist through metamorphosis as unencysted metacercariae in the gut. Adult frogs are infected by eating adult dragonflies, but not by mesocercariae administered experimentally. Kechemir also showed that dragonfly larvae do not become infected when exposed to cercariae. In other words, the dragonfly is an additional obligatory host, not a paratenic host, and four hosts are essential for completion of the life cycle.

The hemiurid *Lecithochirium furcolabiatum* has a similar life cycle, with a mesocercaria in a copepod and a metacercaria in the body cavity of small rockpool fishes, which are eaten by the definitive host, the rockling *Ciliata mustela*.

At the other extreme, the cercariae of *Parahemiurus bennettae* develop caudal vesicles and appendages inside the redia, but these degenerate and the cercariae become sexually mature and gravid inside the redia (Jamieson, 1966). The fate of these inter-redial adults and eggs is unknown but Jamieson suggested that the eggs are released when the snail dies and that these eggs infect other snails directly.

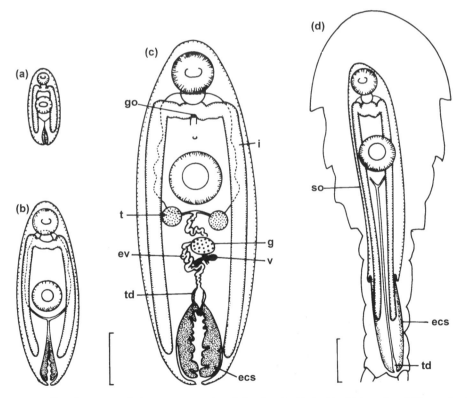

Figure 16.26 Development of the mesocercaria of the hemiurid *Lecithochirium furcolabiatum* in the copepod *Tigriopus brevicornis* at 17°C. (*a*)–(*d*) Mesocercariae at 7, 14, 21 and 21 days post-infection respectively; (*d*) shows mesocercaria with fully extended ecsoma in copepod haemocoel (body outline of copepod only shown). ecs = ecsoma; ev = excretory vesicle; g = germarium; go = genital opening; i = intestine; so = soma; t = testis; td = terminal dilatation of excretory vesicle; v = vitellarium. Scale bars: (*a*)–(*c*) 150 μm; (*d*) 100 μm. Source: reproduced from Matthews and Matthews, 1988a, with permission from Dr B. F. Matthews and CAB International.

An interesting feature of many adult hemiurids is the presence of a telescopic hindbody or ecsoma that can be withdrawn completely into the main body or soma. Matthews and Matthews (1988a) showed that the ecsoma of *L. furcolabiatum* is derived in the mesocercaria by specialization and subsequent eversion of the lining of the posterior region of the bladder (excretory vesicle; Fig. 16.26). The ecsoma appears to be involved in the uptake of amino acids and perhaps larger molecules from the haemolymph of the copepod, and ultrastructural and histochemical differences between the ecsomal and somal tegument in adult hemiurids (Matthews and Matthews, 1988b) point to functional differences in the final host. The somal tegument has features that suggest a protective rather than an absorptive role, and this may reflect the fact that many hemiurids inhabit the relatively harsh environment of the fish stomach. There is some similarity here with the tegument of rumen-inhabiting paramphistomid digeneans (section 15.6.3).

(f)   The dicrocoeliids

The lancet fluke, *Dicrocoelium dendriticum*, has evolved a life cycle independent of the aquatic environment (Fig. 16.27). The adults parasitize herbivores such as sheep

Figure 16.27 Life cycle of *Dicrocoelium dendriticum*. Adult (*a*) in bile duct. Eggs (*b*) voided in faeces, complete development in soil. Eggs eaten by terrestrial snail. Sporocysts (*c*) produce cercariae, which are ejected from snail pneumostome in slime ball (*d*). Slime balls gathered by ants. Most metacercariae (*e*) in haemocoel of gaster. Cyst in suboesophageal ganglion of ant induces behaviour illustrated. Exposed infected ants, locked to vegetation by jaws, consumed by grazing sheep. Source: based on Anokhin, 1966 and Olsen, 1986.

and inhabit the bile ducts, which they reach *via* the opening of the biliary system into the intestine (Olsen, 1986; cf. *Fasciola hepatica*, Chapter 12). Eggs pass down the bile duct and intestinal tract and are voided in the faeces. Most digeneans require water for the maintenance and dispersal of miracidia and cercariae but the fully embryonated eggs of *D. dendriticum* fail to hatch unless eaten by an appropriate land snail and the xiphidiocercariae are not expelled from the snail's lung until enclosed in groups in a protective slimy coat (Fig. 16.27*d*). These packets of cercariae or slime balls are explosively ejected from the pneumostome. Dehydration is retarded by the slime and the balls are collected and eaten, mainly by worker ants.

It is widely assumed that dicrocoeliids manipulate the behaviour of their second intermediate hosts in a similar way to *Microphallus papillorobustus* (section 16.3.4(c)). After consumption by an ant such as *Formica polyctena*, the cercariae of *Dicrocoelium dendriticum* enter the haemocoel by penetrating the wall of the crop and several of them migrate forwards into the head of the ant (Schneider and Hohorst, 1971). Here a single metacercaria enters the brain (suboesophageal ganglion). Following this event, the other cephalic metacercariae rejoin those remaining in the abdomen (gaster) and they all secrete a relatively thick cyst wall (11–12 μm). Romig, Lucius and Frank (1980) found that brain flukes sometimes, but not always, were enclosed in a much thinner cyst wall (1–2 μm). It is assumed that these brain flukes influence the ant's behaviour and they are said to be uninfective to the definitive host.

The behaviour of ants harbouring a brain fluke is strikingly different from that of uninfected ants. Uninfected foraging workers tend to return to their nest at dusk, but infected workers remain in the open, mostly at distances of up to 4 m from the nest (Spindler, Zahler and Loos-Frank, 1986). As the air temperature falls and the humidity increases, they climb up the vegetation and, after adopting a position usually with head downwards (Fig. 16.27*e*), clamp themselves to the plant by means of the jaws (Hohorst and Graefe, 1961; Anokhin, 1966). They maintain this attitude throughout the night and, as the temperature rises and the humidity falls after sunrise, they detach themselves and rejoin their uninfected nest mates. It is assumed that this behaviour enhances the chances of infected ants being eaten by sheep or deer grazing during the cooler hours around dawn and dusk.

The occasional discovery of ants displaying the classical altered behaviour, but with brain flukes in sites other than the suboesophageal ganglia, suggests that the brain fluke may not exert its influence by directly damaging the ganglion (Romig, Lucius and Frank, 1980). Perhaps the host's behaviour is controlled by chemical substances from the parasite diffusing out through the thin cyst wall. In *D. hospes*, a parasite of African ungulates, the ant host harbours two brain flukes, one in each antennal lobe (Romig, Lucius and Frank, 1980) and in *Brachylecithum mosquensis*, a liver fluke from the American robin, *Turdus migratorius*, one or two metacercariae lie in or near the supraoesophageal ganglion (Carney, 1969). Ants infected with *D. hospes* assemble in motionless groups at elevated sites, usually on plants, where they remain for long periods unattached by the mandibles, unresponsive to temperature changes and fed by uninfected nest mates. *B. mosquensis* is particularly interesting because the altered behaviour of its carpenter ant hosts seems likely to attract foraging birds and is linked with a morphological change. Unlike uninfected carpenter ants, which are strongly photophobic, infected workers expose themselves on rocks where they circle slowly or remain motionless for 1–2 h. Moreover, their

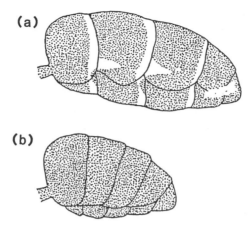

Figure 16.28 The gasters of ants of the genus *Camponotus*, (*a*) parasitized by metacercariae of the dicrocoeliid *Brachylecithum mosquensis* and (*b*) unparasitized. The parasitized condition is distinct from the replete state, in which the intersegmental regions are transparent rather than an opaque white. Source: reproduced from Carney, 1969, with permission from *American Midland Naturalist*.

gasters are greatly distended, not by parasites but by an overgrowth of host tissue (Fig. 16.28). This obesity and the seeking out of exposed places renders them visible to human observers, and presumably also to insectivorous birds, at distances of 8–10 m.

(g)　Paragonimus westermani

In the Far East, freshwater crabs (*Eriocheir* spp.) are important sources of animal protein for humans; the habit of eating these raw or partially cooked provides an opportunity for infection with the troglotrematid *Paragonimus westermani*, a digenean with xiphidiocercariae developing in operculate snails. Other carnivorous animals such as cats and dogs act as reservoir hosts. The xiphidiocercariae of *P. kellicotti*, a relative from North America infecting crayfish-eating mammals, have reduced tails, creep about on the substrate and penetrate crayfish through the thin articular cuticle (see Olsen, 1986), but there is evidence to suggest that cercariae of *P. westermani* rarely if ever leave the snail host and that crabs become infected by eating infected snails (references in Yokogawa, 1965).

After excysting in the intestine of the definitive host, the metacercariae of *P. westermani* bore through into the body cavity, pass through the diaphragm into the pleural space and then enter the lungs (see Yokagawa, 1965). The activities of the parasites in the lungs create a necrotic cavity with a thick fibrous wall, usually referred to incorrectly as a cyst. In dogs and cats, this cavity usually contains two or three parasites, but autopsies on humans with *P. westermani* reveal that isolated individuals are more common. Communication between the cavity and the lung passages is maintained and eggs laid by the flukes are transported to the throat in cilia-propelled mucus, which in humans may be ejected as sputum or swallowed. In humans more eggs are evacuated in the faeces than in the sputum, but it is the latter, which is often blood-stained, and the accompanying light cough, that often lead to an incorrect diagnosis of pulmonary tuberculosis. Detection of eggs in the faeces or sputum provides confirmation of paragonimiasis, but the disease has few symptoms and is rarely a serious threat.

In some endemic areas, the way in which humans acquire *P. westermani* is not obvious and has required some detective work. In Japan, the crabs are usually eaten in soup, but this is cooked for 10–20 min, long enough to kill metacercariae (see Yokogawa, 1965). However, before cooking, the shell and legs are removed from the live crab, the body is cut up on a chopping board and the crab meat is strained through a bamboo basket. Living metacercariae cling to these utensils and to the hands of the cook, and vegetable pickles and spices are often prepared on the chopping board after dicing the crab. This contamination accounts for human infection with lung flukes.

In Kagoshima Prefecture many human cases of paragonimiasis were reported, but there was no tradition of eating crabs (Miyazaki, 1976). The source of infection was identified as wild pigs, which act as paratenic hosts. When these animals eat crabs with metacercariae, the parasites enter the muscles and suspend their development. Humans become infected by eating sliced raw pork. It seems more likely too that tigers, which frequently harbour lung flukes in Sumatra, acquire these *via* a paratenic host rather than from predation on freshwater crabs.

16.3.5   Xiphidiocercariae that penetrate molluscs

Molluscs and arthropods intermingle in many aquatic environments. Given the softness of the molluscan body and ease of penetration into it, and the importance of molluscs as prey for vertebrates, it comes as no surprise that some xiphidiocer-

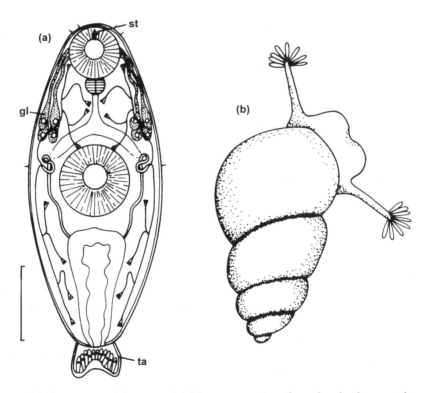

Figure 16.29   (*a*) The cercaria of the opecoelid *Sphaerostoma maius*. gl = stylet glands; st = stylet; ta = tail. (*b*) Cercariae with reduced tails (cercariaea) attached to tentacles of *Bithynia*. Scale bar: (*a*) 50 μm. Source: (*a*) redrawn from Lambert and Lambert, 1974.

cariae have indulged in host-switching. The cercariae of the opecoelid *Sphaerostoma maius* develop in sporocysts in the freshwater prosobranch *Bithynia tentaculata* (see Lambert and Lambert, 1974). The cercaria (so-called cercariaeum) has a greatly reduced tail (Fig. 16.29*a*) and is unable to swim; it attaches itself to the substrate by the reduced tail which is provided with glands. When a snail (*Lymnaea limosa*; *Physa acuta*) passes within reach, the cercariae attach themselves and penetrate, encysting in the mantle or the muscles of the radula. The adult parasite inhabits the intestine of the fish *Leuciscus cephalus*.

The cercariae of the monorchiid *Parasymphylodora markewitschi* have no tail and infect mollusc second intermediate hosts by contagion (Lambert, 1976). The cercariae, attached by their ventral suckers, gather in large numbers on the extremities of the tentacles of the first intermediate host, *B. tentaculata*, giving the tentacles a plumose appearance (Fig. 16.29*b*). The tentacles project in front as the snail progresses and as soon as the larvae touch another mollusc (*Bithynia* or other snails) they attach to it by the free oral sucker, migrate across the surface of the new host and eventually penetrate and encyst. Snails with metacercariae are eaten by the fish *L. cephalus*. Lambert made the interesting discovery that tentacles without cercariae retract when touched with a needle, but do not do so when cercariae are present.

Nasincova and Scholz (1994) found that the tailless cercariae of the monorchiid *Asymphylodora tincae* do not leave the rediae. The definitive host, the freshwater fish *Tinca tinca*, is infected by ingesting pulmonate snails containing these rediae. This secondary reduction from three to two hosts resembles developments in the microphallids and the gorgoderids (see above).

The stylet in *P. markewitschi* is greatly reduced and is lost in *A. tincae*. The stylet is also absent in the monorchiid *Paratimonia gobii*, but the tail is retained. The cercariae emerge from marine bivalves, *Abra ovata*, buried in the sediment (Maillard, 1975). They are photonegative, swim close to the bottom and are drawn in by the inhalant water current of other *Abra ovata*. They penetrate and encyst in their inhalant siphons. Curiously, the definitive host is the small goby *Pomatoschistus microps*, which at 3–4 cm in length is unlikely to ingest a bivalve 1.5 cm in length buried in the sediment. Maillard found that the bivalve is able to shed a damaged siphon and that this autotomy is precipitated by the accumulation of metacercariae in the siphon. The parasitized siphon is rejected at the surface of the sediment and its continuing movements attract foraging gobies.

16.4   THE WAY FORWARD

There is one other group of aquatic organisms with enormous potential as second intermediate hosts for tissue-penetrating cercariae. This is the vertebrates, especially the fishes. Fishes have been exploited by many different digeneans, including the versatile echinostomes, the heterophyids, the opisthorchiids and the strigeids, and they provide the raw material for some remarkable developments, the most exciting of which is the emergence of the schistosomes. The fish-penetrating cercariae will be considered in the next chapter.

CHAPTER 17

Digeneans with aquatic vertebrates, especially fish, as second intermediate hosts

---

17.1 INTRODUCTION

One of the most spectacular evolutionary success stories is the 'explosive' expansion of the osteichthyans (bony fishes) that took place during the period from the Devonian to the late Cretaceous (Pough, Heiser and McFarland, 1990). Of these fishes the teleosteans constitute a bewildering diversity of forms, undoubtedly the largest vertebrate radiation, with close to 21 000 living species in 200–300 families. As we have seen, this expansion provided fertile ground for the diversification of the monogeneans and the cestodes, and it had a similar impact on the digeneans. Fishes not only provided a wealth of potential definitive hosts but also, as prey organisms, supported a great range of piscivorous vertebrates. Penetration and encystation in fishes provided access to these piscivorous vertebrates.

Fishes generally live longer than invertebrates, and digeneans using them as intermediate hosts stand to gain from increased longevity and hence increased chances of being eaten by a suitable definitive host, with the proviso that the parasite is able to circumvent the more sophisticated antiparasite defences of the fishes. Metacercariae of the strigeoids *Uvulifer ambloplitis* and *Posthodiplostomum minimum* were still alive after 4.5 years in captive bluegill, *Lepomis macrochirus* (see Hoffman and Putz, 1965). The evolutionary advantage of recruiting fishes as second intermediate hosts is emphasized by the range of digeneans that have taken this option (Table 17.1).

17.2 MORE ECHINOSTOMES

The versatile echinostomatids have representatives with cercariae capable of penetrating amphibians and fishes. According to Dönges (1980), cercariae of *Isthmiophora melis* penetrate the linings of the mouth and cloaca of adult amphibians, tadpoles and fishes. By experimental infection, Dönges (1962) showed that humans consuming fish with metacercariae can serve as definitive hosts.

Like the azygiids and bivesiculids (section 15.4), some echinostomatid cercariae have greatly enlarged and often coloured tails (Fig. 17.1), indicating that they may be actively ingested by their fish second intermediate hosts. This is consistent with the

Table 17.1 Some digeneans that penetrate fishes (or amphibians)

Parasite	Fish host	Site	Encysted (+) or unencysted (−)	Reference
Echinostomatidae				
Mesorchis denticulatus	Gasterosteus aculeatus	Gills	+	Køie, 1986
	Pungitius pungitius			
	Pomatoschistus microps			
Isthmiophora melis	Phoxinus laevis		+	Dönges, 1962, 1967
	Squalius cephalus			
	(Tadpoles)			
Strigeidae				
Cotylurus erraticus	Salmonids	Pericardial cavity	+	Olson, 1970
Apatemon gracilis	Trout	Pericardial cavity	+	Blair, 1976
	Gasterosteus aculeatus	Eye humours	+	
Cyathocotylidae				
Cyathocotyle orientalis	Pseudorasbora parva*	Muscle	+	Yamaguti, 1975
Holostephanus lühei	Gasterosteus aculeatus	Muscle	+	Erasmus, 1962
Diplostomatidae				
Posthodiplostomum minimum	Freshwater fishes	Various internal organs	+	Hoffman, 1958a
Uvulifer ambloplitis	Freshwater fishes	Muscle	+	Hoffman and Putz, 1965
Diplostomum phoxini	Phoxinus phoxinus	Beneath lining of brain ventricles	−	Rees, 1957
Diplostomum gasterostei	Gasterosteus aculeatus	Retina	−	Williams, M.O. 1966
	Perca fluviatilis	Vitreous humour	−	Kennedy, 1981
Diplostomum spathaceum	Freshwater fishes	Lens	−	Ratanarat–Brockelman, 1974

Table 17.1 (Contd.)

Parasite	Fish host	Site	Encysted (+) or unencysted (−)	Reference
Clinostomatidae				
Clinostomum complanatum	Freshwater fishes	Subcutaneous tissue, muscle	+	Olsen, 1986
Bucephalidae				
Prosorhynchus crucibulum	Marine fishes*	Connective tissue, muscle	+†	Matthews, 1973a
Rhipidocotyle spp.	*Menidia menidia*	Muscle	+	Stunkard, 1976
Bucephalus haimeanus	*Pomatoschistus microps*	Liver	+	El-Mayas and Kearn, 1966
Bucephaloides	*Ciliata mustela**	Crania nerves	±	Matthews, 1974
gracilescens	Whiting	Intracranial fluid, orbit, nasal region	+	Halton and Johnston, 1982
Sanguinicolidae				
Sanguinicola inermis	*Cyprinus carpio*	Blood vessels	−	Kirk and Lewis, 1993
Troglotrematidae				
Nanophyetus salmincola	Salmonid and other fishes	Muscle, fins, kidneys, elsewhere	+	Roberts and Janovy, 1996
Opisthorchiidae				
Opisthorchis viverrini	Cyprinids	Muscle	+	Haswell-Elkins *et al.*, 1992
Heterophyidae				
Cryptocotyle lingua	*Tautogolabrus adspersus* (cunner)	Skin	+	Stunkard, 1930
Cryptogonimidae				
Stemmatostoma pearsoni	Freshwater fishes	Muscle, fins	+	Cribb, 1986

* Experimental infection; † temporary encystment; ± reports of encysted and unencysted metacercariae

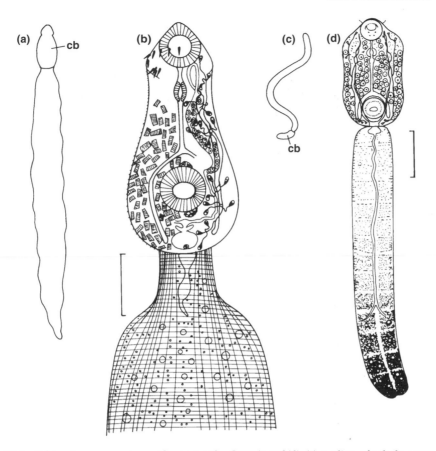

Figure 17.1 Echinostome cercariae with giant tails. *Cercaria rashidi*: (*a*) outline of whole cercaria and (*b*) enlarged view of body and anterior region of tail. Collar spines, tegumental spines, cystogenous glands, gut caeca and flame cell system shown on one side only. Cercaria of *Mesorchis denticulatus*: (*c*) outline of whole cercaria in swimming attitude and (*d*) sketch showing anatomical detail. cb = cercarial body. Scale bars: (*b*) 50 μm; (*d*) 100 μm. Source: (*a*), (*b*) reproduced from Nasir, 1962, with permission from the American Microscopical Society; (*c*), (*d*) reproduced from Køie, 1986, copyright © Springer-Verlag 1986, with permission from Springer-Verlag.

presence of their metacercariae embedded in the gills and beneath the linings of the pharynx and oesophagus of fishes, although such cercariae may also be drawn passively into the gill chamber in the gill-ventilating current.

Nasir (1962) described two giant-tailed echinostomatid cercariae from the freshwater snail *Planorbis carinatus*. One of them, *Cercaria rashidi*, had a total length of up to 1.8 mm with an orange-yellow tail that could exceed 1.6 mm in length (Fig. 17.1*a*, *b*). Nasir failed to infect a variety of snails and the tadpole of *Rana temporaria*, but found cysts in the pharynx and anterior oesophagus of several experimentally exposed fishes, including the stickleback *Gasterosteus aculeatus*, which was also infected naturally.

The tail of the cercaria of *Mesorchis denticulatus* from the brackish-water snail *Hydrobia ulvae* swells when the cercaria leaves the snail and reaches a length of up to 2 mm (body length about 200 μm; Fig. 17.1*c*, *d*). According to Køie (1986), the distal third of the tail is red and the middle third yellowish. Some cercariae passively enter the gill chamber of small fishes but others are ingested actively and are found encysted in the gills. Køie found no cercariae or tails in the stomachs of her fishes, suggesting that they are not swallowed. This may account for the lack of protection from the tail around the bodies of the free-swimming cercariae: unlike azygiids and bivesiculids (see above), echinostomatids may avoid the dangers of a journey through the stomach of their fish hosts.

17.3   *NANOPHYETUS SALMINCOLA*

We saw in the last chapter that xiphidiocercariae are highly specialized for penetrating arthropod cuticle, which they slit open with their stylet. This has not deterred some species from switching to soft-bodied molluscs from their original arthropod hosts (section 16.3.5), and a troglotrematid, *Nanophyetus salmincola*, has recruited fishes as second intermediate hosts. A variety of fish-eating mammals (including humans) and birds serve as definitive hosts, but the parasite has gained notoriety because it transmits a rickettsial disease (*Neorickettsia helminthoeca*) to canid (not human) hosts. Dogs are extremely susceptible and few recover (Roberts and Janovy, 1996).

17.4   THE BUCEPHALIDS

The forked tails of bucephalid cercariae have undergone unique and remarkable modifications to facilitate contact with fish intermediate hosts. *Bucephaloides gracilescens* is associated with the bottom fauna at medium depths on the European continental shelf, where the first intermediate host is the bivalve *Abra alba*, the second intermediate host one of several gadoid fishes and the definitive host the angler fish *Lophius piscatorius* (Fig. 17.2).

The dichotomously branching sporocyst in the bivalve releases cercariae into the exhalant chamber, from which they are expelled. Unlike other furcocercariae, bucephalids have limited swimming ability. However, the tail furcae have circular muscles, which permit them to extend enormously from a retracted length of 0.25 mm to a fully extended length of 2.9 mm, i.e. a 12-fold increase in length. Like the gossamer of small spiders, the filamentous egg appendages of some monogenean eggs (section 5.4) and the mucous threads of some lecithodendriid cercariae (section 16.3.4(a)), the drifting thread-like furcae are swept upwards by water currents and carry the body into suspension (Fig. 17.2). At irregular intervals, partial contractions of the furcae, tending to jerk the body upwards, are the only swimming movements. On stronger contraction of the longitudinal muscles of the furcae they twist into a spiral and the cercaria sinks to the bottom. The tegument covering the posterior edges of the furcae contains secretory granules and is sticky, and the drifting 'fishing lines' adhere readily when a fish makes contact with them. Full contraction of the furcae then brings the tail stem, which is also sticky, into contact with the host and penetration follows. The probability of contact with the host may be increased in *Cercaria pleuromerae* by adhesion between the spread tails of many individuals, forming an extensive drifting net (Wardle, 1988).

The tail stem of the cercaria of *Prosorhynchus squamatus* from *Mytilus edulis* is trilobed (Fig. 17.3*a*) and the adhesive surface of each lobe is invaginated until required. The arrangement provides a stable tripod for attachment (Fig. 17.3*b*) and

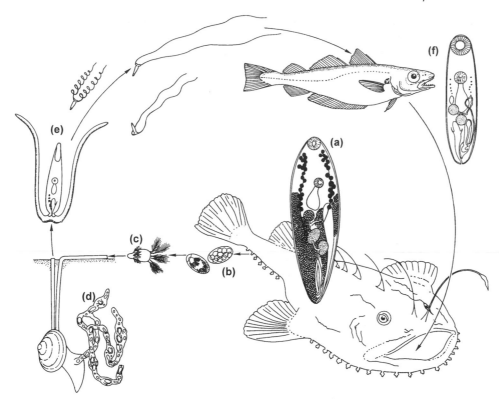

Figure 17.2 The life cycle of a bucephalid. Adult (*a*) in gut of piscivorous fishes, e.g. angler fish (*Lophius piscatorius*). Eggs (*b*) voided in faeces, develop and hatch. Miracidium (*c*) enters sand-dwelling bivalve *via* inhalant siphon, penetrates and produces branched sporocyst (*d*). Cercaria (*e*) leaves *via* exhalant siphon. Tail furcae extensile as shown – adhere to gadoid fish. Cercaria penetrates. Gadoid with metacercaria (*f*) falls prey to angler fish. Source: based mainly on knowledge of *Bucephaloides gracilescens* (see Dawes, 1946 and Matthews, 1974).

elevates the rear of the cercaria ensuring penetration at a favourable angle (Matthews, 1973a). Matthews observed projecting sensilla on the circular rim of each invaginated lobe (Fig. 17.3*c*) and noted that these are vibratile. He suggested that cessation of movement of the sensilla when they make contact with the fish may be the signal for evagination of the adhesive lobes. A similar sensory function for vibratile sensilla on the haptor of the monogenean *Tetraonchus monenteron* has also been proposed (section 6.4.1).

The bucephalids illustrate well the range of sites exploited by metacercariae in the bodies of fish second intermediate hosts and the diversity of parasite–host interactions in these sites, while at the same time highlighting our ignorance of these interactions. Most, if not all, bucephalid metacercariae are able to benefit from the rich source of nutrients available in their fish hosts, in spite of the fact that in some of their chosen sites they are likely to be exposed to antiparasite defences.

Stunkard (1976) observed that the cercaria of *Rhipidocotyle* sp. (body 120–250 μm long and 30–70 μm wide) entered muscle tissue of the fish *Menidia menidia* and within 24–48 h of penetration secreted a thin and flexible cyst wall. In spite of this, the parasite appeared to have access to nutrients from the host and the metacercaria

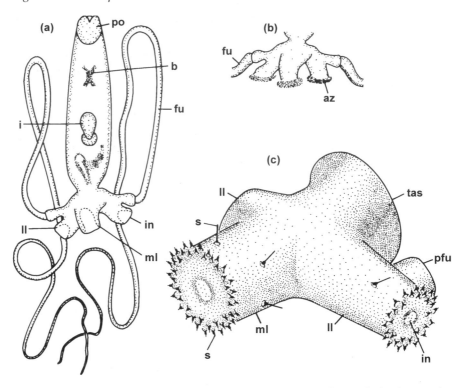

Figure 17.3   (*a*) The cercaria of the bucephalid *Prosorhynchus squamatus*. (*b*) Attached tail stem, showing evaginated adhesive zones. (*c*) Three-dimensional diagram of tail stem, showing arrangement of sensilla. az = adhesive zone; b = 'brain'; fu = tail furca; i = intestine; in = invagination of surface; ll = lateral attachment lobe; ml = median attachment lobe; pfu = point of attachment of furca; po = penetration organ; s = sensillum; tas = tail stem. Scale bar: (*a*) 100 μm. Source: reproduced from Matthews, 1973a, with permission from Dr R. A. Matthews and Cambridge University Press.

grew substantially, the largest cysts measuring 1 mm in length and 500 μm in width. It seems that these cysts did not provoke a host reaction (see also section 17.5).

A different way of gaining access to host resources was described by Matthews (1973a) in *Prosorhynchus crucibulum*, using the turbot, *Scophthalmus maximus*, as second intermediate host. The cercariae enter the connective tissue and musculature, where they secrete a cyst wall within a few hours. However, this cyst wall then breaks down and 1-month-old metacercariae live freely in the tissues. This permits the metacercariae to obtain nutrients by directly ingesting host cells and this diet includes connective tissue cells destined to encapsulate the parasite. Thus, provocation of the host's immune system appears to provide the parasite with a renewable source of food. It is not until the end of the second month, when the metacercaria completes its development and becomes quiescent, that the host connective tissue cells survive in sufficient numbers to form a capsule. There appears to be a different reason for delayed encapsulation by the host in the heterophyid *Stictodora lari* (section 17.5).

The fully grown metacercaria of *P. crucibulum* is about half the size of the adult and, with the exception of the vitellaria, the reproductive system is fully developed. The metacercariae are still alive 10 months after invasion of the turbot.

The metacercaria of *Bucephalus haimeanus* also obtains nutrients from the host by ingesting host tissue (Matthews, 1973b), but this only takes place during the initial migration to the liver of the goby *Pomatoschistus microps*, a journey that takes no more than 1 h and may take as little as 10 min. Encystment occurs on arrival in the liver and the cyst wall persists until the goby is eaten by the final fish host, *Morone labrax*. Nevertheless, the reproductive system appears, although it does not reach the level of development attained by the metacercariae of *P. crucibulum*, and the cysts survive for at least 10 months.

In the bodies of their gadoid second intermediate hosts, the metacercariae of *Bucephaloides gracilescens* (Fig. 17.2) show a predilection for nervous tissue. Matthews (1974) considered the nervous system to be an immunologically inactive site and attributed to this the failure of an experimental and probably unnatural host, the rockling *Ciliata mustela*, to encapsulate the metacercariae. Some, but not all, of the cercariae also lacked a parasite-derived cyst wall. In contrast, in the whiting, a natural and often heavily infected second intermediate host in British waters, Halton and Johnston (1982) found only encysted metacercariae in a range of sites, including the intracranial fluid surrounding the brain, the auditory capsules, the orbits and the connective tissue of the snout. In all of these sites they identified a thin, inner, parasite cyst wall and an outer cellular host capsule. They compared cysts from the brain with those from other sites and discovered that brain cysts have thinner host capsules and better developed male reproductive systems, indicating that the host reaction is less vigorous adjacent to the nervous system and that cysts in this site have perhaps better access to nutrients from the host.

In the hake (*Merluccius merluccius*), cysts of *B. gracilescens* are associated with the brain and cranial and spinal nerves. Crofton and Fraser (1954) offered evidence to suggest that the cercariae reached these sites by entering the canals of the lateral line system.

Halton and Johnston (1982) studied excystment of *B. gracilescens* and found that acid pepsin removes the outer host-derived component of the cyst wall while the inner parasite-derived layer remains intact. The metacercaria is instrumental in breaching the parasite cyst when stimulated by the increased pH in the intestine, an event that is enhanced by bile. Surprisingly, El-Mayas and Kearn (1996) reported excystment of *Bucephalus haimeanus* in acid media, with or without pepsin, while the metacercaria remained totally inactive in the presence of bile. El-Mayas and Kearn suggested that the ability of the parasite to respond to and survive in acid conditions may be related to the relatively long exposure time for food in the stomach of the final host (the bass, *Morone labrax*). How the parasite survives temporary exposure to potentially lethal H^+ ions is unknown, but this achievement may have enabled some bucephalids like *Alcicornis carangis* (Fig. 17.4) to establish themselves permanently in the harsh but relatively unexploited environment of the stomach. The lack of response to bile salts of the thin-walled cysts of *B. haimeanus* was attributed by El-Mayas and Kearn to the fact that they may be exposed to bile before they leave the fish intermediate host, since the cysts are located in the liver.

The bucephalid cercaria, and the adult derived from it, have features that distinguish them from other digeneans. The mouth is not terminal but midventral and communicates *via* a sucker (usually regarded as the pharynx) with a small sac-shaped gut (Figs 17.2, 17.4). Another sucker, which could be regarded as homologous with the ventral sucker of other digeneans, appears at the anterior end of some bucephalids (see Fig. 17.2), but this appears to have a separate origin from the anterior glandular penetration organ of the cercaria (Fig. 17.3*a*). In some adult bucephalids, the anterior region of the body forms a muscular 'rhynchus', which

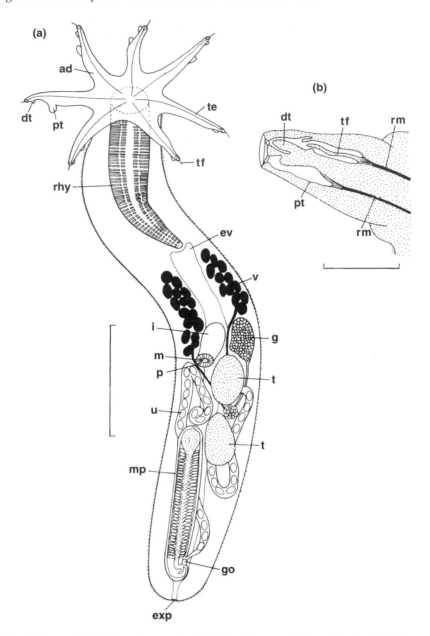

Figure 17.4   The bucephalid *Alcicornis carangis*. (*a*) Adult with expanded anterior disc (tentacles everted). (*b*) Enlarged view of almost fully retracted (inverted) tentacle. ad = anterior disc; dt = distal process of tentacle; ev = excretory vesicle; exp = excretory pore; g = germarium; go = genital opening; i = intestine; m = mouth; mp = male pouch; p = pharynx (?); pt = proximal process of tentacle; rhy = rhynchus; rm = retractor muscle; t = testis; te = tentacle; tf = tentacle filament; u = uterus; v = vitellarium. Scale bars: (*a*) 500 μm; (*b*) 200 μm. Source: redrawn from Rees, 1970.

terminates either in a disc, as in *P. crucibulum*, or in eversible tentacles, as in *A. carangis* (Fig. 17.4).

Matthew's description (1973a) of locomotion in adult *P. crucibulum*, which lives between folds of the intestinal mucosa of the conger eel, offers clues to the reasons for these unusual anatomical features. There appears to have been a shift in locomotion from the more typical leech-like style to progression more reminiscent of an earthworm. The parasite is anchored between the intestinal folds by inflation of the body, aided by backwardly projecting tegumental scales. The rhynchus then extends forwards and anchors the anterior extremity by expanding between the intestinal folds. This permits the body to contract and move forwards.

The overall picture in the bucephalids is one of specialization rather than primitive simplicity. Even the relatively small sac-like intestine, sometimes regarded as a primitive rhabdocoel-like feature, may represent a reduction permitted by the adoption of transtegumental absorption of nutrients.

17.5   THE HETEROPHYIDS AND OPISTHORCHIIDS

The life cycle of the heterophyid *Cryptocotyle concavum* is illustrated in Fig. 17.5. The cercariae of heterophyids and opisthorchiids are strong swimmers with a muscular, single tail, bearing fins (Fig. 17.5*d*). However, they do not swim continuously and exhibit a pattern of alternating active and passive phases that is also a feature of unrelated fish-penetrating cercariae such as *Diplostomum spathaceum* (see below). The cercaria of the heterophyid *Cryptocotyle lingua* has eyes and propels itself

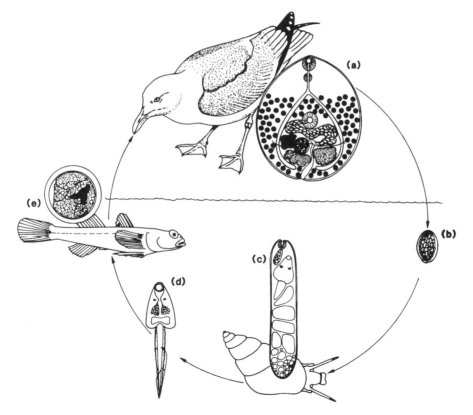

Figure 17.5   The life cycle of the heterophyid *Cryptocotyle concavum*. Adult (*a*) in intestine of bird. Eggs (*b*) enter water in faeces, develop but do not hatch. Eggs probably eaten by snail, *Hydrobia*. Redia (*c*) produces cercariae (*d*) with single tail. Cercaria penetrates and encysts (*e*) in goby, *Pomatoschistus microps*. Bird infected by eating goby. Source: based on Wootton, 1957 and El-Mayas, 1991.

towards the light in a roughly helical path, body first, for periods of 2–3 s, then stops swimming for 15 s, during which period it sinks slowly, tail uppermost (Rothschild, 1939; Chapman and Wilson, 1973).

The host-finding behaviour of these active, fish-penetrating cercariae differs greatly from that of snail-invading cercariae (section 16.2.1). Mechanical stimuli (water turbulence, touch) seem to be highly important for fish-finding, since these stimuli release swimming bursts that are longer than spontaneous bursts and the stimulated parasites are ready to attach to any suitable substrate (Haas, 1992, 1994). There is no indication of directed chemo-orientation, as in some snail-invading cercariae, although exposure to fish extracts and amines leads to a reduction in length of the passive phase in *C. lingua* and a prolongation of the active swimming phase (Chapman, 1974). These responses are consistent with the need to make contact with a host that is likely to create considerable turbulence as it moves about.

According to Haas (1994), contact with the fish host is followed by attachment, enduring contact and penetration, three consecutive behavioural phases that are stimulated by chemical cues of increasing host specificity. It is a feature of fish-finding by cercariae that the specific host is ultimately recognized by its macro-molecules.

When cercariae of *Opisthorchis viverrini* make contact with host skin, attachment is stimulated by glycosaminoglycans (Haas, Granzer and Brockelman, 1990a). It seems that a different chemical signal is required to prevent attached cercariae from detaching and swimming away, but this has not been identified. A third chemical component, probably a protein, is required to initiate penetration behaviour. Haas *et al.* suggested that, for host identification, large molecules may be better than small ones in an environment where the latter may be swamped by small molecules from non-host sources such as mud or decaying materials.

Fish-invading cercariae are also stimulated to swim by shadowing, but Haas (1994) has challenged the assertion that this response enhances the chances of contact with the host and has introduced the idea that such behaviour may serve to avoid predation by planktonic organisms.

Stunkard (1930) expressed his difficulty in understanding how the cercaria of *C. lingua* attaches itself to the skin during the initial stages of penetration since there is no ventral sucker. Nevertheless, he noted that penetrating cercariae resist detachment when vigorously shaken in water. He also emphasized the importance of the backwardly directed tegumental spines for tunnelling through tissue; he suggested that they promote forward progress every time that the larva moves and, at the same time, resist withdrawal from the penetration tunnel. Stunkard noted that gland cells occupying the central region of the body with ducts opening anteriorly (Fig 17.5*d*) were exhausted when the cercariae reached their encystation site beneath the skin, indicating that the glands play an important part in penetration.

If there is a host reaction to the metacercaria of *C. lingua*, this does not prevent the encysted parasite from obtaining nutrients through the thin cyst wall. During the 38-day period required for maturation, the cyst increases in size from $156 \times 70 \,\mu$m to $174 \times 140 \,\mu$m and the metacercaria undergoes considerable development (Rees, 1979). The extent of this development can be judged by the fact that only 3 days are required for maturation of the adult in the seagull definitive host.

Some intriguing relationships between heterophyid cysts and their fish hosts have been reported. Howell (1973) discovered that the cyst of *Stictodora lari*, in the body

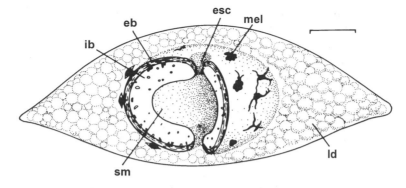

Figure 17.6 Diagrammatic illustration of the host layers encapsulating the metacercaria of the hetero-phyid *Apophallus brevis* in the muscles of the perch *Perca flavescens*. esc = escape canal; ib = inner bony capsule; ld = lipid droplets; mel = melanocyte; ob = outer bony capsule; sm = space occupied by metacer-caria (cyst not shown). Scale bar: 0.25 mm. Source: reproduced from Pike and Burt, 1983, with permission from Cambridge University Press.

cavity of an experimental fish host *Gambusia affinis*, was not encapsulated until 21–23 days after infection, while implanted glass beads were encapsulated within 3 days. Thus, the encysted parasite had some way of avoiding temporarily the host's innate capability to deal with foreign material. Using fluorescent antibody techniques Howell showed that the initial cyst wall did not mimic host tissue but *in vivo* it did become contaminated with material of host origin, which may have provided protection by disguising the cysts as 'self'. Delayed encapsulation by the host corresponded with secretion of an inner cyst layer by cells lining the bladder of the metacercaria.

Pike and Burt (1983) described a host response of great complexity around the thin-walled cysts of *Apophallus brevis* in the muscles of *Perca flavescens*. The cyst becomes enclosed in an elaborate capsule consisting of material similar to or ident-ical with host bone, but at opposite poles there are canals through which the parasite can escape when eaten by the definitive host (Fig. 17.6). Outside the capsule there are additional layers of melanocytes and connective tissue containing lipid droplets. This high degree of organization suggests that capsule formation may be orche-strated by the parasite rather than by the host, but what significance this has for the biology of the parasite is unknown.

Cable (1952, 1963) reported a curious habit in some large-tailed, marine, hetero-phyid cercariae. He observed the emission from the snail of rosette-like aggregates of cercariae, attached to each other by entanglement and/or adhesion of the poster-ior regions of their tails (Fig. 17.7). This so-called zygocercous habit has evolved independently in other digeneans (see Martin, 1968). It differs from the net-like aggregates of bucephalid cercariae (section 17.4) and its significance is not known, although it may enhance conspicuousness and thereby encourage predation by fish intermediate hosts.

In *Aphalloides coelomicola* an interesting abbreviation of the typical opisthorchiatan life cycle has taken place. The cercaria (Fig. 17.8*a*) penetrates the bodies of small brackish water fish, *Gobius minutus*, and encysts in the subepidermal musculature or in the mesenteries (Maillard, 1973). After 8–10 days excystment occurs and the metacercaria enters the body cavity of the fish and becomes mature. The eggs accumulate in a huge uterus (Fig. 17.8*b*) and must be released either by

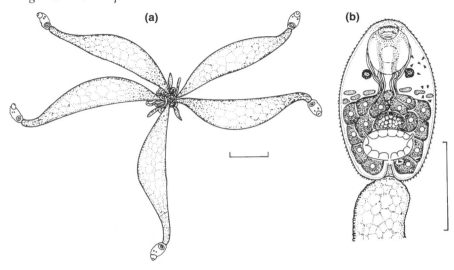

Figure 17.7   (*a*) A small cluster of heterophyid cercariae, *Cercaria caribbea LXX*. (*b*) The body of one of the cercariae. Scale bars: (*a*) 200 μm; (*b*) 50 μm. Source: reproduced from Cable, 1963, with permission from Springer-Verlag.

disintegration following the death of the fish (possibly accelerated by the presence of the parasite), or by digestion of the fish after predation by another vertebrate.

Acetodextra amiuri may have abbreviated its life cycle in a similar way but parasitizes the ovaries of various freshwater fishes (Perkins, 1956). The parasites appear to feed on egg yolk, which young individuals reach by entering the fish egg. Adult parasites packed with their own unlaid eggs are probably released with the host's eggs at spawning. However, early events in the life cycle, including the mode of infection of the fish, remain unknown.

Human consumption of raw fish is widespread, especially in the Far East. I confess to having acquired a taste for it in Japan, although being a parasitologist I should have known better. It is regarded as a delicacy among the affluent strata of society but, for poor people, eating raw fish may be a necessity, since fuel for cooking is often scarce. Fish is the source of human infection with the tapeworm *Diphyllobothrium* (section 10.5.7) and with *Nanophyetus* (section 17.3) and possibly with *Isthmiophora* (section 17.2). It therefore comes as no surprise that a variety of opisthorchiatan digeneans enter the human population by the same route (Table 17.2).

Clonorchis and *Opisthorchis* spp. are liver flukes and, like *Dicrocoelium* (section 16.3.4(f)), enter the small biliary passages from the intestine *via* the common bile duct, not by the devious penetrative route followed by *Fasciola* (Chapter 12). There is an unpleasant association between the incidence of opisthorchiatan liver flukes and cholangiocarcinoma of the liver (Haswell-Elkins, Sithithaworn and Elkins, 1992).

The influence of local customs in the epidemiology of these parasites is well illustrated by reference to *Clonorchis sinensis*. Kobayashi (in Grove, 1990) observed early in this century that the parasite was absent in the human inhabitants of northern China and infrequent in central China, although common in cats and dogs in these areas. Conversely, human infection in southern China was common, the habit of eating raw or undercooked fish being largely restricted to this area. Infection in

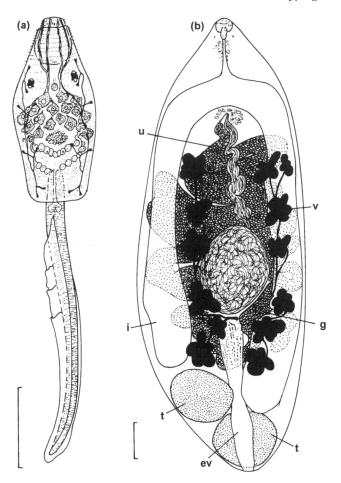

Figure 17.8   The heterophyid *Aphalloides coelomicola*. (*a*) Cercaria in dorsal view. (*b*) Adult in dorsal view. Lettering as in Fig. 17.4. Scale bars: 100 μm. Source: redrawn from Maillard, 1973.

cats and dogs was slight in southern China, probably because fish were too expensive to feed to these animals, and, for the same reason, the parasite was more prevalent in wealthier people. However, the parasite was common in people who raised the fish in freshwater ponds and, in towns, men were more commonly infected than women because raw fish was usually eaten in restaurants not patronized by women. Cats and dogs were identified as significant reservoir hosts in Vietnam.

Fears that clonorchiasis would be introduced to the United States by Chinese immigrants during the First World War led to the exclusion of infected persons from Hawaii. This restriction was rescinded in 1927 when it was realized that the infection would not spread to Americans (see Grove, 1990).

17.6   THE CRYPTOGONIMIDS

The cryptogonimids have a typical opisthorchiatan life cycle, their single-tailed cercariae penetrating and encysting in fishes, with piscivorous fishes as definitive hosts (see, for example, Cribb, 1986). The oral sucker of the metacercaria acquires a

Table 17.2 Some opisthorciatan digeneans parasitizing humans (summarized from Grove, 1990)

Parasite	Mollusc host	Fish host	Site in man	Geographical occurrence
Clonorchis sinensis	Various operculate snails	Freshwater cyprinids	Bile ducts	Eastern Asia (China, Japan, Vietnam)
Opisthorchis felineus	Bithynia leachii	Freshwater cyprinids	Bile ducts	Central and eastern Europe, Russia
Opisthorchis viverrini	Bithynia spp.	Freshwater fishes	Bile ducts	South-east Asia, Thailand
Heterophyes heterophyes	Pirenella conica (Egypt) Cerithidea cingulata (Japan)	Fresh, brackish and saltwater fishes Mugil (Egypt, Japan) Tilapia, Aphanus (Egypt) Acantogobius (Japan)	Intestine	Egypt, Japan
Metagonimus yokogawai	Melania libertina	Freshwater trout (Plectoglossis altivelis)	Intestine	Eastern Asia

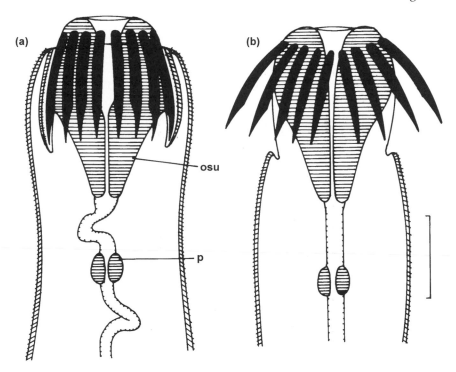

Figure 17.9 Diagrammatic representation of the anterior region of the cryptogonimid *Stemmatostoma pearsoni*, with spiny oral sucker (osu) (*a*) retracted and (*b*) protracted. p = pharynx. Scale bar: approximately 20 μm. Source: based on drawings and scanning electron micrographs by Cribb (1986).

ring of backwardly directed spines similar to, but more substantial than, those of echinostomatids (Fig. 17.9). These spines undoubtedly serve for attachment in the definitive host and can be deployed by protracting the oral sucker (Fig. 17.9).

17.7   THE STRIGEOIDS

The recruitment of fishes as second intermediate hosts has taken place more than once in the strigeoids (strigeids, cyathocephalids, diplostomatids). In these digeneans there is a broad range of parasite–fish interactions which not only parallel those of other groups like the bucephalids and heterophyids but go much further in terms of exploitation of the fish host.

17.7.1   'Black-spot' disease

A feature of many of the strigeoid parasites that encyst in the skin or subepidermal musculature of fishes is the accumulation around them of host chromatophores. Many of these are melanophores containing the black quinone polymer melanin, creating conspicuous 'black-spots' at the site of each parasite. Melanin is also deposited around parasites in arthropods (references in Lackie, 1986), but here appears to be associated with sealing off the intruder within a capsule of sclerotin, the highly inert protein–polyphenol complex occurring in arthropod cuticles and in platyhelminth eggshells. According to Götz (1986), melanin production is always associated with sclerotization and may result from a surplus of quinones. Thus, in arthropods, the association of melanin with parasites may be accidental. In fishes this seems less likely, but what significance it has for the parasite–host relationship is obscure.

In fishes that rely for camouflage on unblemished, reflecting, silver sides (see Denton, 1971), a medium to heavy dose of black-spot might render them more conspicuous to predators. However, this could be disadvantageous for the parasite, since the interference with camouflage might be more significant for aquatic predators such as fishes, which do not serve as hosts, than for aerial predators such as birds, in which the parasites are able to mature. An observation by Hoffman (1956) suggests that, in the diplostomatid *Crassiphiala bulboglossa*, the accumulation of melanophores around cysts may have no adaptive significance in terms of creating conspicuousness. He found that some cysts became surrounded by xanthophores containing yellow pigment, especially cysts located in the less heavily pigmented ventral and ventrolateral areas of the fish. However, there is evidence that these metacercariae interfere with schooling tendencies in their fish hosts and this may increase the danger of predation by kingfishers (Krause and Godin, 1994). A remarkably similar phenomenon has been reported in ligulid cestodes (section 10.5.4).

The conspicuousness of black spots for the human observer and the ease with which the size of the parasite infrapopulation can be determined in the living host have attracted the attention of evolutionary ecologists in their search for parasite–host systems with which to test their hypotheses. A suggestion of great importance is that it is the coevolutionary 'struggle' between parasites and their hosts that is the driving force maintaining sexual reproduction in living organisms ('Red Queen' hypothesis; see also section 14.2). Lively, Craddock and Vrijenhoek (1990) studied the frequency of black-spot (*Uvulifer* sp.) in Mexican pools inhabited by mixed populations of top minnows. In addition to sexually reproducing top minnows, *Poeciliopsis monacha*, two hybrids between *P. monacha* and *P. lucida* occurred; the hybrids could not reproduce sexually, but each female produced clones of herself. Comparing larger fish, parasites were more abundant in the clonal fishes than in the sexual top minnows and, in a pool containing the two hybrids and sexual individuals, the most common hybrid carried most of the parasites. Recolonization by a few fishes of a pool that had previously dried up led to a highly inbred sexual population; in this situation the inbred sexual fishes were more susceptible than the clonal fishes in the same pool. Later, genetically variable sexual females were added to this pool and within 2 years the clonal fishes carried more parasites than the now outbred sexual population.

These results support the assumption that this parasite–host interaction has a genetic basis. They are consistent with the predictions of the Red Queen hypothesis that the genotypic diversity generated by sexual reproduction provides partial relief from the attentions of parasites like *Uvulifer*, and that parasites evolve to disproportionately infect genetically uniform strains as they become common. However, observations on the same parasite–host system by Weeks (1996) do not support these assumptions.

A more specific application of parasites to evolutionary theory is the hypothesis proposed in 1982 by Hamilton and Zuk in relation to sexual selection in vertebrates (see also sections 5.5.1, 7.1.4(d)). They suggested that the development, usually in the male sex, of epigametic traits such as bright colours or long tails, is driven by parasites, in particular by the coevolutionary 'arms race' between parasites and their hosts. They proposed that female vertebrates obtain 'good genes' for their offspring by choosing mates that advertise their freedom from parasites and their resistance to them by displaying the brightest colours or longest tails. This can only operate if the parasites affect the health (viability) of their hosts and if resistance to the parasites is genetically determined.

One of the few tests of the hypothesis conducted on natural populations is that of Fitzgerald, Fournier and Morrissette (1993). They studied a population of three-spined sticklebacks (*Gasterosteus aculeatus*) infected with black-spot (the parasite was not specifically identified), but detected little or no influence on male sexual performance in their study area. There was no association between parasite numbers and the extent of the red nuptial colouring of the male and no relationship between condition and nuptial colour. There was a tendency for males with fully expressed nuptial colour to obtain more eggs. The number of black-spots was similar on males with and without eggs and, among males with eggs, there was no association between numbers of parasites per male and numbers of eggs in the nest.

There was no evidence that parasite load had any effect on parental care – heavily infected males invested as much in their broods as other males and successfully hatched their eggs. However, females with high parasite loads were in poorer condition and produced fewer eggs than less parasitized fishes.

Thus, although the extent of male nuptial colour may sometimes affect female choice in this system, nuptial colour is not influenced by the parasite and the black colour of the parasites in the male stickleback appears to have no effect on choice of a mate by the female. However, choice of female by male pipefish (*Syngnathus typhle*) was found to be influenced by the number of black spots produced by metacercariae of *Cryptocotyle* sp. (see Rosenqvist and Johansson, 1995). Males preferred females with few or no spots and, since high levels of this parasite reduce female fecundity, males may fertilize more eggs by choosing an unparasitized partner.

Uvulifer ambloplitis is a parasite of kingfishers and its relationship with young-of-the-year bluegill sunfish (*Lepomis macrochirus*) inhabiting a pond in North Carolina, USA has been the subject of an important study by Lemly and Esch (1984a, b). Within a few hours of entering the muscle tissue of a fish, the parasite secretes a cyst wall. Then a vigorous host response involving encapsulation and melanization is initiated and may go on for 3 weeks, producing cysts 0.5–1.5 mm in diameter. Lemly and Esch found that the metabolic cost of this to the host is substantial, with severe depletion of the fish's lipid reserves and loss of condition. When encapsulation by the host terminated, no further significant metabolic demands were made by the parasite and, in the laboratory, fishes maintained at temperatures of 20–25°C continued to feed and replenished their lipid reserves. However, in the field, decreasing temperatures in the autumn precluded further feeding and heavily infected fishes (more than 50 cysts per fish) entered winter with insufficient lipid for survival. Thus, 10–20% of the young-of-the-year bluegill population were eliminated during the winter, directly as a consequence of heavy parasitism. This mortality could not be explained by selective predation by kingfishers, which visit the pond most frequently in the spring, or by selective predation by largemouth bass and no evidence was found for any effect on bluegill behaviour that might have rendered them more vulnerable to predation. In older bluegill kept in the laboratory, metacercariae of *U. ambloplitis* may survive for several years.

Other studies have revealed no effect of strigeoid black-spot parasites on the condition of naturally infected fishes (e.g. Baker and Bulow, 1985). Hoffman

(1958a) found that metacercariae of *Posthodiplostomum minimum* lived for at least 16 months in naturally infected and laboratory maintained bluegill (*Lepomis macrochirus*). Hoffman (1958b) claimed that some fishes are capable of eliminating strigeoid cysts by rupture of the body surface, followed by healing after the expulsion of the parasite.

17.7.2   Brain and eye flukes

There is a strong tendency, already foreshadowed in the bucephalids (section 17.4), for strigeoid metacercariae (diplostomatids) to localize in the nervous system (brain, spinal cord, optic nerves, eyes) of their fish intermediate hosts (Table 17.1; Fig. 17.10). The advantages for parasites that colonize the brain or eyes may be twofold. First, the immune system of the host may be less effective in these sites because of isolation of the parasites beyond the blood–brain barrier; this will be especially significant in the eye lens, which is avascular. Secondly, it provides the opportunity for the parasite to interfere with the fish's motor and sensory functions, increasing the chances of the infected fish being caught by predators, including the parasite's

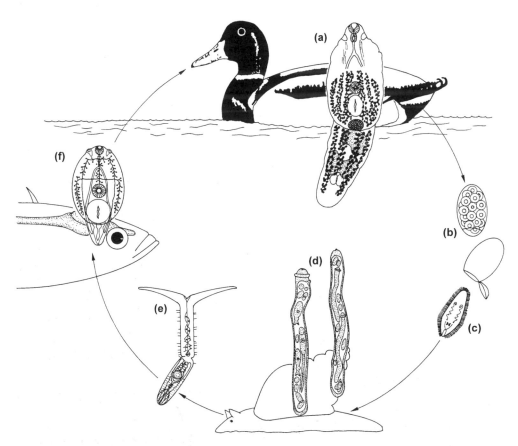

Figure 17.10   The life cycle of *Diplostomum baeri*. Adults (*a*) in intestine of ducks. Eggs (*b*) in faeces, develop and hatch in water. Miracidium (*c*) penetrates snail and gives rise to mother and daughter sporocysts (*d*). Fork-tailed cercaria (*e*) makes contact with stickleback, attaches and penetrates. Travels to optic lobes of brain (rarely eye), where it grows into metacercaria (*f*) without encysting. Bird eats infected stickleback. Source: based on Olsen, 1986.

definitive host, and even to manipulate these functions in favour of selective predation success by the definitive host. A predilection for the eyes might have an additional advantage if some predatory birds selectively remove and eat the eyes of fish, as claimed by Hendrickson (1979).

The preference for immunologically privileged sites will relieve the parasite of the need to create its own barrier (the cyst wall) to isolate it from the fish's defences. It may also reduce the metabolic commitment required by the host to mobilize its defences against the parasite, a demand which can be substantial (see above). In these sheltered sites the exposed parasite could feed and grow without the restrictions imposed by a host or parasite capsule, although these activities and the waste products they are likely to produce are potentially damaging for host nerve tissue (see below). Thus the lives of the host and its parasites could be jeopardized.

The impression is gained from some articles (e.g. review by Chappell, 1995) that significant pathology is a regular feature of infection with diplostomatid metacercariae, and Owen, Barber and Hart (1993) claimed that as few as four metacercariae in the eyes of sticklebacks had a detectable effect on responses to prey. However, it seems to me that the activities of these strigeoids create surprisingly little damage, bearing in mind the delicate nature of the organs occupied and their critical importance to the well-being of the host. It is surprising that lenses like the one shown in Fig. 17.11 can maintain their clarity and ability to focus light in spite of the internal activities of so many parasites, while most invasive experimental procedures directed at lenses lead to the rapid development of opacity. Crowden (1976) found that, although all the dace (*Leuciscus leuciscus*) from the River Thames, England, were infected, pathological conditions such as lens opacity were rarely encountered. Rees (1957) infected minnows (*Phoxinus phoxinus*) with cercariae of

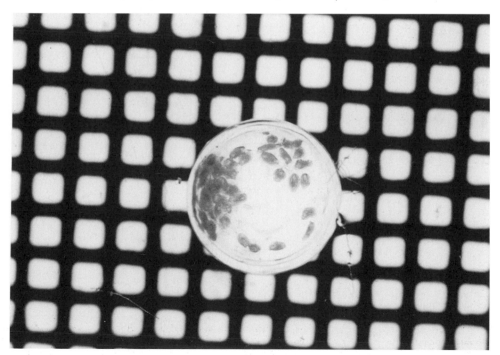

Figure 17.11 An eye lens from the roach, *Rutilus rutilus*, containing several metacercariae of *Diplostomum spathaceum*. Source: photograph by courtesy of Dr G. Duncan.

Diplostomum phoxini and described the pink colour of the brain and the presence of blood in the brain ventricles, which she attributed to minor haemorrhages produced by arriving cercariae. These features were not seen in long-standing infections, although the brains of naturally infected minnows were noticeably swollen (Rees, 1955). Fishes showing no signs of distress contained up to 600 parasites in the brain.

Hendrickson (1978) reported that metacercariae of *D. spathaceum* were still alive in the lens of a grayling (*Thymallus arcticus*) after maintenance in the laboratory for more than 3 years. Hoffman and Hundley (1957) found that metacercariae of *D. baeri eucaliae* were still alive in the brains of sticklebacks 1 year after experimental infection with over 100 parasites.

The metacercariae of *Diplostomum* spp. do not undergo the drastic metamorphosis of strigeids like *Cotylurus brevis* (section 16.3.3(b)), but, nevertheless, grow substantially and develop most of the non-reproductive features of the adult (Fig. 17.12). There is evidence that ingestion and digestion of host tissue provides resources to sustain much if not all of this growth (Rees, 1955; Erasmus, 1958).

A feature that strikes every observer seeing a metacercaria (= diplostomulum) of an eye or brain fluke for the first time is the large number of calcareous bodies scattered through the organism. These bodies are housed in spherical terminations of the branched reserve excretory system (Fig. 17.12c; section 13.1.1). This system appears early in the growing metacercaria (Erasmus, 1958), and the calcareous bodies develop and accumulate as the metacercaria ages (see Tinsley and Sweeting,

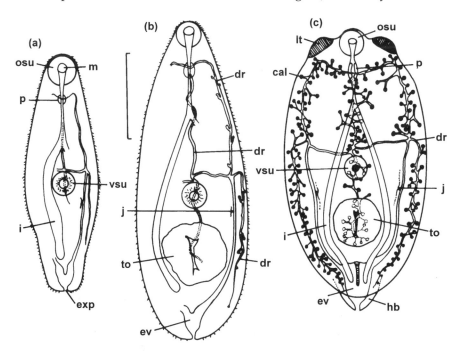

Figure 17.12 Successive stages in ventral view of the development of a diplostomatid in the lens of the fish second intermediate host. After (*a*) 15 days, (*b*) 22 days and (*c*) 544 days in the lens. The reserve excretory system is shown (on one side only in (*a*), (*b*)) but the flame cell system (shown in the cercaria in Fig. 17.13*a*) is omitted, apart from its junction (*j*) with the reserve excretory system. The left gut caecum is also omitted in (*a*), (*b*). cal = calcareous corpuscle; dr = duct of reserve excretory system; ev = excretory vesicle; exp = excretory pore; hb = hind body; i = intestine; lt = lappet; m = mouth; osu = oral sucker; p = pharynx; to = tribocytic organ; vsu = ventral sucker. Scale bar: 100 μm. Source: redrawn from Erasmus, 1958.

1974). Waste materials released by parasites may be toxic to the host in any site and immunologically stimulating in many, but the danger of damage to host tissue is likely to be much greater in particularly sensitive sites such as the brain and the lens. These problems may have been solved in diplostomula by converting some excretory products to insoluble matter and storing this in the reserve excretory system until the opportunity to dispose of it arises in the definitive host. It is interesting in this context that Hoffman and Hundley (1957) found some evidence to suggest that nitrogenous material (urates) may be present in these bodies.

Thus, strigeoids inhabiting the brain and eyes do not exert the lethal effects that might be expected but there is evidence that some of them have a more subtle influence on host behaviour. Brassard, Rau and Curtis (1982) found that guppies (*Lebistes reticulatus = Poecilia reticulata*) exposed experimentally to as few as 25 cercariae of *D. spathaceum* and harbouring as a result of this infection approximately one metacercaria per fish were taken significantly more often than uninfected fishes by the predatory brook trout (*Salvelinus fontinalis*). The behavioural change that was assumed to lead to greater predation was a reduction of activity of the infected fish, but how much of this change was attributable to the migrating cercariae and how much to activities in the lens is not clear. An important point made by Brassard *et al.* is that, in this particular parasite–host system, infected hosts become vulnerable to predators (fishes) that do not serve as definitive hosts. However, the authors suggest that the behavioural change may also channel *D. spathaceum* towards its bird definitive host and that the loss of parasites to fish predators may be an inevitable cost to the parasite to ensure transmission.

Evidence gleaned from a natural population of dace, each infected with from 20 to more than 160 metacercariae of *D. spathaceum*, suggested that, in this system, induced behaviour changes increase vulnerability of the infected fishes to avian predators (Crowden, 1976; Crowden and Broom, 1980). In field data there was no correlation between parasite burden and fish condition, but laboratory studies on fishes from the wild indicated that feeding efficiency of the dace declined as parasite numbers increased. The distance at which the fish detected prey (*Gammarus*) and the proportion of successful attempts at food capture fell, while the time taken to capture a given number of prey increased. The more heavily infected fishes compensated for this by spending more time feeding, so that, over comparable time periods, the number of prey they consumed did not differ from those consumed by less heavily infected fishes; starvation was unlikely and their condition was maintained. However, more heavily infected fishes spent more time close to the surface, either motionless or moving slowly. The definitive host in the study area is thought to be the black-headed gull (*Larus ridibundus*), which, when feeding, submerges only briefly by plunge-diving and is unlikely to take fishes deeper than 12 cm from the surface. Heavily infected dace just below the surface were therefore more likely to be taken by gulls than fishes at deeper levels. This vulnerability was underlined by the ease with which heavily infected fishes were taken in hand nets. Thus, the parasite seems to alter the behaviour of the fish to the parasite's advantage without causing a high level of wasteful fish mortalities.

Crowden and Broom (1980) stressed that the behavioural changes in infected dace manifested themselves only in heavily infected individuals (cf. the effect of *Dicrocoelium* on ants; section 16.3.4(f)) and that this is important in a population where most individuals are infected. The parasite population is usually aggregated (overdispersed) in the host population, with a small number of hosts supporting a disproportionately large number of parasites. Thus potentially harmful effects are confined to relatively few hosts, minimizing the harm to the host population. The activities of gulls and/or the effects of starvation may account for the noticeable loss of heavily infected individuals during the winter.

The occupation of immunologically privileged sites may relieve diplostomatids of the need for a cyst wall, but the retention of a thick capsule around the metacercariae of the strigeid *Apatemon gracilis*, which in sticklebacks are usually sited in the humours of the eye (Blair, 1976), reminds us that the structure has another function. Like its close relative *A. minor*, the cyst wall of *A. gracilis* protects the metacercaria during its passage to the intestine of the bird definitive host and, more importantly, when appropriately stimulated serves to eject the metacercaria in the intestine by expanding rapidly inwards (section 16.3.3(b)). How *Diplostomum* spp. survive the traumatic journey to the intestine of the final host is unknown, but both the brain and the eye have tough outer connective tissue which may offer protection from gizzards and stomach acid for long enough to enable the naked parasites inside to reach the relative safety of the intestine.

Infectivity of the fork-tailed cercariae of *D. spathaceum* declines to zero in just over 20 h at 14°C (Whyte, Secombes and Chappell, 1991). A conspicuous feature of the tail stem is the presence of many pairs of glycogen storage cells (caudal bodies; Fig. 17.13a) and most of this glycogen is used by the cercaria during its free-swimming life, presumably for propulsion (Erasmus, 1958). The swimming pattern of *D. spathaceum* is similar to that of the heterophyids and opisthorchiids (see above) and they respond in much the same way to mechanical stimuli (see Haas, 1992, 1994). However, unlike *Opisthorchis viverrini*, attachment to the substrate is said to be stimulated in *D. spathaceum* by a relatively non-specific factor, host-derived CO_2. This is extremely sensitive, maximal attachment occurring at concentration differences between medium and substrate of as little as 0.1%, and is also very rapid, attachment being completed in 250 ms. The speed of attachment impressed Ferguson (1943), who remarked that large numbers of cercariae of *D. flexicaudum* attached to the tail of a minnow immersed for just 1 s in a suspension of cercariae.

The CO_2 signal has no influence on whether or not the cercaria remains attached, and a combination of other chemical factors is implicated in prolonging its stay (Haas, 1994). Penetration behaviour is released by yet another set of chemical substances, including fatty acids and sialic-acid-containing glycoproteins. Thus, although the cercariae will attach to most animal surfaces because of the ubiquity of the CO_2 signal, the progressively more specific signals required to prolong attachment and induce penetration will restrict the range of organisms into which they penetrate. Nevertheless, *D. spathaceum* was reported by Sweeting (1974) from 23 species of freshwater fish. Perhaps such broad specificity is another spin-off from the occupation of the lens, permitting growth and development in a site sheltered from the host's immune system.

Initial contact of *D. spathaceum* with its host is with the tail furcae and this is then followed by attachment with the anterior organ (Fig. 17.13). Consequently, attachment-releasing CO_2 receptors are likely to be located on the furcae, with chemoreceptors mediating the prolongation of attachment and penetration most probably being located on the body (Haas, 1992). A cysteine proteinase, which probably contributes significantly to penetration, has been identified by Moczon (1994) in the penetration glands of *D. pseudospathaceum*.

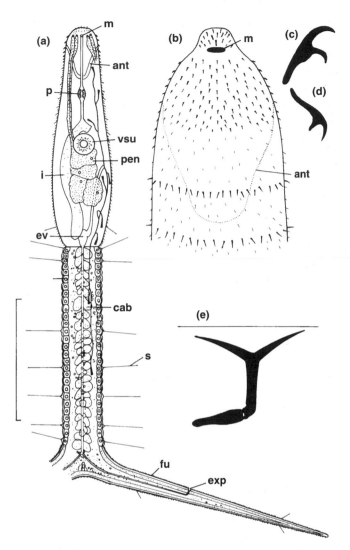

Figure 17.13 A diplostomatid cercaria. (*a*) Ventral view of cercaria (tail furca, excretory system and gut caecum omitted on one side). (*b*) Enlarged anterior end. (*c*) Tegumental hook from ventral region anterior to mouth. (*d*) Hook from ventral region posterior to mouth. (*e*) Attitude of cercaria floating near the surface. ant = anterior organ; cab = caudal body; ev = excretory vesicle; exp = excretory pore; fu = tail furca; i = intestine; m = mouth; p = pharynx; pen = penetration gland cell; s = sensillum; tas = tail stem; vsu = ventral sucker. Scale bar: 100 μm. Source: redrawn from Erasmus, 1958.

The sites occupied by the metacercariae of many strigeoids offer relatively small targets for invading parasites and there is considerable interest in how the cercariae reach sites as small as an eye lens. Although Ferguson (1943) showed experimentally that cercariae of *D. flexicaudum* applied to the eyes in special eye cups were able to pass through the cornea and enter the lens, this is not the usual route. Cercariae may enter the body at any point and the evidence suggests that even those entering the tail may reach the lens, at least in small fishes. Ferguson exposed the tails of individual minnows (*Pimephales promelas*) to a single cercaria of *D. flexicaudum* and in 14 out of 18 of these fishes the cercariae successfully reached one of the eye lenses. In fishes 5–6.5 cm in length, parasites that had invaded the tail reached the lenses in

1.75–2 h (temperature not given); the speed of this movement (about 0.5 mm/min) so impressed Ferguson that he thought it likely that they travelled forward in the blood system rather than by tunnelling through the tissues. In contrast Erasmus (1959) and Ratanarat-Brockelman (1974) discounted the blood system as a migration route for *D. spathaceum* and concluded that parasites travel forwards through the connective tissue and muscle. This conclusion has been questioned by Betterton (1974), who pointed out that the 3% of larvae found in the bloodstream by Erasmus might be sufficient to account for the number recovered from the lens, bearing in mind that cercarial doses of 9000 cercariae per fish were used. Whyte, Secombes and Chappell (1991) repeated Betterton's argument and offered some evidence to suggest that the lymphatic system might also provide a way forward in rainbow trout (*Oncorhynchus mykiss*).

Thus, the route (or routes) taken by migrating diplostomatid metacercariae is still open to question, but it would be remarkable if, as indicated by an observation of Ratanarat-Brockelman, the penetration glands of these small organisms (Fig. 17.13*a*) contained sufficient secretion to sustain an 18 h migration by tunnelling through relatively dense tissues and retain at the end of it enough to enter the lens. Whatever the route taken, the tiny, backwardly-directed, tegumental spines (Fig. 17.13*b–d*) undoubtedly provide purchase in tissue tunnels or in blood or lymph vessels and permit forward progress but not backward movement. The spines on the anterior region of the cercaria illustrated in Fig. 17.13 are more complex than those on the rest of the body, resembling in shape the marginal hooklets of monogeneans. Those in front of the mouth have their blades projecting forwards (Fig. 17.13*c*), while the more numerous spines immediately behind the mouth have their hooks projecting backwards (Fig. 17.13*d*). It is easy to imagine how effective these hooklets would be, in conjunction with changes in shape of the anterior region of the organism, for making forward progress through host tissues.

According to Hendrickson (1979), cercariae of *Ornithodiplostomum ptychocheilus* destined for the brain travel initially through muscle and connective tissue after penetrating the skin. They gain access to the brain cavity either by following the cranial nerves or by following spinal nerves to the spinal cord and then travelling forward along the cord or through the neural canal. Thus, access to the skull and spinal column is *via* the foramina through which the nerves emerge. Cercariae penetrating the tail may reach the brain faster than cercariae entering more anterior regions of the body, because the former have a shorter distance to travel to reach the neural canal, in which they make good forward progress. The tendency for more cercariae of *O. ptychocheilus* to penetrate the head region of the fish was interpreted by Hendrickson not as a preference but as a possible reflection of the greater surface area of this region and the provision of more surface irregularities that may favour penetration. This tendency to penetrate the head is even more pronounced in *D. spathaceum* and may reflect the fact that many cercariae enter *via* the gills (Whyte, Secombes and Chappell, 1991).

There seems little doubt that directed, non-random movements are involved in site location within the fish host in these strigeoids, but we know nothing about the cues that guide migrating parasites. Ferguson (1943) found that metacercariae were still able to find and penetrate eyes from which the lenses had been removed, although more parasites reached eyes with intact lenses. However, he found no parasites in or near lenses implanted in the body muscles or coelom, even in fishes from which eyes or their lenses had been extirpated.

Almost immediately after ingestion of *D. phoxini* by the bird definitive host, mitotic division begins in the relatively undeveloped hindbody (Bell and Hopkins, 1956). This rapidly increases in size as the reproductive organs within it develop, and there are eggs present in the uterus after a mere 3–4 days. Bell and Hopkins pointed out that this pattern of development is different from that found in most tapeworms and many digeneans, where a period of somatic growth precedes sexual maturity in the definitive host. It also differs from the situation in the tapeworm *Schistocephalus solidus*, in which somatic and sexual growth is completed in the fish intermediate host and egg assembly is initiated by the higher body temperature of the bird definitive host (section 10.5.4). A temperature increase is not sufficient to switch on sexual development in *Diplostomum*.

The forebody in diplostomatids is spoon-shaped and the tribocytic organ (section 16.3.3(b)), in its retracted state, communicates with the ventral surface of the forebody by means of a slit-shaped opening behind the ventral sucker (Fig. 17.12). After eversion, the tribocytic organ is spherical. The central region of the exposed surface is spongy and microvillous and the sides are spiny and, as in *Apatemon* (section 16.3.3(b)), there are cisternae of the reserve excretory system closely associated with it. According to Erasmus (1970), the spongy surface may be held in contact with host tissue by the spiny sides, but the functional interrelationships of the tribocytic organ, the lappets and the oral and ventral suckers are no clearer than in other strigeoids.

17.7.3 The extended life cycle – *Alaria*

Many of the predator–prey links indicated in life cycle diagrams published in this and any other textbook of parasitology are selected segments of more complex food webs. The ramifications of these webs may be particularly extensive where predator–prey interrelationships span the interface between aquatic and terrestrial environments. For example, vertebrate predators of tadpoles may include fishes, snakes and small mammals and these in turn may be eaten by larger carnivorous mammals and birds. After metamorphosis, adult frogs are likely to be eaten by a variety of reptiles, mammals and birds, and there are frog-eating frogs and snake-eating snakes. Thus the food web may ascend through several trophic levels and may extend laterally in both aquatic and terrestrial environments. Moreover, the vertebrate participants display a great diversity in anatomy, physiology and behaviour, and few parasites have the adaptability to grow, develop and mature in such a challenging spectrum of hosts. Consequently, these ramifications of the food web must lead to significant mortality in parasites such as digeneans that find themselves in alien hosts.

There are indications that the greater the dependency of the parasite on the host for metabolites the greater the degree of host specificity displayed by the parasite. Conversely, parasites that make few or no demands on the host display a relatively low degree of host specificity (see, for example, the shift in metabolic dependency and specificity in the pseudophyllidean cestode *Schistocephalus solidus* from the definitive to the second intermediate host, section 10.5.4). It is this low specificity option that has been exploited by some strigeids and diplostomatids as a means of surviving in vertebrate hosts in which they are unable to develop to maturity (Fig. 17.14). Parasites finding themselves in the alimentary canals of these inappropriate hosts use their penetration glands to enter the tissues, where they make few demands and are able to survive in a state of suspended development as a so-called mesocercaria (Fig. 17.14*f, g*). Thus, the mesocercaria is able to thread its way through the ramifications of the food web, its development remaining in suspension in inappropriate hosts (so-called paratenic hosts), until, finding itself in a suitable definitive host, development resumes and sexual maturity is reached.

The extent of these ramifications is well illustrated by the work of Johnson (1968) and Shoop and Corkum (1981) on the diplostomatid *Alaria marcianae*, found as an adult in some carnivorous mammals (see below). Fork-tailed cercariae from snails of the genus *Helisoma* penetrate tadpoles but development proceeds to the mesocercarial stage and no further. The mesocercariae survive host metamorphosis and persist in adult frogs. Bullfrogs (*Rana catesbeiana*) have elevated numbers of mesocercariae and may acquire some of these by eating green frogs (*R. clamitans*). Some 98% of aquatic snakes that feed on amphibians contain mesocercariae and a proportion of the high levels of mesocercariae in some snakes (e.g. *Agkistrodon piscivorus*) may be derived from other snakes in their diet. The raccoon (*Procyon lotor*) and the opossum (*Didelphis virginiana*) are naturally infected, and *Alligator mississipiensis*, mice, rats and chicks have been infected experimentally with mesocercariae. The following carnivorous mammals have yielded natural infections of adult *A. marcianae*: cat, striped and spotted skunk, the red fox (*Vulpes fulca*), the grey fox (*Urocyon cinereoargentatus*). There is no evidence that fishes feeding on tadpoles can act as paratenic hosts for *A. marcianae*.

In addition to the frequency of occurrence of *A. marcianae* in the amphibians, reptiles and mammals in the swampy study area of Shoop and Corkum in Louisiana, USA, there is a report of mesocercariae in a local human resident. Shoop and Corkum emphasized the risk of human infection from poorly cooked game animals. They viewed with some concern a new fashion for eating briefly cooked alligator meat and reported that one restaurant regularly purchased local bullfrogs for sale in the establishment.

When mesocercariae of *Alaria* spp. find themselves in the stomach of a definitive host, they undergo a tissue migration as they do in the paratenic host (Shoop and Corkum, 1983a; Fig. 17.14). This takes many mesocercariae to the lungs, where they resume their development and become metacercariae (diplostomula; Fig. 17.14*h*). The diplostomula spend a minimum of 4 days in the lungs and then travel to the intestine, probably *via* the trachea and oesophagus. In the intestine, the folded forebody embraces a villus (Tieszen, Johnson and Dickinson, 1974) and, after development of the reproductive system in the hindbody (Fig. 17.14*a*), egg output begins. Thus, in *Alaria* spp. the life cycle involves a minimum of three hosts (snail, tadpole–frog, carnivorous mammal), with the mesocercaria having the potential to persist after predation as a mesocercaria or to develop into an adult *via* the metacercaria, depending on the host.

The invasiveness of mesocercariae of *Alaria* spp. and their relatives has led to a remarkable new development. Mesocercariae have an affinity for the fat bodies of their hosts, including the mammary glands of mammals. Shoop and Corkum (1983b) showed that there is a change in distribution of mesocercariae of *A. marcianae* in experimentally infected female mice after the birth of their litters. There is a significant increase in the numbers of mesocercariae in the mammary glands. The consequence of this shift, which may be driven by host hormone changes, is that suckling mice ingest mesocercariae from the mother's milk, the parasites going on to establish themselves as mesocercariae in the tissues of the neonates (so-called vertical transmission).

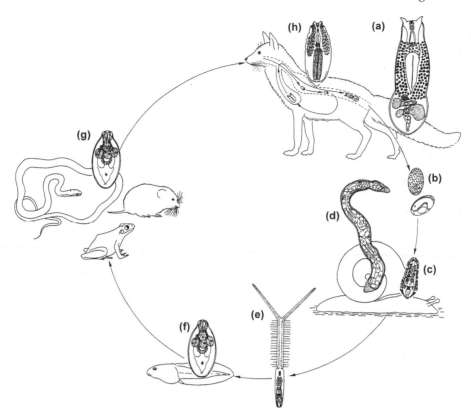

Figure 17.14 The life cycle of *Alaria*. Adult (*a*) in small intestine of fox. Eggs (*b*) in faeces, develop and hatch in water. Miracidium (*c*) penetrates planorbid snails. Sporocyst (*d*) produces fork-tailed cercariae (*e*), which penetrate tadpoles and become mesocercariae (*f*). Mesocercariae (*g*) persist in tissues of tadpole-eating paratenic hosts (frogs, snakes, mammals). When these are eaten by fox, mesocercaria penetrates gut wall and migrates to lung, where it becomes metacercaria (diplostomulum) (*h*). *Diplostomulum* migrates up trachea, is swallowed and achieves maturity in small intestine. Source: based on La Rue and Fallis, 1936, Pearson, 1956, Olsen, 1986.

In other lactating mammals further developments may occur. In a definitive host such as a cat, some mesocercariae mature in the female while others are transmitted to the offspring, where they also reach adulthood. The term 'amphiparatenic host' was coined by Shoop and Corkum (1983b) for those host species that serve as a paratenic host as an adult (the mother) and a definitive host as a juvenile (her offspring).

In cats, vertical transmission of the parasite stops at the F1 generation because the parasites mature in the kittens. Remarkably, this is not so in mice. Female mice of the F1 generation, which had become infected with mesocercariae by nursing from their mothers, were able, in their turn, to pass on mesocercariae to their own offspring (F2) during lactation (see Shoop, 1994).

Vertical transmission *via* the milk occurs also in the related diplostomatid genus *Pharyngostomoides* (see Shoop, 1994), and the phenomenon has been encountered previously in the cestode *Mesocestoides corti* (section 11.7).

There is no evidence for *in utero* transmission of mesocercariae of *Alaria*, but transmission to the offspring before birth has been reported in the viviparous fish second intermediate host of the strigeid *Apatemon graciliformis*. According to Combes and Nassi (1977) cercariae preferentially penetrate female guppies (*Poecilia reticulata*) and find their way to the ovaries. If embryos are present they enter the vitelline vesicles and, provided that no more than two parasites enter each embryo, growth of the young fish is not disturbed and the metacercariae encyst before the young fish are born. Young guppies with one or two cysts survive, but their swimming is often impaired and they spend much time on the bottom. Guppies with more than two cysts usually die within 24 h. The natural definitive host is unknown but Combes and Nassi raised adults by feeding infected guppies to ducks and suggested that infected newborn guppies lying on the bottom may stand a greater chance of being ingested by bottom-feeding birds. The metacercariae persist in the body cavity of *Poecilia* that reach adulthood without being eaten.

Male fish only become infected when there are no females present or when there is a high density of cercariae, but parasites entering male fish degenerate. How the cercariae distinguish between male and female guppies is unknown.

17.8   BLOOD FLUKES

17.8.1   The sanguinicolids

The high degree of immunocompetence of vertebrate second intermediate hosts (fishes, tadpoles) demands the assembly of a cyst wall to protect tissue-dwelling metacercariae from the host's humoral and cellular defences. The combination of host tissue reaction and parasite cyst wall may restrict development of the parasite by preventing access to resources from the host and by limiting the space available for parasite growth. Nevertheless, the versatile digeneans have found ways to satisfy the parasite's conflicting demands for protection and growth. Some, like *Diplostomum spathaceum*, have colonized sites where the immune system is unable to exercise full control; they remain unencysted for the whole of their lives and are able to feed on host tissue and prepare themselves for life in the definitive host by undergoing extensive non-reproductive development (see above). However, in the heterophyids and possibly in some bucephalids, we find hints of a different strategy, one that has had far-reaching consequences. There are indications that some encysted and unencysted parasites (e.g. *Stictodora lari* and *Aphalloides coelomicola* respectively), which inhabit sites that are not known to be immunologically privileged, are not as provocative to the host's immune system as might be expected. The development of this ability to circumvent the immune response of the host appears to be fully expressed in strigeoid mesocercariae, which survive seemingly unmolested in the tissues of a range of vertebrates.

A similar development in an ancestral strigeoid permitted a mesocercaria or a metacercaria to establish itself in the nutrient-rich blood system of a fish second intermediate host. With nothing to prevent the parasite from completing its development and reaching sexual maturity in this environment, the stage was set for loss of the definitive host and the emergence of the sanguinicolids. However, there is no natural escape route for eggs laid by parasites inhabiting the vertebrate circulatory system and the evolution of the sanguinicolids, and the subsequent development of the spirorchids and schistosomes, owes much to the unique features of fish gills. Their capillary beds provide a trap for blood-borne eggs and there are few places in the body where the blood is separated from the outside world by a thinner

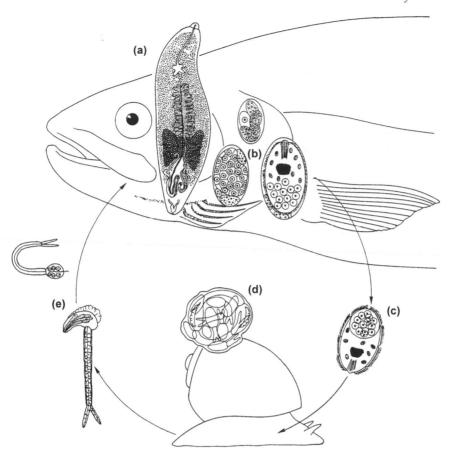

Figure 17.15 The life cycle of *Sanguinicola idahoensis*. Adults (*a*) live in blood vessels and connective tissue of head of trout, *Salmo gairdneri*. Eggs (*b*) accumulate in gill capillaries, expand as they develop. Eggs hatch and miracidium (*c*) leaves gills, penetrates snail, *Lithoglyphus virens*, and produces sporocysts (*d*). Free-swimming cercariae (*e*) (also shown above in inactive floating pose) penetrate skin of trout. Source: stages redrawn from Schell, 1974.

barrier than in the secondary gill lamellae. Most sanguinicolid adults inhabit the ventral arterial system, i.e. the heart, the conus arteriosus, the ventral aorta and the afferent branchial arteries. The eggs are transported in the arterial blood to the gill capillary bed, where they are trapped. Here they develop and hatch *in situ* and the free miracidia readily escape from the gill lamellae into the host's gill-ventilating current (Fig. 17.15).

The eggs have extremely thin, apparently untanned shells, lacking an operculum (Smith, 1972), and Thulin (1982) reported that each egg capsule of *Chimaerohemecus trondheimensis* from the holocephalan *Chimaera monstrosa* is unique in containing nine to 20 miracidia. A single miracidium develops in the egg of *Sanguinicola idahoensis* (see Schell, 1974). However, the dimensions of the egg nearly double as the embryo grows and the volume of the egg nearly quadruples (Fig. 17.15), indicating that this parasite has another innovation – the absorption by the developing embryo of nutrients from the blood of the host. As Tinsley (1983) has pointed out with reference to schistosomes, the blood flukes have extended exploitation of host resources to embryonic development and the egg itself has become parasitic. This

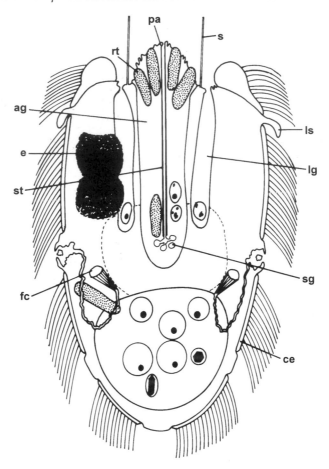

Figure 17.16 The miracidium of *Sanguinicola inermis*. ag = apical gland; ce = ciliated epidermal cell; e = eye; fc = flame cell; lg = lateral gland; ls = lateral sensory papilla; pa = apical papilla; rt = rodlet; s = sensillum; sg = secretory granules; st = stylet. Source: reproduced from McMichael-Phillips, Lewis and Thorndyke, 1992b, with permission from Cambridge University Press.

strategy has the advantages of ovoviviparity but without the disadvantage of reduced rate and discontinuity of egg assembly required to permit embryonic development *in utero*. In fact, because the vitelline cells probably contribute little to embryo nutrition, the metabolic demands of egg assembly on the adult parasite may be less than in other oviparous helminths and may contribute to relatively high rates of egg assembly (section 18.3.1).

Considerable stretching of the eggshell occurs in sanguinicolids and elastin was reported in the shell of *Orchispirium heterovitellatum* by Madhavi and Rao (1971). However, McMichael-Phillips, Lewis and Thorndyke (1992a) found no evidence for the presence of elastin in the eggshell of *Sanguinicola inermis*.

Although miracidia are adept at penetrating the bodies of their molluscan first intermediate hosts, the employment of their penetrative skills to escape from the fish host is a new and unique feature of sanguinicolids. The complexity of the apical gland of miracidia, such as that of *S. inermis* (Fig. 17.16), may be a reflection of this new role. The gland contains rhabdite-like bodies, two kinds of secretion and a tubular stylet-like structure made of microtubules.

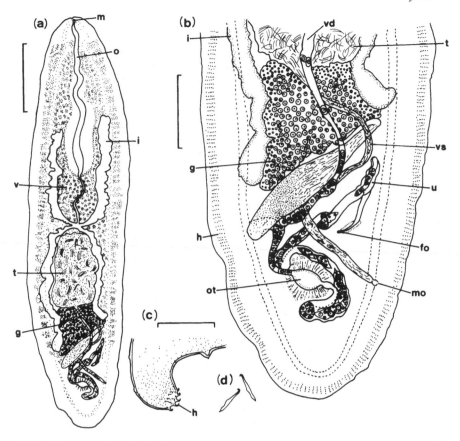

Figure 17.17   The sanguinicolid *Cardicola laruei* from the white trout *Cynoscion arenarius*. (*a*) Whole animal. (*b*) Enlarged view of the posterior region. (*c*) Transverse section of the body margin to show hook deployment. (*d*) Hooks. fo = female genital opening; g = germarium; h = hooks; i = intestine; m = mouth; mo = male genital opening; o = oesophagus; ot = ootype; t = testis; u = uterus; v = vitellarium; vd = vitelline duct; vs = vas deferens. Scale bars: (*a*) 250 µm; (*b*) 100 µm; (*c*) 25 µm. Source: reproduced from Short, 1953, with permission from the American Society of Parasitologists.

Adult sanguinicolids have flattened, elongated bodies ranging from about 1 mm to more than 1 cm in length. They lack oral and ventral suckers (Figs 17.15, 17.17) and most of them rely on lateral tegumental spines or hooks (Fig. 17.17) for attachment to the walls of blood vessels. Undulating swimming movements have been described (e.g. in *S. inermis* by Kirk and Lewis, 1993), and their flattened bodies may increase the effectiveness of this locomotion, enabling the hermaphrodite individuals to make contact for mating. The enlarged surface area created by flattening the body may also reflect the development of transtegumental absorption of nutrients, and another indication of this is the reduction of the intestine to a small lobed sac in *Sanguinicola* spp. (Fig. 17.15*a*).

The fork-tailed cercariae of sanguinicolids are not unlike the fish-penetrating cercariae of their strigeoid relatives, both in anatomy and behaviour. Periods of active swimming and passive sinking alternate. In *Sanguinicola* spp. there is a body fin, which acts like a parachute in the passive phase and delays sinking (Fig. 17.15*e*).

In Chapter 3 reference was made to the scarcity of digeneans in elasmobranch fishes and to the suggestions made by Roberts and Janovy (1996) that high levels of blood urea may have deterred colonization of elasmobranchs by digeneans. Sanguinicolids have successfully established themselves in elasmobranchs and in holocephalans and, if Roberts and Janovy are right, must have developed some way of tolerating urea to make this possible. Sanguinicolids also demonstrate their versatility at the level of the first intermediate host, since some of their marine representatives have switched hosts from molluscs to polychaete annelids (section 3.3).

17.8.2   The spirorchids and turtles

The turtles (testudomorphans) are an ancient group with roots extending back into the Permian period (see Platt, 1992). Their hallmark is the external shell and, according to Pough, Heiser and McFarland (1990), it is this feature that has limited their diversity. These reptiles have changed little since the Triassic period and have shared warm, freshwater environments and tropical, oceanic habitats with the fishes for a very long time.

The turtles have their own group of blood flukes, the spirorchids (Fig. 17.18). These parasites have been reported from seven of the 12 extant families of turtles, including the primitive Pleurodira (side-necked turtles). Platt (1992) compared the

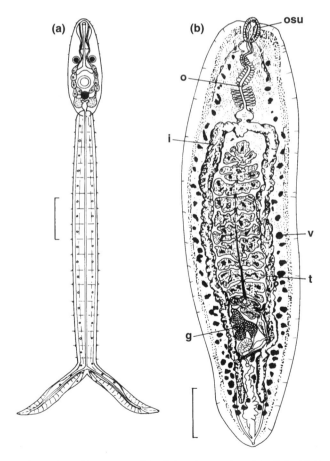

Figure 17.18 *Spirorchis scripta*: (*a*) cercaria; (*b*) adult. osu = oral sucker. Other lettering as in Fig. 17.17. Scale bars: (*a*) 100 μm; (*b*) 200 μm. Source: redrawn from Holliman and Fisher, 1968.

geographical and host distribution of spirorchids and came to the conclusion that the turtles and the spirorchids have had a long coevolutionary history, dating back to the origins of the host group. Whether this parasite–host coevolutionary relationship threads its way back to the fish-like antecedents of the turtles with their sanguinicolid-like parasites, or whether a host-switching event took place between a fish and a cohabiting early turtle remains unknown.

Although the tougher skins of these amphibious reptiles may have deterred cercarial penetration over a large area of the body surface, mucous membranes lining external orifices would still be vulnerable. In the modern *Spirorchis scripta*, Holliman and Fisher (1968) observed the fork-tailed cercariae (Fig. 17.18*a*) attacking the mouth and nares of the painted turtle *Chrysemys picta picta*, and the eyes and eyelids may also be targets, judging by the agitated scraping movements made by the forelimbs of exposed individuals. Platt has suggested that the Testudinidae (tortoises) may have lost their spirorchids when they became terrestrial.

Knowledge of the biology of the spirorchids is limited to a few freshwater species. They are hermaphroditic (Fig. 17.18*b*) and some retain both oral and ventral suckers. Others, like *S. scripta*, have no ventral sucker. *S. parvus*, also from *C. picta picta*, has a cylindrical body and travels through blood vessels by propagating peristaltic waves like an earthworm (Holliman, Fisher and Parker, 1971). *In vitro* the parasite is able to swim by dorso-ventral undulations of the lateral body margins.

Specimens of *Spirorchis scripta* are on average about 2 mm long and are found in arteries and veins of a wide range of body organs, as well as in the heart and in nonvascular tissues (Holliman and Fisher, 1968). Unlike sanguinicolids, they lay operculate eggs with shells that appear to be tanned and, as might be expected from the distribution of adult worms, these eggs lodge in many sites, including the heart, alimentary canal, lungs, brain and liver. Eggs appear in the faeces, but how they reach the gut lumen is unknown. Hatching takes place outside the host and, according to Holliman (1971), the transmission of the parasite is favoured by the preference of both hosts (the snail *Helisoma anceps* and the turtle) for warm, shallow, pond margins.

17.8.3 The origin of the schistosomes

The mammals and birds did not escape the attention of blood flukes when these vertebrates opted for the terrestrial way of life. Many of them retained limited contact with water and they have been exploited by the schistosomes, a group with many new and surprising features. The general consensus is that schistosomes and spirorchids share a common ancestry (see Platt, 1992), but the question remains whether the mammals and birds inherited their blood flukes by the phylogenetic route from their reptilian ancestors or whether they acquired them more recently by host-switching from aquatic reptiles like turtles.

The discovery of a schistosome, *Griphobilharzia amoena*, in a freshwater crocodile (Platt *et al.*, 1991), the only schistosome recorded so far from a cold-blooded vertebrate, does nothing to resolve this question. It could be interpreted as evidence that schistosomes were already well established in reptiles before homeothermic mammals and birds came on the scene, but the isolation of this reptilian schistosome is also consistent with a host-switching event, perhaps from a bird, as suggested by Combes (1991). In fact, there is a high probability that host-switching between definitive hosts has played an important part in the shaping of modern schistosomes (section 18.2), and, if this is the case, we may have underestimated the importance of this phenomenon in the early evolution of blood flukes. In a major study of the phylogenetic relationships of schistosomes, Carmichael (in Basch, 1991) came to the conclusion that birds may be primitive hosts of schistosomes and that there may have been at least two independent transfers to mammals.

A critical factor in the achievement of a successful host-switching event by a blood fluke is its ability to deal effectively with the opposition mobilized against it by the new host's immune system. Elucidating the mechanism of immune evasion by schistosomes has occupied many parasitologists for several decades and our understanding is still far from complete, but an important feature of it is its flexibility, perhaps inherited from a mesocercarial ancestor. Some schistosomes appear to have an innate ability to survive in a range of hosts that are not necessarily related, and others with restricted host specificity appear to have the evolutionary capacity to extend it rapidly when the demand arises (see Chapter 18).

A feature that may have provided more effective opposition to host-switching in schistosomes than the immune system of a potential host may have been the skin of the vertebrate. In response to insolation and dehydration in the terrestrial environment, skin coverings of mammals and birds became much more difficult for soft-bodied cercariae to penetrate, and the more casual and temporary contacts of these vertebrates with water meant that exposure of vulnerable mucous membranes was reduced. That this was not beyond the evolutionary versatility of the antecedents of modern schistosomes is well known, but colonization of the mammals and birds may have had to await enhancement of penetrative ability. Perhaps the aquatic reptiles (turtles, crocodilians) provided the evolutionary platform on which these improvements in skin penetration took place. The schistosomes will be considered in the next chapter.

CHAPTER 18

The schistosomes

18.1 HISTORY AND THE LIFE CYCLE

In 1851, while conducting an autopsy on a young man in Egypt, Theodor Bilharz made an important discovery (see Grove, 1990 for detailed account). He found worms which he first thought were nematodes but later recognized as trematodes (defined in section 3.4). Much to his surprise, these worms were inside tributaries of the hepatic portal vein (Fig. 18.1).

In a letter to von Siebold dated August 1851 (see Bilharz and von Siebold, 1852–1853) Bilharz noted that these worms had an unusual feature – they were not hermaphrodite but were 'something more wonderful, a trematode with divided sex'. His letter went on as follows:

When I investigated the veins of the intestine more carefully than before (and also more effectively by holding the undamaged mesentery in the light), I soon found specimens of the worm that housed a grey thread in the groove of their tail. You can imagine my astonishment when I saw a trematode protruding from the front opening of this groove and moving to and fro, its shape similar to the first, but with everything much finer and more delicate, and, instead of the groove-shaped tail, a string-like abdomen enclosed in the groove-shaped half canal of the male,

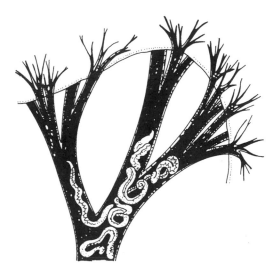

Figure 18.1 *Schistosoma mansoni* in a mesenteric vein of a hamster. Source: reproduced from Baer and Joyeux, 1961, with permission from Masson S. A.

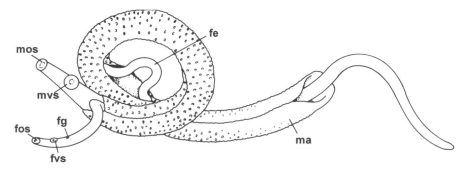

Figure 18.2 A pair of adult schistosomes *in copulo* (based on *Schistosoma mansoni*). fe = female in gynaecophoric canal of male; fg = female gonopore; fos = female oral sucker; fvs – female ventral sucker; ma = male; mos = male oral sucker; mvs = male ventral sucker. Source: reproduced from Whitfield, 1982, with permission from Blackwell Science.

like a sword in its scabbard. It was easy to pull the female out of the male's groove and to have a clear look at its inner structure.

The term 'schistosome' meaning 'split body' refers to the appearance of the couple (Fig. 18.2).

He described the anatomy of the male and the female and coined the term 'gynaecophoric canal' to denote the rolled ventral surface of the male that encloses the filiform female. He named the parasite *Distomum haematobium*. The eggs leave the human body in the urine and Bilharz noted that each egg had a terminal spine. He obtained miracidia from them, but he made no further progress with the life cycle. In fact, it was to be 60 years before it was finally solved, and progress was dogged *en route* by controversy and bitter rivalry. The following account of this unfortunate story is summarized from Farley (1991).

The main protagonists in the scientific row were the eminent German parasitologist Arthur Looss in Egypt and Luigi Sambon in London. The argument stemmed from Sambon's assertion that *Schistosoma* (= *Distomum*) *haematobium* of Bilharz was not the only species of human schistosome. According to Sambon, *S. haematobium* was associated with the blood vessels of the urinogenitals (bladder) and had terminal-spined eggs, while the second species, named *S. mansoni* after Patrick Manson, was associated with the blood vessels of the gut and had lateral-spined eggs (Fig. 18.3*b*, *c*).

Looss regarded the creation of a new species on the basis of eggs as highly improper and claimed that lateral-spined eggs were the abnormal produce of isolated or unimpregnated females of the bladder parasite. The argument continued for several years and was an acrimonious battle in which the scientific training of the British School and the scientific judgement of eminent German professors came under fire. At the time, Looss's position carried the day, although we now know that Sambon was correct. After all, Looss had access to living material while Sambon had only preserved specimens. Leiper, who studied under Looss for a year, wrote in 1911 that Looss had brought the lateral-spined controversy to an end – there was only one schistosome species in Egypt. He would live to regret these words.

Looss's argument with Sambon and Manson extended to the life cycle, knowledge of which in 1906 had advanced no further. The egg released a miracidium that lived for a few hours, but its fate was a mystery. By analogy with the life cycle of *Fasciola*

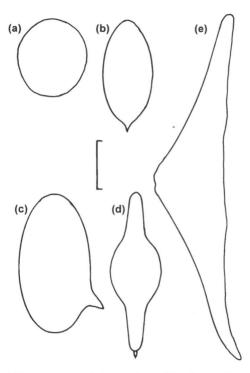

Figure 18.3   The eggs of *Schistosoma* spp. (*a*) *S. japonicum.* (*b*) *S. haematobium.* (*c*) *S. mansoni.* (*d*) *S. bovis.* (*e*) *S. spindale.* Scale bar: 50 μm. Source: reproduced from Sambon, 1909.

hepatica, which had been elucidated at the turn of the century (see Chapter 12), Manson and Sambon considered that the schistosome miracidium must enter an invertebrate such as a mollusc and produce cercariae. They assumed that the free-swimming or encysted cercariae were consumed in drinking water. Looss too suspected that molluscs were involved, but his attempts to infect common Egyptian molluscs and other freshwater organisms failed, as did his attempts to find naturally infected snails. These negative observations led Looss to abandon the idea of an intermediate host and to conclude that the miracidium must directly penetrate humans, a view that may have been partly influenced by his finding that the nematode parasite known as the hookworm enters the human body by penetration.

Looss's high scientific reputation ensured the acceptance of his skin infection theory. In fact, in the first text devoted exclusively to bilharzia (= schistosomiasis), Madden (1907), Professor of Surgery at the Egyptian Medical School, wrote: 'Any small puddle may become defiled with the urine or faeces of a patient suffering from bilharziosis; and, in a very short time, the water or mud is alive with miracidia, which may become applied to the bare feet, legs, or hands, penetrate the skin, and so lead to infection'.

Around this time, the scene moved to Japan. The discovery of a new species, *S. japonicum*, with a spineless egg (Fig. 18.3*a*), was timely because the parasite occurred in domestic animals as well as in humans and this permitted field experiments designed to determine the route of entry into the human body. In 1909, Fujinami and Nakamura conducted a field trial in which disease-free calves were exposed in endemic areas. Calves prevented from drinking and with legs protected by waterproof boots were uninfected (apart from one worm in one calf), while calves with legs exposed, whether their mouths were covered or not, became heavily infected. Many regarded this as support for Looss's theory of skin infection by miracidia.

There followed a debate in Europe as to whether it would be better to pursue research on a focus of *S. haematobium* in Cyprus (supported by Ross at the Liverpool School of Tropical Medicine) or to concentrate on *S. japonicum* in the Far East (supported by Manson at the London School of Tropical Medicine). The decision to support Manson put added strain on relations between the two Schools. Leiper went to China in 1914 but failed to find a Chinese infected with *S. japonicum* and when, at last, eggs were obtained from a dog, the miracidia did not infect any of the local snails that were tested. Moreover, Leiper's failure to infect himself and others by exposure to miracidia cast doubt on Looss's theory.

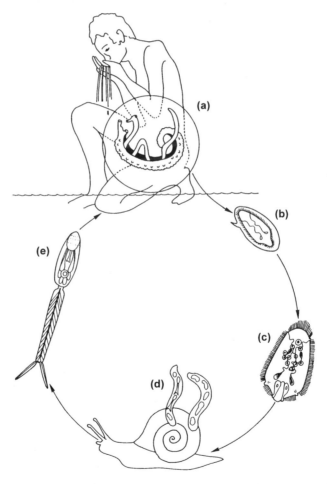

Figure 18.4   Life cycle of *Schistosoma mansoni*. Adults (paired males and females) (*a*) inhabit mesenteric veins of humans. Eggs (*b*) pass through gut wall and are voided in faeces. Each contains fully developed miracidium, which hatches immediately in fresh water. Miracidium (*c*) penetrates snail *Biomphalaria* sp.; daughter sporocyst (*d*) produces fork-tailed cercariae (*e*), which penetrate human skin, leaving tail outside. Cercariae transform to schistosomula and travel *via* blood stream to hepatic portal system, where they pair and mature. Source: based on Lyons, 1978.

This was confirmed by news from Japan. In 1913, Miyairi and Suzuki found that miracidia penetrated local snails and developed into sporocysts. For the first time, the fork-tailed cercariae produced by schistosome sporocysts and released from the snails were seen and described. Leiper then travelled to Japan and collected infected snails. Back in China he released cercariae by dissection and allowed them to make contact with laboratory mice. The outbreak of war precipitated a return voyage to England, during which a lady passenger in a nearby cabin objected to the mice. Leiper was forced to place them in the charge of an Indian butcher, under whose care they all died, although a single male schistosome was found in one of them. On reaching Aden, Leiper still had one infected snail and exposed a single mouse to cercariae from it. Luck at last turned in his favour and, on arrival in London in October 1914, he found male and female worms in the portal veins.

Armed with this knowledge of *S. japonicum*, Leiper was then commissioned to investigate bilharzia in Egypt, with the object of advising British troops on preventative measures. By experiment he demonstrated that the life cycles of Egyptian schistosomes (Fig. 18.4) were similar to that of their Far-Eastern relative and, after discharging his military duties, Leiper finally turned to the species controversy. He showed experimentally that cercariae from *Planorbis* (= *Biomphalaria*) *boissyi* gave rise to lateral-spined eggs, while those from *Bulinus contortus* produce only terminal-spined eggs. He finally laid the controversy to rest by stating that the terminal-spined and lateral-spined eggs found in bilharzial infections are the normal and characteristic products of two distinct species, *B.* (= *Bilharzia*) *haematobium* and *B. mansoni*, and are spread by different intermediate hosts.

18.2   SCHISTOSOME DIVERSITY AND EVOLUTION OF HUMAN SCHISTOSOMES

Depending on the authority consulted, there are between 80 and just over 100 species of schistosomes. Table 18.1 shows a scheme in which 86 species are distributed among 13 genera. Apart from *Griphobilharzia amoena* from a crocodile, they are all parasites of birds (68% of the species) or mammals (32%) and they are found in marine and in freshwater environments. Although males and females of *Schistosoma* spp. differ in shape and size, the sexes of some schistosomes are similar (see Basch, 1991, figs 1–11). The broad interaction between *Schistosoma* spp. and man (Table 18.2) and the impact of these parasites on human health has led to much more intensive research on *Schistosoma* spp. than on other schistosomes. This has the advantage that *Schistosoma* is better known than any other digenean genus, but provides a rather asymmetrical view of the biology of Schistosomatidae.

Combes (1990) considered the possibility that human schistosomes are descended from blood flukes infecting our primate ancestors. He pointed out that no *Schistosoma* sp. specific to a primate other than man has ever been described, and that *Schistosoma* reported from African primates (chimpanzee and baboon) have turned out to be species that also parasitize man. Thus, schistosomes are most likely to have colonized humans relatively recently from non-primate hosts.

In Africa, Combes recognized two lineages of schistosomes, a subdivision that has received support from molecular studies (Barker and Blair, 1996). One of these lineages utilizes pulmonate snails of the genus *Biomphalaria* as first intermediate hosts and includes *Schistosoma mansoni*, a rodent parasite *S. rodhaini*, and two

Table 18.1   The genera of schistosomes and their hosts (from Basch, 1991)

Genus	Definitive host	First intermediate host	Approximate number of species
Griphobilharzia	Freshwater crocodile	Unknown	1
Austrobilharzia	Birds	Marine prosobranchs	6
Ornithobilharzia	Birds	Marine prosobranchs or unknown	4
Trichobilharzia	Ducks, other birds	Freshwater pulmonates or unknown	28
Bilharziella	Ducks (other birds?)	Freshwater pulmonates or unknown	2
Gigantobilharzia	Birds	Freshwater pulmonates, marine opisthobranchs or unknown	12
Dendritobilharzia	Ducks, pelicans	Unknown	2
Macrobilharzia	Birds	Unknown	2
Schistosomatium	Rodents	Freshwater pulmonates	1
Bivitellobilharzia	Elephants	Unknown	2
Heterobilharzia	Carnivores, rodents, lagomorphs	Freshwater pulmonates	1
Orientobilharzia	Ungulates, equines, camels, carnivores	Freshwater pulmonates	4
Schistosoma	Mammals	Freshwater pulmonates and prosobranchs	19

Table 18.2   Species of *Schistosoma* parasitizing humans; details from Grove, 1990 and Greer, Ow-Yang and Yong, 1988 (*S. malayanum*)

Parasite	Geographical location	Snail host	Mammalian host
S. mansoni	Africa, Middle East, South America	*Biomphalaria* spp.	Man, monkeys
S. haematobium	Africa, Middle East	*Bulinus* spp.	Man, other primates
S. japonicum	China, Phillipines, Indonesia (Japan)	*Oncomelania* spp.	Man, dog, cat, buffalo, pig, horse, sheep, rodents
S. intercalatum	Africa	*Bulinus* spp.	Man
S. matthei	Africa	*Bulinus* spp.	Man, sheep
S. mekongi	Laos, Cambodia (Mekong River)	*Tricula operta*	Man
S. malayanum	Malaysia	*Robertsiella kaporensis*	Man, *Rattus* spp.

parasites of the hippopotamus. The other mainly utilizes pulmonate gastropods of the genus *Bulinus* and includes *S. haematobium*, *S. intercalatum* and several parasites of ungulates (*S. bovis*, *S. matthei*, *S. curassoni*, *S. leiperi*, etc.). Combes proposed that humans acquired *S. mansoni* by host-switching from a rodent and *S. haematobium* and *S. intercalatum* by host-switching from ungulates. He suggested that the hippopotamus may have acquired its schistosomes by a separate host-switching event from rodents. However, sequencing data based on ribosomal RNA by Després *et al.* (1995) point to a much more pivotal role for *S. hippopotami*. The indications are that this parasite, with its lateral-spined egg, may be ancestral to both lineages of African schistosomes and the possibility has been raised that the genus *Schistosoma* had its origins in early Suina, since *S. japonicum* commonly occurs in pigs and emerged in this study as the sister-group to all other members of the genus.

S. japonicum in Southeast Asia has also colonized man but, unlike the African schistosomes parasitizing humans, *S. japonicum* has not developed specificity to humans and has been reported from more than 30 different wild animals. Combes (1990) offered some possible explanations for this interesting phenomenon. He pointed out that not only mankind but also modern humans probably originated in Africa, with contact between hominids and schistosomes perhaps dating back to the early emergence of the human lineage and providing ample time for schistosome speciation. Contact between humans and Asian schistosomes would have had to await migration of hominids or modern humans from Africa, thus shortening the period of contact and perhaps accounting for the lack of speciation. If this Asian contact was established after the domestication of mammals by modern man, then the peculiarities of Asian agriculture, particularly the intimate relationship between domesticated buffalo and man and the large-scale irrigation schemes, would favour the maintenance of gene flow between animal and human schistosomes and prevent isolation and speciation. Another possibility is that there are differences in the mechanisms employed by African and Asian schistosomes to evade the host's immune system, such that Asian schistosomes are able to exploit a wide range of mammalian hosts, while African species are restricted to a narrow host range and hence have developed specificity to man.

Human migrations in relatively recent times have provided an opportunity for the spread of schistosomes. Extensive emigration to South America and the Caribbean islands in the 16th and 17th centuries, and particularly the trade in slaves, led to the introduction of *S. mansoni* and undoubtedly also *S. haematobium* (see Combes *et al.*, 1991). *S. mansoni*, but not *S. haematobium*, encountered a snail (*Biomphalaria glabrata*) in the New World that was capable of supporting the parasite. As expected, *S. mansoni* has diverged genetically from its African ancestors since its introduction in America (Imbert-Establet and Combes, 1986). In particular, it has an enhanced affinity for wild rats and other rodents, demonstrating the capacity of schistosomes to adjust rapidly and evade the immune responses of new potential vertebrate hosts.

There has been much human traffic too between countries such as Japan, China and Egypt, where human blood flukes are common, and India, where the parasites have failed to become established in humans. The belief became strong in the first half of this century that there was no danger of establishment of schistosomes in the human population of India because of the absence of suitable snail hosts on the subcontinent (see Southgate and Agrawal, 1990). This belief was shaken a little by the discovery in 1952 of a small focus of urinary schistosomiasis in the human population of Gimvi, in Maharashtra State. The parasite appeared to have established itself in an ancylid snail *Ferrissia tenuis*. *Ferrissia* spp. also occur in Africa but have never been incriminated in the transmission of a mammalian schistosome. This particular Indian focus of the parasite appeared to be in recession, but the experience emphasizes that the establishment of the parasite in a new locality is not necessarily precluded by the absence of the regular snail host.

Nevertheless, *Schistosoma* spp. do appear to be more 'faithful' to their snail intermediate hosts than to their vertebrate definitive hosts, i.e. they are less inclined to indulge in host-switching between snails (see Combes, 1990). In fact, there is agreement that the schistosomes, such as *S. japonicum*, that use prosobranch snails became separated from the African and South American species that use pulmonate hosts early in the Triassic period (see Combes, 1990). Support for this split comes from the molecular data of Barker and Blair (1996).

18.3   GENERAL BIOLOGY

18.3.1   Adult schistosomes

Survival of schistosomes requires that as many eggs as possible escape from the closed circulatory system that they inhabit and enter the aquatic habitat of the snail intermediate host. There are no gills in their land-dwelling hosts, but there are other capillary beds that offer a convenient escape route, namely the well-vascularized walls of the intestine, rectum and bladder, and in the cattle parasite *S. nasale*, the mucous membranes of the nose. Surprisingly perhaps, these blood flukes do not live in the arterial system, from which their eggs would be swept directly into the gut or bladder capillary beds. They inhabit the veins draining these organs, immediately raising the question as to how the eggs reach the capillary bed. In fact, many eggs are swept away from these sites, becoming impacted in internal capillary beds, particularly that of the liver, and it is this aspect of the biology of schistosomes that leads to much of the severe pathology associated with human infections (Warren, 1978; section 18.4.1).

The venous habitat of human schistosomes becomes understandable when we recall that the blood draining from the gut will be rich in nutrients. Adult schistosomes ingest blood and their activities include lysing red blood cells and digesting the haemoglobin that they contain (see review by Dalton *et al.*, 1995). Haematin, the brown–black end product of haemoglobin digestion, accumulates in the gut and is evidence of this activity. Transtegumental uptake of amino acids and glucose has also been demonstrated (Asch and Read, 1975a, b; Rogers and Bueding, 1975) and glucose transporter proteins involved in facilitated diffusion have been identified on the tegumental surface and in the basal tegumental membrane and underlying muscle (Skelly and Shoemaker, 1996). As Chappell (1976) has suggested, it is possible that larger molecules such as peptides are obtained *via* the parasite's gut while the transtegumental route supplies low-molecular-weight nutrients.

Schistosome sex is determined by a chromosomal mechanism, with the female being ZW (usually $2n = 16$) and the male ZZ (see Basch, 1991). Each miracidium will be potentially male or female and snails infected with single miracidia will produce all male or all female cercariae. Naturally occurring snails may produce both male and female cercariae as a result of invasion by more than one miracidium. A vertebrate definitive host must be infected by both male and female cercariae if copulating pairs of adult schistosomes are to become established. In fact, using snails naturally infected with *S. mansoni* and DNA-profiling of parasites, Minchella *et al.* (1995) showed that snails were more likely to carry multiple infections than would be expected by chance, thereby increasing the probability that vertebrate hosts will acquire a dual-sex infection from exposure to cercariae from a single snail.

Why hermaphroditic proto-schistosomes adopted gonochorism (separate sexes or dioecism) is a question that has not been convincingly answered. It is a rare development in platyhelminths, occurring in some fecampiid turbellarians (section 2.3.3), some tapeworms (section 10.6) and some didymozoideans (section 19.1). In schistosomes, it has been linked with the acquisition of warm-blooded hosts, but it is far from clear why this should favour separate sexes (Combes *et al.*, 1991). As suggested by Basch (1990), it could have arisen by selective reinforcement of the tendencies towards protandry and protogyny that we have already met in other platyhelminths (sections 4.15, 5.5.1), with the genes controlling manifestation of the opposite sex becoming non-functional and degenerating. However, the nature of the selection pressures that guided these changes is uncertain. Després and Maurice (1995) have suggested that dimorphism arose before gonochorism in response to the special need to deposit eggs in the smaller venules to facilitate their escape from the body. The development of gonochorism was then favoured by the aggregated distribution of parasites in their hosts.

Whatever the reasons were for investing in separate sexes, this has led to substantial sexual dimorphism in the highly specialized human schistosomes. Females emphasize reproductive efficiency, greatly reducing their body muscles and parenchymatous tissue. It has been estimated that the female of *S. mansoni* produces 300 eggs per day with 38 vitelline cells per egg (Erasmus and Davies, 1979) and it has been claimed that the female converts nearly her own body weight into eggs each day (Becker, 1977; see also section 17.8.1). Reduction of muscles and parenchyma produces a slender female capable of extending into much smaller blood vessels than the bulky male, with the advantage that eggs can be given a head start on their journey to the outside world. The disadvantage is that the weak female is unable to make the journey unaided from the liver to the mesenteric veins, particularly in large hosts, where this distance is considerable, and she must rely on the muscular male for transport to the egg-laying site.

Basch (1990) has taken this concept further and has suggested that the female has also reduced her pharyngeal muscles and is now unable to ingest blood efficiently; he claimed that blood is pumped into the gut of the female by a massaging action of the walls of the gynaecophoric canal of the male. He proposed that unpaired female *S. mansoni* remain stunted and undeveloped because they are unable to eat efficiently, as shown by the lack of black haematin pigment in their gut, and not because, as others have suggested, they require stimulation by substances from the male that promote growth and reproductive maturity (see LoVerde and Chen, 1991). However, Smith and Chappell (1990) pointed out that long-isolated females ('elderly spinsters') have significant quantities of gut haematin, indicating that some females are not entirely incapable of feeding themselves.

In *Schistosomatium douthitti*, a parasite of rodents, females are capable of developing to maturity independently of males and must be capable of feeding themselves (Fig. 18.5). Moreover, females are relatively robust and able to transport themselves to the egg-laying sites (Basch, 1990, 1991). Here they can produce viable eggs without mating and these uniparental, parthenogenetic eggs produce miracidia capable of infecting a snail. This reproductive strategy is clearly of benefit in rodent hosts, which do not always live long enough for male co-infection.

Read and Nee (1990) have pointed out that, provided males contribute genetically to the next generation by fertilizing the eggs of their female companion,

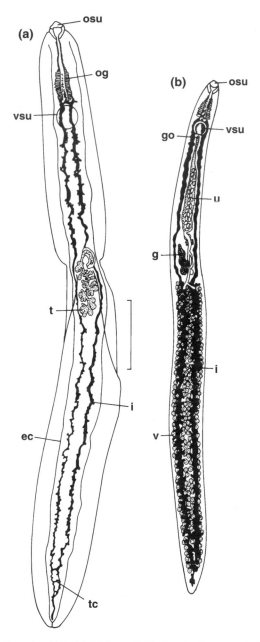

Figure 18.5 *Schistosomatium douthitti.* (*a*) Male and (*b*) female in ventral view. ec = excretory canal; g = germarium; go = genital opening; i = intestine; og = oesophageal glands; osu = oral sucker; t = testis; tc = transverse commissure of intestine; u = uterus containing eggs; v = vitellarium; vsu = ventral sucker. Scale bar: 0.5 mm. Source: reproduced from Price, 1931.

then the division of labour between the two sexes in *Schistosoma* spp. is understandable, with the males catering to the physical needs of the females in which they have a genetic investment. However, if parthenogenesis is the rule, then the elaborate parental care by the male can only be advantageous if the partners are siblings.

18.3.2 Egg development

As discussed above (section 17.8.1), the egg is parasitic – embryonic development takes place in host tissue after laying and is complete before the eggs leave the host. The elevated temperature of the homoiothermic host may permit faster development than in the external environment. There is evidence that the miracidium can take up glucose and amino acids through the eggshell, and the egg of *S. mansoni* approximately doubles its volume during embryogenesis (references in Tinsley, 1983). Although the eggshell is substantial, inward passage of metabolites may be facilitated by the presence of minute branching and anastomosing channels (Swiderski, 1994a).

According to Swiderski (1994a), outer and inner envelopes, both of which are cellular in origin and syncytial in nature, differentiate during embryogenesis and lie between the miracidium and the shell (see also section 13.2). Remnants of the vitelline cells are incorporated in the outer envelope. The structural components of the envelopes and their nutritive reserves (glycogen, lipid) are said to be reabsorbed by the miracidium before hatching. Swiderski (1994b) regarded the envelopes as homologous with the inner and outer egg envelopes of a pseudophyllidean cestode (section 10.8.1).

18.3.3 How eggs escape from the host

How the eggs reach the faeces or urine from the venous system is a perplexing question. In *S. mansoni*, it is necessary for the inert egg to travel by way of the small veins deep into the submucosa and lamina propria (see Fig. 9.3), leave these blood vessels and then find its way out through the mucosal surface into the lumen. Basch (1991) has summarized early views on how this seemingly impossible journey might be achieved.

The suggestion of Lutz (1919, in Basch, 1991) that the female passes through the wall of the venule and deposits her eggs in the perivascular tissue has no support, but the female may give her eggs a head start by pushing her way along one of the small vessels against the blood flow. Whether, during the egg-laying process, the female maintains contact with the male, which is held stationary within the blood vessel by the tegument with its spine-bearing tubercles (Fig. 18.2), is disputed, some workers believing that the female separates completely from the male. Manson-Bahr and Fairley (1920, in Basch, 1991) envisaged the separated female reversing to return to the larger vessel housing her bulky partner, laying eggs at intervals, each egg being trapped by elastic recoil of the venule wall as the female retreats. Because of the orientation of the ootype (Fig. 18.6), the lateral spine of the freshly laid egg of *S. mansoni* will project in the direction of blood flow and is likely to engage in the wall of the vessel, preventing movement of the egg away from the gut.

An old idea of Brumpt (1930, in Basch, 1991) that the male acts like a piston, reversing blood flow and sweeping eggs into more distant venules, was rejected by Faust, Jones and Hoffman (1934, in Basch, 1991), but File (1995) saw retrograde blood flow as a real possibility because of the downstream blockage of small veins by adult worms. Retrograde flow might push eggs back into veins too narrow to admit the female and provides a mechanism for eggs to reach anastomoses with larger veins and thence the liver. In spite of the blockage of vessels by adults and

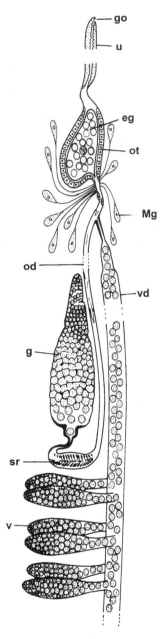

Figure 18.6   Diagrammatic side view of the female reproductive system of *Schistosoma mansoni* with an egg in the ootype. eg = egg; g = germarium; go = genital opening; Mg = Mehlis's gland; od = oviduct; ot = ootype; sr = seminal receptacle; u = uterus; v = vitellarium; vd = common vitelline duct. Source: reproduced from Gönnert, 1955, with permission from Georg Thieme Verlag.

eggs there is no clotting. A serpin (serine protease inhibitor), thought to be secreted by the parasite, is considered to be an inhibitor of thrombin-like enzymes, and may prevent clot formation around the parasite (Modha, Roberts and Kusel, 1996). This serpin may be homologous with an anticoagulating factor detected by Tsang and Damian (1971) in homogenized *S. mansoni*.

Until recently there has been no convincing explanation of how the eggs escape from the venules. The egg spine of *S. mansoni* may be capable of perforating the wall of a blood vessel, but it is not an ideal tool for creating an aperture large enough to accept an egg, and the egg of *S. japonicum* manages to escape into the perivascular tissue without the aid of any spine (Fig. 18.3*a*). There are claims that a collagenase-like enzyme produced by the fully developed miracidium leaks out through the perforations in the eggshell, but, at the time when the egg leaves the venules, the miracidium is not fully developed.

Thus, the traditional view sees the egg as the instrument of escape, while recent work implicates the host. File (1995) showed that schistosome eggs oviposited directly on to a monolayer of endothelial cells from the lining of human umbilical vein were covered by endothelial cells and pushed into the substrate within a 4 h period. Eggs inserted into umbilical veins also became covered by migrating endothelial cells. Uterine secretions known to coat freshly laid eggs, may stimulate endothelial cell activity, since File found that freshly laid eggs interacted more rapidly with these cells than embryonated eggs. Thus, any mechanism (cessation of blood flow, worm behaviour, vascular contraction) that retains eggs against the vessel wall long enough for the lining cells to react is likely to lead to extravasation.

Once in the perivascular tissue the conventional view is that the egg makes further progress towards the lumen by a combination of translocation by muscular movements of the gut wall and histolytic activity of miracidial enzymes leaking out through the shell. However, there are some problems with this concept. The leakage of antigens from eggs into the tissues (intestine wall, liver) attracts host leucocytes, which surround the egg, producing a granuloma. It has been demonstrated that granulomas sequester toxins produced by the egg, reducing damage to adjacent liver cells (section 18.4.1), and there is no reason to believe that granulomas would not be just as effective at sequestering egg-derived enzymes in the gut wall. Another difficulty with the conventional theory is that directionality of egg movement towards the lumen is hard to explain. This has led Damian (1987) to propose the intriguing and radical idea that the host's immune system is exploited by the parasite as a means of promoting egg escape.

Damian (1987) quoted comparative observations by Doenhoff and colleagues that revealed that faecal egg output by *S. mansoni* in immunocompetent mice was significantly higher than in mice in which the immune system had been experimentally suppressed, in spite of similar parasite burdens and fecundity rates in the two groups. The finding by Damian of granulomas free in the crypts of Lieberkühn, which are tubes embedded in the gut wall that communicate with the main gut lumen (see Fig. 9.3), led him to suggest that the granulomas, which are composed of mobile cells, physically transport the eggs to the lumen. There is evidence that granulomas in the crypts are already beginning to dissociate, providing a mechanism for release of the contained eggs into the faeces. The implication is that fewer granulomas are produced by immunosuppressed animals and that, consequently, fewer eggs reach the lumen, but Damian was unable to explain how the mass of cells comprising the granuloma is able to undergo integrated and directional movement.

Some further support for Damian's idea comes from the relationship of *S. mansoni* to guinea pigs and *S. nasale* to buffalo. In both of these relationships, egg escape is poor or fails and there is a corresponding reduced ability to produce granulomas.

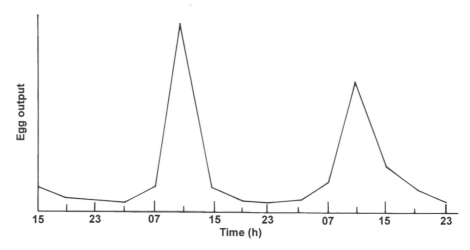

Figure 18.7 Output of eggs of *Schistosoma haematobium* in urine of a single human individual following a normal daily routine. Source: reproduced from McMahon, 1976, with permission from the Australian Society for Parasitology.

Exploitation of host cellular defences to externalize eggs may also have occurred in the turbellarian *Anoplodium hymanae*, which lives in the coelom of a sea cucumber (section 2.3.1), and Damian suggested that primitive host-response-assisted elimination of eggs may have its origins in the widespread ability of invertebrate phagocytes to engulf foreign particles and transport them through the tissues to the outside world. Exploitation of the host's immune response for improving attachment of parasites to their hosts has been mentioned elsewhere (sections 6.4.1, 6.5.2).

Eggs of *S. haematobium* leave the human body in the urine, but the daily pattern of egg output is not uniform. Urine passed during the late morning and early afternoon contains more eggs than at other times (Fig. 18.7). In endemic areas, children are mainly responsible for transmission, and since their water-based activities (playing, urinating) reach a peak at midday, the efflux of eggs at this time will confer a significant advantage on a parasite of an essentially terrestrial host (see McMahon, 1976).

18.3.4 Egg hatching

In spite of the intense research activity that has been focused on the schistosomes, our understanding of egg hatching is still incomplete. There are several unusual features. Although the eggshell is reported to be of tanned protein (Basch, 1991), a distinctive operculum like that of *Fasciola hepatica* (see Fig. 13.5) is absent and the miracidium usually emerges through a vertical or oblique split in the shell (Kassim and Gilbertson, 1976). There is no viscous cushion, as in *F. hepatica* (section 13.2), but there are two sausage-shaped sacs, which incompletely surround the miracidium (Fig. 18.8). The embryological origins of these sacs are unclear. In *S. mansoni*, the miracidium always develops with its anterior region at the opposite end from the spine but when the miracidium gains the ability to move within the shell about half of them reverse this orientation (Blair in Basch, 1991). Some, but not all, larvae become active just prior to hatching and there are

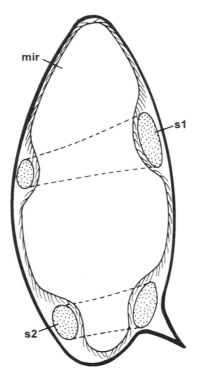

Figure 18.8 The egg of *Schistosoma mansoni* showing the position of the sacs (s1, s2) relative to the miracidium (mir). The sacs incompletely encircle the miracidium. Source: based on Kusel, 1970.

conflicting reports on the effect of increased temperature and of light on hatching (Xu and Dresden, 1990).

Hatching occurs when faeces or urine containing the eggs are diluted, as they would be when sewage enters natural bodies of fresh water, and the idea that osmotic uptake of water leads to increased internal pressure and bursting of the egg is an appealing one. However, as Xu and Dresden (1990) have pointed out, there are observations that are inconsistent with this simple osmotic mechanism. Only eggs containing a clearly defined miracidium hatch successfully. If the osmotic theory were correct, the assumption must be made either that water does not diffuse into immature eggs or that the eggshell becomes more permeable at maturity. Xu and Dresden regarded both of these possibilities as unlikely. Moreover, in order to break an egg of *S. mansoni* they claimed that a pressure of 5000 psi was required and that the osmotic potential of the egg was unable to generate such a pressure.

These observations raised the possibility that there is another dimension to the hatching process and this was reinforced by the report of the hydrolytic enzyme leucine aminopeptidase in the hatching fluid. This enzyme was inhibited by increasing salt concentration in a manner similar to the effect of salt on hatching (Xu and Dresden, 1986). Moreover, hatching was significantly inhibited by bestatin, a competitive inhibitor of leucine aminopeptidase. The implication is that the enzyme has a contributory role in hatching, perhaps weakening the eggshell prior to rupture due to osmotic effects. In this context it is interesting that Zdarska, Giboda and Zenka (1992) have detected in the eggshell of *S. mansoni* a line of weakness, which takes the form of a suture penetrating from the inside but not reaching the outside of the shell.

18.3.5   Host-finding by miracidia

According to Haas *et al.* (1995a), the conflicting results obtained in the 1960s and 1970s by investigators of host-finding in schistosome miracidia can be attributed to use of inappropriate methods. Reinvestigation has revealed that schistosome miracidia respond to macromolecular cues from the snail host, not small molecules, a strategy that may avoid costly responses to the numerous small molecular components of mud in the microhabitats of the snails (Haas *et al.*, 1995a).

The miracidia of *S. mansoni* and *S. haematobium* (like that of *Fasciola hepatica*; section 13.3.2), use chemokinesis – increasing random turning in chemical gradients of increasing strength, with turning back in decreasing gradients (Haas *et al.*, 1995a; Haberl *et al.*, 1995). In contrast, the miracidium of *S. japonicum* uses chemotaxis, swimming up a chemical gradient towards the host (Haas *et al.*, 1995a). Curiously, although *S. mansoni* and *S. haematobium* infect snails of different genera, they both respond to glycoproteins with a molecular mass of more than 30 kDa. However, miracidia of different strains of *S. mansoni* with different susceptible snail hosts, show evidence of selectivity during chemo-orientation, indicating that there may be differences between the glycoproteins from different snails, perhaps in their carbohydrate moieties. So, host specificity may manifest itself as early as during chemo-orientation, and it may be further promoted during attachment to the snail and penetration.

18.3.6   Germinal sacs and interactions with the mollusc

Schistosomes have no rediae. Near the site of entry into the snail, the miracidium transforms into a mother sporocyst, which gives rise to daughter sporocysts. These migrate to the digestive gland/ovotestis, where they grow and produce cercariae. We have already noted that the strategy adopted by schistosomes in relation to avoidance of molluscan defences against parasites is different from that adopted by digeneans that include a redial stage in their intramolluscan development (section 14.1). It seems that, at least in the early stages of parasitism, schistosomes opt for concealment from the mollusc host by mimicking or sequestering host molecules. A high degree of homology has been detected between tropomyosins from *S. mansoni* and from its snail host *Biomphalaria glabrata* by Capron and co-workers (references in Damian, 1991). Of special interest is the finding that the tropomyosin of *S. mansoni* occurs not only in the muscles, as expected, but also in the tegument of the sporocyst. It has been suggested that parasites might preserve highly conserved molecules such as tropomyosin to minimize their foreignness to their host and that the 'choice' of structural proteins by parasites may be constrained by having to adapt to the host's immune system. It has been suggested by Weston and Kemp (1993) that the similarity between schistosome and snail tropomyosins might reflect gene transfer between host and parasite (see also section 20.3).

Infection with germinal sacs of the bird schistosome *Trichobilharzia ocellata* leads to castration and gigantism in the snail host, *Lymnaea stagnalis*, and this gigantism mainly involves an increase in the space occupied by the haemolymph, i.e. the space where the parasites develop and multiply (see review by De Jong-Brink, 1995). Thus, the host is forced to create both resources and space for the production of the enormous numbers of cercariae generated by this parasite. There is evidence to suggest that the physiological changes in the host are brought about by parasitic interference with the host's neuroendocrine system, and a host-derived substance that has been named schistosomin appears to have a central role in regulating these changes. Other kinds of stress have a similar inhibiting effect on host reproduction, so parasites may make use of an already existing host system to curtail energy-consuming activities, thereby favouring the parasite's own development and reproduction.

The strategy of concealment from the host employed by digeneans lacking rediae contrasts with the aggressive immunosuppression employed by redia-producing digeneans. In fact, it has been suggested that this aggressive immunosuppression increases the vulnerability of molluscs to invasion by other opportunistic digeneans, such as schistosomes, creating a need for predatory rediae to remove the competition by consuming the opportunists. Parasites, such as schistosomes, that primarily depend on concealment from the mollusc's internal defences rather than attack no longer require predatory rediae and may have lost them during the course of their evolution.

There is evidence (references in Loker, 1994) that, later in the course of infection, schistosomes develop the capacity to interfere with the activities of host haemocytes, but this seems to be perpetrated in a more subtle way. Although haemocytes are not repelled by *S. mansoni* sporocysts and readily bind to them, in susceptible snail hosts they do not go on to attack the sporocyst surface. Sporocysts of the bird schistosome *Trichobilharzia ocellata* produce factors that induce the central nervous system of the host snail *Lymnea stagnalis* to release substances that reduce the phagocytic activity of the haemocytes (see De Jong-Brink, 1995). The finding that primary infection with a sporocyst-producing digenean can sometimes increase susceptibility to normally incompatible digeneans is also consistent with the idea that sporocyst-producers are capable of interfering with the molluscan defence system.

The bird schistosome *Austrobilharzia terrigalensis* is of special interest because it infects the snail *Velacumantus australis* only if the snail already harbours other digeneans, which are usually redia-producing species (Walker, 1979; Appleton, 1983). The parasite can survive in these situations because it is able to resist the predatory activities of rediae; in fact, the growth and germinal development of redia-bearing species are reduced in the presence of *A. terrigalensis*. Thus, this schistosome appears to have become highly specialized for colonizing and exploiting snails whose internal defence systems have been impaired by primary infections of other digeneans.

18.3.7 Cercarial host-finding and invasion

In the cercaria of *S. mansoni*, an asymmetrically placed pair of anterior gland cells has been identified by Dorsey (1974) as responsible for escape from the snail (Fig. 18.9).

The principal energy reserve of the cercaria is glycogen. Since the free cercaria cannot feed this will be used up during its free-swimming life and the cercaria must have sufficient in reserve when it finds a host to fuel the extremely energetic process of penetration, which consumes 22–35% of the glycogen content of the cercarial body (references in Wilson, 1987). Newly-emerged cercariae may not all be equal in this respect, since a threefold variation has been reported in the glycogen content of different batches, a situation that presumably reflects the nutritional state of the snail host.

The life of the cercaria is prolonged by swimming intermittently, or by spending most of the time attached by the ventral sucker to the water–air interface as in

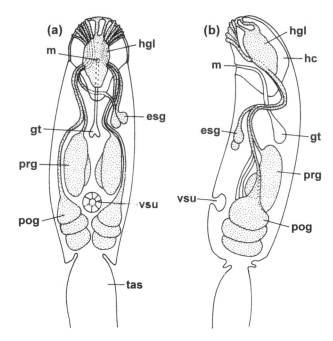

Figure 18.9 Diagrammatic representation of the gland cells in the body of the cercaria of *Schistosoma mansoni* prior to escape from the snail. (*a*) Dorsal view. (*b*) Lateral view. esg = escape glands (one pair; both cells on same side of body); gt = gut; hc = head capsule; hgl = head gland; m = mouth; pog = postacetabular glands (three pairs); prg = preacetabular glands (two pairs); tas = tail stem; vsu = ventral sucker. Source: redrawn from Stirewalt, 1974.

S. japonicum (see Haas, Granzer and Garcia, 1987). The free-swimming cercariae are remarkably manoeuvrable, since they can swim tail or head first, depending on whether the furcae are spread or held together in line with the tail stem.

Since many terrestrial birds and mammals spend relatively short periods of time in contact with water and often restrict these visits to a particular time of day or night, it is important that the cercariae emerge during this contact period. Consequently, many schistosome cercariae have striking daily emergence rhythms, which correspond with the water visitation patterns of their hosts. For example, cercariae of the rodent parasite *S. rodhaini* emerge at night, while those of the cattle parasite *S. bovis* emerge in the early morning (Fig. 18.10). The cercariae of the human parasites *S. mansoni* and *S. haematobium* both emerge in the middle of the day, when human contact with water is most intense, in spite of the fact that *S. mansoni* is related to *S. rodhaini* and *S. haematobium* to *S. bovis*.

Genetic control of shedding rhythms has been demonstrated (see Combes, 1990) and it is essential for the success of any switching of definitive hosts that the cercarial emergence pattern should rapidly adjust to maximize contacts with the new host. There is evidence that *S. mansoni* is capable of this adjustment, since in the Caribbean, at sites where rats are important hosts, cercarial emergence peaks near 16.00h, while at sites where humans are important hosts the peak occurs near 11.00h.

Most observers have failed to detect any directional response of schistosome cercariae to chemical gradients emanating from their vertebrate hosts, although Shiff and Graczyk (1994) claimed to have detected such a response to linoleic acid in *S. mansoni*. Cercariae of *S. mansoni* increase their active swimming phases and

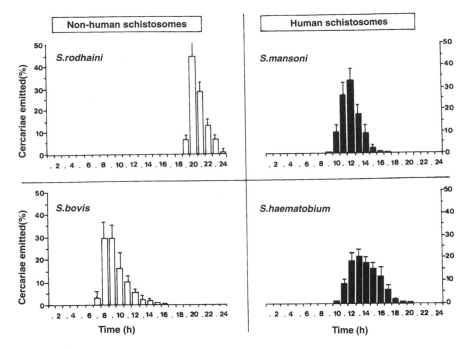

Figure 18.10 The percentage of cercariae emitted in each hour by schistosome-infected vector snails. See text for explanation. Source: reproduced from Combes, 1990, with permission from Elsevier Trends Journals.

perform frequent reversals of swimming direction when stimulated by human skin components, and this may improve their chances of random contacts with a host (see Haas, 1994). Unlike fish-invading cercariae (section 17.5), responses of schistosome cercariae to water turbulence generated by a potential host or to touch are weak or absent, and Haas (1994) interpreted this as adaptive, claiming that it would prevent energy-wasting responses to the many non-host mechanical disturbances in the aquatic environment.

According to Feiler and Haas (1988), responses to shadow stimuli are of great importance for host finding in the duck parasite *Trichobilharzia ocellata*. Shadows induce prolonged forward movements in a downward direction (negative phototaxis), which were thought to lead the cercariae from the water surface to the feet of the duck.

In *S. japonicum*, contact with the host may be entirely passive, cercariae attached to the surface being picked up by any object, including a vertebrate limb, that is immersed and then withdrawn from the water.

The work of Haas and others (see review by Haas, 1994) has demonstrated that, after contact has been made with a potential host, there are considerable differences between schistosome species in the spectrum of host signals to which they respond (Table 18.3). Attachment may be encouraged by skin chemicals and/or by warmth, as in *S. mansoni*, or may require no stimulus, as in *S. japonicum*. Provided that the surface of the potential host has hydrophobic properties, *S. japonicum* will remain attached, but *S. mansoni* will resume swimming if certain skin lipids are absent or if the surface cools. Penetration behaviour in *S. mansoni* requires a completely different chemical stimulus, which takes the form of fatty acids emanating from the skin

Table 18.3   The signals from the host that promote the phases of host-recognition in cercariae of *Schistosoma* spp. (modified from Haas, 1994)

	Attachment	*Enduring contact*	*Directed creeping*	*Penetration*
S. mansoni	L-arginine, warmth [Water turbulence]	Ceramides, acylglycerols, warmth	L-arginine temperature gradients	Fatty acids (not warmth)
S. haematobium	Warmth, [L-arginine]	No stimuli?	L-arginine temperature gradients	Fatty acids (not warmth)
S. spindale	Warmth (not chemical stimuli)	Warmth (not chemical stimuli)	Temperature gradients (not chemical gradients)	Fatty acids (not warmth)
S. japonicum	No stimuli	No stimuli (favoured by solid hydrophobic surfaces)	Temperature gradients (not chemical gradients)	Fatty acids, warmth

[] = stimulus with weak effect

and produced by the action of bacterial esterases on triglycerides. Even if the surface is warm, penetration will not take place in the absence of these chemicals, whereas in *S. japonicum* either fatty acids or warmth will induce penetration behaviour. This broader response of *S. japonicum* is consistent with its wide range of mammalian hosts, which include species with low levels of free fatty acids on the skin surface.

Some schistosome cercariae display a remarkably fine-tuned response to temperature gradients. *S. spindale*, a parasite of ungulates, can detect and move along a gradient as weak as 0.07°C/mm (Haas, Granzer and Brockelman, 1990b) and this was interpreted as an adaptation to reach the skin surface of furry mammals by migrating along the hairs. In this context, it is interesting that the cercaria of *T. ocellata* is stimulated to penetrate duck skin by skin lipids such as cholesterol and ceramides, but fails to respond to the uropygial gland secretion which contains little or none of these lipids and is spread over the feathers during preening. This prevents futile attempts to attach to and penetrate the relatively thick covering of feathers and diverts penetration activity to exposed skin, like that covering the feet (Feiler and Haas, 1988).

What is surprising is that the behaviour of different schistosomes that invade the same hosts (humans) is governed by different sets of stimuli. Haas has suggested that these differences may be related to differences in the aquatic habitats where invasion takes place. He has proposed that the use of chemical cues from the host may be more effective in clear than in muddy, polluted water, and that response to thermal stimuli will be more feasible where the ambient environmental temperatures are relatively low.

King and Higashi (1992) reported that the addition of silver ions to suspensions of *S. mansoni* significantly inhibited their penetration into agar impregnated with a fatty acid (linolenic acid). Washing removed the silver and reversed this inhibition. Since Ag^+ is known to bind to sensilla on the body of the cercaria, they interpreted this as evidence that these sensilla are the chemoreceptors that mediate cercarial response to fatty acids. This was also supported by the fact that the argentophilic sensilla and the cercarial response to fatty acids disappeared 3–4 h after conversion to a schistosomule (see below).

In spite of the sophistication of cercarial behaviour of schistosomes, there appears to be no way for bird schistosomes to avoid penetration and premature death in human skin or, indeed, for human schistosome cercariae to avoid entering bird skin (see Combes, 1990). The frequency of such errors is illustrated by the widespread occurrence of cercarial dermatitis or 'swimmer's itch' in humans, a condition stemming from abortive skin penetration by cercariae of bird or mammal schistosomes (see Grove, 1990).

18.3.8   Skin penetration

Penetrating human skin is a difficult proposition for a small soft-bodied organism like a cercaria. The horny outer layer of the epidermis is a stack of dead cells (stratum corneum), containing the relatively inert protein keratin, and the dermis, from which the epidermis is separated by a basement membrane, contains two other structural proteins, collagen and elastin, which endow it with toughness and flexibility. The young schistosome must traverse the epidermis and enter the dermis to reach the nearest blood or lymph vessels.

Most of the body of a schistosome cercaria is filled with glands and their ducts (Fig. 18.9). Those thought to be concerned with penetration are of two kinds, known in *S. mansoni* as pre-acetabular (two pairs) and post-acetabular (three pairs; Fig. 18.9), distinguished by the appearance and staining properties of their cytoplasm and by reference to the position of the acetabulum or ventral sucker. The ducts of these glands run anteriorly and enter the head capsule (frequently but incorrectly called the oral sucker, which appears later). In *S. mansoni*, their ten separate openings are arranged in two crescentic arrays of five at the anterior extremity of the capsule (Robson and Erasmus, 1970). Some of the secretion of the post-acetabular glands is released before penetration and appears to have an adhesive function as the cercaria undergoes leech-like exploratory movements over the skin surface. Linder (1985) exposed cercariae to lectins bound to fluorochrome. This material had an affinity for post-acetabular secretion, and crescent-shaped fluorescent 'footprints' were clearly revealed at each successive attachment site of the head capsule.

Stirewalt and Hackey (1956) and Stirewalt and Dorsey (1974) have described the penetration of mouse ear skin by the cercariae of *S. mansoni*. After selection of a suitable site, the cercaria foregoes attachment with the ventral sucker and, attached by the anterior extremity, adopts a vertical or oblique attitude relative to the skin surface. The cercariae do not enter *via* a hair shaft, although they may penetrate close to a hair, perhaps because they often migrate along a hair to reach the skin surface. Cercariae negotiate the stratum corneum not by penetrating the dead keratinized epidermal cells (squames) but by passing between them. The squames are disarticulated, perhaps by lysis of their intercellular cement, perhaps by physical force exerted by swelling of the post-acetabular secretion, which is released between them. The muscle protein paramyosin has been reported in the post-acetabular glands of *S. japonicum* by Gobert *et al.* (1997), and they suggested that this protein may play a part in the evasion of host defensive responses (see also section 18.3.9).

In the cercaria of *S. mansoni*, alternate elongation and shortening of the body are important in the penetration process, especially in conjunction with the backwardly directed tegumental spines, which prevent withdrawal from the entry tunnel and provide a purchase for thrusting/probing movements of the anterior region. As the body enters the skin, the tail is set free on the skin surface but before this happens it may contribute to the penetration thrust by virtue of its constant activity, since cercariae without tails either fail to penetrate or do so only very slowly.

Once beneath the stratum corneum there is evidence that the pre-acetabular secretion is released and the cercaria slips more quickly through the tissue, destroying transitional cells (cells converting to keratin production) as it goes. However, it is delayed by the epidermal basement membrane. Once this is perforated, or perhaps bypassed by following a hair canal and its communicating sebaceous gland duct, the parasite must locate a dermal venule or lymph vessel, perforate its wall and gain access to its lumen (see Wilson, 1987).

Proteolytic activity against, among other materials, keratin, glomerular basement membrane, collagen and elastin has been demonstrated in the cercaria of *S. mansoni* (references in Newport *et al.*, 1987; Wilson, 1987), so all the glandular equipment is present for tunnelling through the epidermis and dermis. However, Wilson (1987) pointed out that by the time the schistosomulum reaches the dermis the acetabular glands are exhausted. The only gland that is not exhausted is the head gland, which is located in the head capsule and communicates with the anterior tegument (Fig. 18.9). It seems likely that it plays a part in the final phase of skin migration, in particular perforation of the wall of the blood or lymph vessel.

18.3.9   The cercaria–schistosomulum transition

The intense research effort invested in schistosomes has revealed just how profound the structural, physiological and biochemical changes are that accompany penetration into the vertebrate host. *S. mansoni* takes about 7 min to enter human skin and in those few minutes the nutrient-free freshwater environment is left behind and replaced by host tissue fluids that are warmer, chemically complex and far from hospitable. The conversion of the free-living cercaria to the parasitic schistosomulum that begins as the cercaria penetrates takes less than 1 h to complete *in vivo* (Cousin, Stirewalt and Dorsey, 1981).

The energy metabolism of the cercaria is aerobic, with glycogen as the substrate, but there may be little fuel left after the energy-consuming activity of penetration (see above) and the parasite must then turn to its host as a source of metabolites. The rapid appearance at about this time of a glucose transporter protein on the surface of the schistosomulum may reflect the urgent need for glucose uptake to replenish depleted sugar reserves (Skelly and Shoemaker, 1996).

In spite of the availability of oxygen in the host's blood, adult schistosomes have a fermentative metabolism and produce mainly lactate (Tielens, 1994), and the body of the penetrating cercaria, but not the tail, undergoes a rapid transition from an aerobic to a more anaerobic metabolism. Loss of the tail is not a prerequisite for the transition, and transfer of the cercaria from fresh water to a simple salt solution supplemented with glucose is sufficient to induce the change (Horemans, Tielens and van den Bergh, 1991). Preference for anaerobic energy metabolism in an environment where oxygen seems to be available is an unexplained phenomenon which we have met also in tapeworms (section 9.3.3).

The tegument covering the body of the cercaria is about 0.5 μm thick, with a single trilaminate surface membrane bearing externally a dense, fibrillar, carbohydrate-containing glycocalyx 1–2 μm thick (Fig. 18.11). The glycocalyx is thought to protect

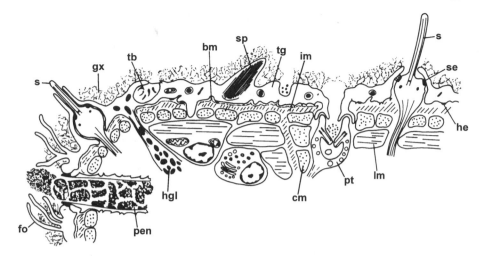

Figure 18.11 Diagrammatic representation of part of the tegument and associated structures of a schistosome cercaria. bm = basal lamina; cm = circular muscle fibre; fo = tegumentary fold; gx = glycocalyx; he = hemidesmosome; hgl = head gland duct; im = interstitial material; lm = longitudinal muscle fibre; pen = penetration gland duct; pt = multiciliated pit sensory organelle; s = sensillum; se = septate desmosome; sp = tegumentary spine; tb = tegumentary vacuole with tadpole-shaped bodies, near head gland opening; tg = tegument. Source: modified from Hockley, 1973.

the cercaria from osmotic stress in fresh water and the schistosomulum is rapidly killed in this medium. During transformation to a schistosomulum the glycocalyx is lost.

The glycocalyx is significant in another way. As a first line of defence against invading pathogens, the blood contains complement, a series of proteins occurring free in the plasma or attached to cell membranes (see Fishelson, 1989, Ramalho-Pinto, 1994). When activated, complement components may be directly cytotoxic or may promote the antiparasite activities of leucocytes. There is evidence that the glycocalyx activates complement, even in individuals not previously exposed to schistosomes; cercariae and schistosomula that still retain elements of the glycocalyx may be killed. Young parasites from the lungs and adults are relatively refractory to lysis by complement, and loss of glycocalyx may be important in the evasion of the complement mechanism. According to Gobert *et al.* (1997), paramyosin, which is present in the post-acetabular glands of the cercaria of *S. japonicum* and on the surface of lung flukes, may play a part in suppressing the complement pathway.

That the glycocalyx is antigenic is demonstrated by the *Cercarienhüllen Reaktion*, a serological test for schistosomiasis in which the glycocalyx interacts with antibodies in the serum of an infected person, forming an envelope around the cercaria (Smyth and Halton, 1983).

An even more remarkable tegumental change accompanies loss of the glycocalyx. Membranous vesicles, each measuring 100–150 nm in diameter, enter the tegument *via* cytoplasmic connections from the tegumental cell bodies (cytons; McLaren, 1980). These vesicles are enclosed by a double membrane, which, by fusion with the surface, is destined to replace the single trilaminate tegumental surface membrane, producing a new covering consisting of an inner and an outer trilaminate layer. Skelly and Shoemaker (1996) identified patches of tegumental immunofluorescence associated with a glucose transporter protein (see above) 15 min after transformation from a cercaria, and interpreted this as upwellings of transporter-containing double membrane from subsurface cytons. If this interpretation is correct, it implies that the schistosomulum has 500–1000 tegumental cytons spaced 2–10 μm apart.

The original trilaminate plasma membrane and the attached glycocalyx are jettisoned in the form of short-lived microvilli; about 40% of the original glycocalyx persists after the surface transformation and is available to interact with the host, as described above, until the last of it is shed a few hours later. This double surface membrane, which is also a feature of the blood-dwelling sanguinicolids and spirorchids (see Chapter 17), but not of digeneans such as *Fasciola hepatica* inhabiting other environments (McLaren and Hockley, 1977), will be considered later. The tegument of the plerocercoid of the tapeworm *Ligula intestinalis* also has a double apical membrane (section 10.5.4).

The external morphology of schistosomula recovered from host skin 3 h after penetration is similar to that of cercariae, except that surface sensilla are lost (see Wilson, 1987). By the time that the schistosomulum leaves the skin, all the midbody tegumental spines have disappeared and the tegument is pitted (Fig. 18.12). The apical area is a pitted spineless knob with no trace of secretory gland ducts. There is a pair of sensory papillae on this knob and others among the anterior spines.

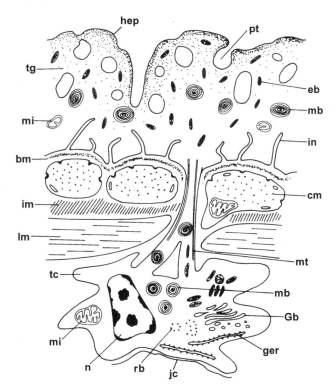

Figure 18.12 Diagrammatic representation of part of the tegument and associated structures of an adult schistosome. eb = elongated body; Gb = Golgi body; ger = granular endoplasmic reticulum; hep = heptalaminate outer membrane; in = invagination of basement membrane; jc = junctional complex with parenchymal cell; mb = membranous body; mi = mitochondrion; mt = microtubule; n = nucleus; pt = surface pit; rb = ribosomes; tc = tegumentary cyton; tg = tegument. Other lettering as in Fig. 18.11. Source: reproduced from Hockley, 1973, with permission from Academic Press.

18.3.10 The journey through the circulatory system

Isotopic labelling of parasites, followed by compressed organ radiography, is a powerful tool for studying the tissue migration of schistosomula after leaving the skin. In the mouse they follow an intravascular route (see Wilson and Coulson, 1989), being carried passively and rapidly between organs in the blood flow, but being obstructed by capillary beds (Fig. 18.13). At least 90% of the penetrants leave the skin and are carried *via* the right side of the heart and the pulmonary artery to the lungs, where they are ensnared by the lung capillary bed.

The lung capillary bed provides an obstacle for the larval stages of other parasitic helminths that enter the circulatory system. Most nematodes that reach this site are unable to progress further; they break out into the alveoli and eventually parasitize the bronchioles or, after travelling up the trachea, are swallowed and parasitize the gut (references in Crabtree and Wilson, 1986). Microfilariae are able to pass through the capillaries without morphological change, but schistosomula must change their shape if they are to reach their unusual vascular environment.

The schistosomula spend a minimum of 2–3 days in the lungs. Initially the parasites totally occlude arterioles 10–20 μm in width, but they more than double their maximum length, which reduces their minimum body diameter to 8 μm (about the same as a red blood corpuscle) and permits passage through distensible pulmonary capillaries 7–10 μm wide. A major developmental step accompanying elongation of the body is the loss of the fibrous interstitial layer that lies between the basal lamina of the tegument and the body wall musculature. These fibres probably oppose extension of the body. A new fibrous layer appears in the adult between the layers of circular and longitudinal muscle.

Migration through pulmonary capillaries in the mouse takes 30–35 h (Wilson and Coulson, 1986). Specimens freshly extracted from mouse lung undergo rhythmic cycles of extension and contraction (Wilson *et al.*, 1978) and make progress when introduced into glass capillaries, as shown in Fig. 18.14. The persistent anterior and posterior spines provide purchase on the walls of the blood vessel: the anterior spines while the body is contracted up behind the head, the posterior spines while the body is extended forward. The absence of spines in the midbody region will facilitate this movement by minimizing friction between the worm surface and the capillary endothelium.

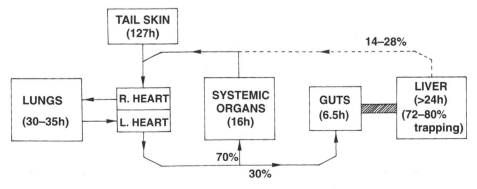

Figure 18.13 Route and dynamics of migration of *Schistosoma mansoni* in the mouse. Percentage distribution of schistosomula is shown, together with half-life of organ transit in parentheses. Shaded area = site of parasitization. Source: reproduced from Wilson and Coulson, 1989, with permission from Elsevier Trends Journals.

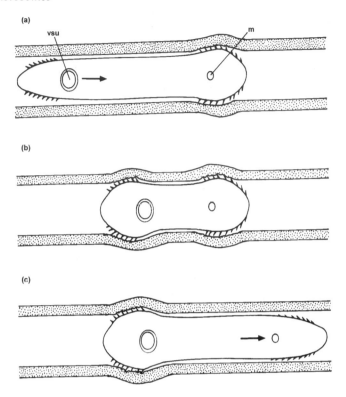

Figure 18.14 Diagrammatic representation of the movement of a schistosomulum along the lumen of a blood vessel, based on *in vitro* observations on schistosomula introduced into glass capillaries and stereoscan micrographs. (*a*) Elongated schistosomulum anchored by head spines and drawing posterior region up towards head. (*b*) Schistosomulum at minimum length. (*c*) Schistosomulum anchored by tail spines, while anterior and midbody are extended forward. m = mouth; vsu = ventral sucker. Source: redrawn from Crabtree and Wilson, 1980.

The barrier between the blood and the air in the mouse lung alveolus may be as thin as 0.1–0.2 μm (Crabtree and Wilson, 1986) and some parasites invariably break through into the alveoli. According to Wilson and Coulson (1989), this is literally the pitfall from which they cannot recover, since their ability to re-enter the tissues diminishes with age.

On leaving the lungs of the mouse, the parasites pass through the left side of the heart and are carried in the arterial blood to the systemic organs in proportion to the fractional distribution of cardiac output (Fig. 18.13). Those entering the splanchnic arteries pass through the gut capillary bed and reach the liver *via* the hepatic portal system.

During the systemic migration the schistosomulum is metabolically quiescent; cell division and growth do not occur (Crabtree and Wilson, 1980). On reaching the liver, new dramatic changes take place. The parasite shortens to the dimensions of a skin parasite and there is marked loss of motility, features that probably operate against passage through the hepatic sinusoids. For the first time the parasite begins to feed on blood cells and growth begins. Why these changes are initiated in the liver capillary bed and not in other organs is unclear; the nutrient-rich composition of hepatic portal blood may be an important factor, or the low pressure in the hepatic portal system.

As they grow they move upstream into progressively larger vessels (see Wilson, 1987). Small numbers of tegumental spines begin to reappear and the subterminal region around the mouth begins to enlarge to form the oral sucker. Between 28 and 35 days post-infection in *S. mansoni*, pairing takes place and each pair migrates along the hepatic portal vessels to the mesenteric or rectal veins. In *S. haematobium*, the paired worms travel *via* the inferior branch of the mesenteric vein and superior rectal vein to the rectal venous plexus. This communicates with the vesical plexus of the bladder wall and this porto-systemic anastomosis is large enough in man to permit the passage of worm pairs.

18.3.11   Circumventing the immune response

The most amazing feature of schistosomes is that they survive immersed in vertebrate blood, a tissue that provides a vehicle for a highly sophisticated immune system. It is even more surprising that this survival is not short-lived. Using a statistical approach, Fulford *et al.* (1995) calculated for *S. mansoni* a maximum likelihood estimate of the mean life span of 8 years, with a 95% confidence limit of 5.7–10.5 years.

In the 1960s, Smithers and Terry began publishing the results of their experimental studies of immunity to *S. mansoni* infection in Rhesus monkeys. This work captivated the imaginations of many parasitologists at the time and stimulated an enormous amount of research and debate that has still not lost its momentum.

Smithers and Terry (1967) focused attention on the fact that monkeys with an initial infection of adult *S. mansoni* were able to destroy a cercarial challenge infection, but at the same time were unable to rid themselves of the adults. Moreover, when they transferred adults surgically from monkeys to the portal system of uninfected monkeys, the recipient monkeys became almost totally resistant to cercarial challenge, in spite of a lack of previous experience of migrating juveniles. These monkeys also failed to reject the transplanted adults. Thus, the adult parasites were exempt from immune responses that they themselves had evoked and the term 'concomitant immunity' was borrowed from tumour biology, to describe this phenomenon (Terry, 1994). It is clear that concomitant immunity might create significant advantages for host and parasite in that, by preventing reinfection, overloading of the host would be avoided, allowing both host and parasite to survive. In the words of Professor George Nelson, quoted by Terry (1994), it is an example of 'the early worm getting the bird'.

Smithers, Terry and Hockley (1969) wondered whether the adult worms evaded the host's immune response by coating their surfaces with host molecules, i.e. by disguising themselves as host tissue. They postulated that newly invaded parasites had not had sufficient time to do this and therefore were vulnerable to host defences. They tested this hypothesis by conducting the surgical transplants of adult worms shown in Fig. 18.15.

The results of these transfers were as follows:

- Monkey worms thrived with uninterrupted egg output when transferred to another monkey.
- Mouse worms transferred to a monkey experienced a 5–6 week interruption of egg laying but then recovered.
- Monkey worms transferred to a monkey that had been immunized against mouse tissue (an 'anti-mouse' monkey) survived with no interruption of egg laying.

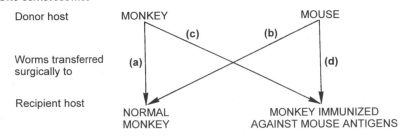

Figure 18.15   Protocol of experiments used to demonstrate presence of host antigens on the tegument of adult *Schistosoma mansoni*. See text for further explanation. Source: reproduced from Wakelin, 1984 (original data from Smithers, Terry and Hockley, 1969), with permission from Cambridge University Press.

- Mouse worms transferred to an 'anti-mouse' monkey ceased egg production and most were dead 25 h after transfer (all dead at 44 h).

These results were consistent with the notion that worms camouflaged their exposed surfaces by acquiring host antigens.

There is considerable evidence that molecules similar to or identical with those of the host are present on the surface of all but the early stages of blood-dwelling schistosomes (references in Terry, 1994). These appear to be from two sources. Some are synthesized by the parasite, for example a molecule closely resembling mouse α_2-macroglobulin. Others are sequestered from the host, for example glycolipids with blood-group specificity and products of the major histocompatibility complex.

The outer of the two tegumental surface membranes is probably the site of these host-like molecules. In newly-transformed schistosomula the inner membrane contains many intramembrane particles, which represent membrane proteins, and resembles the plasma membrane of most mammalian cells, but the outer membrane is protein-poor with few intramembrane particles (Caulfield, 1990). However, the outer membranes of lung worms isolated from mice about a week after infection contain intramembrane particles, whereas outer membranes of worms cultured *in vitro* for the same period do not. Thus, the outer membrane may become charged with host molecules picked up directly from the host, possibly by fusion with adjacent host cells or cell fragments, or manufactured by the parasite as a result of stimulation by the host.

As Butterworth (1990) has pointed out, it is not easy to understand why host-derived molecules freely mobile in the outer lipid membrane of the worm should prevent binding of antibody to a second independent parasite molecule, and a puzzling finding is that resistant parasite stages can still bind antischistosome antibodies (see Ramalho-Pinto, 1994). Important progress in reconciling the paradoxical situation in which adult parasites, but not invading schistosomula, survive in an environment rich in specific antibody and complement is reported by Ramalho-Pinto (1994). It has been shown that schistosomula of *S. mansoni* become resistant to complement by incorporating into their surface a complement-inhibiting protein (so-called decay-accelerating factor) released from human erythrocytes. This factor is expressed in the cell membranes of exposed host cells, such as vascular endothelial cells and erythrocytes, and has an important role in preventing complement activity on these host surfaces.

Other factors may be involved in the loss of susceptibility to immune attack by developing schistosomula. For example, Schmidt (1995) found that the surfaces of male and female *S. mansoni* are entirely covered with glycans, derived it seems from the parasite, and containing a high density of N-acetyllactosamine residues that are immunologically inconspicuous. It has also been suggested that serpins (serine protease inhibitors) may form complexes with parasite proteases, so rendering these enzymes inaccessible to the immune system (Modha, Roberts and Kusel, 1996).

A quite different explanation for the manifestations of concomitant immunity in the mouse has been advanced by Wilson (1990). The observation that concomitant immunity develops after the commencement of oviposition in mice with bisexual infections of *S. mansoni* and does not develop at all in unisexual infections that do not oviposit focuses attention on the key role of the eggs. There is no evidence that eggs produce antigens that stimulate the immune system to attack schistosomula. It appears to be the pathological changes caused by eggs impacted in the liver that are important. Eggs carried astray lodge in the portal venules of the liver and initiate a granulomatous reaction that effectively blocks portal flow to functional units of liver tissue. Portal hypertension develops, leading to the appearance of connections within the liver between portal and hepatic veins and major anastomoses between portal and systemic vascular beds outside the liver, e.g. between the veins of the stomach wall and the oesophagus. Schistosomula of a challenge infection will escape from the portal system *via* these shunts and are likely to be trapped and eliminated in the lungs. Some strains of mice appear to be resistant to schistosome infection because they have a predisposition to form portal-systemic anastomoses in the absence of egg-laying parasites. In humans, the existence of portal shunts in individuals infected with schistosomes is well known, but the contribution that this makes to resistance to challenge infection is undetermined.

Other observations offer a challenge to the concept of concomitant immunity (see Hagan and Wilkins, 1993). Highly irradiated cercariae administered to mice do not mature, but the hosts become immune, especially after multiple doses, without any exposure to adult worms (so-called 'sterile immunity'). Thus, resistance stimulated by adult worms is not the only form of acquired resistance. There is also evidence that adult worms of a primary infection are not entirely resistant to immunological attack. In studies of naturally acquired immunity to *S. bovis* in cattle and experimental work on *S. haematobium* in baboons, it was suspected that antifecundity effects had an immune basis and this notion was supported by the resumption of full egg production by worms surgically transferred from immune to naive hosts. Since the pathology of schistosomiasis is mostly attributable to the effects of eggs that go astray, immune events that act to reduce egg deposition by adult worms would be expected to have a strong selective advantage.

McCullough and Bradley (1973) and Bradley and McCullough (1973) studied the epidemiology of *S. haematobium* in a Tanzanian population and suggested that concomitant immunity played a significant role in the population dynamics of the parasite, especially in the middle years of life. However, a field study carried out in the Gambia pointed to a different interpretation. Wilkins *et al.* (1984) used molluscicide to interrupt transmission of *S. haematobium* in one half of an endemic area. It was found that infection levels in the molluscicide-treated area declined relative to those in the untreated area, where transmission was continuing. This is not consistent with the notion of concomitant immunity, and there is a growing appreciation that the relative stability of egg counts in studies like that of Bradley and McCullough could be a consequence of continued acquisition and loss of worms, i.e. a dynamic equilibrium.

18.3.12   How the immune system kills schistosomula

Within the range of mammalian species that are naturally or experimentally infected with schistosomes, the mechanisms of immunity appear to be multifactorial, with different mechanisms predominating in different hosts and acting at different stages during differentiation of the schistosomulum (Butterworth, 1990). Some of the effective mechanisms have a wide application against invaders such as microorganisms, but others appear to be directed selectively against helminth parasites, including schistosomes, and may represent a special adaptation by the mammalian host for coping with large, tissue-invasive pathogens. Among the various cell types that have been reported to damage schistosomula *in vitro*, eosinophils are particularly effective, especially in the rat and in man, and the elevated levels of immunoglobulin E (IgE) and eosinophils, which are closely associated in helminth infections, may well constitute such a special adaptation (see also section 20.3).

Glauert *et al.* (1978) studied antibody-dependent interactions between human eosinophils and schistosomula of *S. mansoni in vitro*. Human eosinophils initially adhere to the intact schistosomulum and then, in the presence of antibody, spread out intimately over the parasite's surface (Fig. 18.16*a*). The secretory granules of the eosinophil then fuse with the plasma membrane of the cell, thereby ejecting their contents on to the parasite's tegument. This degranulation is followed by vacuolation of the inner layer of the tegument and detachment of the tegument, often in large sheets. The tegument then disintegrates and the fragments are phagocytosed by eosinophils that have not degranulated. Eosinophils then attach themselves to the exposed muscle layers of the parasite and the process of attack and phagocytosis continues. Limited observations on damage to schistosomula *in vivo* were compatible with the view that these *in vitro* studies may reflect the process of destruction in immune hosts.

In the rat, the secretory granules of the eosinophil fuse to form vacuoles, which then eject their contents on to the surface of the schistosomulum (Fig. 18.16*b*; McLaren, Ramalho-Pinto and Smithers, 1978). The lethal cocktail of secretion com-

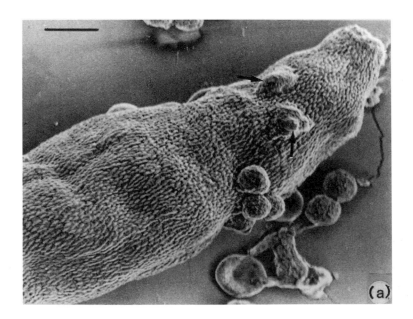

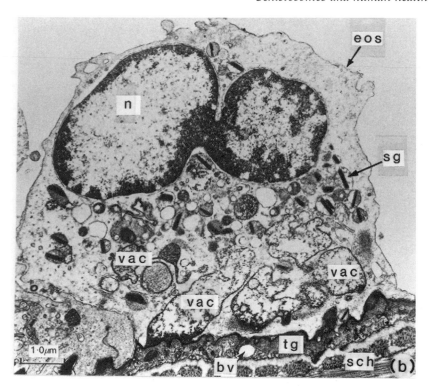

Figure 18.16 Interaction between eosinophils and the schistosomulum of *Schistosoma mansoni*. (*a*) Schistosomulum after incubation with antibody and human eosinophils. Some of the eosinophils are still rounded, others (arrows) are flattened on to the tegument. (*b*) Transmission electron micrograph of a section through a rat eosinophil (eos) attached to the surface of a schistosomulum (sch) after 1 h in culture in fresh normal rat serum. bv = basally situated vacuole; n = nucleus of eosinophil; sg = secretory granule; tg = tegument of schistosomulum; vac = vacuole of eosinophil. Scale bars: (*a*) 5 μm; (*b*) 1 μm. Source: (*a*) permission for reproduction granted to GCK by Drs Audrey Glauert and A. E. Butterworth; (*b*) reproduced from McLaren, Ramalho-Pinto and Smithers, 1978, with permission from Cambridge University Press.

prises hydrolytic enzymes, including peroxidase, and a major basic protein from the crystalloid of the secretion granule.

Human eosinophils from uninfected individuals are capable of inflicting damage on schistosomula *in vitro* but in infected individuals there is an increase in the numbers of circulating eosinophils (eosinophilia) and these cells show a markedly enhanced capacity to kill schistosomula (Butterworth, 1990). This enhancement may be associated with the release by monocytes of substances that stimulate eosinophil functioning.

18.4 SCHISTOSOMES AND HUMAN HEALTH

According to the World Health Organization (WHO) Expert Committee on the Control of Schistosomiasis (1993), 200 million people are infected with schistosomiasis and 20 million of these may be seriously ill. Of the remaining 180 million, 50–60% also have symptoms, generating a public health problem of enormous proportions.

18.4.1   Pathology

Symptoms depend on the species and on the duration and intensity of infection (see Warren, 1978; WHO Expert Committee, 1993). Cercarial densities can reach several thousand per litre but more commonly range from 0.1–1.0 per litre (Butterworth, 1990). Therefore, inhabitants of endemic areas usually acquire parasites by slow trickle exposure, rather than by brief exposure to a large dose as in most experimental infections in laboratory models. Consequently, indigenous populations rarely display early clinical symptoms, which include a temporary dermatitis (swimmer's itch) at the site of cercarial penetration and an acute condition known as Katayama fever, which occurs between 2 and 11 weeks post-infection (p.i.). The latter is associated with *S. japonicum* and less frequently with *S. mansoni* and, as well as fever, involves abdominal pain, nausea and diarrhoea; it abates spontaneously after a few months (Weinstock, 1990). Acute symptoms are of increasing importance among Western visitors to endemic areas (see Jordan and Webbe, 1993).

An epidemic of acute cases involved troops of the Chinese People's Liberation Army after their occupation of Shanghai in the Civil War (see Farley, 1991). The troops were required to learn to swim to prepare for the continuing campaign and several thousand soldiers became infected with *S. japonicum*, leading to a temporary suspension of military training. It was 6 months before the army regained its fighting strength and during that time the United States Seventh Fleet appeared off the coast, deterring any plans to invade Formosa (Taiwan), which was occupied by the Nationalist Chinese. This led to the claim that *S. japonicum* had saved Formosa and possibly altered the course of history in South-East Asia.

Most infected individuals of the indigenous population gradually enter a chronic phase of the disease and the following account is taken from Weinstock (1990) and Butterworth (1990). This involves only minor, although unpleasant, symptoms: blood in the urine (regarded as male menstruation in parts of Africa, since it often coincides with puberty), occasional diarrhoea and cramps. Others go on to develop the classic symptoms of chronic bilharzia. This is brought about not by a virulence factor from the parasite but by the host's reaction (or over-reaction) to the presence of eggs impacted in the tissues. Egg antigens leaking through the shell attract a granulomatous mass of lymphocytes, macrophages and eosinophils in excess extracellular matrix.

Granulomas block blood vessels, contributing to host pathology. However, there is evidence from work on mice that they are not entirely detrimental. Mice deprived of T cells are unable to make granulomas, but they show increased liver cell necrosis and increased mortality. Thus, granulomas may protect the host against toxic parasite products. Moreover, there is evidence that, at about 16 weeks p.i. in mice, the granulomatous reaction is modulated; new granulomas are smaller and therefore do less damage. There is indirect evidence that a similar modulation occurs in chronically infected humans.

In some individuals, but not in others, granuloma formation may be followed by fibrosis. In *S. haematobium* infections fibrosis of the bladder may obstruct the ureters and lead ultimately to kidney failure. In *S. mansoni* liver fibrosis may produce portal hypertension, swelling of the liver and spleen, ascites formation and the development of oesophageal varices. Liver cell function is relatively well maintained, and rupture of the varices and the consequent loss of blood is a major cause of death. It has also been accepted that a causal relationship exists between *S. haematobium* and squamous cell carcinomas of the bladder (Mostafa, Badawi and O'Connor, 1995).

18.4.2 Ecology

Unless stated otherwise, the following account is derived from Jordan and Webbe (1993).

The essential requirement for transmission is human contact with water inhabited by the appropriate snail host, in a community with poor sanitation in which some individuals are infected. Projects such as dam building and irrigation schemes may create new aquatic habitats and contribute to the spread of the disease. The Aswan project in Egypt has not only promoted an increase in the incidence of *Schistosoma* spp. but, by changing water flow, is probably responsible for a change in the relative abundance of the snail hosts. This has led to a replacement of the less pathogenic *S. haematobium* by *S. mansoni* as the commonest species in the Nile Delta, and the spread of the latter species south along the Nile Valley where it was previously absent (see Farley, 1991). Less frequently, dams may lead to a fall in transmission by reducing seasonal flooding, as is the case with the Koka Dam in Ethiopia.

Reasons for human contact with water are many and include domestic, recreational, occupational and religious contacts. Some involve greater risk of infection than others. While collecting water, for example, individuals are exposed only for a short period, frequently early in the morning when cercarial shedding is minimal. In contrast, swimming is a high-risk activity with whole body exposure for longer periods, often during the heat of midday when cercarial shedding is maximal. Bathing may be equally dangerous but the risk is much reduced by the use of soap, which is lethal to cercariae. Ritual washing by Muslim males increases the risk, and infected snails have been found in water containers provided for this purpose in the Yemen.

In China the snail *Oncomelania hupensis* is amphibious and releases cercariae into drops of rain or dew on emergent vegetation. Individuals can be infected by contact with this wet vegetation.

Less is known about how human waste enters water bodies, mainly because of the privacy associated with these functions. However, urination while swimming or bathing is probably of major importance in the transmission of *S. haematobium* and provides direct access to the water for the eggs. It is probably significant that these activities occur around midday and correspond with peak appearance of eggs in the urine (section 18.3.3). Direct defecation into water is less common and many eggs of the intestinal schistosomes probably reach water when faeces deposited in vegetation on the banks of pools and rivers are washed into the water by heavy rain or immersed as the waters rise. Only freshly deposited faeces are likely to maintain transmission, since the viability of ova declines within a few days when exposed to hot sun. Eggs of *S. mansoni* adhere to perianal skin and can be washed off during bathing – an activity observed to be common after defecation in Kenya and the Sudan. Human excreta may be carried to water on the hooves of domestic cattle, but of much greater significance in the propagation of the disease is the role of these animals as carriers of *S. haematobium*.

Processes of sedimentation, decantation and filtration involved in sewage treatment greatly reduce the egg content, but significant numbers of eggs survive in the

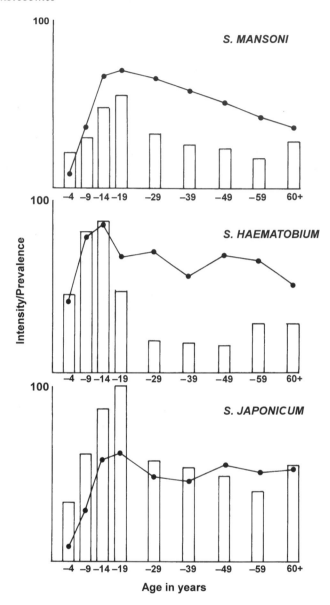

Figure 18.17 General patterns of prevalence (graph) and intensity (histograms – geometric means of egg output) from different well-established areas of schistosomiasis. Source: reproduced from Jordan and Webbe, 1993, with permission from CAB International

effluent. Attempts to prevent faecal contamination of water bodies in Egypt by provision of latrines failed because they rapidly became filthy, fly-infested pits which the population refused to use (Farley, 1991), although modern ventilated pits are more acceptable (Webbe and Jordan, 1993).

In developing countries, children form a high proportion of the population and are responsible for much transmission. In spite of exposure, infection in the first 12 months of life is rare, and it is possible that infants gain protection by transplacental or transmammary transfer of maternal antibodies. In *S. mansoni* and

S. haematobium, prevalence and intensity gradually increase through childhood to peak in the 15–29-year age range, before falling in the older age groups (Fig. 18.17). There seems little doubt that the development of immunity is one component in the reduced infection intensities of older people, with reduced water contact, tissue changes hindering egg escape and reduced egg output by parasites as other possible contributory factors. If the age–intensity pattern is due to developing resistance, then the *S. japonicum* pattern probably indicates that immunity is poorly developed.

Rainfall and temperature profoundly influence transmission patterns, and marked seasonality occurs in some areas. Although rainfall may promote transmission by washing faeces into water bodies, heavy precipitation may wash away snails and have the reverse effect. This is illustrated by the situation in the Gambia discussed by Wright (1971). This small, narrow country is dominated by a main river valley with tributary streams. These become swollen and turbulent during the rainy season and their snail populations and transmission of *S. haematobium* are severely reduced. When the rains cease, water flow eases and water levels fall. The remaining snails multiply, especially in areas of slack water, and human contact increases, leading to dry-season transmission. However, within a 20-mile (32 km) radius, there are sites where there is no transmission (in the main river and adjacent rice fields, where the appropriate snails are absent) and intense wet-season transmission (in temporary pools on the laterite plateau on each side of the river).

18.4.3   Control

The control of schistosomiasis is a topic too large for detailed consideration here. It is a goal that has occupied and eluded parasitologists since the causative agents were identified by Bilharz in the middle of the 19th century. The fascinating story is told in detail by Farley (1991) and the interested reader is referred to this account.

CHAPTER 19

Lesser lights – didymozoideans and aspidogastreans

19.1 THE DIDYMOZOIDEANS

The most remarkable among many unusual features of the endoparasitic didymozoideans is the habitat of the adult parasites. They have established themselves in fishes, not in the gut or in the blood system but in dense tissue sites such as dermal connective tissue and muscle. All have one or more partners and are either encysted in these immunologically competent sites or survive without a protective cyst wall.

A popular view is that didymozoideans are digeneans. Some go as far as to regard them as relatives of the hemiurids (Cable and Nahhas, 1962b) and to describe their unciliated larvae (Fig. 19.1) as miracidia (Self, Peters and Davis, 1963; Williams and Jones, 1994), although no-one has yet demonstrated that there is a mollusc first intermediate host. Self, Peters and Davis (1963) offered embryonated eggs of *Nematobothrium texomensis* to nine species of snail, which, by virtue of their cohabitation with the freshwater fish definitive hosts (buffalo fishes), were potential first intermediate hosts, but, although the eggs were ingested, hatching of a significant number of eggs occurred only in *Succinea avara* and no larval development occurred in any of them.

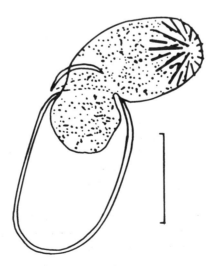

Figure 19.1 Egg of *Didymozoon faciale*, with embryo escaping under pressure of coverglass. Scale bar: 10 μm. Source: reproduced from Baylis, 1938.

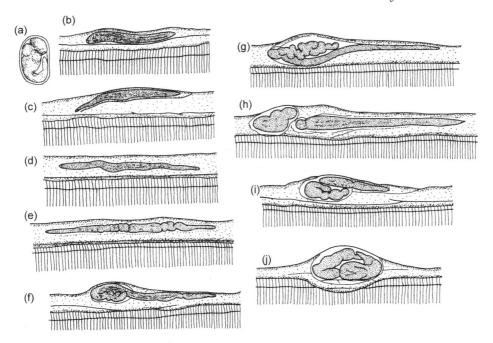

Figure 19.2   *Didymocystis katsuwonicola.* (*a*) Egg with fully formed larva. (*b*)–(**j**) Successive stages in the establishment of two larvae in the host's primary gill lamella. Source: reproduced from Ishii, 1935.

A quite different view was expressed by Ishii (1935) and by Grabda and Siwak-Grabda (1940–1947) and supported by Baer and Joyeux (1961). They considered that the tissue parasite is derived directly from the larva after penetration of the fish host. Grabda and Siwak-Grabda published photographs (their figs 9–11) of consecutive histological sections that they claimed showed a larva of *Nematobothrium sardae* penetrating the host's gill. The development of the tissue stage of *Didymocystis katsuwonicola* was described and illustrated by Ishii (Fig. 19.2). The smallest tissue stage found by Ishii was already about 1 mm in length, and since the egg is merely 17–19 μm by 10–12 μm, there still remains a considerable gap in our knowledge of early events.

Inconsistent with this idea of a direct life cycle are reports of intermediate or paratenic hosts containing immature didymozoideans (Fig. 19.3), implying that the definitive fish host is infected by ingesting infected prey. Parasites regarded as didymozoidean metacercariae have been reported from crustaceans (from goose barnacles by Cable and Nahhas (1962b) and from a copepod by Madhavi (1968); Fig. 19.3*a*), observations that are quoted in support of an affinity with hemiurids. Similar organisms regarded as immature didymozoideans have been found either in the gut lumen or in thin-walled cysts in the gut wall of fishes likely to be preyed upon by hosts of adult didymozoideans (see Cable, 1956 and Lester, 1980; Fig. 19.3*b*). Lester suggested that, after ingestion by the definitive fish host, these young didymozoideans made their way to their definitive sites by migration through the tissues, but feeding experiments did not produce adults.

Tissue-dwelling adult didymozoideans reveal interesting evolutionary trends. In genera such as *Didymozoon* and *Didymocystis* the greatly elongated, filiform, coiled gonads and uterus are accommodated in an inflated, sometimes spherical, posterior body region, from which projects an elongated, more or less flattened anterior

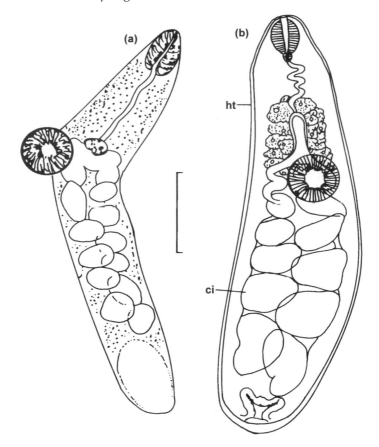

Figure 19.3 Immature parasites assumed to be didymozoideans (*a*) from the crustacean *Paracalanus aculeatus* and (*b*) from the gut wall of the fish *Favonigobius exquisitus*. ci = chamber of intestine; ht = thin sheath of host tissue. Scale bar: 100 μm. Source: reproduced (*a*) from Madhavi, 1968 and (*b*) from Lester, 1980, with permission from the American Society of Parasitologists.

region through which the gut and terminal reproductive ducts thread their way (Fig. 19.4*a*, *b*). Other didymozoideans (*Metanematobothrium*, *Nematobibothrioides*) have a uniformly thread-like or ribbon-like body, often distinguished by its great length (Fig. 19.5). In the sunfish *Mola mola*, Noble (1975) found a specimen of *Nematobibothrioides histoidii* that exceeded 12 m in length, although the width of the worm was a mere 1–2 mm. After stripping the skin from the bodies of sunfish hosts, the thread-like worms were revealed dipping deeply between muscles and traced from one side of the fish to the other under the dorsal fin. The enormous surface area created by the adoption of this shape seems well-suited for uptake of nutrients from the host directly across the tegument and since the elongated parasites lie free in the host tissue there is no physical barrier to such an uptake route. This has probably led to reduction of the pharynx in didymozoideans such as *N. histoidii* (see Noble, 1975) and to loss of all of the digestive tract in *Nematobothrium pelamydis* (see Baer and Joyeux, 1961).

The presence of oral and ventral suckers in some didymozoideans is a digenean feature, but reduction or loss of one or both of these suckers is common (Baer and Joyeux, 1961) and probably reflects the increasing immobility of the tissue-enclosed parasite and lack of the need for attachment. This is expressed further by the

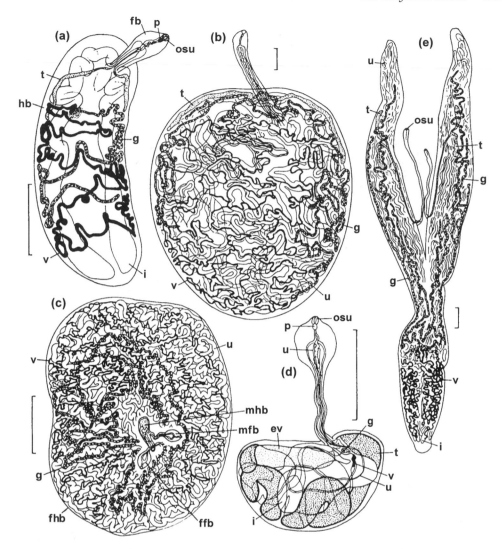

Figure 19.4 Didymozoideans. (*a*) *Didymozoon brevicolle* (uterus omitted in hindbody). (*b*) *Didymocystis alalongae*. (*c*) Pair of individuals of *Coeliotrema thynni*. (*d*) Male individual of *C. thynni*, enlarged. (*e*) *Diplotrema pelamydis*. ev = excretory vesicle; ffb = female forebody; fhb = female hindbody; g = germarium; hb = hindbody; i = intestine; mfb = male forebody; mhb = male hindbody; osu = oral sucker; p = pharynx; t = testis; u = uterus; v = vitellarium. Scale bars: 1 mm. Source: reproduced from Yamaguti, 1938.

apparent absence of muscle in the body of the thread-like *Nematobothrium texomensis* (see Self, Peters and Davis, 1963). Self *et al.* also failed to find flame cells or excretory tubules in adult *N. texomensis*, indicating an even more intimate physiological compatibility between parasite and host tissue.

A second trend concerns the reproductive system. There is a tendency for individuals associated in pairs in cysts or tissue sites to become unisexual (see Baer and Joyeux, 1961), a trend that parallels developments in the fecampiid turbellarians (section 2.3.3), in the tapeworms (section 10.6) and in the schistosomes (section 18.1). In *Didymozoon*, individuals are hermaphrodite (Fig. 19.4*a*). This is also

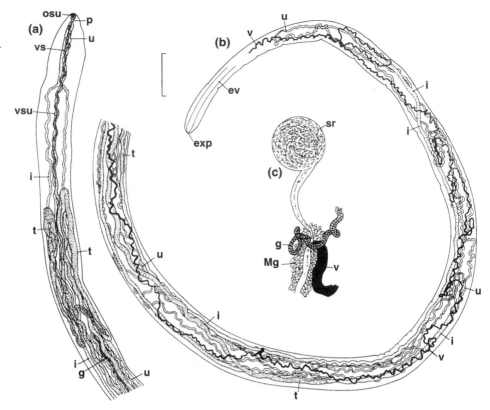

Figure 19.5 A thread-like didymozoidean, *Metanematobothrium guernei*. (*a*) Anterior region. (*b*) Posterior region. (*c*) Genital junction. ev = excretory vesicle; exp = excretory pore; g = germarium; i = intestine; Mg = Mehlis's gland; osu = oral sucker; p = pharynx; sr = seminal receptacle; t = testis; u = uterus; v = vitellarium; vs = vas deferens. Scale bar: (*a*), (*b*) 1 mm. Source: redrawn from Yamaguti, 1938.

the case in *Nematobothrium* and *Didymocystis* (Fig. 19.4*b*), but here there is a slight sexual dimorphism, the female sexual organs in the smaller of the two partners being less developed. This sexual dimorphism goes much further in *Coeliotrema* (Fig. 19.4*c*, *d*), *Wedlia* and *Koellikeria*. In *C. thynni* one individual of the pair is tiny and located in the centre of the larger partner. This small individual has well-developed male organs but rudimentary female ones (Fig. 19.4*d*), while the larger individual is wholly female (Fig. 19.4*c*). Williams (1959) observed rudiments of the organs of the opposite sex both in male and in female individuals of *K. filicollis*. Baer and Joyeux (1961) suggested that the sex of the individuals may be determined by interaction between the partners; assuming that they are basically protandrous hermaphrodites they proposed that the first individual proceeds to development of the female system and then arrests the development of the later arrival at the earlier male phase.

Another development of special interest is the complete fusion of the posterior regions of two hermaphroditic individuals of *Diplotrema pelamydis* (Fig. 19.4*e*), a phenomenon that finds a striking parallel in monogeneans of the genus *Diplozoon* (section 6.5.3).

Uncertainty surrounds the route taken by eggs from the tissue-enclosed adults to the outside world. Lester (1980) claimed that ulceration and rupture of the host tissue enclosing the parasite provided an opportunity for egg release in *Neo-metadidymozoon helicis*, which is located superficially in the dermis of the buccal cavity and gill arches of the marine fish *Platycephalus fuscus*. This is reminiscent of the mechanism of egg release by the nematode *Dracunculus medinensis*. *Nematobo-thrium texomensis* is unique in its occupation of the host's gonads (usually the ovaries), a site that provides unimpeded exit for the eggs of the parasite *via* the gonopore. Self, Peters and Davis (1963) observed worms extending from the genital pore of female hosts and found only fragments of worms in the ovaries after spawning, suggesting that the eggs are liberated by disintegration of the worms. Noble (1975) found small masses of eggs associated with *Nematobibothrioides histoidii* in the connective tissue of the sunfish host and isolated 'strings' of eggs which he believed were left behind as the parent 'snaked' its way though the tissues. He suggested that these eggs were released only when sunfish were attacked by pred-ators. However, Noble also observed that the heads of the worms frequently lay just below the skin, raising the possibility that some eggs may reach the outside world *via* a breach in the skin, perhaps by way of ulceration as in *Neometadidymozoon helicis*. Lester (1980) is of the opinion that release of eggs *via* ulcers may be widespread in didymozoideans.

With regard to release of eggs by predation on the final host, Lester (1980) found that the larvae of *Nematobothrium spinneri* and *Didymozoon brevicolle* were still alive within eggs fed experimentally to fishes and collected from their faeces. Kamegai (1971) suggested that eggs found in human faeces and associated with the consump-tion of flying fish (*Cypselurus* sp.) were derived from muscle-dwelling didymozoi-deans.

19.2   THE ASPIDOGASTREANS

The aspidogastreans are regarded as close relatives of the digeneans. Indeed, mod-ern DNA sequencing techniques point to a common ancestry, a conclusion that is not readily reconciled with their biology (Chapter 3). If they are as ancient as the digeneans, then there is no better example of how different the rate of evolution can be in two related groups parasitizing the same group of hosts, the molluscs. The spectacular expansion of the digeneans is arguably unequalled in the parasitic platyhelminths, while the aspidogastreans are relatively insignificant, with a mere 11 genera listed by Yamaguti (1963).

Nevertheless, although there are few aspidogastreans, some aspects of their biology are intriguing. In particular, they make up for their small numbers in the remarkable versatility of their life cycles. Like the graffillids, some of them inhabit the gut of gastropod or lamellibranch molluscs; others live in the renal and pericar-dial cavities of lamellibranchs (Fig. 19.6). We recall that the gut contents of these molluscs are propelled by cilia and gut flow is relatively weak (section 2.3.2); similarly any flow from the pericardial cavity through the kidney is likely to be feeble. Consequently, endosymbiotes of molluscs have little need for a strong attachment organ and in graffillids this is entirely lacking. It is therefore surprising to learn that aspidogastreans have a remarkably complex multiloculate ventral attachment organ (see Figs 19.6, 19.10).

The incongruity of this elaborate holdfast was recognized by Huehner, Hannan and Garvin (1989), who suggested that it may have a secondary role in feeding. In

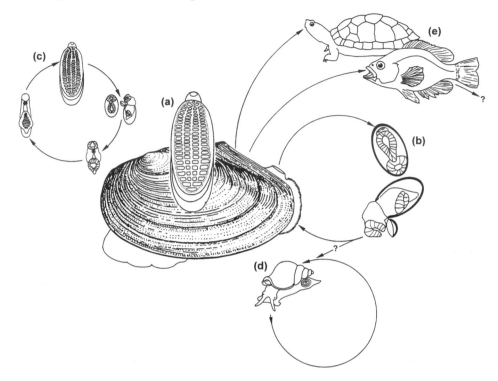

Figure 19.6   The life cycle of *Aspidogaster conchicola*. Adult (*a*) in pericardial cavity or kidney of bivalve mollusc. Eggs (*b*) leave host *via* exhalant siphon. Eggs or freshly hatched larvae drawn into new hosts *via* inhalant siphon. Autoinfection (*c*) may occur. Parasites also cycle between gastropod molluscs (*d*) and survive in the intestines of mollusc-feeding fish and turtles (*e*).

support of this they reported that so-called marginal bodies (Fig. 19.7), located between the marginal loculi of the holdfast, are gland reservoirs containing esterases, and they directed attention to the identification by Trimble, Bailey and Nelson (1971) and Trimble, Bailey and Sheppard (1972) of phosphatases and esterases in the ventral attachment organ of *Aspidogaster conchicola*. Huehner, Hannan and Garvin (1989) also found evidence that cells of the host digestive epithelium were ingested by *A. conchicola* and Gentner (1971) claimed that they ingested host blood cells. Hence the contribution to feeding, if any, made by the ventral attachment organ remains obscure.

The larvae of *A. conchicola* (see below and Fig. 19.8) have an oral sucker and a single postero-ventral sucker (see, for example, Williams, 1942). The multiloculate ventral holdfast of the adult appears during larval development and, according to Williams (1942) and to Huehner and Etges (1977), incorporates the larval postero-ventral sucker in *A. conchicola*. Strong positive allometric growth characterizes development of the holdfast and, according to Huehner and Etges (1977), the first step is the appearance of a transverse septum which divides off a new anterior loculus from the larval postero-ventral sucker. New transverse septa develop in sequence one behind the other, separating off new transverse loculi as the holdfast grows in an anterior direction (Fig. 19.8). When 11 or 12 transverse loculi have developed (parasite length 0.9–1.1 mm), a median longitudinal septum appears and subdivides the anterior three to five loculi. In parasites with 20 transverse loculi,

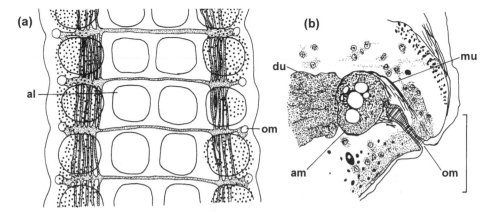

Figure 19.7 The marginal glands of *Multicotyle purvisi*. (*a*) Part of the holdfast organ of the adult showing the extent of the glands (course stipple) and gland ducts (fine stipple). (*b*) Section through the opening of a marginal gland. al = alveolus of holdfast organ; am = ampulla; du = gland duct; mu = muscle; om = opening of marginal gland. Scale bar: (*b*) 50 μm. Source: (*a*) reproduced from Rohde, 1971, with permission from Gustav Fischer Verlag, Jena; (*b*) reproduced from Rohde, 1966, with permission from Springer-Verlag.

secondary longitudinal septa appear on each side of the median longitudinal septum. By the time the parasite has reached a length of 1.4–1.5 mm at 180 days the rest of the 22–26 transverse loculi are completed, as is the longitudinal septation. At 1.6 mm positive allometric growth of the holdfast ceases and growth rates of the holdfast and the body are then equal.

The observations of Williams (1942) on *A. conchicola* suggest that there is a fundamental change in the mode of locomotion as development proceeds. Young post-larvae move like leeches using the oral sucker and the, as yet undeveloped, postero-ventral sucker, but, after differentiation of the holdfast, the oral sucker no longer plays a part in locomotion and creeping is presumably achieved by a shuffling

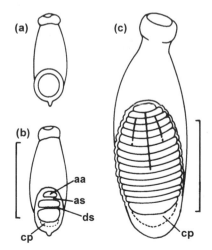

Figure 19.8 Consecutive stages in the development of *Aspidogaster conchicola*. (*a*) Prior to septum formation. (*b*) New anterior alveolus (aa) formed by a completed anterior septum (as); second developing septum (ds) with crescentic pit (cp) of posterior alveolus. (*c*) Juvenile, showing onset of longitudinal septum formation. Scale bars: 0.5 mm. Source: reproduced from Huehner and Etges, 1977, with permission from the American Society of Parasitologists.

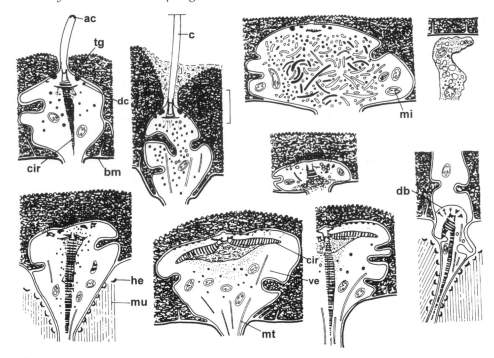

Figure 19.9   A few of the sensilla on the body of a single juvenile *Lobatostoma manteri* from the marine prosobranch mollusc *Cerithium moniliferum*. ac = apical cap of cilium; bm = basal lamina; c = cilium; cir = ciliary rootlet; db = dense bar; dc = dense collar; he = hemidesmosome; mi = mitochondrion; mu = muscle fibre; mt = microtubule; tg = tegument; ve = vesicle. Scale bar: 1 μm. Source: reproduced from Rohde, 1989, copyright © Springer-Verlag 1989, with permission from Springer-Verlag.

motion of the holdfast loculi. Thoney and Burreson (1987) described waves of undulating motion, independent of the dorsal portion of the body, passing down the holdfast of *Multicalyx cristata*. Rohde (1973) observed specimens of *Lobatostoma manteri* creeping from the digestive gland of its snail host into the stomach and back again.

Another unusual feature of aspidogastreans is the great abundance of sensilla on their bodies (Fig. 19.9) and the corresponding complexity of the nervous system (Rohde, 1968). According to Rohde (1989), the sensilla exceed, both in variety and number, those displayed by other parasitic platyhelminths and indeed by free-living flatworms, and this feature of aspidogastreans is totally inconsistent with the idea that parasitism leads to a reduction in anatomical complexity (so-called sacculinization). Rohde believed that the diversity and abundance of sensilla is related to the avoidance of damage to the snail host, but how such damage control would operate is unclear. It is more likely that this degree of sensory complexity is linked with behavioural complexities as yet unknown. At its simplest level, locomotion involving the coordination of the many suctorial loculi of the holdfast would place a heavy demand on the nervous system.

Perhaps the most interesting features of the aspidogastreans concern their life cycle and host range. *A. conchicola* has been reported from freshwater bivalves belonging to the families Unionidae, Mutelidae, Sphaeriidae and Corbicularidae (see Rohde, 1972), where it most commonly inhabits the pericardial cavity or kidney (Fig. 19.6). In addition, *A. conchicola* is said to parasitize gastropods of the genus

Viviparus (see Michelson, 1970), but in these molluscs it inhabits a different site – the ducts of the digestive gland. The eggs of aspidogastreans are operculate with tanned shells (Fried and Haseeb, 1991), and Huehner and Etges (1977) found embryonated eggs in the faeces of infected gastropods, but these failed to hatch when kept in pond water and were all dead at 35 days. On the other hand, uninfected snails became infected when they ingested embryonated eggs from the faeces of infected snails and the unciliated larvae emerging from these eggs migrated to the digestive gland from the intestine. The life cycle from egg to egg was completed in 270 days at 20°C.

How *A. conchicola* reaches the kidney of bivalves is still uncertain. The finding of young parasites in the kidneys led Bakker and Davids (1973) to suggest that fully developed eggs hatch after being drawn into the mantle cavity in the inhalant current and that the larvae then find their way to the kidney openings. Since eggs within the uterus of adults may contain active embryos (Williams, 1942) autoinfection is also possible. Voeltzkow (in Rohde, 1972) found young parasites in the intestine and believed that the pericardial cavity was reached by penetration from the adjacent intestine. Whichever route the parasites take, it is truly remarkable that they have the behavioural flexibility to follow different routes and to occupy very different microhabitats depending on whether they encounter a snail or a clam. It is tempting to suspect that the snail- and clam-infecting parasites are genetically distinct but phenotypically indistinguishable. However, this has not been demonstrated and there is an even more surprising dimension to their life history.

Unlike the digeneans, aspidogastreans have colonized sites in their molluscan hosts that communicate with the outside world and there is no problem in disseminating their eggs and infecting other molluscs. However, there is another problem. Molluscs are abundant and ubiquitous and, provided that their shells can be crushed, they provide nutritious food for aquatic vertebrates. The threat to aspidogastreans from this predation must have been considerable and has been countered by overcoming in some way the digestive processes of the vertebrate, permitting the flatworms to survive and to continue to reproduce after the destruction of their molluscan host. In fact, *A. conchicola* has been reported from the intestines of freshwater teleosts belonging to several families, and from freshwater turtles. Thus, the parasite's versatility encompasses the intestines of vertebrates as well as the gut and excretory systems of molluscs. How aspidogastreans are capable of functioning in such widely different hosts and habitats is a challenging question that has not been addressed, and the suggestion of Rohde (1972, 1973) that aspidogastreans are ill-adapted to parasitism is clearly untenable.

This amazing development may not be entirely the preserve of the aspidogastreans. Freeman and Llewellyn (1958) discovered an adult fellodistomatid digenean, *Proctoeces subtenuis*, in the kidney of the mud-dwelling bivalve *Scrobicularia plana* at a site in the Thames estuary, England. Adult *P. subtenuis* had previously been recorded from the hind gut of labrid and sparid fishes at localities elsewhere in the world, and the similarity with the life cycle of an aspidogastrean is striking. However, it remains to be demonstrated that adult digeneans in the bivalves are able to survive in the gut of mollusc-feeding fishes and, so far, no infected fishes have been found in the Thames estuary. Another possibility is that elsewhere in the world fishes serve as hosts for the adult parasites and become infected by consuming bivalve molluscs harbouring immature metacercariae; the subpopulation in the Thames area may have lost the fish host and achieved sexual maturity in the bivalve second intermediate host (the first intermediate host is unknown).

Once the ability to survive in vertebrate predators had become established as a safety net for aspidogastrean parasites reproducing and cycling between molluscs, the stage was set for further evolutionary change. *Cotylogaster occidentalis* (Fig. 19.10*a*) is an intestinal parasite of freshwater clams and snails (see Dickerman, 1948; Fredericksen, 1978). The parasite spreads to new molluscan hosts by way of eggs which, unlike those of *A. conchicola*, liberate a ciliated, free-swimming larva (= cotylocidium; Fig. 19.11). The parasites also survive if the molluscan hosts are eaten by a freshwater fish, the sheepshead, *Aplodinotus grunniens*, but, according to Fredericksen (1972, 1978), parasites from the intestine of the fish are larger than those from the mollusc and are more likely to be gravid. Moreover, fish parasites lay more eggs and more of them hatch. Thus, in the fish host there has been a shift towards enhanced growth and increased reproductive capacity, while maintaining the ability to reproduce in the molluscan host.

Cable (1974) records what may be another small step in this trend. The aspido-gastreans that he identified as *C. occidentalis* in the digestive gland of the snail *Pleurocera acuta* rarely contained eggs; he never found eggs or larvae in water in which many thousands of *P. acuta* had been kept at all seasons over a period of more than 30 years. These observations, coupled with the finding that cross-fertilization

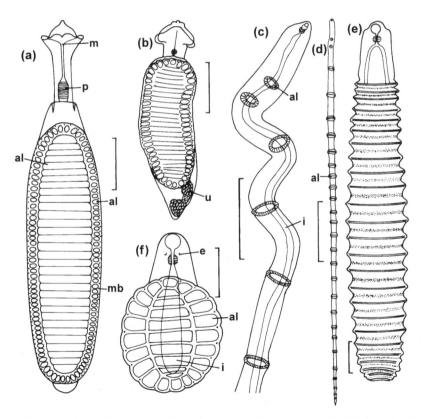

Figure 19.10 Some adult aspidogastreans, showing, in particular, the ventral holdfast organ. (*a*) *Cotylogaster occidentalis*. (*b*) *Lobatostoma manteri* (specimen from a fish). (*c*), (*d*) Anterior region and whole individual respectively of *Stichocotyle nephropis*. (*e*) *Rugogaster hydrolagi*. (*f*) *Cotylaspis insignis*. al = alveolus of holdfast organ; i = intestine; m = mouth; mb = marginal body; p = pharynx; u = uterus containing eggs. Scale bars: (*a*), (*b*), (*e*) 1 mm; (*c*) 5 mm; (*d*) 1 cm; (*f*) 0.5 mm. Source: redrawn from: (*a*) Nickerson, 1902; (*b*) Rohde, 1973; (*c*), (*d*) Odhner, 1910; (*e*) Schell, 1973; (*f*) Osborn, 1903.

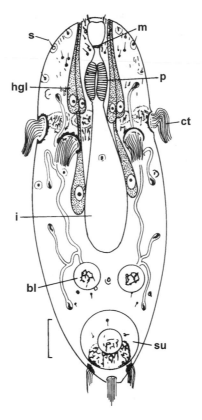

Figure 19.11 The cotylocidium (larva) of *Cotylogaster occidentalis*. bl = bladder containing concretions; ct = ciliary tuft; hgl = head glands; i = intestine; m = mouth; p = pharynx; s = sensillum; su = sucker. Scale bar: 20 μm. Source: redrawn from Fredericksen, 1978.

was impossible because only single parasites occurred in the snails, convinced him that the life cycle is not normally completed in *P. acuta*; the implication is that a fish host is required.

The shift of sexual maturation from the mollusc to the fish, perhaps favoured because more abundant resources in the fish boost egg production of the parasite, is completed in *Lobatostoma manteri* (see Rohde, 1973, 1975, 1994b). Sexually mature parasites (Fig. 19.10*b*) occur in the intestine of the snub-nosed dart (*Trachinotus blochi*), a marine teleost with special adaptations for the handling and crushing of the thick shells of the gastropod *Cerithium moniliferum*. The eggs of the parasite contain fully developed unciliated larvae when laid and these are eaten by *C. moniliferum* and other snails and hatch in the stomach. The larvae migrate to the digestive gland and differentiate into pre-adults; the gonads contain young spermatozoa and egg cells but these do not mature until the snail is eaten by the fish.

Some aspidogastreans have not been deterred by high levels of urea in elasmobranch fishes. For example, *Stichocotyle nephropis* lives in the bile ducts and *Multicalyx cristata* in the gall bladder of rays. The biliary system of mammals is occupied by digeneans (e.g. by *Fasciola hepatica*; section 13.1) and by the tapeworm *Vampirolepis macrostoma* (section 9.6). The biliary system has imposed on the aspidogastreans a worm-like shape and appears to have led to the reduction of the holdfast to a single longitudinal row of suckers (see Dollfus, 1958 and Thoney and Burreson,

1987, respectively). In *S. nephropis* this holdfast has fragmented and the suckers are separate (Fig. 19.10*c, d*). The most exciting feature of *S. nephropis*, however, is the occurrence of encysted stages of the parasite in crustaceans such as the Norway lobster, *Nephrops norvegicus*. Roberts and Janovy (1996) voiced the possibility that the intervention of the lobster is accidental, but this is hard to reconcile with prevalences as high as 20% in some areas off the coast of Scotland (Symonds, 1972) and the observation of MacKenzie (1964) that sucker numbers increase in encysted larvae and that larvae probably live for at least a year. Cysts containing up to eight white larvae, each measuring 3–8 mm in length, were found by MacKenzie on the outside of the gut, mainly around the rectum but occasionally as far forward as the anterior part of the abdomen. So, it seems likely that *Stichocotyle* has gained access to its fish host by recruiting a new invertebrate intermediate host, a crustacean, on which the fish preys. No additional molluscan host has been reported, and the crustacean intermediate host may have entirely replaced the molluscan phase of the life cycle.

The rectal gland of holocephalans has also been colonized by aspidogastreans. This is associated with significant changes in the holdfast organ. *Rugogaster hydrolagi* has lost the alveoli, apart from a rudimentary (?), anteriorly situated, ventral sucker (see Schell, 1973), and the holdfast is represented by a longitudinal series of transverse ridges (Fig. 19.10*e*). Schell experienced difficulty in removing the parasites from the rectal gland and observed that the lining of the gland at the attachment site was corrugated, but how such an organ achieves such firm attachment or, indeed, why firm attachment is needed in such a sheltered environment is unexplained. It will be recalled that the rectal gland of elasmobranchs is a site that is occupied by some monogeneans (see Table 7.1). The rectal gland provides a habitat in elasmobranchs, and possibly also in holocephalans, where parasites can avoid high concentrations of urea, but to survive here they must tolerate elevated levels of sodium chloride (see Burger and Hess, 1960).

The versatile aspidogastreans have one other surprise. *Cotylaspis insignis* (see Fig. 19.10*f*) was found by Osborn (1903) living in the mantle cavity of freshwater mussels of the genus *Anodonta*. Most of the parasites were found attached by the ventral holdfast to the outer surface of the kidney and others were found adhering to the free edge of the inner gill or to the adjoining surfaces of the visceral mass. Osborn made a careful search of the interior of the kidney and of the pericardial chamber, but did not find any specimens of *Cotylaspis* in these sites. Thus *C. insignis* is ectoparasitic or possibly ectocommensal. Osborn described how the animal gathered material from the lining of the mantle cavity by expanding the region around the mouth, applying the expanded disc to the host and then contracting the disc so as to sweep the material, which was not identified by Osborn, into the mouth. This procedure was then repeated in another place. Embryonic development takes place after the eggs are laid but the rest of the life cycle is unknown.

CHAPTER 20

Summary and conclusions

20.1 ORIGINS AND DEVELOPMENT OF SYMBIOSIS IN THE PLATYHELMINTHS

Although a relatively small number of turbellarians live in permanent symbiotic associations (about 200 species from 35 families, according to Jennings (1997), attributed to Cannon, 1986), the allocation of only one chapter of this book to the symbiotic turbellarians in no way reflects their importance relative to the other symbiotic platyhelminths. A relationship that has attracted special interest is the mutualistic association between acoel turbellarians (*Convoluta* spp.) and intracellular algal cells (section 2.1), and the turbellarians associated symbiotically with multi-cellular animals are nutritionally diverse, spanning the entire spectrum from simple facultative sheltering (inquilinism) to obligate endoparasitism (Jennings, 1997). According to Jennings, the majority of permanent symbiotes are rhabdocoels and the sites they occupy are usually the branchial and body cavities, gills and external body surfaces of arthropods (mainly crustaceans) and the guts and body cavities of echinoderms, molluscs and sipunculids. Less commonly, turbellarians associate with polychaetes, elasmobranch and teleost fishes and freshwater chelonians, being found mainly on the body surfaces.

Defining the relationships between turbellarian symbiotes and their hosts is often difficult because many of the nutritional strategies of the symbiotes contain elements of more than one life style. Different degrees of emphasis upon these elements produce different nutritional patterns for each symbiote and determine the extent of metabolic dependence on the host (Jennings, 1997). Symbiosis has arisen and developed several times in the turbellarians and we can recognize evolutionary pathways in which the nutritional emphasis shifts from predation, supplemented by commensalism, to parasitism. For example, in the umagillid rhabdocoels found in the alimentary tract of echinoderms (section 2.3.1) we can trace a transition from a benign relationship to a parasitic one: there is a gradual shift in the diet of the platyhelminth partner from predominantly co-symbiotic protozoans to host intestinal tissue and gut contents, leading to loss of the gut and absorption of nutrients through the body surface, as in *Fallacohospes* (section 2.3.1). Parasitic graffillids in molluscs (section 2.3.2) and pterastericolids in starfish (section 2.3.4) may have had similar origins. Lost in obscurity are the origins of gutless turbellarians such as the fecampiids (section 2.3.3), the pre-adults of which are endoparasites of crustaceans, and the rhabdocoel *Acholades* (section 2.3.5), which digests the tissue of starfish tube feet with enzymes released from and through the epidermis.

Ectosymbiotic turbellarians are epizoic predators, which use their hosts as feeding platforms from which they capture either free-living invertebrates that come within reach and/or co-symbiotic organisms (Jennings, 1997). As in endosymbiotes, the same tendency to supplement this diet by opportunistic commensalism can be identified. A significant feature of the hosts (crustaceans, xiphosurans, chelonians) that harbour these commensals (temnocephalans and triclads, sections 2.2.2, 2.2.3) is that they crush, shred or tear their food before ingestion and inevitably lose food particles, which can be picked up by appropriately stationed symbiotes. As Jennings (1997) has indicated, this exploitation of untidy eating habits is widespread in invertebrates, occurring in protozoans, rotifers, annelid worms and the recently-discovered cycliophorans as well as in the turbellarians. So successful is this habit that these relationships seem to be stable and are not evolving further towards parasitism (Jennings, 1997). Therefore, the ectoparasitism displayed by the mono-geneans and by turbellarians such as *Micropharynx* (section 2.4) may have evolved by an entirely different route. These fish parasites may have developed from opportunistic, free-living, predatory platyhelminths that directly developed a taste for fish epidermal cells without an intervening stage involving commensalism. The ancestors of the trematodes (digeneans and aspidogastreans) may have established themselves as parasites of molluscs in a similar way.

The suggestion has been made in this book (Chapter 3) that endoparasitism (gut parasitism) of vertebrates as exemplified by the cestodes and by the digeneans evolved in different ways from the endoparasitism of turbellarians. The proposal is that cestodes evolved not from endocommensal ancestors but from ancestors that were already well established as fish ectoparasites. These were the oncophoreans, which also gave rise to modern monogeneans. On the other hand, the colonization of vertebrate guts by digeneans may have stemmed from the serendipitous survival of free-living reproductive individuals released from their larval life as endoparasites of molluscs and ingested by predatory fishes. Predation by fishes was probably important also in the establishment of aspidogastreans in vertebrates, but this involved direct predation on the mollusc and survival of ingested intramolluscan parasitic stages.

The evolutionary potentials of the oncophorean–fish and trematode–mollusc associations were enormous. The monogenean skin and gill parasites became estab-lished on all the major groups of fishes, and monogeneans profited in particular from the explosive evolutionary expansion of the teleosts. Some monogeneans survive in largely terrestrial amphibians, but have not yet succeeded in breaking their reproductive link with the aquatic environment. However, this has been achieved by some cestodes by changes in the egg (section 11.1 and has facilitated the invasion of the terrestrial higher vertebrates, the reptiles, the birds and the mammals.

Trematode exploitation of the vertebrates is a secondary development in a tem-poral sense, arising from platyhelminths already well established as endoparasites of molluscs. Ties with the aquatic environment remain strong in both the digeneans and the aspidogastreans because of the importance of molluscs in the life cycle. Even those that successfully employ terrestrial gastropods as hosts can rarely dispense entirely with a moist environment for transmission. Nevertheless, this did not prevent the digeneans from colonizing the terrestrial vertebrates; they achieved this either by exploiting the fact that most terrestrial vertebrates visit water bodies or by acquiring intermediate hosts that have both aquatic and terrestrial phases in their life histories and are, therefore, able to carry the parasite across the water–air interface.

All of these parasites, monogeneans, cestodes, digeneans and aspidogastreans, took advantage of the rapidly diversifying vertebrates and gave rise to a significant proportion of modern platyhelminths.

Multiple-host life cycles are a significant feature of cestodes and trematodes. Students often express surprise that such parasites with two, three or even four hosts survive in the harsh natural world. How can apparently tenuous links between two or more often disparate hosts be indefinitely maintained by such seemingly delicate platyhelminths in such an unstable world? That this conception grossly underestimates the robustness of platyhelminths and their life cycles is demonstrated by calling to mind the failure so far of all human attempts to break the life cycles of parasites like *Schistosoma*. Life cycles involving three or four hosts are likely to be more robust and stable than we might imagine. However, weaknesses or gaps in these extended food chains are likely to arise from time to time and such changes have undoubtedly led to extinctions of platyhelminth symbiotes. This book contains many examples of others that appear to have had the flexibility to bypass the weak link and survive with fewer hosts, although it is not always easy to determine whether life-cycle curtailment is the correct interpretation of past events or whether a particular short life cycle is primitive.

20.2 THE PLATYHELMINTH FLAIR FOR SYMBIOSIS

The platyhelminths are traditionally regarded as relatively low on the evolutionary ladder and yet their evolutionary potential may have been enormous. According to Willmer (1990), the platyhelminths may have provided the evolutionary springboard from which higher metazoans of many kinds originated (section 1.3). This potential may also be reflected in the extensive and varied commitment of the platyhelminths to symbiosis, a commitment that, arguably, other groups of animals fail to match. We need to know the reasons for this commitment and whether the platyhelminth level of body organization has special features that encouraged the development of the symbiotic way of life.

The morphological freedom provided by the triploblastic, acoelomate body-construction of platyhelminths, and the possession of a well developed nervous system coupled with antagonistic body muscles and a hydrostatic skeleton (Chapter 1), allowed the development of complex locomotor and behaviour patterns and probably had significant benefits for the symbiotic life style. With no cuticle to restrict body shape changes, platyhelminths can insinuate themselves through narrow openings into the bodies of their hosts and, once inside, negotiate tubes, ducts and apertures. Different growth patterns can create virtually any body shape, so that platyhelminths can mould themselves to fit their microenvironment or adjust their surface area to match their physiological requirements. The nematodes are much less flexible in terms of locomotion and shape change because they have adopted a high pseudocoelomic pressure and a thick cuticle, and use only longitudinal muscle.

At first sight, the lack of a cuticle in the platyhelminths might be regarded as a hindrance for an endoparasite, but the naked flatworms are at least as successful as the externally-protected nematodes as inhabitants of the harsh environment of the vertebrate digestive system. In fact, the continuous, syncytial, cytoplasmic layer or tegument of the major parasitic platyhelminths provides highly effective protection against host bile and digestive enzymes and against the effector arm of the host's immune system and, at the same time, at least in tapeworms, is capable of actively absorbing nutrients in direct competition with the host. This versatility could only be achieved by a living surface layer.

The lack of cell boundaries in the tegument may be an important contributory factor to its role as a selective barrier, although gutless endoparasitic turbellarians such as *Acholades* presumably absorb nutrients through a ciliated cellular epidermis. In platyhelminths with a tegument, the subsurface location of the nucleated cell bodies (cytons), including the associated machinery for macromolecular synthesis, provides ready access to nutrients from the parenchyma, and, at least in cestodes, provides an inward route for nutrients absorbed from the host's gut lumen. Wilson and Webster (1974) have suggested that the development of the protonephridial system, permitting a relatively permeable body surface, may have been a factor in the shaping of the kind of endosymbiosis that has developed in the platyhelminths.

The presence of a syncytial tegument seems linked to parasitism but cannot be attributed exclusively to the demands of an endoparasitic life style, since endoparasitic turbellarians have a cellular epidermis and a syncytial tegument occurs in monogeneans. According to those who regard the major parasitic platyhelminths as monophyletic neodermatans, the syncytial tegument or neodermis has evolved only once, but the possibility of an independent origin of this feature in oncophoreans and in trematodes should not be overlooked.

The major parasitic platyhelminths have not entirely lost their cellular ciliated epidermis, even though it is replaced in adult life by the syncytial tegument. This ciliated epidermis has an important role in the propulsion of the larvae and provides a means of disseminating small organisms through the three-dimensional aquatic environment. The effectiveness and importance of this is illustrated by the fact that ciliated larvae are retained by most monogeneans (oncomiracidia), many tapeworms (coracidia), gyrocotylideans (lycophore larvae), most digeneans (miracidia) and some aspidogastreans (cotylocidia).

Adhesion of symbiotes to their host is an almost universal necessity, although the strength required of the bond varies depending upon the microhabitat of the symbiote and the forces acting to dislodge it. Adhesive structures in the form of glandular secretions and suckers are widespread in free-living turbellarians (see Jennings, 1997), and similar organs in ancestral turbellarians must have been important preadaptations favouring the development of symbiosis.

Jennings (1997) identified another possible preadaptation with useful applications in both ecto- and endosymbiotes. This is the bulbous pharynx, containing prominent circular and radial muscle fibres, often with protrusible capability. As Jennings has pointed out, this organ is versatile. It is adept at all of the following:

- seizing small prey organisms in free-living and ectosymbiotic turbellarians;
- penetrating the surface of larger prey species;
- ingesting food particles lost by the host in ectocommensals;
- taking up host gut contents or gut tissue in endosymbiotes like digeneans, or host epidermal cells or blood in ectoparasites such as monogeneans.

In egestion, the pharynx appears to have an important function as a pump for flushing out the gut (section 4.5) and this may be a widespread feature in free-living and symbiotic flatworms.

Features of platyhelminth reproduction may also have favoured the establishment and development of symbiotic relationships. Unlike the dioecious nematodes most flatworms are hermaphrodite. The ability of each individual to assemble eggs maximizes reproductive exploitation of the virtually unlimited resources available to symbiotes. At the same time, it does not prevent cross-fertilization and genetic interchange, and holds open the option of self-fertilization, permitting reproductive contributions even from isolated symbiotes. Hermaphroditism and self-fertilization may also have encouraged the rapid establishment and spread of favourable mutations.

The chemically resistant, tanned eggshell/capsule (section 1.4.5) is a feature of most platyhelminths, including free-living forms. It probably predates the development of endoparasitism and is likely to have been particularly important as a preadaptation for an endosymbiotic (commensal or parasitic) existence. Any feature, such as a resistant eggshell, that prolongs embryonic life might be valuable to any commensal or parasite awaiting contact with a host. Moreover, the embryos in eggs or capsules released by endosymbiotes newly established in the digestive systems of animal hosts would be protected from the host's digestive fluids as they passed out of the gut, in the same way as such eggs or capsules are protected when they are eaten by predators (section 1.4.5). However, some oncophoreans were flexible enough to discard the protection afforded by the quinone-tanned eggshell when it became advantageous to do so, permitting important changes to take place in the cestode life cycle (section 11.1).

Two features of the biology of symbiotic flatworms are interrelated and have figured frequently in discussions of adaptation to parasitism; these are fecundity and food reserves. High fecundity has long been regarded as a prerequisite for endosymbiosis, the argument being that large numbers of offspring are needed to compensate for the high mortality during transmission between hosts. As for food reserves, glycogen predominates in endosymbiotes of all types, including turbellarians, cestodes and digeneans, while lipid is the major storage product in ectosymbiotic turbellarians and, indeed, in free-living predatory turbellarians (Jennings, 1997). The emphasis on glycogen accumulation in endosymbiotes, creating the potential for releasing energy by anaerobic glycolytic pathways, has been interpreted in the past as a preadaptation permitting colonization of host internal sites where oxygen levels are low. Bryant (1994) proposed that this preadaptation could be traced back to an ancient world where the free-living ancestors of modern platyhelminths were exposed to an environment with little or no oxygen. Bryant supported the notion that haem-based respiratory electron transfer systems evolved at first in relation to the nullification of the toxic effects of oxygen, as this new pollutant appeared and began to accumulate in the atmosphere. According to Bryant, anoxic environments, including internal sites in potential hosts, persisted alongside oxygen-rich ones; the inhabitants of the anoxic sites retained their emphasis on anaerobic energy generation along with cytochrome systems concerned primarily with oxygen detoxification.

Jennings (1997) has offered an alternative explanation for these features of endosymbiotes. He pointed out that high glycogen levels are not exclusively a feature of endosymbiotes that inhabit anaerobic sites, and he quoted as examples umagillids inhabiting the perivisceral coelom of echinoids and the gut near the bases of the respiratory trees of holothurians, and digeneans living in the vertebrate lung. His argument starts with the premise that endosymbiotic organisms have a food supply that is virtually assured and unlimited. Since animals will always reproduce to the fullest extent possible, symbiotes will channel resources surplus to growth and maintenance into reproduction. Thus, endosymbiotes will have so-called 'r'-type survival and reproductive strategies, a feature of which is high fecundity. High fecundity is likely to be highly energy-dependent and will need, therefore, an energy source such as glycogen, which is easy to mobilize and constantly renewable, whatever the nature and source of the food. Jennings offered this as an explanation for the presence of glycogen at relatively high levels in virtually all endosymbiotes, whether they live in oxygen-rich environments or not, but, as Jennings pointed out, once glycogen dependence is established it would permit endosymbiotes to occupy oxygen-poor habitats by way of anaerobic glycolysis.

The food supply of predatory free-living turbellarians is less reliable. The same can be said for ectosymbiotic turbellarians with elements of predation and commensalism in their nutritional strategies, since both of these components are subject to the vagaries of prey availability. Therefore, any food surplus in these turbellarians is likely to be stored as lipid, which is energetically the most efficient reserve when mobilized and acts as a buffer to tide the symbiote over lean periods in food availability. Further surpluses are used for production of eggs, which are few. These eggs are well-provisioned, largely with lipid, and hatch at an advanced developmental stage. Lipid storage and low fecundity are typical of food-limited predators and are features of so-called 'K'-type survival and reproductive strategies (see Jennings, 1997).

Thus, glycogen storage and high fecundity may be consequences of the endosymbiotic life style rather than prerequisites for it. Nevertheless, as Jennings pointed out, once established, high fecundity in endosymbiotes will contribute greatly to the survival chances of the species by compensating for losses inherent in transmission between hosts.

The monogeneans are ectoparasites but their exclusively parasitic life style is likely to provide an assured food supply. However, the expectation of high glycogen levels is not supported by those measurements that have been made, with the exception of the bladder parasite *Polystoma integerrimum* (see Halton, 1967). Factors that will also have a bearing are the relatively low fecundity of most monogeneans and generally well aerated habitats. *Polystoma* is atypical in occupying an oxygen-poor environment (according to Halton, 1967) and in concentrating egg production in a limited spring season (section 7.1.5), hence creating a demand for extensive accumulated reserves. Intermittent exposure to low ambient oxygen levels is inflicted on the skin parasite *Entobdella soleae* by the daily activity pattern and anoxic environment of its bottom-dwelling flat-fish host, but the parasite appears to cope with anoxic periods by increased breathing movements (section 4.6) rather than by resorting to anaerobic metabolic pathways.

The discovery that parasitic flatworms are unable to synthesize fatty acids and sterols *de novo* used to be regarded as an adaptation to parasitism, but Meyer, Meyer and Bueding (1970) found that *Schistosoma mansoni* and *Hymenolepis diminuta* share this inability with the free-living triclad *Dugesia dorotocephala*. Meyer *et al.* suggested that this inability may be common to the entire phylum and it may have been another important factor predisposing platyhelminths to a symbiotic way of life.

20.3   LOOKING FORWARD

The symbiotic relationships of turbellarians deserve more consideration than they have received to date, but there is little doubt that future research priorities will remain basically unchanged, focusing on the major groups of parasitic platyhelminths and especially on those of medical and veterinary importance. There are several research areas in which we can expect exciting discoveries.

There are many challenging immunological questions relating to platyhelminth parasites. With regard to the mammalian immune system, Bell (1996) has proposed that helminth parasites are as important as any other infectious agent in shaping its evolution, ontogeny and function. He suggested that infection with helminths has led to the evolution in vertebrates of the Th2 system, which expresses when activated elevated levels of IgE, eosinophils and mast cells (see also section 18.3.12). Even more intriguing is Bell's proposal that activation of the Th2 system by helminths during ontogeny and adulthood in humans may be essential for the proper homoeostasis of the developing and adult immune systems. In the absence of such a stimulus, as in modern Western populations in which helminth parasites have been virtually eliminated during the last 100 years or so, this arm of the immune system may malfunction, leading to the clinical manifestation of allergic disorders such as asthma. Bell pointed out that the allergies that plague Western populations are rarely encountered in developing countries where helminth parasites are prevalent.

No group of platyhelminths has attracted more interest from immunologists than the blood-dwelling digeneans of the genus *Schistosoma* and yet, in spite of this prolonged and intense activity, many questions remain to be answered (Chapter 18) and the parasite–host relationships of other schistosomes and of sanguinicolid and spirorchid blood parasites (Chapter 17) are virtually unexplored.

Interactions between monogeneans and fishes raise interesting questions about the role of fish skin in responding to ectoparasites. We know from work on capsalids and gyrodactylids that the skin is not an immunologically inert organ (sections 5.3, 5.5.1), but little is known about the mechanism, or about how most sedentary monogeneans like polyopisthocotyleans escape any local host response (section 6.5.2).

A relatively new area of immunological exploration is the interaction between platyhelminths and invertebrate hosts. Although they lack some of the more spectacular features of the vertebrate internal defence system (see review by Loker, 1994), especially long-term and highly specific immunological memory, invertebrates are capable of mounting efficient humoral and cellular defences against symbiotes. Moreover, these 'invertebrates' are diverse, many with lineages that have been distinct since the Cambrian period, and we would expect that they have produced many different solutions to the problems posed by symbiotes. Loker illustrated this by reference to the molluscs and the arthropods; he pointed out that both are protostome phyla and yet their internal defence systems are remarkably different. The internal defence systems of some phyla have scarcely been investigated, so we can expect even greater diversity. Since platyhelminths have symbiotic relationships with members of most of the major groups of metazoans, they must be very good at avoiding the consequences of different kinds of defence systems. However, our understanding of how this is achieved is rudimentary.

Work on digenean germinal sacs and their interactions with molluscan defence systems has already brought into focus unexpected differences between redia-producing and sporocyst-producing digeneans and has highlighted the predatory nature of rediae (section 14.1). This has raised fundamental questions about how rediae recognize germinal sacs of their own species.

Of special interest are those few parasitic platyhelminths that are capable of infecting a range of hosts. This broad host specificity involves not only metabolically inactive stages, such as the mesocercariae of digeneans of the genus *Alaria* (section 17.7.3), but also fully mature adult parasites, such as the skin-parasitic monogenean *Neobenedenia melleni* (section 5.3) and the adult cestode *Bothriocephalus acheilognathi* (section 10.5.3). The basis of such wide host tolerance is unknown, but is likely to involve immunological and physiological factors.

One of the possible strategies for evading host defences that has attracted great interest is the notion that parasites might avoid recognition by displaying host molecules sequestered from the host or synthesized by the parasite. The genomes of parasites with the capacity to produce macromolecules identical or similar to those of the host can be assumed to include DNA sequences identical or similar to those of the host. Such DNA homology could have come about by gene conservation during the long evolutionary history of the metazoans or by more recent adaptive change based on natural selection. Howell (1985) has raised another exciting possibility. He has suggested that DNA homology may have arisen by the direct incorporation of host genetic material into the parasite genome. The similarity between schistosome and snail tropomyosins (section 18.3.6) and between sparganum (cestode) growth factor and human growth hormone (section 10.5.8) may reflect such a transfer. Events of this kind would provide the opportunity for more rapid evolution of the parasite–host relationship and are not without precedent: there is evidence for gene capture from vertebrate hosts by a symbiotic bacterium and by several viruses (Lewin, 1985 and Barinaga, 1992 respectively). Howell has suggested that retroviruses may provide a vehicle for gene transfer between host and parasite.

Howell went further to suggest that gene transfer from parasite to host and even from parasite to parasite within the same host should be considered. With regard to the latter possibility, he raised the question of non-reciprocal cross-immunity, in which one parasite species generates immunity against another species inhabiting the same host (section 9.3.4). According to Howell, this could represent a situation where one parasite species acquires the DNA of another, expresses the antigens for which it encodes and generates immunity to the potentially competing parasite. It remains to be seen whether platyhelminth parasites indulge in these transgenic activities with their hosts or with other parasites.

Largely as a result of the application of immunocytochemical techniques, there has been considerable recent expansion in our knowledge of the patterns of platyhelminth nervous systems and particularly in the identities and distributions of neuroactive substances (see, for example, Halton *et al.*, 1994 and Reuter and Gustafsson, 1995). However, apart from observations linking regulatory peptides with egg assembly (section 7.1.5), our understanding of the function of the platyhelminth nervous system remains largely speculative (see Halton *et al.*, 1994). The possibility has been raised that endoparasites may gain an advantage by releasing peptides that influence the physiological systems of their hosts (Halton *et al.*, 1994). Such exogenous peptides could improve the parasite's environment by interacting with the host's immune system or by inducing pathological changes. There are also strong indications that parasites may gain advantage by responding to host-derived chemical messages in the shape of host hormones (sections 7.1.5, 17.7.3). All of these questions offer substantial challenges for the future.

Few features of parasitism have excited the imagination more in recent years than the alleged manipulation of morphological and behavioural features of the host by parasites. Parasitologists have been quick to regard such changes as advantageous for the parasite, but more recent considerations have introduced uncertainty about whether some of these effects are adaptive, in the sense that the survival of the parasite is promoted by them (Moore and Gotelli, 1990; Poulin, 1995), and this area is likely to evoke more critical studies in the future. There is a link here too with neurophysiology, since parasite-derived exogenous peptides may be involved in modulating host behaviour (Halton *et al.*, 1994).

In evolutionary biology and ecology, the symbiotic platyhelminths have a great deal to offer and some of the areas deserving special attention have been highlighted by Esch and Fernandez (1993). Many believe that the influence of parasites on the evolution of their hosts has been greatly underestimated. It has been suggested that parasites may be the driving force behind the evolution of secondary sexual characters in vertebrates and even the evolution of sex itself. Zuk and McKean (1996) argue that selection acting differently on male and female hosts has been an important contributor to the widespread host sexual differences in parasitic infections (section 7.1.4(c)), and this evolutionary approach is of great interest since it links mating systems, secondary sexual characters, hormones and the immune system of vertebrates. Sex differences in parasitic infections also occur in invertebrate hosts and are virtually unexplored. Parasitic platyhelminths have already contributed to the evolutionary debate (sections 7.1.4(d), 17.7.1) and will undoubtedly continue to do so.

The impact of modern biotechnology on our knowledge of the parasitic platyhelminths is hard to predict but will certainly be great. It is already heralded as likely to provide the final answers to phylogenetic questions. However, there is a need for caution. If gene transfer has indeed taken place between host and parasite and/or between one parasite species and another (see above), then phylogenetic patterns based on molecular studies may be compromised. This matter was raised by Willmer (1990) and she also made the point that molecular convergence may be as real as anatomical convergence, the latter being a widespread phenomenon in the parasitic platyhelminths, ranging from loss of whole systems such as the gut to independent development of complex organs such as attachment suckers and cirruses. Certainly, some of the conclusions drawn from studies of molecular phylogeny of parasitic platyhelminths are inconsistent with intuitive views based on knowledge of their biology. Nevertheless, future studies on gene sequencing and genetics are bound to provide important information on the affinities and evolution of the platyhelminths.

The account in this book of symbiosis in the platyhelminths has involved the synthesis of material ranging from field studies to the special techniques of molecular science and has illustrated a variety of biological concepts but particularly evolutionary adaptation at all levels. In recent times more traditional zoological studies of an academic nature have been displaced by more fashionable research. The importance of the latter cannot be denied but the precipitate decline in the former approach is, I believe, unfortunate because, as demonstrated in this book, our knowledge of the general biology of platyhelminth symbiotes is far from complete. Even basic information about life cycles is often lacking and in areas such as physiology and functional morphology there is much to do. These gaps in our knowledge will not be filled unless the biological community has the will to maintain some support for this kind of research. It would be pleasing to me if this book stimulates even to a small degree a heightened awareness of the continuing importance and fascination of this approach.

Appendix

A classification scheme compiled from several sources showing the possible relationships between the symbiotic platyhelminths described in this book. Genera featuring significantly in the text are shown and some additional group names in common use are also given.

Turbellarians
1) Acoela
        Convolutidae   *Convoluta*
2) Prolecithophora
        Urostomidae   *Ichthyophaga*
3) Polycladida
4) Tricladida   *Bdellasimilis, Micropharynx*
5) Rhabdocoela
    Dalyelliida
        Umagillidae   *Syndisyrinx, Wahlia, Anoplodium, Desmote, Fallacohospes*
        Graffillidae   *Paravortex*
        Pterastericolidae   *Pterastericola, Triloborhynchus*
        Acholadidae   *Acholades*
        Fecampiidae   *Fecampia, Kronborgia*
    Temnocephalida
        Temnocephalidae   *Temnocephala, Craspedella, Diceratocephala*
6) Udonellida   *Udonella*

Oncophoreans
1) Monogenea
    Monopisthocotylea
        Enoplocotylidae   *Enoplocotyle*
        Acanthocotylidae   *Acanthocotyle*
        Microbothriidae   *Leptocotyle*
        Monocotylidae   *Dendromonocotyle, Horricauda*
        Capsalidae   *Entobdella, Benedenia, Trimusculotrema, Trochopus, Montchadskyella (?)*
        Gyrodactylidae   *Gyrodactylus, Phanerothecium, Isancistrum*
        Dactylogyridae
            Dactylogyrinae   *Dactylogyrus, Neodactylogyrus*
Dactylogyroideans   Ancyrocephalinae   *Haliotrema, Chauhanellus, Hamatopeduncularia, Neodiplectanotrema, Paradiplectanotrema, Enterogyrus*
        Calceostomatinae   *Neocalceostomoides*
        Amphibdellidae   *Amphibdelloides*
        Tetraonchidae   *Tetraonchus*
        Diplectanidae   *Lamellodiscus, Telegamatrix*
    Polyopisthocotylea
        Chimaericolidae   *Callorhynchicola*
        Hexabothriidae   *Hexabothrium*

Mazocraeideans	{ Diclidophoridae	*Diclidophora, Cyclocotyla, Heterobothrium*
	Gastrocotylidae	*Gastrocotyle*
	Plectanocotylidae	*Plectanocotyle*
	Mazocraeidae	*Kuhnia, Grubea*
	Microcotylidae	*Microcotyle, Axine, Caballeraxine*
	Anthocotylidae	*Anthocotyle*
	{ Diplozoidae	*Diplozoon*
	Polystomatidae	*Polystoma, Eupolystoma, Pseudodiplorchis, Concinnocotyla*
	Sphyranuridae	

2) Gyrocotylidea — *Gyrocotyle*

3) Cestoda

Caryophyllidea	*Biacetabulum, Archigetes, Glaridacris*
Spathebothriidea	*Cyathocephalus*
Pseudophyllidea	*Bothriocephalus, Triaenophorus, Diphyllobothrium, Ligula, Schistocephalus, Spirometra*
Trypanorhyncha	*Lacistorhynchus, Grillotia, Gilquinia*
Amphilinidea(?)	*Amphilina, Austramphilina, Nesolecithus*
Tetraphyllidea	*Acanthobothrium, Phyllobothrium*
Nippotaenidea	*Amurotaenia*
Proteocephalidea	*Proteocephalus*
Tetrabothriidea	*Tetrabothrius*
Cyclophyllidea	
Mesocestoididae	*Mesocestoides*
Taeniidae	*Taenia, Echinococcus*
Anoplocephalidae	*Moniezia, Oochoristica, Thysanosoma, Cittotaenia*
Dilepididae	*Dipylidium, Valipora*
Paruterinidae	*Paruterina*
Nematotaeniidae	
Hymenolepididae	*Hymenolepis, Vampirolepis, Ditestolepis*

Trematodes

1) Digenea

Echinostomata	
Echinostomatidae	*Aporchis, Mesorchis, Echinostoma, Echinoparyphium, Hypoderaeum, Himasthla, Isthmiophora*
Notocotylidae	*Notocotylus*
Philophthalmidae	*Philophthalmus*
Haploporidae	*Saccocoelium*
Mesometridae	*Mesometra*
Fasciolidae	*Fasciola*
Paramphistomidae	*Cotylophoron, Diplodiscus*
Psilostomidae	
Haplosplanchnidae	
Heronimidae	*Heronimus*

	Strigeata	
Strigeoids	{ Strigeidae	*Cotylurus, Apatemon*
	Cyathocotylidae	
	{ Diplostomatidae	*Diplostomum, Posthodiplostomum, Uvulifer, Crassiphiala, Ornithodiplostomum, Alaria*
	Brachylaimidae	*Postharmostomum, Brachylaima, Hasstilesia*
	Leucochloridiidae	*Leucochloridium*
	Azygiidae	*Azygia, Proterometra*
	Bivesiculidae	*Bivesicula, Paucivitellosus*
	Gymnophallidae	*Gymnophallus, Parvatrema*

Bucephalidae	*Bucephalus, Bucephaloides, Rhipidocotyle, Prosorhynchus, Alcicornis*
Clinostomatidae	
Fellodistomatidae	*Monascus, Steringophorus, Lintonium, Proctoeces*
Transversotrematidae	*Transversotrema, Prototransversotrema*
Cyclocoelidae	*Cyclocoelum*
Sanguinicolidae	*Sanguinicola, Aporocotyle*
Spirorchiidae	*Spirorchis*
Schistosomatidae	*Schistosoma, Schistosomatium, Griphobilharzia, Trichobilharzia, Austrobilharzia*

Plagiorchata

Dicrocoelidae	*Dicrocoelium, Brachylecithum*
Lecithodendriidae	*Allassogonoporus, Acanthatrium*
Plagiorchiidae	*Plagiorchis*
Monorchiidae	*Asymphylodora, Parasymphylodora, Paratimonia*
Gorgoderidae	*Gorgodera, Phyllodistomum*
Opecoelidae	*Sphaerostoma*
Microphallidae	*Microphallus, Maritrema, Levinseniella*
Troglotrematidae	*Paragonimus, Nanophyetus*
Lepocreadiidae	*Lepocreadium, Opechona*

Opisthorchiata

Opisthorchiidae	*Opisthorchis, Clonorchis*
Hemiuridae	*Lecithochirium, Genarchopsis, Parahemiurus*
Halipegidae	*Halipegus*
Heterophyidae	*Apophallus, Stictodora, Cryptocotyle, Acetodextra*
Cryptogonimidae	*Stemmatostoma (? Aphalloides)*

2) ? Didymozoidea	*Didymozoon, Didymocystis, Metanematobothrium, Nematobothrium, Nematobibothrioides, Coeliotrema, Koellikeria, Diplotrema, Neometadidymozoon, Wedlia*
3) Aspidogastrea	*Aspidogaster, Multicalyx, Lobatostoma, Cotylogaster, Stichocotyle, Rugogaster, Cotylaspis*

References

Adam, M. E. and **Lewis, J. W.** (1992) Sites of encystment by the metacercariae of *Echinoparyphium recurvatum* in *Lymnaea peregra*. *Journal of Helminthology*, **66**, 96–99.

Alexander, R. M. (1979) *The Invertebrates*, Cambridge University Press, Cambridge.

Allen, K. M. and **Tinsley, R. C.** (1989) The diet and gastrodermal ultrastructure of polystomatid monogeneans infecting chelonians. *Parasitology*, **98**, 265–273.

Allison, F. R. (1980) Sensory receptors of the rosette organ of *Gyrocotyle rugosa*. *International Journal for Parasitology*, **10**, 341–353.

Allison, L. N. (1943) *Leucochloridiomorpha constantiae* (Mueller) (Brachylaemidae), its life cycle and taxonomic relationships among digenetic trematodes. *Transactions of the American Microscopical Society*, **62**, 127–168.

Alpers, D. H. (1994) Digestion and absorption of carbohydrates and proteins, in *Physiology of the Gastrointestinal Tract*, 3rd edn, (ed. L. R. Johnson), Raven Press, New York, pp. 1723–1749.

Al-Sehaibani, M. A. (1990) Studies on dactylogyroidean monogeneans. MPhil Thesis, University of East Anglia, UK.

Amin, O. M. (1978) On the crustacean hosts of larval acanthocephalan and cestode parasites in southwestern Lake Michigan. *Journal of Parasitology*, **64**, 842–845.

Andersen, K. (1975) The functional morphology of the scolex of *Diphyllobothrium* Cobbold (Cestoda, Pseudophyllidea). A scanning electron and light microscopical study on scoleces of adult *D. dendriticum* (Nitzsch), *D. latum* (L). and *D. ditremum* (Creplin). *International Journal for Parasitology*, **5**, 487–493.

Anderson, M. G. (1962) *Proterometra dickermani*, sp. nov. (Trematoda: Azygiidae). *Transactions of the American Microscopical Society*, **81**, 279–282.

Anderson, M. G. and **Anderson, F. M.** (1963) Life history of *Proterometra dickermani* Anderson, 1962. *Journal of Parasitology*, **49**, 275–280.

Anderson, M. G. and **Anderson, F. M.** (1967) The life histories of *Proterometra albacauda* and *Proterometra septimae*, spp. n. (Trematoda: Azygiidae) and a redescription of *Proterometra catenaria* Smith, 1934. *Journal of Parasitology*, **53**, 31–37.

Anderson, R. M. and **Whitfield, P. J.** (1975) Survival characteristics of the free-living cercarial population of the ectoparasitic digenean *Transversotrema patialensis* (Soparker, 1924). *Parasitology*, **70**, 295–310.

Andreassen, J. (1981) Immunity to adult cestodes. *Parasitology*, **82**, 153–159.

Andreassen, J. and **Hopkins, C. A.** (1980) Immunologically mediated rejection of *Hymenolepis diminuta* by its normal host, the rat. *Journal of Parasitology*, **66**, 898–903.

Anokhin, I. A. (1966) Diurnal cycle of activity and the behaviour of ants (*Formica pratensis* Retz.), invaded by the metacercaria of *Dicrocoelium lanceolatum*, during the grazing period. *Zoologicheskii Zhurnal*, **45**, 687–691 (in Russian).

Apelt, G. (1969) Fortpflanzungsbiologie, Entwicklungszyklen und vergleichende Frühentwicklung acoeler Turbellarien. *Marine Biology*, **4**, 267–325.

Appleton, C. C. (1983) Studies on *Austrobilharzia terrigalensis* (Trematoda: Schistosomatidae) in the Swan estuary, Western Australia: frequency of infection of the intermediate host population. *International Journal for Parasitology*, **13**, 51–60.

Arai, H. P. (1980) Migratory activity and related phenomena in *Hymenolepsis diminuta*, in *Biology of the Tapeworm* Hymenolepsis diminuta, (ed. H. P. Arai), Academic Press, London, pp. 615–637.

Archer, D. M. and **Hopkins, C. A.** (1958) Studies on cestode metabolism, III. Growth pattern of *Diphyllobothrium* sp. in a definitive host. *Experimental Parasitology*, **7**, 125–144.

Arme, C. (1984) Points in question. The terminology of parasitology: the need for uniformity. *International Journal for Parasitology*, **14**, 539–540.

Arme, C. (1993) A day in the life of a tapeworm. *Helminthologia*, **30**, 3–7.

Arme, C. and **Read, C. P.** (1970) A surface enzyme in *Hymenolepis diminuta* (Cestoda). *Journal of Parasitology*, **56**, 514–516.

Armstrong, E. P., Halton, D. W., Tinsley, R. C. *et al.* (1997) Immunocytochemical evidence for the involvement of an FMRFamide-related peptide in egg production in the flatworm parasite *Polystoma nearcticum*. *Journal of Comparative Neurology*, **377**, 41–48.

Asch, H. L. and **Read, C. P.** (1975a) Transtegumental absorption of amino acids by male *Schistosoma mansoni*. *Journal of Parasitology*, **61**, 378–379.

Asch, H. L. and **Read, C. P.** (1975b) Membrane transport in *Schistosoma mansoni*: transport of amino acids by adult males. *Experimental Parasitology*, **38**, 123–135.

Audousset, J. C., Rondelaud, D., Dreyfuss, G. and **Vareille-Morel, C.** (1989) Les émissions cercariennes de *Fasciola hepatica* L. chez le mollusque *Lymnaea truncatula* Müller. A propos de quelques observations chronobiologiques. *Bulletin de la Société française de Parasitologie*, **7**, 217–224.

Baer, J. and **Joyeux, C.** (1961) Classe des Trématodes, in *Traité de Zoologie*, (ed. P. Grassé), Masson, Paris, pp. 561–692.

Baker, S. C. and **Bulow, F. J.** (1985) Effects of black-spot disease on the condition of stone-rollers *Campostoma anomalum*. *American Midland Naturalist*, **114**, 198–199.

Bakke, T. A., Harris, P. D., Jansen, P. A. and **Hansen, L. P.** (1992) Host specificity and dispersal strategy in gyrodactylid monogeneans, with particular reference to *Gyrodactylus salaris* (Platyhelminthes, Monogenea). *Diseases of Aquatic Organisms*, **13**, 63–74.

Bakker, K. E. and **Davids, C.** (1973) Notes on the life history of *Aspidogaster conchicola* Baer, 1826 (Trematoda; Aspidogastridae). *Journal of Helminthology*, **47**, 269–276.

Ball, C. S. (1916) The development of *Paravortex gemellipara* (*Graffilla gemellipara* Linton). *Journal of Morphology*, **27**, 453–557.

Ball, I. R. and **Khan, R. A.** (1976) On *Micropharynx parasitica* Jägerskiöld, a marine planarian ectoparasitic on thorny skate, *Raja radiata* Donovan, from the North Atlantic Ocean. *Journal of Fish Biology*, **8**, 419–426.

Ball, I. R. and **Reynoldson, T. B.** (1981) British planarians. Platyhelminthes: Tricladida, in *Synopses of the British Fauna 19*, (eds D. M. Kermack and R. S. K. Barnes), Cambridge University Press, Cambridge.

Barber, I. and **Huntingford, F. A.** (1996) Parasite infection alters schooling behaviour: deviant positioning of helminth-infected minnows in conspecific groups. *Proceedings of the Royal Society of London B*, **263**, 1095–1102.

Bargues, M. D. and **Mas-Coma, S.** (1991) Intracellular development in digenean sporocyst early stages. *Research and Reviews in Parasitology*, **51**, 111–124.

Barinaga, M. (1992) Viruses launch their own 'Star Wars'. *Science*, **258**, 1730–1731.

Barker, S. C. and **Blair, D.** (1996) Molecular phylogeny of *Schistosoma* species supports traditional groupings within the genus. *Journal of Parasitology*, **82**, 292–298.

Barnes, R. D. (1987) *Invertebrate Zoology*, 5th edn, Saunders College Publishing, New York.

Barrett, J. (1981) *Biochemistry of Parasitic Helminths*, Macmillan, London.

Barrett, J. and **Precious, W. Y.** (1994) The effect of sodium ions on the excystment of *Hymenolepis diminuta* cysticercoids. *Folia Parasitologica*, **41**, 44.

Barrington, E. J. W. (1979) *Invertebrate Structure and Function*, 2nd edn, Nelson, Sunbury-on-Thames, UK.

Bartoli, P. (1972) Les cycles biologiques de *Gymnophallus nereicola* J. Rebecq et G. Prévot, 1962 et *G. fossarum* P. Bartoli, 1965, espèces jumelles parasites d'oiseaux de rivages marins (Trematoda, Digenea, Gymnophallidae). *Annales de Parasitologie*, **47**, 193–223.

Bartoli, P. (1973) La pénétration et l'installation des cercaires de *Gymnophallus fossarum* P. Bartoli, 1965 (Digenea, Gymnophallidae) chez *Cardium glaucum* Bruguière. *Bulletin du Muséum national d'Histoire naturelle*, Série 3, Zoologie, **91**, 319–334.

Bartoli, P. (1974) Un cas d'exclusion compétitive chez les Trématodes: l'élimination de *Gymnophallus choledochus* T. Odhner, 1900 par *G. nereicola* J. Rebecq et G. Prévot, 1962 en Camargue (France) (Digenea, Gymnophallidae). *Bulletin de la Société Zoologique de France*, **99**, 551–559.

Bartoli, P. (1976–1978) Modification de la croissance et du comportement de *Venerupis aurea* parasité par *Gymnophallus fossarum* P. Bartoli, 1965 (Trematoda, Digenea). *Haliotis*, **7**, 23–28.

Bartoli, P. (1981) Les particularités biologiques et écologiques favorisant le recrutement de *Gymnophallus nereicola* J. Rebecq et G. Prévot, 1962 (Digenea, Gymnophallidae) par l'annélide polychète *Nereis diversicolor* O. F. Muller. *Annales de Parasitologie*, **56**, 261–270.

Bartoli, P. (1983) Populations ou espèces? Recherches sur la signification de la transmission de trématodes Lepocreadiinae (T. Odhner, 1905) dans deux écosystèmes marins. *Annales de Parasitologie,* **58,** 117–139.

Bartoli, P. (1984) Redescription de *Cercaria setifera* F. S. Monticelli, 1914 (nec J. Müller) (Trematoda) parasite de *Conus ventricosus* Hwass; comparaison avec quelques cercaires optalmotrichocerques de Méditerranée occidentale. *Annales de Parasitologie,* **59,** 161–176.

Bartoli, P. (1987) Caractères adaptifs originaux des digènes intestinaux de *Sarpa salpa* (Teleostei, Sparidae) et leur interprétation en termes d'évolution. *Annales de Parasitologie,* **62,** 542–576.

Bartoli, P. and **Combes, C.** (1986) Stratégies de dissémination des cercaires de trématodes dans un écosystème marin littoral. *Acta Ecologica,* **7,** 101–114.

Bartoli, P. and **Prévot, G.** (1986) Stratégies d'infestation des hôtes cibles chez les trématodes marins parasites de *Larus cachinnans michaellis* de Provence. *Annales de Parasitologie,* **61,** 533–552.

Basch, P. F. (1990) Why do schistosomes have separate sexes? *Parasitology Today,* **6,** 160–163.

Basch, P. F. (1991) *Schistosomes – Development, Reproduction and Host Relations,* Oxford University Press, Oxford.

Bashiruddin, M. and **Karling, T. G.** (1970) A new entocommensal turbellarian (Fa. Pterastericolidae) from the sea star *Astropecten irregularis. Zeitschrift für Morphologie und Ökologie der Tiere,* **67,** 16–28.

Bauer, O. N. (1958) Fishes as carriers of human helminthoses, in *Parasitology of Fishes,* (eds V. A. Dogiel, G. K. Petrushevski and Y. I. Polyanski), Leningrad University Press. English translation by Z. Kabata (1961), Oliver & Boyd, London, pp. 320–334.

Baylis, H. A. (1938) On two species of the trematode genus *Didymozoon* from the mackerel. *Journal of the Marine Biological Association of the United Kingdom,* **22,** 485–492.

Baylis, H. A. and **Jones, E. I.** (1933) Some records of parasitic worms from marine fishes at Plymouth. *Journal of the Marine Biological Association of the United Kingdom,* **18,** 627–634.

Bayssade-Dufour, C., Albaret, J., Samnaliev, P. *et al.* (1980) Les structures argyrophiles tégumentaires des stades larvaires (miracidium, rédie, cercaire) de *Fasciola hepatica.* Comparaison avec *F. gigantica. Annales de Parasitologie,* **55,** 553–564.

Beauchamp, P. de (1961) Classe des Turbellariés, in *Traité de Zoologie, vol. IV,* (ed. P. Grassé), Masson, Paris, pp. 35–212.

Beaver, P. C. (1937) Experimental studies on *Echinostoma revolutum* (Froelich) a fluke from birds and mammals. *Illinois Biological Monographs,* **15,** 7–96.

Becker, W. (1977) Zur Stoffwechselphysiologie der Miracidien von *Schistosoma mansoni* während ihrer Aktivierung innerhalb der Eischale. *Zeitschrift für Parasitenkunde,* **52,** 69–79.

Bednarz, J. (1962) The developmental cycle of germ-cells in *F. hepatica* L. 1758 (Trematodes, Digenea). *Zoologica Poloniae,* **12,** 439–466.

Befus, A. D. and **Podesta, R. B.** (1976) Intestine, in *Ecological Aspects of Parasitology,* (ed. C. R. Kennedy), North-Holland, Amsterdam, pp. 303–325.

Bell, E. J. and **Hopkins, C. A.** (1956) The development of *Diplostomum phoxini* (Strigeida, Trematoda). *Annals of Tropical Medicine and Parasitology,* **50,** 275–282.

Bell, R. G. (1996) IgE, allergies and helminth parasites: a new perspective on an old conundrum. *Immunology and Cell Biology,* **74,** 337–345.

Belopolskaya, M. M. (1958) On the structure of eggs of some cestodes. *Nauchnye Doklady vysshei Shkoly,* **4,** 7–10 (in Russian).

Beneden, P. J. van. (1858) *Mémoires sur les Vers Intestinaux,* Baillière et fils, Paris.

Beneden, P. J. van and **Hesse, C. E.** (1864) Recherches sur les bdellodes (hirudinées) et les trématodes marins. *Mémoires de l'Académie royale des sciences, des lettres et des beaux-arts de Belgique,* **34,** 1–142.

Benjamin, L. R. and **James, B. L.** (1987) The development of the metacercaria of *Maritrema linguilla* Jäg., 1908 (Digenea: Microphallidae) in the intermediate host, *Ligia oceanica* (L.). *Parasitology,* **94,** 221–231.

Bennett, C. E. (1975) *Fasciola hepatica*: development of caecal epithelium during migration in the mouse. *Experimental Parasitology,* **37,** 426–441.

Bennett, C. E. and **Threadgold, L. T.** (1975) *Fasciola hepatica*: development of tegument during migration in the mouse. *Experimental Parasitology,* **38,** 38–55.

Bertin, L. (1958) Organes de la respiration aquatique, in *Traité de Zoologie,* (ed. P. Grassé), vol. 13, fasc. 2, Masson, Paris, pp. 1303–1341.

Betterton, C. (1974) Studies on the host-specificity of the eyefluke, *Diplostomum spathaceum* in brown and rainbow trout. *Parasitology*, **69**, 11–29.

Bijukumar, A. and **Kearn, G. C.** (1996) *Furcohaptor cynoglossi* n. g., n. sp., an ancyrocephaline monogenean gill parasite with a bifurcate haptor and a note on its adhesive attitude. *Systematic Parasitology*, **34**, 71–76.

Bilharz, T. and **von Siebold, C. T.** (1852–1853) Ein Beitrag zur Helminthographia humana, aus brieflichen Mittheilungen des Dr. Bilharz in Cairo, nebst Bemerkungen von Prof. C. Th v. Siebold in Breslau. *Zeitschrift für wissenschaftliche Zoologie*, **4**, 53–76.

Blair, D. (1976) Observations on the life-cycle of the strigeoid trematode, *Apatemon* (*Apatemon*) *gracilis* (Rudolphi, 1819) Szidat, 1928. *Journal of Helminthology*, **50**, 125–131.

Blair, D. (1993) The phylogenetic position of the Aspidobothrea within the parasitic flatworms inferred from ribosomal RNA sequence data. *International Journal for Parasitology*, **23**, 169–178.

Blankespoor, C. L., Pappas, P. W. and **Eisner, T.** (1997) Impairment of the chemical defence of the beetle, *Tenebrio molitor*, by metacestodes (cysticercoids) of the tapeworm, *Hymenolepis diminuta*. *Parasitology*, **115**, 105–110.

Boeger, W. A., Kritsky, D. C. and **Belmont-Jégu, E.** (1994) Neotropical Monogenoidea. 20. Two new species of oviparous Gyrodactylidea (Polyonchoinea) from loricariid catfishes (Siluriformes) in Brazil and the phylogenetic status of Ooegyrodactylidae Harris, 1983. *Journal of the Helminthological Society of Washington*, **61**, 34–44.

Bona, F. V. (1994) Family Dilepididae Railliet & Henry, 1909, in *Key to the Cestode Parasites of Vertebrates*, (eds L. F. Khalil, A. Jones and R. A. Bray), CAB International, Wallingford, UK, pp. 443–554.

Bondad-Reantaso, M. G., Ogawa, K., Yoshinaga, T. and **Wakabayashi, H.** (1995) Acquired protection against *Neobenedenia girellae* in Japanese flounder. *Fish Pathology*, **30**, 233–238.

Bonham, K. and **Guberlet, J. E.** (1938) Ectoparasitic trematodes of Puget Sound fishes – *Acanthocotyle*. *American Midland Naturalist*, **20**, 590–602.

Borradaile, L. A., Potts, F. A., Eastham, L. E. S. and **Saunders, J. T.** (1961) *The Invertebrata*, Cambridge University Press, Cambridge.

Borucinska, J. and **Caira, J. N.** (1993) A comparison of mode of attachment and histopathogenicity of four tapeworm species representing two orders infecting the spiral intestine of the nurse shark, *Ginglymostoma cirratum*. *Journal of Parasitology*, **79**, 238–246.

Bourgat, R. and **Kulo, S.-D.** (1977) Recherches sur le cycle biologique d'un Paramphistomatidae (Trematoda) d'amphibiens en Afrique. *Annales de Parasitologie*, **52**, 7–12.

Bradley, D. J. and **McCullough, F. S.** (1973) Egg output stability and the epidemiology of *Schistosoma haematobium*. Part II. An analysis of the epidemiology of endemic *S. haematobium*. *Transactions of the Royal Society of Tropical Medicine and Hygiene*, **67**, 491–500.

Brassard, P., Rau, M. E. and **Curtis, M. A.** (1982) Parasite-induced susceptibility to predation in diplostomiasis. *Parasitology*, **85**, 495–501.

Bråten, T. and **Hopkins, C. A.** (1969) The migration of *Hymenolepis diminuta* in the rat's intestine during normal development and following surgical transplantation. *Parasitology*, **59**, 891–905.

Brooks, D. R., O'Grady, R. T. and **Glen, D. R.** (1985) Phylogenetic analysis of the Digenea (Platyhelminthes: Cercomeria) with comments on their adaptive radiation. *Canadian Journal of Zoology*, **63**, 411–443.

Bryant, C. (1994) Ancient biochemistries and the evolution of parasites. *International Journal for Parasitology*, **24**, 1089–1097.

Bryant, C. and **Behm, C. A.** (1989) *Biochemical Adaptation in Parasites*, Chapman & Hall, London.

Buchsbaum, R. (1957) *Animals without Backbones*, Penguin Books, Harmondsworth, UK.

Bundy, D. A. P. (1981) Periodicity in the hatching of digenean eggs: a possible circadian rhythm in the life-cycle of *Transversotrema patialense*. *Parasitology*, **83**, 13–22.

Burger, J. W. and **Hess, W. N.** (1960) Function of the rectal gland in the spiny dogfish. *Science*, **131**, 670–671.

Burns, W. C. (1961a) Penetration and development of *Allassogonoporus vespertilionis* and *Acanthatrium oregonense* (Trematoda: Lecithodendriidae) cercariae in caddis fly larvae. *Journal of Parasitology*, **47**, 927–932.

Burns, W. C. (1961b) Six virgulate xiphidiocercariae from Oregon, including redescriptions of *Allassogonoporus vespertilionis* and *Acanthatrium oregonense*. *Journal of Parasitology*, **47**, 919–925.

Burren, C. H., Ehrlich, I. and **Johnson, P.** (1967) Excretion of lipids by the liver fluke (*Fasciola hepatica* L.). *Lipids*, **2**, 353–356.

Butterworth, A. E. (1990) Immunology of schistosomiasis, in *Modern Parasite Biology: Cellular, Immunological and Molecular Aspects*, (ed. D. J. Wyler), W. H. Freeman & Co., New York, pp. 262–288.

Bychowsky, B. E. (1957) *Monogenetic Trematodes, their Classification and Phylogeny*, Academy of Sciences, USSR, Moscow, Leningrad (in Russian; English translation by W. J. Hargis and P. C. Oustinov, 1961, American Institute of Biological Sciences, Washington).

Bychowsky, B. E., Korotaeva, V. D. and **Nagibina, L. F.** (1970) [*Montchadskyella intestinale* gen. et sp. n. – a new member of endoparasitic monogeneans (Monogenoidea)]. *Parazitologiya*, **4**, 451–457 (in Russian).

Bychowsky, B. E. and **Nagibina, L. F.** (1976) [New species of Monogenea of the genus *Telegamatrix*]. *Biologiya Morya*, **2**, 10–15 (in Russian).

Cable, J., Harris, P. D. and **Tinsley, R. C.** (1996) Ultrastructural adaptations for viviparity in the female reproductive system of gyrodactylid monogeneans. *Tissue and Cell*, **28**, 515–526.

Cable, J. and **Tinsley, R. C.** (1991) Intra-uterine larval development of the polystomatid monogeneans, *Pseudodiplorchis americanus* and *Neodiplorchis scaphiopodis*. *Parasitology*, **103**, 253–266.

Cable, J. and **Tinsley, R. C.** (1992a) Unique ultrastructural adaptations of *Pseudodiplorchis americanus* (Polystomatidae: Monogenea) to a sequence of hostile conditions following host infection. *Parasitology*, **105**, 229–241.

Cable, J. and **Tinsley, R. C.** (1992b) Tegumental ultrastructure of *Pseudodiplorchis americanus* larvae (Monogenea: Polystomatidae). *International Journal for Parasitology*, **22**, 819–829.

Cable, R. M. (1952) Studies on marine digenetic trematodes of Puerto Rico. Four species of magnocercous heterophyid cercariae with zygocercous aggregation in one. *Journal of Parasitology*, **38**, 28.

Cable, R. M. (1954a) A new marine cercaria from the Woods Hole region and its bearing on the interpretation of larval types in the Fellodistomatidae (Trematoda: Digenea). *Biological Bulletin*, **106**, 15–20.

Cable, R. M. (1954b) Studies on marine digenetic trematodes of Puerto Rico. The life cycle in the family Haplosplanchnidae. *Journal of Parasitology*, **40**, 71–76.

Cable, R. M. (1956) Marine cercariae of Puerto Rico, in *Scientific Survey of Porto Rico and the Virgin Islands*, New York Academy of Sciences, vol. 16, pp. 493–577.

Cable, R. M. (1963) Marine cercariae from Curaçao and Jamaica. *Zeitschrift für Parasitenkunde*, **23**, 429–469.

Cable, R. M. (1974) Phylogeny and taxonomy of trematodes with reference to marine species, in *Symbiosis in the Sea*, (ed. W. B. Vernberg), University of South Carolina Press, pp. 173–193.

Cable, R. M. (1985) The partial life cycle and affinities of an unusual xiphidiocercaria from *Amnicola limosa* (Say) in Indiana, U. S. A. (Digenia: Lecithodendriidae). *Journal of Parasitology*, **71**, 342–344.

Cable, R. M. and **Crandall, R. B.** (1956) Larval stages and phylogeny as exemplified by the lung fluke of turtles. *Science*, **124**, 890.

Cable, R. M. and **Nahhas, F. M.** (1962a) *Bivesicula caribbensis* sp. n. (Trematoda: Digenea) and its life history. *Journal of Parasitology*, **48**, 536–538.

Cable, R. M. and **Nahhas, F. M.** (1962b) *Lepas* sp., second intermediate host of a didymozoid trematode. *Journal of Parasitology*, **48**, 34.

Calentine, R. L. (1964) The life cycle of *Archigetes iowensis* (Cestoda: Caryophyllaeidae). *Journal of Parasitology*, **50**, 454–458.

Caley, J. (1974) The functional significance of scolex retraction and subsequent cyst formation in the cysticercoid larva of *Hymenolepis microstoma*. *Parasitology*, **68**, 207–227.

Caley, J. (1975a) *In vitro* hatching of the tapeworm *Moniezia expansa* (Cestoda: Anoplocephalidae) and some properties of the egg membranes. *Zeitschrift für Parasitenkunde*, **45**, 335–346.

Caley, J. (1975b) A comparative study of the two alternative larval forms of *Hymenolepis nana*, the dwarf tapeworm, with special reference to the process of excystment. *Zeitschrift für Parasitenkunde*, **47**, 217–235.

Campbell, D. (1985) The life cycle of *Gymnophallus rebecqui* (Digenea: Gymnophallidae) and the response of the bivalve *Abra tenuis* to its metacercariae. *Journal of the Marine Biological Association of the United Kingdom*, **65**, 589–601.

Campbell, R. A. and **Beveridge, I.** (1994) Order Trypanorhyncha Diesing, 1863, in *Keys to the Cestode Parasites of Vertebrates*, (eds L. F. Khalil, A. Jones and R. A. Bray), CAB International, Wallingford, UK, pp. 51–148.

Cannon, L. R. G. (1990) *Apidioplana apluda* n. sp., a turbellarian symbiote of gorgonian corals from the Great Barrier Reef, with a review of the family Apidioplanidae (Polycladida: Acotylea). *Memoirs of the Queensland Museum*, **28**, 435–442.

Cannon, L. R. G. and **Grygier, M. J.** (1991) The turbellarian *Notoplana comes* n. sp. (Leptoplanidae: Acotylea: Polycladida) found with the intertidal brittlestar *Ophiocoma scolopendrina* (Ophiocomidae: Ophiuroidea) in Okinawa, Japan. *Galaxea*, **10**, 23–33.

Cannon, L. R. G. and **Jennings, J. B.** (1987) Occurrence and nutritional relationships of four ectosymbiotes of the freshwater crayfishes *Cherax dispar* Riek and *Cherax punctatus* Clark (Crustacea: Decapoda) in Queensland. *Australian Journal of Marine and Freshwater Research*, **38**, 419–427.

Carmona, C., Dowd, A. J., Smith, A. M. and **Dalton, J. P.** (1993) Cathepsin L proteinase secreted by *Fasciola hepatica* in vitro prevents antibody-mediated eosinophil attachment to newly excysted juveniles. *Molecular and Biochemical Parasitology*, **62**, 9–18.

Carney, W. P. (1969) Behavioral and morphological changes in carpenter ants harboring dicrocoeliid metacercariae. *American Midland Naturalist*, **82**, 605–611.

Caulfield, J. P. (1990) Cell biology of schistosomes. II. Tegumental membranes and their interaction with human blood cells, in *Modern Parasite Biology: Cellular, Immunological and Molecular Aspects*, (ed. D. J. Wyler), W. H. Freeman & Co., New York, pp. 107–125.

Caullery, M. and **Mesnil, F.** (1903) Recherches sur les 'Fecampia' Giard, Turbellariés Rhabdocèles, parasites internes des crustacés. *Annales de la Faculté des Sciences de Marseille*, **13**, 131–168.

Caveny, B. A. and **Etges, F. J.** (1971) Life history studies of *Microphallus opacus* (Trematoda: Microphallidae). *Journal of Parasitology*, **57**, 1215–1221.

Chandler, A. C. (1939) The effects of numbers and age of worms on development of primary and secondary infections with *Hymenolepis diminuta* in rats, and an investigation into the true nature of 'premunition' in tapeworm infections. *American Journal of Hygiene*, **29D**, 105–114.

Chapman, H. D. (1974) The behaviour of the cercaria of *Cryptocotyle lingua*. *Zeitschrift für Parasitenkunde*, **44**, 211–226.

Chapman, H. D. and **Wilson, R. A.** (1973) The propulsion of the cercariae of *Himasthla secunda* (Nicoll) and *Cryptocotyle lingua*. *Parasitology*, **67**, 1–15.

Chappell, L. H. (1976) The nutrition of *Schistosoma* and *Fasciola*. *Parasitology*, **73**, xxii.

Chappell, L. H. (1995) The biology of diplostomatid eyeflukes of fishes. *Journal of Helminthology*, **69**, 97–101.

Chauvin, A., Bouvet, G. and **Boulard, C.** (1995) Humoral and cellular immune responses to *Fasciola hepatica* experimental primary and secondary infection in sheep. *International Journal for Parasitology*, **25**, 1227–1241.

Cheng, T. C. and **Thakur, A. S.** (1967) Thermal activation and inactivation of *Philophthalmus gralli* metacercariae. *Journal of Parasitology*, **53**, 212–213.

Christensen, A. M. and **Kanneworff, B.** (1964) *Kronborgia amphipodicola* gen. et sp. nov., a dioecious turbellarian parasitizing ampeliscid amphipods. *Ophelia*, **1**, 147–166.

Christensen, A. M. and **Kanneworff, B.** (1965) Life history and biology of *Kronborgia amphipodicola* Christensen & Kanneworff (Turbellaria, Neorhabdocoela). *Ophelia*, **2**, 237–251.

Christensen, A. M. and **Kanneworff, B.** (1967) On some cocoons belonging to undescribed species of endoparasitic turbellarians. *Ophelia*, **4**, 29–42.

Clark, W. C. (1974) Interpretation of life history pattern in the Digenea. *International Journal for Parasitology*, **4**, 115–123.

Clarke, M. R. and **Merrett, N.** (1972) The significance of squid, whale and other remains from the stomachs of bottom-living deep-sea fish. *Journal of the Marine Biological Association of the United Kingdom*, **52**, 599–603.

Cleveland, G. and **Kearn, G. C.** (1989) The function of the paraoesophageal glands in an echinostome (digenean) cercaria (?*Cercaria spinifera* La Valette, 1855). *Journal of Helminthology*, **63**, 231–238.

Cohen, C., Reinhardt, B., Castellani, L. *et al.* (1982) Schistosome surface spines are 'crystals' of actin. *Journal of Cell Biology*, **95**, 987–988.

Cohen, J. (1977) *Reproduction*, Butterworth, London.

Combes, C. (1968) Biologie, écologie des cycles et biogéographie de digènes et monogènes d'amphibiens dans l'est des Pyrénées. *Mémoires du Muséum national d'Histoire naturelle*, Série A, Zoologie, **51**, 1–195.

Combes, C. (1972) Ecologie des Polystomatidae (Monogenea): facteurs influençant le volume et le rythme de la ponte. *International Journal for Parasitology*, **2**, 233–238.

Combes, C. (1990) Where do human schistosomes come from? An evolutionary approach. *Trends in Ecology and Evolution*, **5**, 334–337.

Combes, C. (1991) The schistosome scandal. *Acta oecologica*, **12**, 165–173.

Combes, C., Bourgat, R. and **Salami-Cadoux, M.-L.** (1973) Découverte d'un cycle interne direct intervenant comme mode habituel dans la reproduction d'un Plathelminthe. *Comptes rendus de l'Académie des Sciences (Paris)*, Série D, **276**, 2005–2006.

Combes, C., Bourgat, R. and **Salami-Cadoux, M.-L.** (1976) Valeur adaptative du mode de transmission chez les Polystomatidae (Monogenea). *Bulletin d'Ecologie*, **7**, 207–214.

Combes, C., Després, D., Establet, D. *et al.* (1991) Schistosomatidae (Trematoda): some views on their origin and evolution. *Research and Reviews in Parasitology*, **51**, 25–28.

Combes, C. and **Nassi, H.** (1977) Metacercarial dispersion and intracellular parasitism in a strigeid trematode. *International Journal for Parasitology*, **7**, 501–503.

Cone, D. K., Gratzek, J. B. and **Hoffman, G. L.** (1987) A study of *Enterogyrus* sp. (Monogenea) parasitizing the foregut of captive *Pomacanthus paru* (Pomacanthidae) in Georgia. *Canadian Journal of Zoology*, **65**, 312–316.

Conn, D. B. and **Etges, F. J.** (1983) Maternal transmission of asexually proliferative *Mesocestoides corti* tetrathyridia (Cestoda) in mice. *Journal of Parasitology*, **69**, 922–925.

Conn, D. B. and **Etges, F. J.** (1984) Fine structure and histochemistry of the parenchyma and uterine egg capsules of *Oochoristica anolis* (Cestoda: Linstowiidae). *Zeitschrift für Parasitenkunde*, **70**, 769–779.

Conn, D. B., Etges, F. J. and **Sidner, R. A.** (1984) Fine structure of the gravid paruterine organ and embryonic envelopes of *Mesocestoides lineatus* (Cestoda). *Journal of Parasitology*, **70**, 68–77.

Conradt, U. and **Peters, W.** (1989) Investigations on the occurrence of pinocytosis in the tegument of *Schistocephalus solidus*. *Parasitology Research*, **75**, 630–635.

Conradt, U. and **Schmidt, J.** (1992) A double surface membrane in plerocercoids of *Ligula intestinalis* (Cestoda: Pseudophyllidea). *Parasitology Research*, **78**, 123–129.

Cooper, A. R. (1920) *Glaridacris catostomi* gen. nov., sp. nov.: a cestodarian parasite. *Transactions of the American Microscopical Society*, **39**, 5–24.

Cort, W. W., Brackett, S., Olivier, L. and **Nolf, L. O.** (1945) Influence of larval trematode infections in snails on their second intermediate host relations to the strigeid trematode, *Cotylurus flabelliformis* (Faust, 1917). *Journal of Parasitology*, **31**, 61–78.

Cousin, C. E., Stirewalt, M. A. and **Dorsey, C. H.** (1981) *Schistosoma mansoni*: ultrastructure of early transformation of skin- and shear-pressure derived schistosomules. *Experimental Parasitology*, **51**, 341–365.

Cowell, L. E., Watanabe, W. O., Head, W. D. *et al.* (1993) Use of tropical cleaner fish to control the ectoparasite *Neobenedenia melleni* (Monogenea: Capsalidae) on seawater-cultured Florida red tilapia. *Aquaculture*, **113**, 189–200.

Cox, W. T. and **Hendrickson, G. L.** (1991) Observations on the life cycle of *Proteocephalus tumidocollus* (Cestoda: Proteocephalidae) in steelhead trout *Oncorhynchus mykiss*. *Journal of the Helminthological Society of Washington*, **58**, 39–42.

Crabtree, J. E. and **Wilson, R. A.** (1980) *Schistosoma mansoni*: a scanning electron microscope study of the developing schistosomulum. *Parasitology*, **81**, 553–564.

Crabtree, J. E. and **Wilson, R. A.** (1986) *Schistosoma mansoni*: an ultrastructural examination of pulmonary migration. *Parasitology*, **92**, 343–354.

Craig, P. S., Deshan, L., Macpherson, C. N. L. *et al.* (1992) A large focus of alveolar echinococcosis in central China. *Lancet*, **340**, 826–831.

Crandall, R. B. (1960) The life history and affinities of the turtle lung fluke, *Heronimus chelydrae* MacCallum, 1902. *Journal of Parasitology*, **46**, 289–307.

Cribb, T. H. (1986) The life cycle and morphology of *Stemmatostoma pearsoni*, gen. et sp. nov., with notes on the morphology of *Telogaster opisthorchis* Macfarlane (Digenea: Cryptogonimidae). *Australian Journal of Zoology*, **34**, 279–304.

Cribb, T. H. (1988) Life cycle and biology of *Prototransversotrema steeri* Angel, 1969 (Digenea: Transversotrematidae). *Australian Journal of Zoology*, **36**, 111–129.

Crofton, H. D. and **Fraser, P. G.** (1954) The mode of infection of the hake, *Merluccius merluccius* (L.) by the trematode *Bucephalopsis gracilescens* (Rud.). *Proceedings of the Zoological Society of London*, **124**, 105–109.

Crompton, D. W. T. (1970) *An Ecological Approach to Acanthocephalan Physiology*, Cambridge University Press, London.

Crowden, A. E. (1976) *Diplostomum spathaceum* in the Thames; occurrence and effects on fish behaviour. *Parasitology*, **73**, vii.

Crowden, A. E. and **Broom, D. M.** (1980) Effects of the eyefluke, *Diplostomum spathaceum*, on the behaviour of dace (*Leuciscus leuciscus*). *Animal Behaviour*, **28**, 287–294.

Cunningham, J. T. (1890) *A Treatise on the Common Sole*, Marine Biological Association of the United Kingdom, Plymouth, UK.

Czaplinski, B. and **Vaucher, C.** (1994) Family Hymenolepididae Ariola, 1899, in *Keys to the Cestode Parasites of Vertebrates*, (eds L. F. Khalil, A. Jones and R. A. Bray), CAB International, Wallingford, UK, pp. 595–663.

Dailey, M. D. (1985) Diseases of Mammalia: Cetacea, in *Diseases of Marine Animals, vol. IV, part 2*, (ed. O. Kinne), Biologische Anstalt, Helgoland, pp. 805–847.

Dalton, J. P. and **Heffernan, M.** (1989) Thiol proteases released in vitro by *Fasciola hepatica*. *Molecular and Biochemical Parasitology*, **35**, 161–166.

Dalton, J. P., Smith, A. M., Clough, K. A. and **Brindley, P. J.** (1995) Digestion of haemoglobin by schistosomes: 35 years on. *Parasitology Today*, **11**, 299–303.

Damian, R. T. (1987) The exploitation of host immune responses by parasites. *Journal of Parasitology*, **73**, 3–13.

Damian, R. T. (1991) Tropomyosin and molecular mimicry. *Parasitology Today*, **7**, 96.

Dawes, B. (1946) *The Trematoda*, Cambridge University Press, Cambridge.

Dawes, B. (1962) On the growth and maturation of *Fasciola hepatica* L. in the mouse. *Journal of Helminthology*, **36**, 11–38.

Dawes, B. (1963a) *Fasciola hepatica* L., a tissue feeder. *Nature*, **198**, 1011–1012.

Dawes, B. (1963b) Hyperplasia of the bile duct in fascioliasis and its relation to the problem of nutrition in the liver-fluke, *Fasciola hepatica* L. *Parasitology*, **53**, 123–133.

Deblock, S. (1980) Inventaires des trématodes larvaires parasites des mollusques *Hydrobia* (Prosobranches) des côtes de France. *Parassitologia*, **22**, 1–105.

Deblock, S. and **Maillard, C.** (1975) Contribution à l'étude des Microphallidae Travassos, 1920 (Trematoda). XXXII. – *Microphallus breviatus* n. sp., espèce à cycle évolutif abrégé originaire d'un étang méditerranéen du Languedoc. *Acta Tropica*, **32**, 317–326.

Denton, E. (1971) Reflectors in fishes. *Scientific American*, **224**, No. 1, 65–72.

Desowitz, R. S. (1981) *New Guinea Tapeworms and Jewish Grandmothers: Tales of Parasites and People*, W. W. Norton, New York.

Després, L., Kruger, F. J., Imbert-Establet, D. and **Adamson, M. L.** (1995) ITS2 ribosomal RNA indicates *Schistosoma hippopotami* is a distinct species. *International Journal for Parasitology*, **25**, 1509–1514.

Després, L. and **Maurice, S.** (1995) The evolution of dimorphism and separate sexes in schistosomes. *Proceedings of the Royal Society (London) B*, **262**, 175–180.

Dickerman, E. E. (1948) On the life cycle and systematic position of the aspidogastrid trematode, *Cotylogaster occidentalis* Nickerson, 1902. *Journal of Parasitology*, **34**, 164.

Dike, S. C. and **Read, C. P.** (1971) Relation of tegumentary phosphohydrolases and sugar transport in *Hymenolepis diminuta*. *Journal of Parasitology*, **57**, 1251–1255.

Dixon, K. E. (1964) Excystment of metacercariae of *Fasciola hepatica* L. in vitro. *Nature*, **202**, 1240–1241.

Dixon, K. E. (1965) The structure and histochemistry of the cyst wall of the metacercaria of *Fasciola hepatica* L. *Parasitology*, **55**, 215–226.

Dixon, K. E. (1966) The physiology of excystment of the metacercaria of *Fasciola hepatica* L. *Parasitology*, **56**, 431–456.

Dixon, K. E. (1968) Encystment of the cercaria of *Fasciola hepatica*. *Wiadomosci Parazytologiczne*, **14**, 689–701.

Dogiel, V. A. (1962) *General Parasitology*, (English translation by Z. Kabata, 1964), Oliver & Boyd, London.

Dollfus, R. P. (1958) Cours d'helminthologie. I. – Trématodes sous-classe Aspidogastrea. *Annales de Parasitologie*, **33**, 305–395.

Dollfus, R. P. (1966) Organismes dont la présence dans le plancton marin était, jusqu'à présent, ignorée: larves et postlarves de cestodes tétrarhynques. *Compte rendu hebdomadaire des séances de l'Académie des sciences.* Paris, **262**, 2612–2615.

Dolmen, D. (1987) *Gyrodactylus salaris* (Monogenea) in Norway; infestations and management, in *Proceedings of a Symposium on Parasites and Diseases in Natural Waters and Aquaculture in Nordic Countries,* (eds A. Stenmark and G. Malmberg), Zoo-tax, Naturhistoriska Riksmuseet, University of Stockholm, Stockholm, pp. 63–69.

Dönges, J. (1962) Entwicklungszyklus einer fakultativ humanpathogenen Echinostomatide-nart. *Medizinische Welt,* **27**, 1499–1500.

Dönges, J. (1964) Der Lebenszyklus von *Posthodiplostomum cuticola* (v. Nordmann 1832) Dubois 1936 (Trematoda, Diplostomatidae). *Zeitschrift für Parasitenkunde,* **24**, 169–248.

Dönges, J. (1967) Der modifizierende Einfluss des Endwirtes auf die Entwicklung des Darmegels *Isthmiophora melis* (Schrank 1788). Zugleich ein Beitrag zur taxonomischen Klärung des Genus *Isthmiophora* Lühe 1909 (Trematoda, Echinostomatidae). *Zeitschrift für Parasitenkunde,* **29**, 1–14.

Dönges, J. (1969) Entwicklungs- und Lebensdauer von Metacercarien. *Zeitschrift für Parasitenkunde,* **31**, 340–366.

Dönges, J. (1980) *Parasitologie. Mit besonderer Berücksichtigung humanpathogener Formen,* Georg Thieme, Stuttgart.

Dönges, J. and **Götzelmann, M.** (1988) Digenetic trematodes: multiplication of the intramolluscan stages in some species is potentially unlimited. *Journal of Parasitology,* **74**, 884–885.

Dorsey, C. H. (1974) *Schistosoma mansoni:* ultrastructure of cercarial escape glands. *Experimental Parasitology,* **36**, 386–396.

Dreyfuss, G. and **Rondelaud, D.** (1994) *Fasciola hepatica:* a study of the shedding of cercariae from *Lymnaea truncatula* raised under constant conditions of temperature and photoperiod. *Parasite,* **1**, 401–404.

Dubinina, M. N. (1964) *Ligulidae (Cestoda: Pseudophyllidea) and their Evolution,* 'Nauka' Publishing House, Leningrad.

Dubinina, M. N. (1985) Class Amphilinida, in [*Keys to Parasites of the Freshwater Fish Fauna of the USSR*], volume 2, [*Parasitic Metazoa*], (ed. O. N. Bauer), Leningrad Publishing House 'Nauka', Leningrad, pp. 388–394 (in Russian).

Ebino, K. Y., Yoshinaga, K., Suwa, T. *et al.* (1989) Effects of coprophagy on pregnant mice – is coprophagy beneficial on a balanced diet? *Experimental Animals,* **38**, 245–252.

Eckert, J., von Brand, T. and **Voge, M.** (1969) Asexual multiplication of *Mesocestoides corti* (Cestoda) in the intestine of dogs and skunks. *Journal of Parasitology,* **55**, 241–249.

Eckert, J., Pawlowski, F. K., Vuitton, D. A. *et al.* (1995) Medical aspects of echinococcosis. *Parasitology Today,* **11**, 273–276.

Ehlers U. (1986) Comments on a phylogenetic system of the Platyhelminthes. *Hydrobiologia* **132**, 1–12.

El-Mayas, H. (1991) A study on digenean parasites from the salt marshes of north Norfolk, PhD Thesis, University of East Anglia, UK.

El-Mayas, H. and **Kearn, G. C.** (1996) *In vitro* excystment of the metacercaria of *Bucephalus haimeanus* from the common goby (*Pomatoschistus microps*). *Journal of Helminthology,* **70**, 193–200.

El-Naggar, M. M. and **Kearn, G. C.** (1983) Glands associated with the anterior adhesive areas and body margins in the skin-parasitic monogenean *Entobdella soleae. International Journal for Parasitology,* **13**, 67–81.

Erasmus, D. A. (1958) Studies on the morphology, biology and development of a strigeid cercaria (*Cercaria X* Baylis 1930). *Parasitology,* **48**, 312–335.

Erasmus, D. A. (1959) The migration of *Cercaria X* Baylis (Strigeida) within the fish intermediate host. *Parasitology,* **49**, 173–190.

Erasmus, D. A. (1962) Studies on the adult and metacercaria of *Holostephanus lühei* Szidat, 1936. *Parasitology,* **52**, 353–374.

Erasmus, D. A. (1969a) Studies on the host–parasite interface of strigeoid trematodes. V. Regional differentiation of the adhesive organ of *Apatemon gracilis minor* Yamaguti, 1933. *Parasitology,* **59**, 245–256.

Erasmus, D. A. (1969b) Studies on the host–parasite interface of strigeoid trematodes. IV. The ultrastructure of the lappets of *Apatemon gracilis minor* Yamaguti, 1933. *Parasitology,* **59**, 193–201.

Erasmus, D. A. (1970) The host–parasite interface of strigeoid trematodes. VII. Ultrastructural observations on the adhesive organ of *Diplostomum phoxini* Faust, 1918. *Zeitschrift für Parasitenkunde*, **33**, 211–224.

Erasmus, D. A. (1972) *The Biology of the Trematodes*, Edward Arnold, London.

Erasmus, D. A. and **Davies, T. W.** (1979) *Schistosoma mansoni* and *S. haematobium*: calcium metabolism of the vitelline cell. *Experimental Parasitology*, **47**, 91–106.

Ergens, R. (1988) *Paraquadriacanthus nasalis* gen. et sp. n. (Monogenea: Ancyrocephalidae) from *Clarias lazera* Cuvier et Valenciennes. *Folia Parasitologica*, **35**, 189–191.

Esch, G. W. and **Fernandez, J. C.** (1993) *A Functional Biology of Parasitism*, Chapman & Hall, London.

Euzet, L. and **Combes, C.** (1967) Présence au Nord-Tchad de *Eupolystoma alluaudi* (de Beauchamp, 1913) (Monogenea, Polystomatidae). *Annales de Parasitologie humaine et comparée*, **42**, 403–406.

Euzet, L. and **Combes, C.** (1980) Les problèmes de l'espèces congénériques de Monogènes Monopisthocotylea, in *Les problèmes de l'espèces dans le règne animal*, (eds C. Bocquet, J. Genermont and M. Lamotte). *Mémoires de la Société Zoologique de France*, **40**, 239–285.

Euzet, L. and **Maillard, C.** (1976) [The mechanism of attachment to the host of some Hexabothriidae (Monogenea)] (in Russian). *Proceedings of the Institute of Biology and Pedology, Far-East Science Centre, Vladivostok*, **34**, 115–122.

Euzet, L. and **Marc, A.** (1963) *Microcotyle donavini* van Beneden et Hesse 1863, espèce type du genre *Microcotyle* van Beneden et Hesse 1863. *Annales de Parasitologie humaine et comparée*, **38**, 875–886.

Euzet, L. and **Wahl, E.** (1970) Biologie de *Rhinecotyle crepitacula* Euzet et Trilles, 1960 (Monogenea) parasite de *Sphyraena piscatorum* Cadenat, 1964 (Teleostei) dans la lagune Ebrié (Côte d'Ivoire). *Revue Suisse de Zoologie*, **77**, 687–703.

Evans, N. A., Whitfield, P. J. and **Dobson, A. P.** (1981) Parasite utilization of a host community: the distribution and occurrence of metacercarial cysts of *Echinoparyphium recurvatum* (Digenea: Echinostomatidae) in seven species of mollusc at Harting Pond, Sussex. *Parasitology*, **83**, 1–12.

Evans, W. S., Hardy, M. C., Singh, R. *et al.* (1992) Effect of the rat tapeworm, *Hymenolepis diminuta*, on the coprophagic activity of its intermediate host, *Tribolium confusum*. *Canadian Journal of Zoology*, **70**, 2311–2314.

Fairweather, I. and **Threadgold, L. T.** (1983) *Hymenolepis nana*: the fine structure of the adult nervous system. *Parasitology*, **86**, 89–103.

Fares, A. and **Maillard, C.** (1974) Recherches sur quelques Haploporidae (Trematoda) parasites des muges de Méditerranée occidentale: systématique et cycles évolutifs. *Zeitschrift für Parasitenkunde*, **45**, 11–43.

Farley, J. (1991) *Bilharzia. A History of Imperial Tropical Medicine*, Cambridge University Press, Cambridge.

Feiler, W. and **Haas, W.** (1988) Host-finding in *Trichobilharzia ocellata* cercariae: swimming and attachment to the host. *Parasitology*, **96**, 493–505.

Ferguson, M. S. (1943) Migration and localization of an animal parasite within the host. *Journal of Experimental Zoology*, **93**, 375–399.

Ferraz, E., Shinn, A. P. and **Sommerville, C.** (1994) *Gyrodactylus gemini* n. sp. (Monogenea: Gyrodactylidae), a parasite of *Semaprochilodus taeniurus* (Steindachner) from the Venezuelan Amazon. *Systematic Parasitology*, **29**, 217–222.

File, S. (1995) Interaction of schistosome eggs with vascular endothelium. *Journal of Parasitology*, **81**, 234–238.

Fischer, H. and **Freeman, R. S.** (1969) Penetration of parenteral plerocercoids of *Proteocephalus ambloplitis* (Leidy) into the gut of smallmouth bass. *Journal of Parasitology*, **55**, 766–774.

Fischer, H. and **Freeman, R. S.** (1973) The role of plerocercoids in the biology of *Proteocephalus ambloplitis* (Cestoda) maturing in smallmouth bass. *Canadian Journal of Zoology*, **51**, 133–141.

Fishelson, Z. (1989) Complement and parasitic trematodes. *Parasitology Today*, **5**, 19–25.

Fitzgerald, G. J., Fournier, M. and **Morrissette, J.** (1993) Sexual selection in an anadromous population of threespine sticklebacks – no role for parasites. *Evolutionary Ecology*, **8**, 348–356.

Flisser, A. and **Larralde, C.** (1986) Cysticercosis, in *Immunodiagnosis of Parasitic Diseases*, vol. 1, *Helminthic Diseases*, (eds K. W. Walls and P. M. Schantz), Academic Press, New York, pp. 109–161.

Fournier A. and **Combes, C.** (1978) Structure of photoreceptors of *Polystoma integerrimum* (Platyhelminths, Monogenea). *Zoomorphologie*, **91**, 147–155.

Fournier, A. and **Combes, C.** (1979) Démonstration d'une dualité évolutive des embryons chez *Eupolystoma alluaudi* (Monogenea: Polystomatidae) et de son rôle dans la genèse du cycle interne. *Comptes rendus de l'Académie des Sciences (Paris)*, Série D, **289**, 745–747.

Fredericksen, D. W. (1972) Morphology and taxonomy of *Cotylogaster occidentalis* (Trematoda: Aspidogastridae). *Journal of Parasitology*, **58**, 1110–1116.

Fredericksen, D. W. (1978) The fine structure and phylogenetic position of the cotylocidium larva of *Cotylogaster occidentalis* Nickerson 1902 (Trematoda: Aspidogastridae). *Journal of Parasitology*, **64**, 961–976.

Freeman, R. F. H. and **Llewellyn, J.** (1958) An adult digenetic trematode from an invertebrate host: *Proctoeces subtenuis* (Linton) from the lamellibranch *Scrobicularia plana* (Da Costa). *Journal of the Marine Biological Association of the United Kingdom*, **37**, 435–457.

Freeman, R. S. (1952) The biology and life history of *Monoecocestus* Beddard, 1914 (Cestoda: Anoplocephalidae) from the porcupine. *Journal of Parasitology*, **38**, 111–129.

Freeman, R. S. (1964) On the biology of *Proteocephalus parallacticus* Maclulich (Cestoda) in Algonquin Park, Canada. *Canadian Journal of Zoology*, **42**, 387–408.

Freeman, R. S. (1973) Ontogeny of cestodes and its bearing on their phylogeny and systematics. *Advances in Parasitology*, **11**, 481–557.

Fried, B. and **Haseeb, M. A.** (1991) Platyhelminthes: Aspidogastrea, Monogenea and Digenea, in *Microscopic Anatomy of Invertebrates*, vol. 3, *Platyhelminthes and Nemertinea*, Wiley-Liss, New York, pp. 141–209.

Fried, B. and **King, B. W.** (1989) Attraction of *Echinostoma revolutum* cercariae to *Biomphalaria glabrata* dialysate. *Journal of Parasitology*, **75**, 55–57.

Fuhrmann, O. (1931) *Cestoidea, Handbuch der Zoologie de Kukenthal*, vol. 2, pp. 141–416.

Fujino, T., Fried, B. and **Hosier, D. W.** (1994) The expulsion of *Echinostoma trivolvis* (Trematoda) from ICR mice: extension–retraction mechanisms and ultrastructure of the collar spines. *Parasitology Research*, **80**, 581–587.

Fulford, A. J. C., Butterworth, A. E., Ouma, J. H. and **Sturrock, R. F.** (1995) A statistical approach to schistosome population dynamics and estimation of the life-span of *Schistosoma mansoni* in man. *Parasitology*, **110**, 307–316.

Gajdusek, D. C. (1978) Introduction of *Taenia solium* into West New Guinea with a note on an epidemic of burns from cysticercus epilepsy in the Ekari people of the Wissel Lakes area. *Papua New Guinea Medical Journal*, **21**, 329–342.

Gallien, L. (1935) Recherches expérimentales sur le dimorphisme évolutif et la biologie de *Polystoma integerrimum* Fröhl. *Travaux de la Station zoologique de Wimereux*, **12**, 1–181.

Garstang, W. (1954) *Larval Forms and other Zoological Verses*, Blackwell, Oxford.

Gentner, H. W. (1971) Notes on the biology of *Aspidogaster conchicola* and *Cotylaspis insignis*. *Zeitschrift für Parasitenkunde*, **35**, 263–269.

Gerasev, P. I., Gayevskaya, A. V. and **Kovaleva, A. A.** (1987) New genera of monogeneans of the diplectanotreme group (Ancyrocephalinae). *Parazitologicheskii Sbornik*, **34**, 192–210.

Gibson, D. I. (1987) Questions in digenean systematics and evolution. *Parasitology*, **95**, 429–460.

Gibson, D. I. and **Bray, R. A.** (1994) The evolutionary expansion and host–parasite relationships of the Digenea. *International Journal for Parasitology*, **24**, 1213–1226.

Ginetsinskaya, T. A. (1958) The life cycles of fish helminths and the biology of their larval stages, in *Parasitology of Fishes*, (eds V. A. Dogiel, G. K. Petrushevski and Y. I. Polyanski), English translation by Z. Kabata (1961), Oliver & Boyd, London, pp. 140–179.

Glauert, A. M., Butterworth, A. E., Sturrock, R. F. and **Houba, V.** (1978) The mechanism of antibody-dependent, eosinophil-mediated damage to schistosomula of *Schistosoma mansoni in vitro*: a study by phase-contrast and electron microscopy. *Journal of Cell Science*, **34**, 173–192.

Gobert, G. N., Stenzel, D. J., Jones, M. K. *et al.* (1997) *Schistosoma japonicum*: immunolocalization of paramyosin during development. *Parasitology*, **114**, 45–52.

Gönnert, R. (1955) Schistosomiasis-Studien. I. Beiträge zur Anatomie und Histologie von *Schistosoma mansoni*. *Zeitschrift für Tropenmedizin und Parasitologie*, **6**, 18–33.

Gönnert, R. (1970) Hatching and chemical nature of membranes of taeniid eggs. *Journal of Parasitology*, **56**, 559–560.

Goodrich, E. S. (1946) The study of nephridia and genital ducts since 1895. *Quarterly Journal of Microscopical Science*, **86**, 113–301.

Goto, S. (1894) Studies on the ectoparasitic trematodes of Japan. *Journal of the College of Science, Imperial University of Tokyo*, **8**, 1–273.

Götz, P. (1986) Mechanisms of encapsulation in dipteran hosts, in *Immune Mechanisms in Invertebrate Vectors* (ed. A. M. Lackie), Symposia of the Zoological Society of London, no. 56, Clarendon Press, Oxford, pp. 1–19.

Grabda, E. and **Siwak-Grabda, J.** (1940–1947) Recherches sur *Nematobothrium sardae* G. A. et W. G. MacCallum 1916 (Didymozoonidae Monticelli 1888) parasite de la cavité branchiale du poisson *Sarda sarda* Bloch provenant de la Mer Noire. *Zoologica Poloniae*, **4**, 11–33.

Grabda-Kazubska, B. (1980) Observations on the life cycle of *Diplodiscus subclavatus* (Pallas, 1760) (Trematoda, Diplodiscidae). *Acta Parasitologica Polonica*, **27**, 261–271.

Greer, G. J., Ow-Yang, C. K. and **Yong, H.** (1988) *Schistosoma malayensis* n. sp.: a *Schistosoma japonicum*-complex schistosome from Peninsular Malaysia. *Journal of Parasitology*, **74**, 471–480.

Grove, D. I. (1990) *A History of Human Helminthology*, CAB International, Wallingford, UK.

Gusev, A. V. (1985) Class Monogenea, in [*Keys to Parasites of the Freshwater Fish Fauna of the USSR*], *vol. 2*, [*Parasitic Metazoa*], (ed. O. N. Bauer), Leningrad Publishing House 'Nauka', Leningrad, pp. 10–253 (in Russian).

Gusev, A. V. (1995) Some pathways and factors of monogenean microevolution. *Canadian Journal of Fisheries and Aquatic Sciences*, **52**, 52–56.

Gusev, A. V. and **Strelkov, Y. A.** (1960) [*Ancylodiscoides* (Monogenoidea) of Far-East sheat-fishes (*Silurus* and *Parasilurus*)] (in Russian). *Trudy Zoologicheskogo Instituta Leningrad*, **28**, 197–255.

Gustafsson, M. K. S. (1992) The neuroanatomy of parasitic flatworms. *Advances in Neuroimmunology*, **2**, 267–286.

Haas, W. (1992) Physiological analysis of cercarial behavior. *Journal of Parasitology*, **78**, 243–255.

Haas, W. (1994) Physiological analyses of host-finding behaviour in trematode cercariae: adaptations for transmission success. *Parasitology*, **109**, S15–S29.

Haas, W., Granzer, M. and **Brockelman, C. R.** (1990a) *Opisthorchis viverrini*: finding and recognition of the fish host by the cercariae. *Experimental Parasitology*, **71**, 422–431.

Haas, W., Granzer, M. and **Brockelman, C. R.** (1990b) Finding and recognition of the bovine host by the cercariae of *Schistosoma spindale*. *Parasitology Research*, **76**, 343–350.

Haas, W., Granzer, M. and **Garcia, E. G.** (1987) Host identification by *Schistosoma japonicum* cercariae. *Journal of Parasitology*, **73**, 568–577.

Haas, W., Haberl, B., Kalbe, M. and **Körner, M.** (1995a) Snail-host-finding by miracidia and cercariae: chemical host cues. *Parasitology Today*, **11**, 468–472.

Haas, W., Körner, M., Hutterer, E. *et al.* (1995b) Finding and recognition of the snail intermediate hosts by three species of echinostome cercariae. *Parasitology*, **110**, 133–142.

Haberl, B., Kalbe, M., Fuchs, H. *et al.* (1995) *Schistosoma mansoni* and *S. haematobium*: miracidial host-finding behaviour is stimulated by macromolecules. *International Journal for Parasitology*, **25**, 551–560.

Hagan, P. and **Wilkins, H. A.** (1993) Concomitant immunity in schistosomiasis. *Parasitology Today*, **9**, 3–6.

Haight, M., Davidson, D. and **Pasternak, J.** (1977a) Cell cycle analysis in developing cercariae of *Trichobilharzia ocellata* (Trematoda: Schistosomatidae). *Journal of Parasitology*, **63**, 274–281.

Haight, M., Davidson, D. and **Pasternak, J.** (1977b) Relationship between nuclear morphology and the phases of the cell cycle during cercarial development of the digenetic trematode *Trichobilharzia ocellata*. *Journal of Parasitology*, **63**, 267–273.

Hall, M. C. (1934) The discharge of eggs from the segments of *Thysanosoma actinioides*. *Proceedings of the Helminthological Society of Washington*, **1**, 6–7.

Halton, D. W. (1967) Studies on glycogen deposition in Trematoda. *Comparative Biochemistry and Physiology*, **23**, 113–120.

Halton, D. W. (1975) Intracellular digestion and cellular defaecation in a monogenean, *Diclidophora merlangi*. *Parasitology*, **70**, 331–340.

Halton, D. W. (1976) *Diclidophora merlangi*: sloughing and renewal of haematin cells. *Experimental Parasitology*, **40**, 41–47.

Halton, D. W. and **Johnston, B. R.** (1982) Functional morphology of the metacercarial cyst of *Bucephaloides gracilescens* (Trematoda: Bucephalidae). *Parasitology*, **85**, 45–52.

Halton, D. W., Shaw, C., Maule, A. G. and **Smart, D.** (1994) Regulatory peptides in helminth parasites. *Advances in Parasitology*, **34**, 163–227.

Halton, D. W. and **Stranock, S. D.** (1976) The fine structure and histochemistry of the caecal epithelium of *Calicotyle kroyeri* (Monogenea: Monopisthocotylea). *International Journal for Parasitology* **6**, 253–263.

Hamilton, W. D. and **Zuk, M.** (1982) Heritable true fitness and bright birds: a role for parasites? *Science*, **218**, 384–387.

Hanna, R. E. B. (1980) *Fasciola hepatica*: glycocalyx replacement in the juvenile as a possible mechanism for protection against host immunity. *Experimental Parasitology*, **50**, 113–114.

Harris, P. D. (1983) The morphology and life-cycle of the oviparous *Oögyrodactylus farlowellae* gen. et sp. nov. (Monogenea, Gyrodactylidea). *Parasitology*, **87**, 405–420.

Harris, P. D. (1986) Species of *Gyrodactylus* von Nordmann, 1832 (Monogenea Gyrodactylidae) from poeciliid fishes, with a description of *G. turnbulli* sp. nov. from the guppy, *Poecilia reticulata* Peters. *Journal of Natural History*, **20**, 183–191.

Harris, P. D. (1993) Interactions between reproduction and population biology in gyrodactylid monogeneans – a review. *Bulletin Français de la Pêche et de la Pisciculture*, **328**, 47–65.

Harris, P. D., Jansen, P. A. and **Bakke, T. A.** (1994) The population age structure and reproductive biology of *Gyrodactylus salaris* Malmberg (Monogenea). *Parasitology*, **108**, 167–173.

Haswell-Elkins, M. R., Sithithaworn, P. and **Elkins, D.** (1992) *Opisthorchis viverrini* and cholangiocarcinoma in northeast Thailand. *Parasitology Today*, **8**, 86–89.

Heath, D. D. (1971) The migration of oncospheres of *Taenia pisiformis*, *T. serialis* and *Echinococcus granulosus* within the intermediate host. *International Journal for Parasitology*, **1**, 145–152.

Hedges, S. B., Hass, C. A. and **Maxson, L. R.** (1993) Relations of fish and tetrapods. *Nature*, **363**, 501–502.

Helluy, S. (1982) Relations hôtes–parasite du trématode *Microphallus papillorobustus* (Rankin, 1940). I. Pénétration des cercaires et rapports des métacercaires avec le tissu nerveux des *Gammarus*, hôtes intermédiaires. *Annales de Parasitologie*, **57**, 263–270.

Helluy, S. (1983) Relations hôtes–parasite du trématode *Microphallus papillorobustus* (Rankin, 1940). II. – Modifications du comportement des *Gammarus* hôtes intermédiaires et localisation des métacercaires. *Annales de Parasitologie*, **58**, 1–17.

Helluy, S. (1984) Relations hôtes–parasites du trématode *Microphallus papillorobustus* (Rankin, 1940). III – Facteurs impliqués dans les modifications du comportement des *Gammarus* hôtes intermédiaires et tests de prédation. *Annales de Parasitologie*, **59**, 41–56.

Hendrickson, G. L. (1978) Observations on strigeoid trematodes from the eyes of southeastern Wyoming fish. I. *Diplostomulum spathaceum* (Rudolphi, 1819). *Proceedings of the Helminthological Society of Washington*, **45**, 60–64.

Hendrickson, G. L. (1979) *Ornithodiplostomum ptychocheilus*: migration to the brain of the fish intermediate host, *Pimephales promelas*. *Experimental Parasitology*, **48**, 245–258.

Herber, E. C. (1942) Life history studies on two trematodes of the subfamily Notocotylinae. *Journal of Parasitology*, **28**, 179–196.

Hertel, L. A. (1993) Excretion and osmoregulation in the flatworms. *Transactions of the American Microscopical Society*, **112**, 10–17.

Hess, E. (1972) Contribution à la biologie larvaire de *Mesocestoides corti* Hoeppli, 1925 (Cestoda, Cyclophyllidea). Note préliminaire. *Revue Suisse de Zoologie*, **79**, 1031–1037.

Hess, E. (1980) Ultrastructural study of the tetrathyridium of *Mesocestoides corti* Hoeppli, 1925: tegument and parenchyma. *Zeitschrift für Parasitenkunde*, **61**, 135–159.

Heyneman, D. (1962) Studies on helminth immunity. II. Influence of *Hymenolepis nana* (Cestoda: Hymenolepididae) in dual infections with *H. diminuta* in white mice and rats. *Experimental Parasitology*, **12**, 7–18.

Hickman, J. L. (1963) The biology of *Oochoristica vacuolata* Hickman (Cestoda). *Papers and Proceedings of the Royal Society of Tasmania*, **97**, 81–104.

Hine, P. M. (1977) New species of *Nippotaenia* and *Amurotaenia* (Cestoda: Nippotaeniidae) from New Zealand freshwater fishes. *Journal of the Royal Society of New Zealand*, **7**, 143–155.

Hoberg, E. P. (1994) Order Tetrabothriidea Baer, 1954, in *Key to the Cestode Parasites of Vertebrates*, (eds L. F. Khalil, A. Jones and R. A. Bray), CAB International, Wallingford, UK, pp. 295–304.

Hockley, D. J. (1973) Ultrastructure of the tegument of *Schistosoma*. *Advances in Parasitology*, **11**, 233–305.

Hoffman, G. L. (1956) The life cycle of *Crassiphiala bulboglossa* (Trematoda: Strigeida). Development of the metacercaria and cyst, and effect on the fish hosts. *Journal of Parasitology*, **42**, 435–444.

Hoffman, G. L. (1958a) Experimental studies on the cercaria and metacercaria of a strigeoid trematode, *Posthodiplostomum minimum*. *Experimental Parasitology*, **7**, 23–50.

Hoffman, G. L. (1958b) Studies on the life-cycle of *Ornithodiplostomum ptychocheilus* (Faust) (Trematoda: Strigeoidea) and the 'self cure' of infected fish. *Journal of Parasitology*, **44**, 416–421.

Hoffman, G. L. and **Hundley, J. B.** (1957) The life-cycle of *Diplostomum baeri eucaliae* n. subsp. (Trematoda: Strigeida). *Journal of Parasitology*, **43**, 613–627.

Hoffman, G. L. and **Putz, R. E.** (1965) The black-spot (*Uvulifer ambloplitis*: Trematoda: Strigeoidea) of centrarchid fishes. *Transactions of the American Fisheries Society*, **94**, 143–151.

Hohorst, W. and **Graefe, G.** (1961) Ameisen – obligatorische Zwischenwirte des Lanzettegels (*Dicrocoelium dendriticum*). *Naturwissenschaften*, **48**, 229–230.

Holliman, R. B. (1971) Ecological observations on two species of spirorchid trematodes. *American Midland Naturalist*, **86**, 509–512.

Holliman, R. B. and **Fisher, J. E.** (1968) Life cycle and pathology of *Spirorchis scripta* Stunkard, 1923 (Digenea: Spirorchiidae) in *Chrysemys picta picta*. *Journal of Parasitology*, **54**, 310–318.

Holliman, R. B., **Fisher, J. E.** and **Parker, J. C.** (1971) Studies on *Spirorchis parvus* (Stunkard, 1923) and its pathological effects on *Chrysemys picta picta*. *Journal of Parasitology*, **57**, 71–77.

Hopkins, C. A. (1980) Immunity and *Hymenolepis diminuta*, in *Biology of the Tapeworm* Hymenolepis diminuta, (ed. H. P. Arai), Academic Press, London, pp. 551–614.

Hopkins, C. A. and **Allen, L. M.** (1979) *Hymenolepis diminuta*: the role of the tail in determining the position of the worm in the intestine of the rat. *Parasitology*, **79**, 401–410.

Hopkins, C. A., **Law, L. M.** and **Threadgold, L. T.** (1978) *Schistocephalus solidus*: pinocytosis by the plerocercoid tegument. *Experimental Parasitology*, **44**, 161–172.

Hopkins, C. A. and **Smyth, J. D.** (1951) Notes on the morphology and life history of *Schistocephalus solidus* (Cestoda: Diphyllobothriidae). *Parasitology*, **41**, 283–291.

Horemans, A. M. C., **Tielens, A. G. M.** and **van den Bergh, S. G.** (1991) The transition from an aerobic to an anaerobic energy metabolism in transforming *Schistosoma mansoni* cercariae occurs exclusively in the head. *Parasitology*, **102**, 259–265.

Horsfall, M. W. (1934) Studies on the life history and morphology of the cystocercous cercariae. *Transactions of the American Microscopical Society*, **53**, 311–347.

Horton, M. A. and **Whittington, I. D.** (1994) A new species of *Metabenedeniella* (Monogenea: Capsalidae) from the dorsal fin of *Diagramma pictum* (Perciformes: Haemulidae) from the Great Barrier Reef, Australia with a revision of the genus. *Journal of Parasitology*, **80**, 998–1007.

Houde, A. E. and **Torio, A. J.** (1992) Effect of parasitic infection on male colour pattern and female choice in guppies. *Behavioral Ecology*, **3**, 346–351.

Howell, M. J. (1973) The resistance of cysts of *Stictodora lari* (Trematoda: Heterophyidae) to encapsulation by cells of the fish host. *International Journal for Parasitology*, **3**, 653–659.

Howell, M. J. (1985) Gene exchange between hosts and parasites. *International Journal for Parasitology*, **15**, 597–600.

Huehner, M. K. and **Etges, F. J.** (1977) The life cycle and development of *Aspidogaster conchicola* in the snails, *Viviparus malleatus* and *Goniobasis livescens*. *Journal of Parasitology*, **63**, 669–674.

Huehner, M. K., **Hannan, K.** and **Garvin, M.** (1989) Feeding habits and marginal organ histochemistry of *Aspidogaster conchicola* (Trematoda: Aspidogastrea). *Journal of Parasitology*, **75**, 848–852.

Hughes, D. L. and **Harness, E.** (1973a) Attempts to demonstrate a 'host antigen' effect by the experimental transfer of adult *Fasciola hepatica* into recipient animals immunised against the donor. *Research in Veterinary Science*, **14**, 151–154.

Hughes, D. L. and **Harness, E.** (1973b) The experimental transfer of immature *Fasciola hepatica* from donor mice and hamsters to rats immunised against the donors. *Research in Veterinary Science*, **14**, 220–222.

Hughes, G. M. (1984) General anatomy of the gills, in *Fish Physiology*, vol. 10, (eds W. S. Hoar and D. J. Randall), Academic Press, London.

Hunter, G. C. and **Kille, R. A.** (1950) Some observations on *Dictyocotyle coeliaca* Nybelin, 1941 (Monogenea). *Journal of Helminthology*, **24**, 15–22.

Hurd, H. (1993) Reproductive disturbances induced by parasites and pathogens of insects, in *Parasites and Pathogens of Insects, Volume 1, Parasites*, (eds N. E. Beckage, S. N. Thompson and B. A. Federici), Academic Press, London, pp. 87–105.

Hurd, H. and **Fogo, S.** (1991) Changes induced by *Hymenolepis diminuta* (Cestoda) in the behaviour of the intermediate host *Tenebrio molitor*. *Canadian Journal of Zoology*, **69**, 2291–2294.

Hyman, L. H. (1951) *The Invertebrates: Platyhelminthes and Rhynchocoela. The Acoelomate Bilateria*, McGraw-Hill, New York.

Imbert-Establet, D. and **Combes, C.** (1986) *Schistosoma mansoni*: comparison of a Caribbean and African strain and experimental crossing based on compatibility with intermediate hosts and *Rattus rattus*. *Experimental Parasitology*, **61**, 210–218.

Insler, G. D. and **Roberts, L. S.** (1980) Developmental physiology of tapeworms. XV. A system for testing possible crowding factors *in vitro*. *Journal of Experimental Zoology*, **211**, 45–54.

Ishida, S. and **Teshirogi, W.** (1986) Eggshell formation in polyclads. *Hydrobiologia*, **132**, 127–135.

Ishii, N. (1935) Studies on the family Didymozooidae (Monticelli, 1888). *Japanese Journal of Zoology*, **6**, 279–335.

Isseroff, H. and **Cable, R. M.** (1968) Fine structure of photoreceptors in larval trematodes. A comparative study. *Zeitschrift für Zellforschung*, **86**, 511–534.

Isseroff, H. and **Read, C. P.** (1969) Studies on membrane transport – VI. Absorption of amino acids by fascioliid trematodes. *Comparative Biochemistry and Physiology*, **30**, 1153–1159.

Isseroff, H. and **Read, C. P.** (1974) Studies on membrane transport – VIII. Absorption of monosaccharides by *Fasciola hepatica*. *Comparative Biochemistry and Physiology*, **47A**, 141–152.

Ivanov, A. V. (1952) Morphology of *Udonella caligorum* Johnston, 1835, and the position of Udonellidae in the systematics of platyhelminths. *Parazitologicheskii Sbornik*, **14**, 112–163 (English translation, 1981, edited by Simmons, J. E., Hargis, W. J. and Zwerner, D. E. University of California at Berkeley, Berkeley, CA).

James, B. L. (1964) The life cycle of *Parvatrema homoeotecnum* sp. nov. (Trematoda: Digenea) and a review of the family Gymnophallidae Morozov, 1955. *Parasitology*, **54**, 1–41.

James, B. L. (1968) Studies on the life-cycle of *Microphallus pygmaeus* (Levinsen, 1881) (Trematoda: Microphallidae). *Journal of Natural History*, **2**, 155–172.

Jamieson, B. G. M. (1966) Larval stages of the progenetic trematode *Parahemiurus bennettae* Jamieson, 1966 (Digenea, Hemiuridae) and the evolutionary origin of cercariae. *Proceedings of the Royal Society of Queensland*, **77**, 81–92.

Jarecka, L. (1961) Morphological adaptations of tapeworm eggs and their importance in the life cycles. *Acta Parasitologica Polonica*, **9**, 409–426.

Jarecka, L. (1970) Life cycle of *Valipora campylancristrota* (Wedl, 1855) Baer and Bona 1958–1960 (Cestoda–Dilepididae) and the description of cercoscolex – a new type of cestode larva. *Bulletin de l'Académie Polonaise des Sciences*, Série des Sciences Biologiques, Cl. II, **18**, 99–102.

Jarecka, L. and **Burt, M. D. B.** (1984) The cercoid larvae of *Pseudanthobothrium hanseni* Baer, 1956 and *Pseudanthobothrium* sp. (Cestoda, Tetraphyllidea) from experimentally infected harpacticoid copepods. *Acta Parasitologica Polonica*, **29**, 23–26.

Jennings, F. W., Mulligan, W. and **Urquhart, G. M.** (1956) Radioisotope studies on the anaemia produced by infection with *Fasciola hepatica*. *Experimental Parasitology*, **5**, 458–468.

Jennings, J. B. (1968a) Feeding, digestion and food storage in two species of temnocephalid flatworms (Turbellaria: Rhabdocoela). *Journal of Zoology (London)*, **156**, 1–8.

Jennings, J. B. (1968b) A new temnocephalid flatworm from Costa Rica. *Journal of Natural History*, **2**, 117–120.

Jennings, J. B. (1981) Physiological adaptations to entosymbiosis in three species of graffillid rhabdocoels. *Hydrobiologia*, **84**, 147–153.

Jennings, J. B. (1985) Feeding and digestion in the aberrant planarian *Bdellasimilis barwicki* (Turbellaria: Tricladida: Procerodidae): an ectosymbiote of freshwater turtles in Queensland and New South Wales. *Australian Journal of Zoology*, **33**, 317–327.

Jennings, J. B. (1988) Nutrition and respiration in symbiotic Turbellaria. *Progress in Zoology*, **36**, 3–13.

Jennings, J. B. (1989) Epidermal uptake of nutrients in an unusual turbellarian parasitic in the starfish *Coscinasterias calamaria* in Tasmanian waters. *Biological Bulletin*, **176**, 327–336.

Jennings, J. B. (1997) Nutritional and respiratory pathways to parasitism exemplified in the Turbellaria. *International Journal for Parasitology*, **27**, 679–691.

Jennings, J. B. and **Calow, P.** (1975) The relationship between high fecundity and the evolution of entoparasitism. *Oecologia*, **21**, 109–115.

Jennings, J. B. and **Cannon, L. R. G.** (1985) Observations on the occurrence, nutritional physiology and respiratory pigment of three species of flatworms (Rhabdocoela: Pterastericolidae) entosymbiotic in starfish from temperate and tropical waters. *Ophelia*, **24**, 199–215.

Jennings, J. B. and **Cannon, L. R. G.** (1987) The occurrence, spectral properties and probable rôle of haemoglobins in four species of entosymbiotic turbellarians (Rhabdocoela: Umagillidae). *Ophelia*, **27**, 143–154.

Jennings, J. B. and **LeFlore, W. B.** (1979) Occurrence and possible adaptive significance of some histochemically demonstrable dehydrogenases in two entosymbiotic rhabdocoels (Platyhelminthes: Turbellaria). *Comparative Biochemistry and Physiology*, **62B**, 301–304.

Jennings, J. B. and **Phillips, J. I.** (1978) Feeding and digestion in three entosymbiotic graffillid rhabdocoels from bivalve and gastropod molluscs. *Biological Bulletin*, **155**, 542–562.

Jespersen, A. and **Lützen, J.** (1972) *Triloborhynchus psilastericola* n. sp., a parasitic turbellarian (Fam. Pterastericolidae) from the starfish *Psilaster andromeda* (Müller and Troschel). *Zeitschrift für Morphologie und Ökologie der Tiere*, **71**, 290–298.

Jimenez-Albarran, M. and **Guevara-Pozo, D.** (1980a) Estudios experimentales sobre biologia de *Fasciola hepatica*: 4 – Influencia de la naturaleza del soporte (papel de celofan o vidrio) en la fijación de las metacercarias de *Fasciola hepatica*. *Revista Ibérica de Parasitologia*, **40**, 251–256.

Jimenez-Albarran, M. and **Guevara-Pozo, D.** (1980b) Estudios experimentales sobre biologia de *Fasciola hepatica*: 5 – Influencia del color de la luz en la fijación de las metacercarias sobre la superficie del vidrio. *Revista Ibérica de Parasitologia*, **40**, 443–452.

Johnsen, B. O. and **Jensen, A. J.** (1991) The *Gyrodactylus* story in Norway. *Aquaculture*, **98**, 289–302.

Johnson, A. D. (1968) Life history of *Alaria marcianae* (La Rue, 1917) Walton, 1949 (Trematoda: Diplostomatidae). *Journal of Parasitology*, **54**, 324–332.

Johnson, J. C. (1920) The life cycle of *Echinostoma revolutum* (Froelich). *University of California Publications in Zoology*, **19**, 335–388.

Johnson, S. G., Lively, C. M. and **Schrag, S. J.** (1995) Evolution and ecological correlates of uniparental reproduction in freshwater snails. *Experientia*, **51**, 498–509.

Jones, M. F. and **Alicata, J. E.** (1935) Development and morphology of the cestode *Hymenolepis cantaniana*, in coleopteran and avian hosts. *Journal of the Washington Academy of Sciences*, **25**, 237–247.

Jones, M. K. (1985) Morphology of *Baerietta hickmani* n. sp. (Cestoda, Nematotaeniidae) from Australian scincid lizards. *Journal of Parasitology*, **71**, 4–9.

Jones, M. K. (1987) A taxonomic revision of the Nematotaeniidae Lühe, 1910 (Cestoda: Cyclophyllidea). *Systematic Parasitology*, **10**, 165–245.

Jones, M. K. (1988) Formation of the paruterine capsules and embryonic envelopes in *Cylindrotaenia hickmani* (Jones, 1985) (Cestoda: Nematotaeniidae). *Australian Journal of Zoology*, **36**, 545–563.

Jones, T. C. and **Lester, J. G.** (1992) The life history and biology of *Diceratocephala boschmai* (Platyhelminthes; Temnocephalida), an ectosymbiont on the redclaw crayfish *Cherax quadricarinatus*. *Hydrobiologia*, **248**, 193–199.

Jong-Brink, M. De. (1995) How schistosomes profit from the stress responses they elicit in their hosts. *Advances in Parasitology*, **35**, 177–256.

Jordan, P. and **Webbe, G.** (1993) Epidemiology, in *Human Schistosomiasis*, (eds P. Jordan, G. Webbe and R. F. Sturrock), CAB International, Wallingford, UK, pp. 87–158.

Joyeux, C. and **Baer, J.** (1961) Classe des cestodes, in *Traité de Zoologie, vol. IV*, (ed. P. Grassé), Masson, Paris, pp. 347–560.

Kamegai, S. (1971) The determination of a generic name of flying fishes' muscle parasite, a didymozoid, whose ova have occasionally been found in human faeces in Japan. *Japanese Journal of Parasitology*, **20**, 170–176.

Kanneworff, B. and **Christensen, A. M.** (1966) *Kronborgia caridicola* sp. nov., an endoparasitic turbellarian from north Atlantic shrimps. *Ophelia*, **3**, 65–80.

Kassim, O. and **Gilbertson, D. E.** (1976) Hatching of *Schistosoma mansoni* eggs and observations on motility of miracidia. *Journal of Parasitology*, **62**, 715–720.

Kawana, H. (1940) Study on the development of the excretory system of *Fasciola hepatica* L., with special reference of its first intermediate host in central China. *Journal of the Shanghai Science Institute, Sect. IV*, **5**, 13–34.

Kearn, G. C. (1962) Breathing movements in *Entobdella soleae* (Trematoda, Monogenea) from the skin of the common sole. *Journal of the Marine Biological Association of the United Kingdom*, **42**, 93–104.

Kearn, G. C. (1963a) Feeding in some monogenean skin parasites: *Entobdella soleae* on *Solea solea* and *Acanthocotyle* sp. on *Raia clavata. Journal of the Marine Biological Association of the United Kingdom*, **43**, 749–766.

Kearn, G. C. (1963b) The egg, oncomiracidium and larval development of *Entobdella soleae*, a monogenean skin parasite of the common sole. *Parasitology*, **53**, 435–447.

Kearn, G. C. (1963c) The life cycle of the monogenean *Entobdella soleae*, a skin parasite of the common sole. *Parasitology*, **53**, 253–263.

Kearn, G. C. (1964) The attachment of the monogenean *Entobdella soleae* to the skin of the common sole. *Parasitology*, **54**, 327–335.

Kearn, G. C. (1965) The biology of *Leptocotyle minor*, a skin parasite of the dogfish, *Scyliorhinus canicula. Parasitology*, **55**, 473–480.

Kearn, G. C. (1966) The adhesive mechanism of the monogenean parasite *Tetraonchus monenteron* from the gills of the pike (*Esox lucius*). *Parasitology*, **56**, 505–511.

Kearn, G. C. (1967a) Experiments on host-finding and host-specificity in the monogenean *Entobdella soleae. Parasitology*, **57**, 585–605.

Kearn, G. C. (1967b) The life-cycles and larval development of some acanthocotylids (Monogenea) from Plymouth rays. *Parasitology*, **57**, 157–167.

Kearn, G. C. (1970) The production, transfer and assimilation of spermatophores by *Entobdella soleae*, a monogenean skin parasite of the common sole. *Parasitology*, **60**, 301–311.

Kearn, G. C. (1971a) The physiology and behaviour of the monogenean skin parasite *Entobdella soleae* in relation to its host (*Solea solea*), in *Ecology and Physiology of Parasites, a Symposium*, (ed. A. M. Fallis), University of Toronto Press, Toronto, Ontario, pp. 161–187.

Kearn, G. C. (1971b) The attachment site, invasion route and larval development of *Trochopus pini*, a monogenean from the gills of *Trigla hirundo. Parasitology*, **63**, 513–525.

Kearn, G. C. (1971c) The attachment of the ancyrocephalid monogenean *Haliotrema balisticus* to the gills of the trigger fish, *Balistes capriscus* (=*carolinensis*). *Parasitology*, **63**, 157–162.

Kearn, G. C. (1973) An endogenous circadian hatching rhythm in the monogenean skin parasite *Entobdella soleae*, and its relationship to the activity rhythm of the host (*Solea solea*). *Parasitology*, **66**, 101–122.

Kearn, G. C. (1974a) A comparative study of the glandular and excretory systems of the oncomiracidia of the monogenean skin parasites *Entobdella hippoglossi, E. diadema* and *E. soleae. Parasitology*, **69**, 257–269.

Kearn, G. C. (1974b) The effects of fish skin mucus on hatching in the monogenean parasite *Entobdella soleae* from the skin of the common sole, *Solea solea. Parasitology*, **68**, 173–188.

Kearn, G. C. (1974c) Nocturnal hatching in the monogenean skin parasite *Entobdella hippoglossi* from the halibut, *Hippoglossus hippoglossus. Parasitology*, **68**, 161–172.

Kearn, G. C. (1975) The mode of hatching of the monogenean *Entobdella soleae*, a skin parasite of the common sole (*Solea solea*). *Parasitology*, **71**, 419–431.

Kearn, G. C. (1976a) Body surface of fishes, in *Ecological Aspects of Parasitology*, (ed. C. R. Kennedy), North-Holland, Amsterdam, pp. 185–208.

Kearn, G. C. (1976b) [Observations on monogenean parasites from the nasal fossae of European rays: *Empruthotrema raiae* (MacCallum, 1916) Johnston and Tiegs, 1922 and *E. torpedinis* sp. nov. from *Torpedo marmorata*]. *Proceedings of the Institute of Biology and Pedology, Far-East Science Centre, Vladivostok*, **34**, 45–54 (in Russian with English summary).

Kearn, G. C. (1978) Early development and microhabitat of the monogenean *Horricauda rhinobatidis*, with observations on the related *Troglocephalus rhinobatidis*, from *Rhinobatos batillum* from Queensland, Australia. *International Journal for Parasitology*, **8**, 305–311.

Kearn, G. C. (1980) Light and gravity responses of the oncomiracidium of *Entobdella soleae* and their role in host location. *Parasitology*, **81**, 71–89.

Kearn, G. C. (1981) Behaviour of oncomiracidia. *Parasitology*, **82**, 57–59.

Kearn, G. C. (1982) Rapid hatching induced by light intensity reduction in the monogenean *Entobdella diadema. Journal of Parasitology*, **68**, 171–172.

Kearn, G. C. (1984) The migration of the monogenean *Entobdella soleae* on the surface of its host, *Solea solea*. *International Journal for Parasitology*, **14**, 63–69.

Kearn, G. C. (1985) Observations on egg production in the monogenean *Entobdella soleae*. *International Journal for Parasitology*, **15**, 187–194.

Kearn, G. C. (1986) The eggs of monogeneans. *Advances in Parasitology*, **25**, 175–273.

Kearn, G. C. (1987a) Locomotion in the gill-parasitic monogenean *Tetraonchus monenteron*. *Journal of Parasitology*, **73**, 224–225.

Kearn, G. C. (1987b) The site of development of the monogenean *Calicotyle kroyeri*, a parasite of rays. *Journal of the Marine biological Association of the United Kingdom*, **67**, 77–87.

Kearn, G. C. (1988a) Orientation and locomotion in the monogenean parasite *Entobdella soleae* on the skin of its host (*Solea solea*). *International Journal for Parasitology*, **18**, 753–759.

Kearn, G. C. (1988b) The monogenean skin parasite *Entobdella soleae*: movement of adults and juveniles from host to host (*Solea solea*). *International Journal for Parasitology*, **18**, 313–319.

Kearn, G. C. (1990) The rate of development and longevity of the monogenean skin parasite *Entobdella soleae*. *Journal of Helminthology*, **64**, 340–342.

Kearn, G. C. (1993) A new species of the genus *Enoplocotyle* (Platyhelminthes: Monogenea) parasitic on the skin of the moray eel *Gymnothorax kidako* in Japan, with observations on hatching and the oncomiracidium. *Journal of Zoology*, **229**, 533–544.

Kearn, G. C. (1994) Evolutionary expansion of the Monogenea. *International Journal for Parasitology*, **24**, 1227–1271.

Kearn, G. C., Al-Sehaibani, M. A., Whittington, I. D. *et al.* (1996) Swallowing of sea water and its role in egestion in the monogenean *Entobdella soleae*, a skin parasite of the common sole (*Solea solea*), with observations on other monogeneans and on a freshwater temnocephalan. *Journal of Natural History*, **30**, 637–646.

Kearn, G. C. and **Baker, N. O. B.** (1973) Ultrastructural and histochemical observations on the pigmented eyes of the oncomiracidium of *Entobdella soleae*, a monogenean skin parasite of the common sole, *Solea solea*. *Zeitschrift für Parasitenkunde*, **41**, 239–254.

Kearn, G. C., Cleveland, G. and **Wilkins, S.** (1989) An *in vitro* study of the forcible ejection of the metacercaria of the strigeid digenean *Apatemon* (*Australapatemon*) *minor* Yamaguti, 1933 by rapid inward expansion of the cyst wall. *Journal of Helminthology*, **63**, 63–71.

Kearn, G. C. and **Evans-Gowing, R.** (In press) Attachment and detachment of the anterior adhesive pads of the monogenean (platyhelminth) parasite *Entobdella soleae* from the skin of the common sole (*Solea solea*). *International Journal for Parasitology*.

Kearn, G. C. and **Gowing, R.** (1989) Glands and sensilla associated with the haptor of the gill-parasitic monogenean *Tetraonchus monenteron*. *International Journal for Parasitology*, **19**, 673–679.

Kearn, G. C. and **Gowing, R.** (1990) Vestigial marginal hooklets in the oncomiracidium of the microbothriid monogenean *Leptocotyle minor*. *Parasitology Research*, **76**, 406–408.

Kearn, G. C., James, R. and **Evans-Gowing, R.** (1993) Insemination and population density in *Entobdella soleae*, a monogenean skin parasite of the common sole, *Solea solea*. *International Journal for Parasitology*, **23**, 891–899.

Kearn, G. C. and **Macdonald, S.** (1976) The chemical nature of host hatching factors in the monogenean skin parasites *Entobdella soleae* and *Acanthocotyle lobianchi*. *International Journal for Parasitology*, **6**, 457–466.

Kearn, G. C. and **Vasconcelos, M. E.** (1979) Preliminary list of monogenean parasites of Portuguese marine fishes, with a note on *Enoplocotyle minima* (Tagliani, 1912). *Boletim Instituto Nacional de Investigação das Pescas, Lisbõa*, **1**, 25–36.

Kearn, G. C. and **Whittington, I. D.** (1991) Swimming in a sub-adult monogenean of the genus *Entobdella*. *International Journal for Parasitology*, **21**, 739–741.

Kearn, G. C. and **Whittington, I. D.** (1992a) Diversity of reproductive behaviour in platyhelminth parasites: insemination in some benedeniine (capsalid) monogeneans. *Parasitology*, **104**, 489–496.

Kearn, G. C. and **Whittington, I. D.** (1992b) A response to light in an adult encotyllabine (capsalid) monogenean from the pharyngeal tooth pads of some marine teleost fishes. *International Journal for Parasitology*, **22**, 119–121.

Kearn, G. C. and **Whittington, I. D.** (1994) Ancyrocephaline monogeneans of the genera *Chauhanellus* and *Hamatopeduncularia* from the gills of the blue catfish, *Arius graeffei*, in the Brisbane River and Moreton Bay, Queensland, Australia, with descriptions of four new species. *International Journal for Parasitology*, **24**, 569–588.

Kearn, G. C., Whittington, I. D. and **Evans-Gowing, R.** (1995) Use of cement for attachment in *Neocalceostomoides brisbanensis*, a calceostomatine monogenean from the gill chamber of the blue catfish, *Arius graeffei*. *International Journal for Parasitology*, **25**, 299–306.

Kechemir, N. (1978) Démonstration expérimentale d'un cycle biologique à quatres hôtes obligatoires chez les trématodes hémiurides. *Annales de Parasitologie*, **53**, 75–92.

Keeble, F. (1910) *Plant-Animals, a Study in Symbiosis*, Cambridge University Press, London.

Kennedy, C. R. (1965) The mode of hatching of the egg of the cestode *Archigetes hepatica* (Yamaguti). *Parasitology*, **55**, 18P.

Kennedy, C. R. (1981) Long term studies on the population biology of two species of eyefluke, *Diplostomum gasterostei* and *Tylodelphys clavata* (Digenea: Diplostomatidae), concurrently infecting the eyes of perch, *Perca fluviatilis*. *Journal of Fish Biology*, **19**, 221–236.

Kent, M. L. and **Olson, A. C. Jr** (1986) Interrelationships of a parasitic turbellarian, (*Paravortex* sp.) (Graffillidae, Rhabdocoela) and its marine fish hosts. *Fish Pathology*, **21**, 65–72.

Keymer, A. (1980) The influence of *Hymenolepis diminuta* on the survival and fecundity of the intermediate host, *Tribolium confusum*. *Parasitology*, **81**, 405–421.

Keymer, A. (1982) The dynamics of infection of *Tribolium confusum* by *Hymenolepis diminuta*: the influence of exposure time and host density. *Parasitology*, **84**, 157–166.

Khalil, G. M. and **Cable, R. M.** (1968) Germinal development in *Philophthalmus megalurus* (Cort, 1914) (Trematoda: Digenea). *Zeitschrift für Parasitenkunde*, **31**, 211–231.

Khalil, L. F. (1970) Further studies on *Macrogyrodactylus polypteri*, a monogenean on the African freshwater fish *Polypterus senegalus*. *Journal of Helminthology*, **44**, 329–348.

Kim, S. and **Fried, B.** (1989) Pathological effects of *Echinostoma caproni* (Trematoda) in the domestic chick. *Journal of Helminthology*, **63**, 227–230.

King, C. L. and **Higashi, G. I.** (1992) *Schistosoma mansoni*: silver ion (Ag^+) stimulates and reversibly inhibits lipid-induced cercarial penetration. *Experimental Parasitology*, **75**, 31–39.

Kirk, R. S. and **Lewis, J. W.** (1993) The life-cycle and morphology of *Sanguinicola inermis* Plehn, 1905 (Digenea: Sanguinicolidae). *Systematic Parasitology*, **25**, 125–133.

Kisielewska, K. (1961) Circulation of tapeworms of *Sorex araneus araneus* L. in biocenosis of Bialowieza National Park. *Acta Parasitologica Polonica*, **9**, 331–369.

Knapp, W. P. W. van der and **Loker, E. S.** (1990) Immune mechanisms in trematode–snail interactions. *Parasitology Today*, **6**, 175–182.

Køie, M. (1975) On the morphology and life-history of *Opechona bacillaris* (Molin, 1859) Looss, 1907 (Trematoda, Lepocreadiidae). *Ophelia*, **13**, 63–86.

Køie, M. (1979) On the morphology and life-history of *Monascus* (= *Haplocladus*) *filiformis* (Rudolphi, 1819) Looss, 1907 and *Steringophorus furciger* (Olsson, 1868) Odhner, 1905 (Trematoda, Fellodistomidae). *Ophelia*, **18**, 113–132.

Køie, M. (1982) The redia, cercaria and early stages of *Aporocotyle simplex* Odhner, 1900 (Sanguinicolidae) – a digenetic trematode which has a polychaete annelid as the only intermediate host. *Ophelia*, **21**, 115–145.

Køie, M. (1985) On the morphology and life-history of *Lepidapedon elongatum* (Lebour, 1908) Nicoll, 1910 (Trematoda, Lepocreadiidae). *Ophelia*, **24**, 135–153.

Køie, M. (1986) The life-history of *Mesorchis denticulatus* (Rudolphi, 1802) Dietz, 1909 (Trematoda, Echinostomatidae). *Zeitschrift für Parasitenkunde*, **72**, 335–343.

Køie, M., Nansen, P. and **Christensen, N.** (1977) Stereoscan studies of rediae, cercariae, cysts, excysted metacercariae and migratory stages of *Fasciola hepatica*. *Zeitschrift für Parasitenkunde*, **54**, 289–297.

Kong, Y., Chung, Y.-B., Cho, S.-Y. and **Kang, S.-Y.** (1994) Cleavage of immunoglobulin G by excretory–secretory cathepsin S-like protease of *Spirometra mansoni* plerocercoid. *Parasitology*, **109**, 611–621.

Kozloff, E. N. (1965) *Desmote inops* sp. n. and *Fallacohospes inchoatus* gen. and sp. n., umagillid rhabdocoels from the intestine of the crinoid *Florometra serratissima* (A. H. Clark). *Journal of Parasitology*, **51**, 305–312.

Kozloff, E. N. and **Shinn, G. L.** (1987) *Wahlia pulchella* n. sp., a turbellarian flatworm (Neorhabdocoela: Umagillidae) from the intestine of the sea cucumber *Stichopus californicus*. *Journal of Parasitology*, **73**, 194–202.

Krause, J. and **Godin, J.** (1994) Influence of parasitism on the shoaling of banded killifish, *Fundulus diaphanus*. *Canadian Journal of Zoology*, **72**, 1775–1779.

Kritsky, D. C. and **Boeger, W. A.** (1991) Neotropical Monogenea. 16. New species of oviparous Gyrodactylidea with proposal of *Nothogyrodactylus* gen. n. (Oogyrodactylidae). *Journal of the Helminthological Society of Washington*, **58**, 7–15.

Kritsky, D. C. and **Fritts, T. H.** (1970) Monogenetic trematodes from Costa Rica with the proposal of *Anacanthocotyle* gen. n. (Gyrodactylidae: Isancistrinae). *Proceedings of the Helminthological Society of Washington*, **37**, 63–68.

Kruidenier, F. J. (1951) The formation and function of mucoids in virgulate cercariae, including a study of the virgula organ. *American Midland Naturalist*, **46**, 660–683.

Krull, W. H. (1939) Observations on the distribution and ecology of the oribatid mites. *Journal of the Washington Academy of Sciences*, **29**, 519–528.

Kumazawa, H. (1992) A kinetic study of egg production, fecal egg output, and the rate of proglottid shedding in *Hymenolepis nana*. *Journal of Parasitology*, **78**, 498–504.

Kuris, A. M. (1980) Effect of exposure to *Echinostoma liei* miracidia on growth and survival of young *Biomphalaria glabrata* snails. *International Journal for Parasitology*, **10**, 303–308.

Kusel, J. R. (1970) Studies on the structure and hatching of the eggs of *Schistosoma mansoni*. *Parasitology*, **60**, 79–88.

LaBeau, M. R. and **Peters, L. E.** (1995) *Proterometra autraini* n. sp. (Digenea: Azygiidae) from Michigan's upper peninsula and a key to species of *Proterometra*. *Journal of Parasitology*, **81**, 442–445.

Lackie, A. M. (1975) The activation of infective stages of endoparasites of vertebrates. *Biological Reviews*, **50**, 285–323.

Lackie, A. M. (1976) Evasion of the haemocytic defence reaction of certain insects by larvae of *Hymenolepis diminuta* (Cestoda). *Parasitology*, **73**, 97–107.

Lackie, A. M. (ed.) (1986) *Immune Mechanisms in Invertebrate Vectors*, Symposia of the Zoological Society of London, no. 56, Clarendon Press, Oxford.

Lafferty, K. D. (1993a) Effects of parasitic castration on growth, reproduction and population dynamics of the marine snail *Cerithidia californica*. *Marine Ecology Progress Series*, **96**, 229–237.

Lafferty, K. D. (1993b) The marine snail, *Cerithidia californica*, matures at smaller sizes when parasitism is high. *Oikos*, **68**, 3–11.

Lambert, M. (1976) Cycle biologique de *Parasymphylodora markewitschi* (Kulakovskaya, 1947) (Trematoda Digenea, Monorchiidae). *Bulletin du Muséum national d'Histoire naturelle*, Série 3, Zoologie, **284**, 1107–1114.

Lambert, M. and **Lambert, A.** (1974) Cycle biologique de *Sphaerostoma maius* Janiszwiska, 1949 (Digenea, Opecoelidae), parasite du chevaine *Leuciscus cephalus* L. (Cyprinidae). *Bulletin du Muséum national d'Histoire naturelle*, Série 3, Zoologie, **158**, 885–897.

La Rue, G. R. and **Fallis, A. M.** (1936) Morphological study of *Alaria canis* n. sp. (Trematoda: Alariidae), a trematode parasite of the dog. *Transactions of the American Microscopical Society*, **55**, 340–351.

Lawson, J. R. and **Gemmell, M. A.** (1983) Hydatidosis and cysticercosis: the dynamics of transmission. *Advances in Parasitology*, **22**, 261–308.

Lebour, M. V. (1916) Medusae as hosts for larval trematodes. *Journal of the Marine Biological Association of the United Kingdom*, **11**, 57–59.

Lebour, M. V. (1917) Some parasites of *Sagitta bipunctata*. *Journal of the Marine Biological Association of the United Kingdom*, **11**, 201–206.

Lemly, A. D. and **Esch, G. W.** (1984a) Effects of the trematode *Uvulifer ambloplitis* on juvenile bluegill sunfish, *Lepomis macrochirus*: ecological implications. *Journal of Parasitology*, **70**, 475–492.

Lemly, A. D. and **Esch, G. W.** (1984b) Population biology of the trematode *Uvulifer ambloplitis* (Hughes, 1927) in juvenile bluegill sunfish, *Lepomis macrochirus*, and largemouth bass, *Micropterus salmoides*. *Journal of Parasitology*, **70**, 466–474.

Lengy, J. (1960) Study on *Paramphistomum microbothrium* Fischoeder, 1901 a rumen parasite of cattle in Israel. *Bulletin of the Research Council of Israel*, **9B**, 71–130.

Lester, R. J. G. (1971) The influence of *Schistocephalus* plerocercoids on the respiration of *Gasterosteus* and a possible resulting effect on the behaviour of the fish. *Canadian Journal of Zoology*, **49**, 361–366.

Lester, R. J. G. (1972) Attachment of *Gyrodactylus* to *Gasterosteus* and host response. *Journal of Parasitology*, **58**, 717–722.

Lester, R. J. G. (1980) Host–parasite relations in some didymozoid trematodes. *Journal of Parasitology*, **66**, 527–531.

Lethbridge, R. C. (1971a) The chemical composition and some properties of the egg layers in *Hymenolepis diminuta* eggs. *Parasitology*, **63**, 275–288.

Lethbridge, R. C. (1971b) The hatching of *Hymenolepis diminuta* eggs and penetration of the hexacanths in *Tenebrio molitor* beetles. *Parasitology*, **62**, 445–456.

Leuckart, R. (1886) *The Parasites of Man and the Diseases which Proceed from them. A Textbook for Students and Practitioners*, (trans. W. E. Hoyle and J. Young), Pentland, Edinburgh.

Lewin, R. (1985) Fish to bacterium gene transfer. *Science*, **227**, 1020.

Lewis, M. C., Welsford, I. G. and **Uglem, G. L.** (1989) Cercarial emergence of *Proterometra macrostoma* and *P. edneyi* (Digenea: Azygiidae): contrasting responses to light:dark cycling. *Parasitology*, **99**, 215–223.

Lewis, P. D. (1974) Helminths of terrestrial molluscs in Nebraska. II. Life cycle of *Leucochloridium variae* McIntosh, 1932 (Digenea: Leucochloridiidae). *Journal of Parasitology*, **60**, 251–255.

Le Zotte, L. A. (1954) Studies on marine digenetic trematodes of Puerto Rico: the family Bivesiculidae, its biology and affinities. *Journal of Parasitology*, **40**, 148–162.

Lie, K. J. (1973) Larval trematode antagonism: principles and possible application as a control method. *Experimental Parasitology*, **33**, 343–349.

Linder, E. (1985) *Schistosoma mansoni*: visualization with fluorescent lectins of secretions and surface carbohydrates of living cercariae. *Experimental Parasitology*, **59**, 307–312.

Litchford, R. G. (1963) Observations on *Hymenolepis microstoma* in three laboratory hosts: *Mesocricetus auratus, Mus musculus*, and *Rattus norvegicus. Journal of Parasitology*, **49**, 403–410.

Little, P. A. (1929) The anatomy and histology of *Phyllonella soleae* Ben. & Hesse, an ectoparasitic trematode of the sole, *Solea vulgaris* Quensel. *Parasitology*, **21**, 324–337.

Lively, C. M., Craddock, C. and **Vrijenhoek, R. C.** (1990) Red Queen hypothesis supported by parasitism in sexual and clonal fish. *Nature*, **344**, 864–866.

Llewellyn, J. (1954) Observations on the food and the gut pigment of the Polyopisthocotylea (Trematoda: Monogenea). *Parasitology*, **44**, 428–437.

Llewellyn, J. (1956a) The adhesive mechanisms of monogenetic trematodes: the attachment of *Plectanocotyle gurnardi* (v. Ben. & Hesse) to the gills of *Trigla. Journal of the Marine Biological Association of the United Kingdom*, **35**, 507–514.

Llewellyn, J. (1956b) The host-specificity, micro-ecology, adhesive attitudes and comparative morphology of some trematode gill parasites. *Journal of the Marine Biological Association of the United Kingdom*, **35**, 113–127.

Llewellyn, J. (1957a) The mechanism of the attachment of *Kuhnia scombri* (Kuhn, 1829) (Trematoda: Monogenea) to the gills of its host *Scomber scombrus* L., including a note on the taxonomy of the parasite. *Parasitology*, **47**, 30–39.

Llewellyn, J. (1957b) Host specificity in monogenetic trematodes, in *First Symposium on Host Specificity among Parasites of Vertebrates*, Paul Attinger, Neuchâtel, pp. 199–212.

Llewellyn, J. (1958) The adhesive mechanisms of monogenetic trematodes: the attachment of species of the Diclidophoridae to the gills of gadoid fishes. *Journal of the Marine Biological Association of the United Kingdom*, **37**, 67–79.

Llewellyn, J. (1960) Amphibdellid (monogenean) parasites of electric rays (Torpedinidae). *Journal of the Marine Biological Association of the United Kingdom*, **39**, 561–589.

Llewellyn, J. (1962) The life histories and population dynamics of monogenean gill parasites of *Trachurus trachurus* (L.). *Journal of the Marine Biological Association of the United Kingdom*, **42**, 587–600.

Llewellyn, J. (1963) Larvae and larval development of monogeneans. *Advances in Parasitology*, **1**, 287–326.

Llewellyn, J. (1965) The evolution of parasitic platyhelminths, in *Evolution of Parasites, Third Symposium of the British Society for Parasitology* (ed. A. E. R. Taylor), Blackwell Scientific Publications, Oxford, pp. 47–78.

Llewellyn, J. (1966) The effects of fish hosts upon the body shapes of their monogenean parasites. *Proceedings of the First International Congress of Parasitology*.

Llewellyn, J. (1970) Taxonomy, genetics and evolution of parasites: Monogenea. *Journal of Parasitology*, **56**, 493–504.

Llewellyn, J. (1972) Phylum Platyhelminthes, in *Textbook of Zoology. Invertebrates*, 7th edn, (eds A. J. Marshall and W. D. Williams), Macmillan, London, pp. 188–226.

Llewellyn, J. (1983) Sperm transfer in the monogenean gill parasite *Gastrocotyle trachuri*. *Proceedings of the Royal Society of London B*, **219**, 439–446.

Llewellyn, J. (1984) The biology of *Isancistrum subulatae* n. sp., a monogenean parasitic on the squid, *Alloteuthis subulata*, at Plymouth. *Journal of the Marine Biological Association of the United Kingdom*, **64**, 285–302.

Llewellyn, J. (1986) Phylogenetic inference from platyhelminth life-cycle stages, in *Proceedings of the Sixth International Congress of Parasitology*, (ed. M. Howell), Australian Academy of Sciences, Canberra, pp. 281–289.

Llewellyn, J., Macdonald, S. and **Green, J. E.** (1980) Host-specificity and speciation in diclidophoran (monogenean) gill parasites of trisopteran (gadoid) fishes at Plymouth. *Journal of the Marine Biological Association of the United Kingdom*, **60**, 73–79.

Llewellyn, J. and **Simmons, J. E.** (1984) The attachment of the monogenean parasite *Callorhynchicola multitesticulatus* to the gills of its holocephalan host *Callorhynchus milii*. *International Journal for Parasitology*, **14**, 191–196.

LoBue, C. P. and **Bell, M. A.** (1993) Phenotypic manipulation by the cestode parasite *Schistocephalus solidus* of its intermediate host, *Gasterosteus aculeatus*, the threespine stickleback. *American Naturalist*, **142**, 725–735.

Loker, E. S. (1994) On being a parasite in an invertebrate host: a short survival course. *Journal of Parasitology*, **80**, 728–747.

Loos-Frank, B. (1980) The common vole, *Microtus arvalis* Pall. as intermediate host of *Mesocestoides* (Cestoda) in Germany. *Zeitschrift für Parasitenkunde*, **63**, 129–136.

LoVerde, P. T. and **Chen, L.** (1991) Schistosome female reproductive development. *Parasitology Today*, **7**, 303–308.

Lowenberger, C. A. and **Rau, M. E.** (1994a) *Plagiorchis elegans*: emergence, longevity and infectivity of cercariae, and host behavioural modifications during cercarial emergence. *Parasitology*, **109**, 65–72.

Lowenberger, C. A. and **Rau, M. E.** (1994b) Selective oviposition by *Aedes aegypti* (Diptera: Culicidae) in response to a larval parasite, *Plagiorchis elegans* (Trematoda: Plagiorchiidae). *Environmental Entomology*, **23**, 1269–1276.

Lumsden, R. D. and **Specian, R.** (1980) The morphology, histology, and fine structure of the adult stage of the cyclophyllidean tapeworm *Hymenolepis diminuta*, in *Biology of the Tapeworm* Hymenolepis diminuta, (ed. H. P. Arai), Academic Press, London, pp. 157–280.

Lynch, J. E. (1945) Redescription of the species of *Gyrocotyle* from the ratfish, *Hydrolagus colliei* (Lay and Bennet), with notes on the morphology and taxonomy of the genus. *Journal of Parasitology*, **31**, 418–446.

Lyons, K. M. (1966) The chemical nature and evolutionary significance of monogenean attachment sclerites. *Parasitology*, **56**, 63–100.

Lyons, K. M. (1969a) Sense organs of monogenean skin parasites ending in a typical cilium. *Parasitology*, **59**, 611–623.

Lyons, K. M. (1969b) Compound sensilla in monogenean skin parasites. *Parasitology*, **59**, 625–636.

Lyons, K. M. (1969c) The fine structure of the body wall of *Gyrocotyle urna*. *Zeitschrift für Parasitenkunde*, **33**, 95–109.

Lyons, K. M. (1970) The fine structure and function of the adult epidermis of two skin parasitic monogeneans, *Entobdella soleae* and *Acanthocotyle elegans*. *Parasitology*, **60**, 39–52.

Lyons, K. M. (1972) Sense organs of monogeneans, in *Behavioural Aspects of Parasite Transmission*, (eds E. U. Canning and C. A. Wright), Supplement 1 to the *Zoological Journal of the Linnean Society*, **51**, 19–30.

Lyons, K. M. (1978) *The Biology of Helminth Parasites*, Edward Arnold, London.

McCarthy, A. M. (1990a) Speciation of echinostomes: evidence for the existence of two sympatric sibling species in the complex *Echinoparyphium recurvatum* (von Linstow 1873) (Digenea: Echinostomatidae). *Parasitology*, **101**, 35–42.

McCarthy, A. M. (1990b) The influence of second intermediate host dispersion pattern upon the transmission of cercariae of *Echinoparyphium recurvatum* (Digenea: Echinostomatidae). *Parasitology*, **101**, 43–47.

McCullough, F. S. and **Bradley, D. J.** (1973) Egg output stability and the epidemiology of *Schistosoma haematobium*. Part I. Variation and stability in *Schistosoma haematobium* egg counts. *Transactions of the Royal Society of Tropical Medicine and Hygiene*, **67**, 475–490.

McDaniel, J. S. and **Dixon, K. E.** (1967) Utilization of exogenous glucose by the rediae of *Parorchis acanthus* (Digenea: Philophthalmidae) and *Cryptocotyle lingua* (Digenea: Heterophyidae). *Biological Bulletin*, **133**, 591–599.

Macdonald, S. (1974) Host skin mucus as a hatching stimulant in *Acanthocotyle lobianchi*, a monogenean from the skin of *Raja* spp. *Parasitology*, **68**, 331–338.

Macdonald, S. and **Caley, J.** (1975) Sexual reproduction in the monogenean *Diclidophora merlangi*: tissue penetration by sperms. *Zeitschrift für Parasitenkunde*, **45**, 323–334.

Macdonald, S. and **Combes, C.** (1978) The hatching rhythm of *Polystoma integerrimum*, a monogenean from the frog *Rana temporaria*. *Chronobiologia*, **5**, 277–285.

Macdonald, S. and **Llewellyn, J.** (1980) Reproduction in *Acanthocotyle greeni* n. sp. (Monogenea) from the skin of *Raia* spp. at Plymouth. *Journal of the Marine Biological Association of the United Kingdom*, **60**, 81–88.

McGonigle, S. and **Dalton, J. P.** (1995) Isolation of *Fasciola hepatica* haemoglobin. *Parasitology*, **111**, 209–215.

MacKenzie, K. (1964) *Stichocotyle nephropis* Cunningham, 1887 (Trematoda) in Scottish waters. *Annals and Magazine of Natural History*, **6**, 505–506.

MacKenzie, K. (1965) The plerocercoid of *Gilquinia squali* Fabricius, 1794. *Parasitology*, **55**, 607–615.

MacKenzie, K. (1975) Some aspects of the biology of the plerocercoid of *Gilquinia squali* Fabricius 1794 (Cestoda: Trypanorhyncha). *Journal of Fish Biology*, **7**, 321–327.

Mackiewicz, J. S. (1981) Caryophyllidea (Cestoidea): evolution and classification. *Advances in Parasitology*, **19**, 139–206.

Mackiewicz, J. S. (1988) Cestode transmission patterns. *Journal of Parasitology*, **74**, 60–71.

Mackiewicz, J. S. (1994) Order Caryophyllidea van Beneden in Carus, 1863, in *Key to the Cestode Parasites of Vertebrates*, (eds L. F. Khalil, A. Jones and R. A. Bray), CAB International, Wallingford, UK, pp. 21–43.

McKindsey, C. W. and **McLaughlin, J. D.** (1995a) Species- and size-specific infection of snails by *Cyclocoelum mutabile* (Digenea: Cyclocoelidae). *Journal of Parasitology*, **81**, 513–519.

McKindsey, C. W. and **McLaughlin, J. D.** (1995b) Field studies on the transmission and survival of *Cyclocoelum mutabile* (Digenea) infections in natural snail populations in southern Manitoba, Canada. *Journal of Parasitology*, **81**, 520–525.

MacKinnon, B. M. (1982a) The development of the ventral papillae of *Notocotylus triserialis* (Digenea: Notocotylidae). *Zeitschrift für Parasitenkunde*, **68**, 279–293.

MacKinnon, B. M. (1982b) The structure and possible function of the ventral papillae of *Notocotylus triserialis* Diesing, 1839. *Parasitology*, **84**, 313–332.

McLaren, D. J. (1980) Schistosoma mansoni: *the Parasite Surface in Relation to Host Immunity*, Research Studies Press, Letchworth, UK.

McLaren, D. J. and **Hockley, D. J.** (1977) Blood flukes have a double outer membrane. *Nature*, **269**, 147–149.

McLaren, D. J., **Ramalho-Pinto, F. J.** and **Smithers, S. R.** (1978) Ultrastructural evidence for complement and antibody-dependent damage to schistosomula of *Schistosoma mansoni* by rat eosinophils *in vitro*. *Parasitology*, **77**, 313–324.

McLaughlin, J. D. (1976) Experimental studies on the life cycle of *Cyclocoelum mutabile* (Zender) (Trematoda: Cyclocoelidae). *Canadian Journal of Zoology*, **54**, 48–54.

McMahon, J. E. (1976) Circadian rhythm in *Schistosoma haematobium* egg excretion. *International Journal for Parasitology*, **6**, 373–377.

McManus, D. P. and **James, B. L.** (1975) The absorption of sugars and organic acids by the daughter sporocysts of *Microphallus similis* (Jäg.). *International Journal for Parasitology*, **5**, 33–38.

McMichael-Phillips, D. F., **Lewis, J. W.** and **Thorndyke, M. C.** (1992a) Ultrastructure of the egg of *Sanguinicola inermis* Plehn, 1905 (Digenea: Sanguinicolidae). *Journal of Natural History*, **26**, 895–904.

McMichael-Phillips, D. F., **Lewis, J. W.** and **Thorndyke, M. C.** (1992b) Ultrastructural studies on the miracidium of *Sanguinicola inermis* (Digenea: Sanguinicolidae). *Parasitology*, **105**, 435–443.

McPhail, J. D. and **Peacock, S. D.** (1983) Some effects of the cestode (*Schistocephalus solidus*) on reproduction in the threespine stickleback (*Gasterosteus aculeatus*): evolutionary aspects of a host–parasite interaction. *Canadian Journal of Zoology*, **61**, 901–908.

Macpherson, C. N. L. (1983) An active intermediate host role for man in the life cycle of *Echinococcus granulosus* in Turkana, Kenya. *American Journal of Tropical Medicine and Hygiene*, **32**, 397–404.

Macy, R. W. and **Bell, W. D.** (1968) The life cycle of *Astacatrematula macrocotyla* gen. et sp. n. (Trematoda: Psilostomidae) from Oregon. *Journal of Parasitology*, **54**, 319–323.

Madden, F. C. (1907) *Bilharziosis*, Cassell & Co., London.

Madhavi, R. (1968) A didymozoid metacercaria from the copepod, *Paracalanus aculeatus* Giesbrecht, from Bay of Bengal. *Journal of Parasitology*, **54**, 629.

Madhavi, R. (1978) Life history of *Genarchopsis goppo* Ozaki, 1925 (Trematoda: Hemiuridae) from the freshwater fish *Channa punctata*. *Journal of Helminthology*, **52**, 251–259.

Madhavi, R. and **Anderson, R. M.** (1985) Variability in the susceptibility of the fish host, *Poecilia reticulata*, to infection with *Gyrodactylus bullatarudis* (Monogenea). *Parasitology*, **91**, 531–544.

Madhavi, R. and **Rao, K. H.** (1971) *Orchispirium heterovitellatum*: chemical nature of the eggshell. *Experimental Parasitology*, **30**, 345–348.

Maillard, C. (1973) Mise en évidence du cycle évolutif abrégé d'*Aphalloides coelomicola* Dollfus, Chabaud ct Golvan, 1957 (Trematoda). Notion d'«hôte historique». *Compte rendu hebdomadaire des séances de l'Académie des sciences. Paris*. Série D, **277**, 317–320.

Maillard, C. (1975) Cycle évolutif de *Paratimonia gobii* Prévot et Bartoli 1967 (Trematoda – Monorchiidae). *Acta Tropica*, **32**, 327–333.

Malmberg, G. (1970) The excretory systems and the marginal hooks as a basis for the systematics of *Gyrodactylus* (Trematoda, Monogenea). *Archiv för Zoologi*, **23**, 1–235.

Mamaev, Y. L. and **Kurochkin, Y. V.** (1976) [The monogenean *Caballeraxine chainanica*, its method of attachment to the host, morphology and systematic position]. *Biologiya Morya*, **2**, 16–21 (in Russian).

Manter, H. W. (1951) Studies on *Gyrocotyle rugosa* Diesing, 1850, a cestodarian parasite of the elephant fish, *Callorhynchus milii*. *Zoology Publications from Victoria University College*, **17**, 1–11.

Manter, H. W. and **Prince, D. F.** (1953) Some monogenetic trematodes of marine fishes from Fiji. *Proceedings of the Helminthological Society of Washington*, **20**, 105–112.

Margolis, L., Esch, G. W., Holmes, J. C. *et al.* (1982) The use of ecological terms in Parasitology (report of an *ad hoc* committee of the American Society of Parasitologists). *Journal of Parasitology*, **68**, 131–133.

Martin, W. E. (1968) *Cercaria gorgonocephala* Ward, 1916, a zygocercous species in northwestern United States. *Transactions of the American Microscopical Society*, **87**, 472–476.

Mas-Coma, S. and **Montoliu, I.** (1987) The life cycle of *Dollfusinus frontalis*, a brachylaimid trematode of small mammals (Insectivora and Rodentia). *International Journal for Parasitology*, **17**, 1063–1079.

Mathias, P. (1924) Sur le cycle évolutif d'un trématode de la famille des Psilostomidae (*Psilotrema spiculigerum* Mühling). *Comptes rendus de l'Académie des Sciences (Paris)*, **178**, 1217–1219.

Matthews, B. F. (1980) *Cercaria vaullegeardi* Pelseneer, 1906 (Digenea: Hemiuridae); the daughter sporocyst and emergence of the cercaria. *Parasitology*, **81**, 61–69.

Matthews, B. F. (1981a) *Cercaria vaullegeardi* Pelseneer, 1906 (Digenea: Hemiuridae); development and ultrastructure. *Parasitology*, **83**, 575–586.

Matthews, B. F. (1981b) *Cercaria vaullegeardi* Pelseneer, 1906 (Digenea: Hemiuridae); the infection mechanism. *Parasitology*, **83**, 587–593.

Matthews, B. F. (1982) *Cercaria calliostomae* Dollfus, 1923 (Digenea: Hemiuridae); development, morphology and emergence. *Zeitschrift für Parasitenkunde*, **67**, 45–53.

Matthews, B. F. and **Matthews, R. A.** (1988a) The ecsoma in Hemiuridae (Digenea: Hemiuroidea): tegumental structure and function in the mesocercaria of *Lecithochirium furcolabiatum* (Jones, 1933) Dawes, 1947. *Journal of Helminthology*, **62**, 317–330.

Matthews, B. F. and **Matthews, R. A.** (1988b) The tegument in Hemiuridae (Digenea: Hemiuroidea): structure and function in the adult. *Journal of Helminthology*, **62**, 305–316.

Matthews, R. A. (1973a) The life-cycle of *Prosorhynchus crucibulum* (Rudolphi, 1819) Odhner, 1905, and a comparison of its cercaria with that of *Prosorhynchus squamatus* Odhner, 1905. *Parasitology*, **66**, 133–164.

Matthews, R. A. (1973b) The life-cycle of *Bucephalus haimeanus* Lacaze-Duthiers, 1854 from *Cardium edule* L. *Parasitology*, **67**, 341–350.

Matthews, R. A. (1974) The life-cycle of *Bucephaloides gracilescens* (Rudolphi, 1819) Hopkins, 1954 (Digenea: Gasterostomata). *Parasitology*, **68**, 1–12.

Mattison, R. G., Hanna, R. E. B. and **Nizami, W. A.** (1994) Ultrastructure and histochemistry of the tegument of juvenile paramphistomes during migration in Indian ruminants. *Journal of Helminthology*, **68**, 211–221.

Measures, L. N., Beverley-Burton, M. and **Williams, A.** (1990) Three new species of *Monocotyle* (Monogenea: Monocotylidae) from the stingray, *Himantura uarnak* (Rajiformes: Dasyatidae) from the Great Barrier Reef: phylogenetic reconstruction, systematics and emended diagnoses. *International Journal for Parasitology*, **20**, 755–767.

Meggitt, F. J. (1914) The structure and life-history of a tapeworm (*Ichthyotaenia filicollis* Rud.) parasitic in the stickleback. *Proceedings of the Zoological Society of London*, **8**, 113–138.

Mehlhorn, H., Becker, B., Andrews, P. and **Thomas, H.** (1981) On the nature of the proglottids of cestodes: a light and electron microscopic study on *Taenia*, *Hymenolepis*, and *Echinococcus*. *Zeitschrift für Parasitenkunde*, **65**, 243–259.

Menitskii, Y. L. (1963) [Structure and systematic position of the turbellarian *Ichthyophaga subcutanea* Syromjatnizova 1949, parasitizing fish]. *Parazitologicheskii Sbornik*, **21**, 245–258 (In Russian).

Mettrick, D. F. (1980) The intestine as an environment for *Hymenolepis diminuta*, in *Biology of the Tapeworm* Hymenolepis diminuta, (ed. H. P. Arai), Academic Press, London, pp. 281–356.

Meyer, F., Meyer, H. and **Bueding, E.** (1970) Lipid metabolism in the parasitic and free-living flatworms, *Schistosoma mansoni* and *Dugesia dorotocephala*. *Biochimica et Biophysica Acta*, **210**, 257–266.

Meyer, M. C. (1972) The pattern of circulation of *Diphyllobothrium sebago* (Cestoda: Pseudophyllidea) in an enzootic area. *Journal of Wildlife Diseases*, **8**, 215–220.

Meyerhof, E. and **Rothschild, M.** (1940) A prolific trematode. *Nature*, **146**, 367–368.

Michelson, E. H. (1970) *Aspidogaster conchicola* from freshwater gastropods in the United States. *Journal of Parasitology*, **56**, 709–712.

Miller, A. (1961) The mouthparts and digestive tract of adult dung beetles (Coleoptera: Scarabaeidae), with reference to the ingestion of helminth eggs. *Journal of Parasitology*, **50**, 735–744.

Mills, C. A. (1979a) Attachment and feeding of the adult ectoparasitic digenean *Transversotrema patialense* (Soparkar, 1924) on the zebra fish *Brachydanio rerio* (Hamilton-Buchanan). *Journal of Fish Diseases*, 1979, 443–447.

Mills, C. A. (1979b) The influence of differing ionic environments on the cercarial, post-cercarial and adult stages of the ectoparasitic digenean *Transversotrema patialense*. *International Journal for Parasitology*, **9**, 603–608.

Mills, C. A., Anderson, R. M. and **Whitfield, P. J.** (1979) Density-dependent survival and reproduction within populations of the ectoparasitic digenean *Transversotrema patialense* on the fish host. *Journal of Animal Ecology*, **48**, 383–399.

Minchella, D. J., Sollenberger, K. M. and **Pereira De Souza, C.** (1995) Distribution of Schistosome genetic diversity within molluscan intermediate hosts. *Parasitology*, **111**, 217–220.

Minchin, D. (1991) *Udonella caligorum* Johnston (Trematoda) from the Celtic Sea. *Irish Naturalists' Journal*, **23**, 509–510.

Mitchell, J. B. (1973) *Gorgoderina vitelliloba* (Trematoda: Gorgoderidae) in its definitive host, *Rana temporaria*. *International Journal for Parasitology*, **3**, 539–544.

Mitchell, J. B. and **Mason, A. R.** (1980) Emergence of the cercariae of *Gorgoderina vitelliloba* from daughter sporocysts. *International Journal for Parasitology*, **10**, 75–80.

Miyazaki, I. (1976) A newly recognized mode of human infection with the lung fluke, *Paragonimus westermani* (Kerbert 1878). *Journal of Parasitology*, **62**, 646–648.

Moczon, T. (1977) Penetration of *Hymenolepis diminuta* oncospheres across the intestinal tissues of *Tenebrio molitor* beetles. *Bulletin de l'Académie Polonaise des Sciences*, **25**, 531–535.

Moczon, T. (1994) Histochemistry of proteinases in the cercariae of *Diplostomum pseudospathaceum* (Trematoda, Diplostomatidae). *Parasitology Research*, **80**, 684–686.

Moczon, T. (1996a) A serine proteinase in the penetration glands of the hexacanths of *Hymenolepis diminuta* (Cestoda, Cyclophyllidea). *Parasitology Research*, **82**, 67–71.

Moczon, T. (1996b) A serine proteinase in the penetration glands of the cercariae of *Plagiorchis elegans* (Trematoda, Plagiorchiidae). *Parasitology Research*, **82**, 72–76.

Modha, J., Roberts, M. C. and **Kusel, J. R.** (1996) Schistosomes and serpins: a complex business. *Parasitology Today*, **12**, 119–121.

Monaco, L. H., Wood, R. A. and **Mizelle, J. D.** (1954) Studies on monogenetic trematodes. XVI. Rhamnocercinae, a new subfamily of Dactylogyridae. *American Midland Naturalist*, **52**, 129–132.

Moog, F. (1981) The lining of the small intestine. *Scientific American*, **245**, 116–125.

Moore, J. (1981) Asexual reproduction and environmental predictability in cestodes (Cyclophyllidea: Taeniidae). *Evolution*, **35**, 723–741.

Moore, J. and **Gotelli, N. J.** (1990) A phylogenetic perspective on the evolution of altered host behaviours: a critical look at the manipulation hypothesis, in *Parasitism and Host Behaviour*, (eds C. J. Barnard and J. M. Behnke), Taylor & Francis, London, pp. 193–233.

Moore, M. M., Kaattari, S. L. and **Olson, R. E.** (1994) Biologically active factors against the monogenetic trematode *Gyrodactylus stellatus* in the serum and mucus of infected juvenile English soles. *Journal of Aquatic Animal Health*, **6**, 93–100.

Morand, S., Robert, F. and **Connors, V. A.** (1995) Complexity in parasite life cycles: population biology of cestodes in fish. *Journal of Animal Ecology*, **64**, 256–264.

Morris, G. P. and **Halton, D. W.** (1975) The occurrence of bacteria and mycoplasma-like organisms in a monogenean parasite *Diclidophora merlangi*. *International Journal for Parasitology*, **5**, 495–498.

Moss, G. D. (1970) The excretory metabolism of the endoparasitic digenean *Fasciola hepatica* and its relationship to its respiratory metabolism. *Parasitology*, **60**, 1–19.

Mostafa, M. H., Badawi, A. F. and **O'Connor, P. J.** (1995) Bladder cancer associated with schistosomiasis. *Parasitology Today*, **11**, 87–89.

Mount, P. M. (1970) Histogenesis of the rostellar hooks of *Taenia crassiceps* (Zeder, 1800) (Cestoda). *Journal of Parasitology*, **56**, 947–961.

Mudry, D. R. and **Dailey, M. D.** (1971) Postembryonic development of certain tetraphyllidean and trypanorhynchan cestodes with a possible alternative life cycle for the order Trypanorhyncha. *Canadian Journal of Zoology*, **49**, 1249–1253.

Mueller, J. F. (1966) Host–parasite relationships as illustrated by the cestode *Spirometra mansonoides*, in *Host Parasite Relationships*, (ed. J. E. McCauley), Proceedings of the 26th Annual Biology Colloquium, Oregon State University Press, Corvallis, OR, pp. 15–58.

Mueller, J. F. (1972) Survival and longevity of *Mesocestoides* tetrathyridia under adverse conditions. *Journal of Parasitology*, **58**, 228.

Mueller, J. F. (1974) The biology of *Spirometra*. *Journal of Parasitology*, **60**, 3–13.

Murrills, R. J., Reader, T. A. J. and **Southgate, V. R.** (1985a) Studies on the invasion of *Notocotylus attenuatus* (Notocotylidae: Digenea) into its snail host *Lymnaea peregra*: the contents of the fully embryonated egg. *Parasitology*, **91**, 397–405.

Murrills, R. J., Reader, T. A. J. and **Southgate, V. R.** (1985b) Studies on the invasion of *Notocotylus attenuatus* (Notocotylidae: Digenea) into its snail host, *Lymnaea peregra*. *In vitro* observations on the hatching mechanism of the egg. *Parasitology*, **91**, 545–554.

Nasincova, V. and **Scholz, T.** (1994) The life cycle of *Asymphylodora tincae* (Modeer 1790) (Trematoda: Monorchiidae): a unique development in monorchiid trematodes. *Parasitology Research*, **80**, 192–197.

Nasir, P. (1960) Trematode parasites of snails from Edgbaston Pool: the life history of the strigeid *Cotylurus brevis* Dubois & Rausch, 1950. *Parasitology*, **50**, 551–575.

Nasir, P. (1962) Two new species of giant-tailed echinostome cercariae from *Planorbis carinatus* (Müll). *Transactions of the American Microscopical Society*, **81**, 132–137.

Nelson, G. S. and **Rausch, R. L.** (1963) *Echinococcus* infections in man and animals in Kenya. *Annals of Tropical Medicine and Parasitology*, **57**, 136–149.

Newport, G., McKerrow, J., Hedstrom, R. *et al.* (1987) Schistosome elastases: biological importance, structure, function and stage-specific expression. *Biochemical Society Symposia*, **53**, 115–121.

Nichols, K. C. (1975) Observations on little-known flatworms: *Udonella*. *International Journal for Parasitology*, **5**, 475–482.

Nickerson, W. S. (1902) *Cotylogaster occidentalis* n. sp. and a revision of the family Aspidobothridae. *Zoologische Jahrbücher, Systematik*, **15**, 597–624.

Noble, E. R. and **Noble G. A.** (1971) *Parasitology: The Biology of Animal Parasites*, Lea & Febiger, Philadelphia, PA.

Noble, G. A. (1975) Description of *Nematobibothrioides histoidii* (Noble, 1974) (Trematoda: Didymozoidae) and comparison with other genera. *Journal of Parasitology*, **61**, 224–227.

Nollen, P. M. (1975) Studies on the reproductive system of *Hymenolepis diminuta* using autoradiography and transplantation. *Journal of Parasitology*, **61**, 100–104.

Oaks, J. A. and **Holy, J. M.** (1994) *Hymenolepis diminuta*: two morphologically distinct tegumental secretory mechanisms are present in the cestode. *Experimental Parasitology*, **79**, 292–300.

Odhner, T. (1910) *Stichocotyle nephropis* J. T. Cunningham, ein aberranter Trematode der Digenienfamilie Aspidogastridae. *Kungliga Svenska Vetenskapsakademiens Handlingar*, **45**, 3–16.

Ohman, C. (1966) The structure and function of the adhesive organ in strigeid trematodes. Part III. *Apatemon gracilis minor* Yamaguti, 1933. *Parasitology*, **56**, 209–226.

Oliver, G. (1969) Recherches sur les Diplectanidae (Monogenea) parasites de téléostéens du Golfe du Lion. II. – Lamellodiscinae nov. sub-fam. *Vie et milieu, Série A, Biologie Marine*, **20**, 43–72.

Olsen, O. W. (1986) *Animal Parasites, their Life Cycles and Ecology*, Dover, New York.

Olson, R. E. (1970) The life cycle of *Cotylurus erraticus* (Rudolphi, 1809) Szidat, 1928 (Trematoda: Strigeidae). *Journal of Parasitology*, **56**, 55–63.

Osborn, H. L. (1903) On the habits and structure of *Cotylaspis insignis* Leidy, from Lake Chautauqua, New York. *Journal of Morphology*, **18**, 1–44.

Owen, S. F., Barber, I. and **Hart, P. J. B.** (1993) Low level infection by eye fluke, *Diplostomum* spp., affects the vision of three-spined sticklebacks, *Gasterosteus aculeatus*. *Journal of Fish Biology*, **42**, 803–806.

Paperna, I. (1964) Competitive exclusion of *Dactylogyrus extensus* by *Dactylogyrus vastator* (Trematoda, Monogenea) on the gills of reared carp. *Journal of Parasitology*, **50**, 94–98.

Pappas, P. W. (1978) Tryptic and protease activities in the normal and *Hymenolepis diminuta*-infected rat small intestine. *Journal of Parasitology*, **64**, 562–564.

Pappas, P. W. (1980) Enzyme interactions at the host–parasite interface, in *Cellular Interactions in Symbiosis and Parasitism*, (eds C. B. Cook, P. W. Pappas and E. D. Rudolph), Ohio State University Press, Columbus, OH, pp. 145–172.

Pappas, P. W. (1983) Host–parasite interface, in *Biology of the Eucestoda, vol. 2*, (eds C. Arme and P. W. Pappas), Academic Press, London, pp. 297–334.

Pappas, P. W. (1987) The tapeworm trap and other esoteric remedies for intestinal helminths. *Parasitology Today*, **3**, 282–283.

Pappas, P. W., Marschall, E. A., Morrison, S. E. *et al.* (1995) Increased coprophagic activity of the beetle *Tenebrio molitor*, on feces containing eggs of the tapeworm, *Hymenolepis diminuta*. *International Journal for Parasitology*, **25**, 1179–1184.

Pappas, P. W. and **Read, C. P.** (1972a) Trypsin inactivation by intact *Hymenolepis diminuta*. *Journal of Parasitology*, **58**, 864–871.

Pappas, P. W. and **Read, C. P.** (1972b) Inactivation of α-and β-chymotrypsin by intact *Hymenolepis diminuta* (Cestoda). *Biological Bulletin*, **143**, 605–616.

Pariselle, A., Lambert, A. and **Euzet, L.** (1991) A new type of haptor in mesoparasitic monogeneans of the genus *Enterogyrus* Paperna, 1963, with a description of *Enterogyrus foratus* n. sp. and *E. coronatus* n. sp., stomach parasites of cichlids in West Africa. *Systematic Parasitology*, **20**, 211–220.

Parshad, V. R. and **Guraya, S. S.** (1978) Morphological and histochemical observations on the digestive system of *Cotylophoron cotylophoron*. *Journal of Helminthology*, **52**, 327–333.

Pearson, A. G. M., Fincham, A. G., Waters, H. and **Bundy, D. A. P.** (1985) Differences in composition between *Fasciola hepatica* spines and cestode hooks. *Comparative Biochemistry and Physiology*, **81B**, 373–376.

Pearson, J. C. (1956) Studies on the life cycles and morphology of the larval stages of *Alaria arisaemoides* Augustine and Uribe, 1927 and *Alaria canis* LaRue and Fallis, 1936 (Trematoda: Diplostomidae). *Canadian Journal of Zoology*, **34**, 295–387.

Pearson, J. C. (1968) Observations on the morphology and life-cycle of *Paucivitellosus fragilis* Coil, Reid & Kuntz, 1965 (Trematoda: Bivesiculidae). *Parasitology*, **58**, 769–788.

Pearson, J. C. (1972) A phylogeny of life-cycle patterns of the Digenea. *Advances in Parasitology*, **10**, 153–189.

Pearson, J. C. (1986) The paranephridial system in the Digenea: occurrence and possible phylogenetic significance, in *Parasite Lives*, (eds M. Cremin, C. Dobson and D. E. Moorhouse), University of Queensland Press, Brisbane, Queensland, pp. 56–68.

Pearson, J. C. (1992) On the position of the digenean family Heronimidae: an inquiry into a cladistic classification of the Digenea. *Systematic Parasitology*, **21**, 81–166.

Pence, D. B. (1967) The fine structure and histochemistry of the infective eggs of *Dipylidium caninum*. *Journal of Parasitology*, **53**, 1041–1054.

Perkins, K. W. (1956) Studies on the morphology and biology of *Acetodextra amiuri* (Stafford) (Trematoda: Heterophyidae). *American Midland Naturalist*, **55**, 139–161.

Petrushevski, G. K. and **Shulman, S. S.** (1958) The parasitic diseases of fishes in the natural waters of the USSR, in *Parasitology of Fishes*, (eds V. A. Dogiel, G. K. Petrushevski and Y. I. Polyanski), Leningrad University Press. English translation by Z. Kabata (1961), Oliver & Boyd, London, pp. 299–319.

Phares, C. K. (1987) Plerocercoid growth factor: a homologue of human growth hormone. *Parasitology Today*, **3**, 346–349.

Phares, K. and **Kubik, J.** (1996) The growth factor from plerocercoids of *Spirometra mansonoides* is both a growth hormone agonist and a cysteine proteinase. *Journal of Parasitology*, **82**, 210–215.

Phillips, J. I. (1978) The occurrence and distribution of haemoglobin in the entosymbiotic rhabdocoel *Paravortex scrobiculariae* (Graff) (Platyhelminthes: Turbellaria). *Comparative Biochemistry and Physiology*, **61A**, 679–683.

Pichelin, S., Whittington, I. and **Pearson, J.** (1991) *Concinnocotyla* (Monogenea: Polystomatidae), a new genus for the polystome from the Australian lungfish *Neoceratodus forsteri*. *Systematic Parasitology*, **18**, 81–93.

Pike, A. W. (1969) Observations on the life cycles of *Notocotylus triserialis* Diesing, 1839, and *N. imbricatus* (Looss, 1893) *sensu* Szidat, 1935. *Journal of Helminthology*, **43**, 145–165.

Pike, A. W. and **Burt, M. D. B.** (1983) The tissue response of yellow perch, *Perca flavescens* Mitchill to infections with the metacercarial cyst of *Apophallus brevis* Ransom, 1920. *Parasitology*, **87**, 393–404.

Plateaux, L. (1972) Sur les modifications produites chez une fourmi par la présence d'un parasite cestode. *Annales des Sciences naturelles, Zoologie*, **14**, 203–220.

Platt, T. R. (1992) A phylogenetic and biogeographic analysis of the genera of Spirorchinae (Digenea: Spirorchidae) parasitic in freshwater turtles. *Journal of Parasitology*, **78**, 616–629.

Platt, T. R., Blair, D., Purdie, J. and **Melville, L.** (1991) *Griphobilharzia amoena* n. gen., n. sp. (Digenea: Schistosomatidae), a parasite of the freshwater crocodile *Crocodylus johnstoni* (Reptilia: Crocodylia) from Australia, with the erection of a new subfamily, Griphobilharziinae. *Journal of Parasitology*, **77**, 65–68.

Platzer, E. G. and **Roberts, L. S.** (1969) Developmental physiology of cestodes. V. Effects of vitamin deficient diets and host coprophagy prevention on development of *Hymenolepis diminuta*. *Journal of Parasitology*, **55**, 1143–1152.

Podesta, R. B. and **Mettrick, D. F.** (1974) Pathophysiology of cestode infections: effect of *Hymenolepis diminuta* on oxygen tensions, pH and gastrointestinal function. *International Journal for Parasitology*, **4**, 277–292.

Polzer, M. and **Conradt, U.** (1994) Identification and partial characterization of the proteases from different developmental stages of *Schistocephalus solidus* (Cestoda: Pseudophyllidae). *International Journal for Parasitology*, **24**, 967–973.

Popiel, I. and **James, B. L.** (1976) The effect of glycogen and glucose on oxygen consumption in the daughter sporocysts of *Cercaria linearis* Stunkard, 1932 and *Cercaria stunkardi* Palombi, 1934 (Digenea: Opecoelidae). *Zeitschrift für Parasitenkunde*, **51**, 71–77.

Popova, L. B. and **Davydov, V. G.** (1988) Studies of localization of *Amphilina foliacea* (Amphilinidae Dubinina, 1974) in definitive hosts. *Helminthologia*, **25**, 129–138.

Pough, F. H., Heiser, J. B. and **McFarland, W. N.** (1990) *Vertebrate Life*, 3rd edn, Macmillan, New York.

Poulin, R. (1995) 'Adaptive' changes in the behaviour of parasitized animals: a critical review. *International Journal for Parasitology*, **25**, 1371–1383.

Prévot, G. (1971) Cycle évolutif d'*Aporchis massiliensis* Timon-David, 1955 (Digenea Echinostomatidae), parasite du Goéland *Larus argentatus michaellis* Naumann. *Bulletin de la Société Zoologique de France*, **96**, 197–208.

Price, E. W. (1943) A new trematode of the genus *Polystoma* (Monogenea: Polystomatidae) from *Xenopus laevis* Daud. *Proceedings of the Helminthological Society of Washington*, **10**, 83–85.

Price, H. F. (1931) Life history of *Schistosomatium douthitti* (Cort). *American Journal of Hygiene*, **13**, 685–727.

Pronin, N. M., Timoshenko, T. M. and **Sanzhieva, S. D.** (1989) Dynamics of egg production of the cestode *Diphyllobothrium dendriticum* (Nitzsch, 1824) (Cestoda: Pseudophyllidea) and the concept of fecundity in helminths. *Folia Parasitologica*, **36**, 49–57.

Prudhoe, S. (1968) A new polyclad turbellarian associating with a hermit crab in the Hawaiian Islands. *Pacific Science*, **22**, 408–411.

Prudhoe, S. (1985) *A Monograph on Polyclad Turbellaria*, British Museum (Natural History), Oxford University Press, Oxford.

Pugh, R. E. (1987) Effects on the development of *Dipylidium caninum* and on the host reaction to this parasite in the adult flea (*Ctenocephalides felis felis*). *Parasitology Research*, **73**, 171–177.

Ramalho-Pinto, F. J. (1994) How *Schistosoma mansoni* evades complement-mediated immune attack. *Ciência e Cultura*, **46**, 433–440.

Ratanarat-Brockelman, C. (1974) Migration of *Diplostomum spathaceum* (Trematoda) in the fish intermediate host. *Zeitschrift für Parasitenkunde*, **43**, 123–134.

Ratcliffe, L. H. (1968) Hatching of *Dicrocoelium lanceolatum* egg. *Experimental Parasitology*, **23**, 67–78.

Rau, M. E. and **Caron, F. R.** (1979) Parasite-induced susceptibility of moose to hunting. *Canadian Journal of Zoology*, **57**, 2466–2468.

Rausch, R. L. (1993) The biology of *Echinococcus granulosus*, in *Compendium on Cystic Echinococcosis with Special Reference to the Xinjiang Uygur Autonomous Region, the People's Republic of China*, (eds F. L. Andersen, Chai, J.-J. and Liu, F.-J.), Brigham Young University, Provo, UT, pp. 27–56.

Rausch, R. L. (1994) Family Taeniidae Ludwig, 1886, in *Key to the Cestode Parasites of Vertebrates*, (eds L. F. Khalil, A. Jones and R. A. Bray), CAB International, Wallingford, UK, pp. 665–672.

Rauther, M. (1937) Kiemen der Anamnier. Fische, in *Handbuch der Vergleichende Anatomie der Wirbeltiere, de Bolk, etc., vol. 3*, Berlin and Vienna, pp. 224–251.

Read, A. F. and **Nee, S.** (1990) Male schistosomes: more than just muscle? *Parasitology Today*, **6**, 297.

Read, C. P. (1950) The vertebrate small intestine as an environment for parasitic helminths. *Rice Institute Pamphlet*, **37**, pp. 94.

Read, C. P. (1966) Nutrition of intestinal helminths, in *Biology of Parasites*, (ed. E. J. L. Soulsby), Academic Press, New York, pp. 101–126.

Read, C. P. (1967) Longevity of the tapeworm *Hymenolepis diminuta*. *Journal of Parasitology*, **53**, 1055–1056.

Read, C. P. (1971) The microcosm of intestinal helminths, in *Ecology and Physiology of Parasites*, (ed. A. M. Fallis), University of Toronto Press, Toronto, Ontario, pp. 188–200.

Read, C. P. and **Kilejian, A. Z.** (1969) Circadian migratory behavior of a cestode symbiote in the rat host. *Journal of Parasitology*, **55**, 574–578.

Read, C. P., Rothman, A. H. and **Simmons, J. E. Jr** (1963) Studies on membrane transport, with special reference to parasite-host integration. *Annals of the New York Academy of Sciences*, **113**, 154–205.

Rees, F. G. (1979) The morphology and ultrastructure of the female reproductive ducts in the metacercaria and adult of *Cryptocotyle lingua* (Creplin) (Digenea: Heterophyidae). *Zeitschrift für Parasitenkunde*, **60**, 157–176.

Rees, F. G. (1981) The ultrastructure of the epidermis of the redia of *Parorchis acanthus* Nicoll (Digenea; Philophthalmidae). *Zeitschrift für Parasitenkunde*, **65**, 19–30.

Rees, F. G. (1983a) The ultrastructure of the intestine of the redia of *Parorchis acanthus* Nicoll (Digenea: Philophthalmidae) from the digestive gland of *Nucella lapillus* L. *Parasitology*, **87**, 159–166.

Rees, F. G. (1983b) The ultrastructure of the fore-gut of the redia of *Parorchis acanthus* Nicoll (Digenea: Philophthalmidae) from the digestive gland of *Nucella lapillus* L. *Parasitology*, **87**, 151–158.

Rees, G. (1939) Studies on the germ cell cycle of the digenetic trematode *Parorchis acanthus* Nicoll. Part I. Anatomy of the genitalia and gametogenesis in the adult. *Parasitology*, **31**, 417–433.

Rees, G. (1940) Studies on the germ cell cycle of the digenetic trematode *Parorchis acanthus* Nicoll. Part II. Structure of the miracidium and germinal development in the larval stages. *Parasitology*, **32**, 372–391.

Rees, G. (1950) The plerocercoid larva of *Grillotia heptanchi* (Vaullegeard). *Parasitology*, **40**, 265–272.

Rees, G. (1952) The structure of the adult and larval stages of *Plagiorchis (Multiglandularis) megalorchis* n. nom. from the turkey and an experimental demonstration of the life history. *Parasitology*, **42**, 92–113.

Rees, G. (1955) The adult and diplostomulum stage (*Diplostomulum phoxini* (Faust)) of *Diplostomum pelmatoides* Dubois and an experimental demonstration of part of the life cycle. *Parasitology*, **45**, 295–312.

Rees, G. (1956) The scolex of *Tetrabothrius affinis* (Lönnberg), a cestode from *Balaenoptera musculus* L., the blue whale. *Parasitology*, **46**, 425–442.

Rees, G. (1957) *Cercaria diplostomi phoxini* (Faust), a furcocercaria which develops into *Diplostomulum phoxini* in the brain of the minnow. *Parasitology*, **47**, 126–137.

Rees, G. (1967) The histochemistry of the cystogenous gland cells and cyst wall of *Parorchis acanthus* Nicoll, and some details of the morphology and fine structure of the cercaria. *Parasitology*, **57**, 87–110.

Rees, G. (1970) Some helminth parasites of fishes of Bermuda and an account of the attachment organ of *Alcicornis carangis* MacCallum, 1917 (Digenea: Bucephalidae). *Parasitology*, **60**, 195–221.

Rees, G. and **Williams, H. H.** (1965) The functional morphology of the scolex and the genitalia of *Acanthobothrium coronatum* (Rud.) (Cestoda: Tetraphyllidea). *Parasitology*, **55**, 617–651.

Rees, J. A. (1986) Studies on adhesive gland systems of monogeneans. PhD thesis, University of East Anglia, UK.

Reith, E. J. and **Ross, M. H.** (1965) *Atlas of Descriptive Histology*, Harper & Row, New York.

Reuter, M. and **Gustafsson, M. K. S.** (1995) The flatworm nervous system, in *The Nervous Systems of Invertebrates: an Evolutionary and Comparative Approach*, (eds O. Breidbach and W. Kutsch), Birkhäuser Verlag, Basel, pp. 25–59.

Richards, G. R. and **Chubb, J. C.** (1996) Host response to initial and challenge infections, following treatment, of *Gyrodactylus bullatarudis* and *G. turnbulli* (Monogenea) on the guppy (*Poecilia reticulata*). *Parasitology Research*, **82**, 242–247.

Richardson, L. R. (1968) A new bdellourid-like triclad turbellarian ectoconsortic on Murray River Chelonia. *Proceedings of the Linnean Society of New South Wales*, **93**, 90–97.

Rickard, M. D. and **Howell, M. J.** (1982) Comparative aspects of immunity in fascioliasis and cysticercosis in domesticated animals, in *Biology and Control of Endoparasites* (eds L. E. A. Symons, A. D. Donald and J. K. Dineen), Academic Press, Sydney, NSW, pp. 343–373.

Riegel, J. A. (1972) *Comparative Physiology of Renal Excretion*, Oliver & Boyd, Edinburgh.

Rieger, R. M., Tyler, S., Smith, J. P. S. III and **Rieger, G. E.** (1991) Platyhelminthes: Turbellaria, in *Microscopic Anatomy of Invertebrates, vol. 3, Platyhelminthes and Nemertinea*, (eds F. W. Harrison and B. J. Bogitsh), Wiley-Liss, New York, pp. 7–140.

Riley, M. W. and **Uglem, G. L.** (1995) *Proterometra macrostoma* (Digenea: Azygiidae): variations in cercarial morphology and physiology. *Parasitology*, **110**, 429–436.

Robert, F. and **Gabrion, C.** (1991) Cestodoses de l'avifaune Camarguaise. Rôle d'*Artemia* (Crustacea, Anostraca) et stratégies de rencontre hôte–parasite. *Annales de Parasitologie Humaine et Comparée*, **66**, 226–235.

Robert, F., Renaud, F., Mathieu, E. and **Gabrion, C.** (1988) Importance of the paratenic host in the biology of *Bothriocephalus gregarius* (Cestoda, Pseudophyllidea), a parasite of the turbot. *International Journal for Parasitology*, **18**, 611–621.

Roberts, L. S. (1980) Development of *Hymenolepis diminuta* in its definitive host, in *Biology of the Tapeworm* Hymenolepis diminuta, (ed. H. P. Arai), Academic Press, London, pp. 357–423.

Roberts, L. S. (1983) Carbohydrate metabolism, in *Biology of the Eucestoda, vol. 2*, (eds C. Arme and P. W. Pappas), Academic Press, London, pp. 343–390.

Roberts, L. S. and **Janovy, J. Jr** (1996) *Foundations of Parasitology*, 5th edn, Wm C. Brown, Dubuque, IO.

Robinson, G. and **Threadgold, L. T.** (1975) Electron microscope studies of *Fasciola hepatica*. XII. The fine structure of the gastrodermis. *Experimental Parasitology*, **37**, 20–36.

Robson, R. T. and **Erasmus, D. A.** (1970) The ultrastructure, based on stereoscan observations, of the oral sucker of the cercaria of *Schistosoma mansoni* with special reference to penetration. *Zeitschrift für Parasitenkunde*, **35**, 76–86.

Rodgers, L. O. and **Kuntz, R. E.** (1940) A new polystomatid monogenean fluke from a spadefoot. *Wasmann Collector*, **4**, 37–40.

Rogers, S. H. and **Bueding, E.** (1975) Anatomical localization of glucose uptake by *Schistosoma mansoni* adults. *International Journal for Parasitology*, **5**, 369–371.

Rohde, K. (1966) Die Funktion der Randkörper der Aspidobothria (Trematoda). *Naturwissenschaften*, **53**, 587–588.

Rohde, K. (1968) The nervous systems of *Multicotyle purvisi* Dawes, 1941 (Aspidogastrea) and *Diaschistorchis multitesticularis* Rohde, 1962 (Digenea). Implications for the ecology of the parasites. *Zeitschrift für Parasitenkunde*, **30**, 78–94.

Rohde, K. (1971) Untersuchungen an *Multicotyle purvisi* Dawes, 1941 (Trematoda: Aspidogastrea). V. Licht- und elektronenmikroskopischer Bau der Randkörper. *Zoologische Jahrbücher, Anatomie*, **88**, 387–398.

Rohde, K. (1972) The Aspidogastrea, especially *Multicotyle purvisi* Dawes, 1941. *Advances in Parasitology*, **10**, 77–151.

Rohde, K. (1973) Structure and development of *Lobatostoma manteri* sp. nov. (Trematoda: Aspidogastrea) from the Great Barrier Reef, Australia. *Parasitology*, **66**, 63–83.

Rohde, K. (1975) Early development and pathogenesis of *Lobatostoma manteri* Rohde (Trematoda: Aspidogastrea). *International Journal for Parasitology*, **5**, 597–607.

Rohde, K. (1976) Monogenean gill parasites of *Scomberomorus commersoni* Lacépède and other mackerel on the Australian east coast. *Zeitschrift für Parasitenkunde*, **51**, 49–69.

Rohde, K. (1989) At least eight types of sense receptors in an endoparasitic flatworm: a counter-trend to sacculinization. *Naturwissenschaften*, **76**, 383–385.

Rohde, K. (1994a) The origins of parasitism in the Platyhelminthes. *International Journal for Parasitology*, **24**, 1099–1115.

Rohde, K. (1994b) The minor groups of parasitic Platyhelminthes. *Advances in Parasitology*, **33**, 145–234.

Rohde, K. and **Georgi, M.** (1983) Structure and development of *Austramphilina elongata* Johnston, 1931 (Cestodaria: Amphilinidea). *International Journal for Parasitology*, **13**, 273–287.

Rohde, K., Hefford, C., Ellis, J. T. *et al.* (1993) Contributions to the phylogeny of Platyhelminthes based on partial sequencing of 18S ribosomal DNA. *International Journal for Parasitology*, **23**, 705–724.

Rohde, K., Johnson, A. M., Baverstock, P. R. and **Watson, N. A.** (1995) Aspects of the phylogeny of Platyhelminthes based on 18S ribosomal DNA and protonephridial ultrastructure. *Hydrobiologia*, **305**, 27–35.

Rohde, K., Luton, K., Baverstock, P. R. and **Johnson, A. M.** (1994) The phylogenetic relationships of *Kronborgia* (Platyhelminthes, Fecampiida) based on comparison of 18S ribosomal DNA sequences. *International Journal for Parasitology*, **24**, 657–669.

Romig, T., Lucius, R. and **Frank, W.** (1980) Cerebral larvae in the second intermediate host of *Dicrocoelium dendriticum* (Rudolphi, 1819) and *Dicrocoelium hospes* Looss, 1907 (Trematodes, Dicrocoeliidae). *Zeitschrift für Parasitenkunde*, **63**, 277–286.

Rondelaud, D. and **Barthe, D.** (1978) Arguments et propositions pour une nouvelle interprétation de l'évolution de *Fasciola hepatica* L. dans *Lymnaea (Galba) truncatula* Müller. *Annales de Parasitologie*, **53**, 201–213.

Rondelaud, D. and **Barthe, D.** (1981) Les générations rédiennes de *Fasciola hepatica* L. chez *Lymnaea truncatula* Müller, à propos des effets de plusieurs facteurs. *Annales de Parasitologie* **57**, 245–262.

Rondelaud, D. and **Barthe, D.** (1982) Les générations rédiennes de *Fasciola hepatica* L. chez *Lymnaea truncatula* Müller. Pluralité des schémas de développement. *Annales de Parasitologie*, **57**, 639–642.

Rosenqvist, G. and **Johansson, K.** (1995) Male avoidance of parasitized females explained by direct benefits in a pipefish. *Animal Behaviour*, **49**, 1039–1045.

Rothschild, M. (1939) A note on the life cycle of *Cryptocotyle lingua* (Creplin) 1825 (Trematoda). *Novitates Zoologicae*, **41**, 178–180.

Roubal, F. R. (1994a) Attachment of eggs by *Lamellodiscus acanthopagri* (Monogenea: Diplectanidae) to the gills of *Acanthopagrus australis* (Pisces: Sparidae), with evidence for autoinfection and postsettlement migration. *Canadian Journal of Zoology*, **72**, 87–95.

Roubal, F. R. (1994b) Observations on the eggs and fecundity of dactylogyrid and diplectanid monogeneans from the Australian marine sparid fish, *Acanthopagrus australis*. *Folia Parasitologica*, **41**, 220–222.

Rowan, W. B. (1955) The life cycle and epizootiology of the rabbit trematode, *Hasstilesia tricolor* (Stiles and Hassall, 1894) Hall, 1916 (Trematoda: Brachylaemidae). *Transactions of the American Microscopical Society,* **74**, 1–21.

Ruff, M. D. and **Read, C. P.** (1973) Inhibition of pancreatic lipase by *Hymenolepis diminuta. Journal of Parasitology,* **59**, 105–111.

Rybicka, K. (1966) Embryogenesis in cestodes. *Advances in Parasitology,* **4**, 107–186.

Rybicka, K. (1972) Ultrastructure of embryonic envelopes and their differentiation in *Hymenolepis diminuta* (Cestoda). *Journal of Parasitology,* **58**, 849–863.

Sakanari, J. and **Moser, M.** (1985a) Salinity and temperature effects on the eggs, coracidia, and procercoids of *Lacistorhynchus tenuis* (Cestoda: Trypanorhyncha) and induced mortality in a first intermediate host. *Journal of Parasitology,* **71**, 583–587.

Sakanari, J. and **Moser, M.** (1985b) Infectivity of, and laboratory infection with, an elasmobranch cestode, *Lacistorhynchus tenuis* (van Beneden, 1858). *Journal of Parasitology,* **71**, 788–791.

Sakanari, J. A. and **Moser, M.** (1989) Complete life cycle of the elasmobranch cestode, *Lacistorhynchus dollfusi* Beveridge and Sakanari, 1987 (Trypanorhyncha). *Journal of Parasitology,* **75**, 806–808.

Salami-Cadoux, M.-L. (1975) Transmission et développement du monogène réno-vésical *Eupolystoma alluaudi* (de Beauchamp, 1913) Euzet et Combes, 1967. *Bulletin de la Société Zoologique de France,* **100**, 283–292.

Sambon, L. W. (1909) What is '*Schistosoma mansoni*' Sambon, 1907? *Journal of Tropical Medicine and Hygiene,* **12**, 1–11.

Samnaliev, P., Kanev, I, and **Vassilev, I.** (1978) Interactions between larval and parthenite stages of some trematodes in one and the same intermediate host. *Proceedings of the Fourth International Congress of Parasitology,* Warsaw, pp. 45–46.

Sawma, J. T., Isseroff, H. and **Reino, D.** (1978) Proline in fascioliasis – IV. Induction of bile duct hyperplasia. *Comparative Biochemistry and Physiology,* **61A**, 239–243.

Schacher, J. F., Khalil, G. M. and **Salman, S.** (1965) A field study of *Halzoun* (parasitic pharyngitis) in Lebanon. *Journal of Tropical Medicine and Hygiene,* **68**, 226–230.

Schad, G. A. (1966) Immunity, competition, and natural regulation of helminth populations. *American Naturalist,* **100**, 359–364.

Schantz, P. M., Chai, J., Craig, P. S. *et al.* (1995) Epidemiology and control of hydatid disease, in *Echinococcus and Hydatid Disease*, (eds R. C. A. Thompson and A. J. Lymbery), CAB International, Wallingford, UK, pp. 233–331.

Schantz, P. M., Gottstein, B., Ammann, R. and **Lanier, A.** (1991) Hydatid and the Arctic. *Parasitology Today,* **7**, 35–36.

Schell, S. C. (1973) *Rugogaster hydrolagi* gen. et sp. n. (Trematoda: Aspidobothrea: Rugogastridae fam. n.) from the ratfish, *Hydrolagus colliei* (Lay and Bennett, 1839). *Journal of Parasitology,* **59**, 803–805.

Schell, S. C. (1974) The life history of *Sanguinicola idahoensis* sp. n. (Trematoda: Sanguinicolidae), a blood parasite of steelhead trout, *Salmo gairdneri* Richardson. *Journal of Parasitology,* **60**, 561–566.

Schmidt, G. D. (1969) *Dioecotaenia cancellata* (Linton, 1890) gen. et comb. n., a dioecious cestode (Tetraphyllidea) from the cow-nosed ray, *Rhinoptera bonasus* (Mitchell), in Chesapeake Bay, with the proposal of a new family, Dioecotaeniidae. *Journal of Parasitology,* **55**, 271–275.

Schmidt, G. D. (1986) *CRC Handbook of Tapeworm Identification*, CRC Press, Boca Raton, FL.

Schmidt, G. D. and **Roberts, L. S.** (1989) *Foundations of Parasitology*, 4th edn, Times Mirror–Mosby College Publishing , St. Louis, MO.

Schmidt, J. (1995) Glycans with *N*-acetyllactosamine type 2-like residues covering adult *Schistosoma mansoni*, and glycomimesis as a putative mechanism of immune evasion. *Parasitology,* **111**, 325–336.

Schmidt, P. J. (1942) A fish with feather-like gill-leaflets. *Copeia,* **1942, No. 2**, 98–100.

Schneider, G. and **Hohorst, W.** (1971) Wanderung der Metacercarien des Lanzett-Egels in Ameisen. *Naturwissenschaften,* **58**, 327–328.

Schroeder, L. L., Pappas, P. W. and **Means, G. E.** (1981) Trypsin inactivation by intact *Hymenolepis diminuta* (Cestoda): some characteristics of the inactivated enzyme. *Journal of Parasitology,* **67**, 378–385.

Schwabe, C. W. and **Daoud, K. A.** (1961) Epidemiology of *Echinococcus* in the Middle East. I. Human infection in Lebanon, 1949 to 1959. *American Journal of Tropical Medicine and Hygiene,* **10**, 374–381.

Scott, M. E. (1985a) Experimental epidemiology of *Gyrodactylus bullatarudis* (Monogenea) on guppies (*Poecilia reticulata*): short- and long-term studies, in *Ecology and Genetics of Host-Parasite Interactions*, (eds D. Rollinson and R. M. Anderson), Academic Press, London, pp. 21–38.

Scott, M. E. (1985b) Dynamics of challenge infections of *Gyrodactylus bullatarudis* Turnbull (Monogenea) on guppies, *Poecilia reticulata* (Peters). *Journal of Fish Diseases*, **8**, 495–503.

Scott, M. E. and **Anderson, R. M.** (1984) The population dynamics of *Gyrodactylus bullatarudis* (Monogenea) within laboratory populations of the fish host *Poecilia reticulata*. *Parasitology*, **89**, 159–194.

Self, J. T., Peters, L. E. and **Davis, C. E.** (1963) The egg, miracidium, and adult of *Nematobothrium texomensis* (Trematoda: Digenea). *Journal of Parasitology*, **49**, 731–736.

Sewell, K. B. and **Whittington, I. D.** (1995) A light microscope study of the attachment organs and their role in locomotion of *Craspedella* sp. (Platyhelminthes: Rhabdocoela: Temnocephalidae), an ectosymbiont from the branchial chamber of *Cherax quadricarinatus* (Crustacea: Parastacidae) in Queensland, Australia. *Journal of Natural History*, **29**, 1121–1141.

Sharma, P. N. and **Hanna, R. E. B.** (1988) Ultrastructure and cytochemistry of the tegument of *Orthocoelium scoliocoelium* and *Paramphistomum cervi* (Trematoda: Digenea). *Journal of Helminthology*, **62**, 331–343.

Shaw, C., Maule, A. G. and **Halton, D. W.** (1996) Platyhelminth FMRFamide-related peptides. *International Journal for Parasitology*, **26**, 335–345.

Shiff, C. J. and **Graczyk, T. K.** (1994) A chemokinetic response in *Schistosoma mansoni* cercariae. *Journal of Parasitology*, **80**, 879–883.

Shinn, G. L. (1981) The diet of three species of umagillid neorhabdocoel turbellarians inhabiting the intestine of echinoids. *Hydrobiologia*, **84**, 155–162.

Shinn, G. L. (1983a) The life history of *Syndisyrinx franciscanus*, a symbiotic turbellarian from the intestine of echinoids, with observations on the mechanism of hatching. *Ophelia*, **22**, 57–79.

Shinn, G. L. (1983b) *Anoplodium hymanae* sp. n., an umagillid turbellarian from the coelom of *Stichopus californicus*, a Northeast Pacific holothurian. *Canadian Journal of Zoology*, **61**, 750–760.

Shinn, G. L. (1985a) Reproduction of *Anoplodium hymanae*, a turbellarian flatworm (Neorhabdocoela, Umagillidae) inhabiting the coelom of sea cucumbers; production of egg capsules, and escape of infective stages *without* evisceration of the host. *Biological Bulletin*, **169**, 182–198.

Shinn, G. L. (1985b) Infection of new hosts by *Anoplodium hymanae*, a turbellarian flatworm (Neorhabdocoela, Umagillidae) inhabiting the coelom of the sea cucumber *Stichopus californicus*. *Biological Bulletin*, **169**, 199–214.

Shinn, G. L. (1986a) Life history and function of the secondary uterus of *Wahlia pulchella*, an umagillid turbellarian from the intestine of a northeastern Pacific sea cucumber (*Stichopus californicus*). *Ophelia*, 59–74.

Shinn, G. L. (1986b) Spontaneous hatching of *Fallacohospes inchoatus*, an umagillid flatworm from the northeastern Pacific crinoid *Florometra serratissima*. *Canadian Journal of Zoology*, **64**, 2068–2071.

Shinn, G. L. (1993) Formation of egg capsules by flatworms (Phylum Platyhelminthes). *Transactions of the American Microscopical Society*, **112**, 18–34.

Shinn, G. L. and **Christensen, A. M.** (1985) *Kronborgia pugettensis* sp. nov. (Neorhabdocoela: Fecampiidae), an endoparasitic turbellarian infesting the shrimp *Heptacarpus kincaidi* (Rathbun), with notes on its life-history. *Parasitology*, **91**, 431–447.

Shinn, G. L. and **Cloney, R. A.** (1986) Egg capsules of a parasitic turbellarian flatworm: ultrastructure of hatching sutures. *Journal of Morphology*, **188**, 15–28.

Shoop, W. L. (1994) Vertical transmission in the Trematoda. *Journal of the Helminthological Society of Washington*, **61**, 153–161.

Shoop, W. L. and **Corkum, K. C.** (1981) Epidemiology of *Alaria marcianae* mesocercariae in Louisiana. *Journal of Parasitology*, **67**, 928–931.

Shoop, W. L. and **Corkum, K. C.** (1983a) Migration of *Alaria marcianae* (Trematoda) in domestic cats. *Journal of Parasitology*, **69**, 912–917.

Shoop, W. L. and **Corkum, K. C.** (1983b) Transmammary infection of paratenic and definitive hosts with *Alaria marcianae* (Trematoda) mesocercariae. *Journal of Parasitology*, **69**, 731–735.

Short, R. B. (1953) A new blood fluke, *Cardicola laruei* n. g., n. sp., (Aporocotylidae) from marine fishes. *Journal of Parasitology*, **39**, 304–309.

Siebold, C. T. von (1857) *On Tape and Cystic Worms, with an Introduction on the Origin of Intestinal Worms*, (trans. T. H. Huxley), Sydenham Society, London.

Sillman, E. I. (1962) The life history of *Azygia longa* (Leidy 1851) (Trematoda: Digenea), and notes on *A. acuminata* Goldberger 1911. *Transactions of the American Microscopical Society*, **81**, 43–65.

Silverman, P. H. (1954a) Studies on the biology of some tapeworms of the genus *Taenia*. II. – The morphology and development of the taeniid hexacanth embryo and its enclosing membranes, with some notes on the state of development and propagation of gravid segments. *Annals of Tropical Medicine and Parasitology*, **48**, 356–366.

Silverman, P. H. (1954b) Studies on the biology of some tapeworms of the genus *Taenia*. I. – Factors affecting hatching and activation of taeniid ova, and some criteria of their viability. *Annals of Tropical Medicine and Parasitology*, **48**, 207–215.

Silverman, P. H. and **Griffiths, R. B.** (1955) A review of methods of sewage disposal in Great Britain, with special reference to the epizootiology of *Cysticercus bovis*. *Annals of Tropical Medicine and Parasitology*, **49**, 436–450.

Simmons, J. E. (1974) *Gyrocotyle*: a century-old enigma, in *Symbiosis in the Sea*, (ed. W. B. Vernberg), University of South Carolina Press, pp. 195–218.

Simpkin, K. G., Chapman, C. R. and **Coles, G. C.** (1980) *Fasciola hepatica*: a proteolytic digestive enzyme. *Experimental Parasitology*, **49**, 281–287.

Sinitsin, D. F. (1931) A glimpse into the life history of the tapeworm of sheep, *Moniezia expansa. Journal of Parasitology*, **17**, 223–227.

Skelly, P. J. and **Shoemaker, C. B.** (1996) Rapid appearance and asymmetric distribution of glucose transporter SGTP4 at the apical surface of intramammalian-stage *Schistosoma mansoni. Proceedings of the National Academy of Sciences of the USA*, **93**, 3642–3646.

Sluys, R. (1990) On *Bdellasimilis barwicki* (Platyhelminthes: Tricladida) and its phyletic position. *Invertebrate Taxonomy*, **4**, 149–158.

Smith, A. M., Dowd, A. J., Heffernan, M. *et al.* (1993) *Fasciola hepatica*: a secreted cathepsin L-like proteinase cleaves host immunoglobulin. *International Journal for Parasitology*, **23**, 977–983.

Smith, D. C. and **Douglas, A. E.** (1987) *The Biology of Symbiosis*, Edward Arnold, London.

Smith, J. E., Carthy, J. D., Chapman, G. *et al.* (1971) *The Invertebrate Panorama*, Weidenfeld & Nicolson, London.

Smith, J. W. (1972) The blood flukes of cold-blooded vertebrates and some comparison with the schistosomes. *Helminthological Abstracts*, **41**, 161–204.

Smith, S. W. G. and **Chappell, L. H.** (1990) Single sex schistosomes and chemical messengers. *Parasitology Today*, **6**, 297–298.

Smithers, S. R. and **Terry, R. J.** (1967) Resistance to experimental infection with *Schistosoma mansoni* in Rhesus monkeys induced by the transfer of adult worms. *Transactions of the Royal Society of Tropical Medicine and Hygiene*, **61**, 517–533.

Smithers, S. R., Terry, R. J. and **Hockley, D. J.** (1969) Host antigens in schistosomiasis. *Proceedings of the Royal Society B*, **171**, 483–494.

Smyth, J. D. (1954) A technique for the histochemical demonstration of polyphenol oxidase and its application to egg-shell formation in helminths and byssus formation in *Mytilus*. *Quarterly Journal of Microscopical Science*, **95**, 139–152.

Smyth, J. D. (1962) Some aspects of the evolution of the host–parasite relationship, in *The Evolution of Living Organisms*, (ed. G. W. Leeper), University Press, Melbourne, pp. 136–148.

Smyth, J. D. (1964) The biology of the hydatid organisms. *Advances in Parasitology*, **2**, 169–219.

Smyth, J. D. (1969) *The Physiology of Cestodes*, Oliver & Boyd, Edinburgh.

Smyth, J. D. and **Halton, D. W.** (1983) *The Physiology of Trematodes*, 2nd edn, Cambridge University Press, Cambridge.

Snelson, F. F. Jr, Gruber, S. H., Murru, F. L. and **Schmid, T. H.** (1990) Southern stingray, *Dasyatis americana*: host for a symbiotic cleaner wrasse. *Copeia*, **1990**, 961–965.

Sommer, F. (1880) Die Anatomie des Leberegels *Distomum hepaticum* L. *Zeitschrift für wissenschaftliche Zoologie*, **34**, 539–640.

Sousa, W. P. (1992) Interspecific interactions among larval trematode parasites of freshwater and marine snails. *American Zoologist*, **32**, 583–592.

Sousa, W. P. (1993) Interspecific antagonism and species coexistence in a diverse guild of larval trematode parasites. *Ecological Monographs*, **63**, 103–128.

Southgate, V. R. (1971) Observations on the fine structure of the cercaria of *Notocotylus attenuatus* and formation of the cyst wall of the metacercaria. *Zeitschrift für Zellforschung,* **120**, 420–449.

Southgate, V. R. and **Agrawal, M. C.** (1990) Human schistosomiasis in India? *Parasitology Today,* **6**, 166–168.

Specht, D. and **Voge, M.** (1965) Asexual multiplication of *Mesocestoides* tetrathyridia in laboratory animals. *Journal of Parasitology,* **51**, 268–272.

Spindler, E.-M., Zahler, M. and **Loos-Frank, B.** (1986) Behavioural aspects of ants as intermediate hosts of *Dicrocoelium dendriticum. Zeitschrift für Parasitenkunde,* **72**, 689–692.

Sproston, N. G. (1946) A synopsis of the monogenetic trematodes. *Transactions of the Zoological Society of London,* **25**, 185–600.

Stabrowski, A. and **Nollen, P. M.** (1985) The responses of *Philophthalmus gralli* and *P. megalurus* miracidia to light, gravity and magnetic fields. *International Journal for Parasitology,* **15**, 551–555.

Starr, M. P. (1975) A generalized scheme for classifying organismic associations, in *Symbiosis, Symposia of the Society for Experimental Biology,* no. 29, (eds D. H. Jennings and D. L. Lee), Cambridge University Press, Cambridge, pp. 1–20.

Steenstrup, J. J. S. (1845) *On the Alternation of Generations; or, the Propagation and Development of Animals through Alternate Generations: a Peculiar Form of Fostering the Young in the Lower Classes of Animals,* translated from German by G. Busk, Ray Society, London.

Stehr-Green, J. K., Stehr-Green, P. A., Schantz, P. M. *et al.* (1988) Risk factors for infection with *Echinococcus multilocularis* in Alaska. *American Journal of Tropical Medicine and Hygiene,* **38**, 380–385.

Stiles, C. W. (1906) *Illustrated Key to the Cestode Parasites of Man, Hygienic Laboratory – Bulletin No. 25,* United States Treasury Dept, Washington, DC.

Stirewalt, M. A. (1974) *Schistosoma mansoni*: cercaria to schistosomule. *Advances in Parasitology,* **12**, 115–182.

Stirewalt, M. A. and **Dorsey, C. H.** (1974) *Schistosoma mansoni*: cercarial penetration of host epidermis at the ultrastructural level. *Experimental Parasitology,* **35**, 1–15.

Stirewalt, M. A. and **Hackey, J. R.** (1956) Penetration of host skin by cercariae of *Schistosoma mansoni*. I. Observed entry into skin of mouse, hamster, rat, monkey and man. *Journal of Parasitology,* **42**, 565–580.

Stunkard, H. W. (1924) A new trematode, *Oculotrema hippopotami* n. g., n. sp., from the eye of the hippopotamus. *Parasitology,* **16**, 436–441.

Stunkard, H. W. (1930) The life history of *Cryptocotyle lingua* (Creplin), with notes on the physiology of the metacercariae. *Journal of Morphology and Physiology,* **50**, 143–190.

Stunkard, H. W. (1938) The development of *Moniezia expansa* in the intermediate host. *Parasitology,* **30**, 491–501.

Stunkard, H. W. (1956) The morphology and life-history of the digenetic trematode *Azygia sebago* Ward, 1910. *Biological Bulletin,* **111**, 248–268.

Stunkard, H. W. (1958) The morphology and life-history of *Levinseniella minuta* (Trematoda: Microphallidae). *Journal of Parasitology,* **44**, 225–230.

Stunkard, H. W. (1976) The life cycles, intermediate hosts, and larval stages of *Rhipidocotyle transversale* Chandler, 1935 and *Rhipidocotyle lintoni* Hopkins, 1954: life-cycles and systematics of bucephalid trematodes. *Biological Bulletin,* **150**, 294–317.

Stunkard, H. W. (1978) The life-cycle and taxonomic relations of *Lintonium vibex* (Linton, 1900) Stunkard and Nigrelli, 1930 (Trematoda: Fellodistomidae). *Biological Bulletin,* **155**, 383–394.

Stunkard, H. W. (1980) Successive hosts and developmental stages in the life history of *Neopechona cablei* sp. n. (Trematoda: Lepocreadiidae). *Journal of Parasitology,* **66**, 636–641.

Sukhdeo, M. V. K. (1992) Intestinal contractions and migration behaviour in *Hymenolepis diminuta. International Journal for Parasitology,* **22**, 813–817.

Sukhdeo, M. V. K., Hsu, S. C., Thompson, C. S. and **Mettrick, D. F.** (1984) *Hymenolepis diminuta*: behavioral effects of 5-hydroxytryptamine, acetylcholine, histamine and somatostatin. *Journal of Parasitology,* **70**, 682–688.

Sukhdeo, M. V. K. and **Kerr, M. S.** (1992) Behavioural adaptation of the tapeworm *Hymenolepis diminuta* to its environment. *Parasitology,* **104**, 331–336.

Sukhdeo, M. V. K. and **Mettrick, D. F.** (1986) The behavior of juvenile *Fasciola hepatica. Journal of Parasitology,* **72**, 492–497.

Sukhdeo, M. V. K. and **Mettrick, D. F.** (1987) Parasite behaviour: understanding platyhelminth responses. *Advances in Parasitology*, **26**, 73–144.

Sukhdeo, M. V. K., Sangster, N. C. and **Mettrick, D. F.** (1988) Permanent feeding sites of adult *Fasciola hepatica* in rabbits? *International Journal for Parasitology*, **18**, 509–512.

Sukhdeo, M. V. K. and **Sukhdeo, S. C.** (1989) Gastrointestinal hormones: environmental cues for *Fasciola hepatica*? *Parasitology*, **98**, 239–243.

Sukhdeo, M. V. K., Sukhdeo, S. C. and **Mettrick, D. F.** (1987) Site-finding behaviour of *Fasciola hepatica* (Trematoda), a parasitic flatworm. *Behaviour*, **103**, 174–186.

Sweeting, R. A. (1974) Investigations into natural and experimental infections of freshwater fish by the common eye-fluke *Diplostomum spathaceum* Rud. *Parasitology*, **69**, 291–300.

Swiderski, Z. (1994a) Origin, differentiation and ultrastructure of egg envelopes surrounding the miracidia of *Schistosoma mansoni*. *Acta Parasitologica*, **39**, 64–72.

Swiderski, Z. (1994b) Homology and analogy in egg envelopes surrounding miracidia of *Schistosoma mansoni* and coracidia of *Bothriocephalus clavibothrium*. *Acta Parasitologica*, **39**, 123–130.

Symonds, D. J. (1972) Infestation of *Nephrops norvegicus* (L.) by *Stichocotyle nephropis* Cunningham in British waters. *Journal of Natural History*, **6**, 423–426.

Szidat, L. (1932) Über Cysticerke Riesencercarien, Insbesondere *Cercaria mirabilis* M. Braun und *Cercaria splendens* n. sp., und ihre Entwicklung im Magen von Raubfischen zu Trematoden der Gattung *Azygia* Looss. *Zeitschrift für Parasitenkunde*, **4**, 477–505.

Tappenden, T. (1989) Studies on the reproductive system of monogeneans. PhD thesis, University of East Anglia, UK.

Tappenden, T., Kearn, G. C. and **Evans-Gowing, R.** (1993) Fertilization and the functional anatomy of the germarium in the monogenean *Entobdella soleae*. *International Journal for Parasitology*, **23**, 901–911.

Terry, R. J. (1994) Human immunity to schistosomes: concomitant immunity? *Parasitology Today*, **10**, 377–378.

Theron, A. (1976) Le cycle biologique de *Plagiorchis neomidis* Brendow, 1970, digène parasite de *Neomys fodiens* dans les Pyrénées. Chronobiologie de l'émission cercarienne. *Annales de Parasitologie*, **51**, 329–340.

Thomas, A. P. (1883) The life history of the liver-fluke (*Fasciola hepatica*). *Quarterly Journal of Microscopical Science*, **23**, 99–133.

Thomas, J. S. and **Pascoe, D.** (1973) The digestion of exogenous carbohydrate by the daughter sporocysts of *Cercaria linearis* Lespés, 1857 and *Cercaria stunkardi* Palombi, 1938, *in vitro*. *Zeitschrift für Parasitenkunde*, **43**, 17–23.

Thompson, R. C. A. (1979) Biology and speciation of *Echinococcus granulosus*. *Australian Veterinary Journal*, **55**, 93–98.

Thompson, R. C. A. (1995) Biology and systematics of *Echinococcus*, in Echinococcus *and Hydatid Disease*, (eds R. C. A. Thompson and A. J. Lymbery), CAB International, Wallingford, UK, pp. 1–50.

Thoney, D. A. and **Burreson, E. M.** (1987) Morphology and development of the adult and cotylocidium of *Multicalyx cristata* (Aspidocotylea), a gall bladder parasite of elasmobranchs. *Proceedings of the Helminthological Society of Washington*, **54**, 96–104.

Thornhill, J. A., Jones, T. and **Kusel, K. R.** (1986) Increased oviposition and growth in immature *Biomphalaria glabrata* after exposure to *Schistosoma mansoni*. *Parasitology*, **93**, 443–450.

Threadgold, L. T. (1963) The tegument and associated structures of *Fasciola hepatica*. *Quarterly Journal of Microscopical Science*, **104**, 505–512.

Threadgold, L. T. (1967) Electron microscope studies of *Fasciola hepatica*. III. Further observations on the tegument and associated structures. *Parasitology*, **57**, 633–637.

Threadgold, L. T. (1978) *Fasciola hepatica*: a transmission and scanning electron microscopical study of the apical surface of the gastrodermal cells. *Parasitology*, **76**, 85–90.

Threadgold, L. T. and **Hopkins, C. A.** (1981) *Schistocephalus solidus* and *Ligula intestinalis*: pinocytosis by the tegument. *Experimental Parasitology*, **51**, 444–456.

Thulin, J. (1982) The morphology of the miracidium of *Chimaerohemecus trondheimensis* Van der Land, 1967 (Digenea: Sanguinicolidae). *Parasitology*, **85**, ix–x.

Thurston, J. P. (1968) The frequency distribution of *Oculotrema hippopotami* (Monogenea: Polystomatidae) on *Hippopotamus amphibius*. *Journal of Zoology (London)*, **154**, 481–485.

Thurston, J. P. (1970) Haemoglobin in a monogenean, *Oculotrema hippopotami*. *Nature*, **228**, 578–579.

Thurston, J. P. and **Laws, R. M.** (1965) *Oculotrema hippopotami* (Trematoda: Monogenea) in Uganda. *Nature*, **205**, 1127.

Tielens, A. G. M. (1994) Energy generation in parasitic helminths. *Parasitology Today*, **10**, 346–352.

Tierney, J. F., Huntingford, F. A. and **Crompton, D. W. T.** (1993) The relationship between infectivity of *Schistocephalus solidus* (Cestoda) and anti-predator behaviour of its intermediate host, the three-spined stickleback, *Gasterosteus aculeatus*. *Animal Behaviour*, **46**, 603–605.

Tieszen, J. E., Johnson, A. D. and **Dickinson, J. P.** (1974) Structure and function of the holdfast organ and lappets of *Alaria mustelae* Bosma, 1931, with further studies on esterases of *A. marcianae* (La Rue, 1917) (Trematoda: Diplostomatidae). *Journal of Parasitology*, **60**, 567–573.

Timofeeva, T. A. (1977) [The distribution of *Udonella caligorum* Johnston in the waters of eastern Murman], in *Biology of the Northern Waters of the European Part of the USSR*, Apatity, Publishing House of the Kol'skiy Branch of the Academy of Sciences of the USSR, pp. 54–59 (in Russian).

Tinsley, R. C. (1978) Oviposition, hatching and the oncomiracidium of *Eupolystoma anterorchis* (Monogenoidea). *Parasitology*, **77**, 121–132.

Tinsley, R. C. (1981) The evidence from parasite relationships for the evolutionary status of *Xenopus* (Anura Pipidae). *Monitore zoologico italiano, N. S. Suppl.*, **15**, 367–385.

Tinsley, R. C. (1983) Ovoviviparity in platyhelminth life-cycles. *Parasitology*, **86**, 161–196.

Tinsley, R. C. (1989) The effects of host sex on transmission success. *Parasitology Today*, **5**, 190–195.

Tinsley, R. C. (1990a) The influence of parasite infection on mating success in spadefoot toads, *Scaphiopus couchii*. *American Zoologist*, **30**, 313–324.

Tinsley, R. C. (1990b) Host behaviour and opportunism in parasite life cycles, in *Parasitism and Host Behaviour*, (eds C. J. Barnard and J. M. Behnke), Taylor & Francis, London, pp. 158–192.

Tinsley, R. C. (1991) The control of egg production in the monogenean, *Polystoma*. *Abstracts of the Spring Meeting of the British Society for Parasitology*. April 1991, p. 46.

Tinsley, R. C. and **Earle, C. M.** (1983) Invasion of vertebrate lungs by the polystomatid monogeneans *Pseudodiplorchis americanus* and *Neodiplorchis scaphiopodis*. *Parasitology*, **86**, 501–517.

Tinsley, R. C. and **Jackson, H. C.** (1986) Intestinal migration in the life-cycle of *Pseudodiplorchis americanus* (Monogenea). *Parasitology*, **93**, 451–469.

Tinsley, R. C. and **Jackson, H. C.** (1988) Pulsed transmission of *Pseudodiplorchis americanus* (Monogenea) between desert hosts (*Scaphiopus couchii*). *Parasitology*, **97**, 437–452.

Tinsley, R. C. and **Owen, R. W.** (1975) Studies on the biology of *Protopolystoma xenopodis* (Monogenoidea): the oncomiracidium and life-cycle. *Parasitology*, **71**, 445–463.

Tinsley, R. C. and **Sweeting, R. A.** (1974) Studies on the biology and taxonomy of *Diplostomulum* (*Tylodelphylus*) *xenopodis* from the African clawed toad, *Xenopus laevis*. *Journal of Helminthology*, **48**, 247–263.

Tirard, C., Berrebi, P., Raibaut, A. and **Frenaud, F.** (1992) Parasites as biological markers: evolutionary relationships in the heterospecific combination of helminths (monogeneans) and teleosts (Gadidae). *Biological Journal of the Linnean Society*, **47**, 173–182.

Tocque, K. and **Tinsley, R. C.** (1991a) Asymmetric reproductive output by the monogenean *Pseudodiplorchis americanus*. *Parasitology*, **102**, 213–220.

Tocque, K. and **Tinsley, R. C.** (1991b) The influence of desert temperature cycles on the reproductive biology of *Pseudodiplorchis americanus* (Monogenea). *Parasitology*, **103**, 111–120.

Tocque, K. and **Tinsley, R. C.** (1992) Ingestion of host blood by the monogenean *Pseudodiplorchis americanus*: a quantitative analysis. *Parasitology*, **104**, 283–289.

Todd, J. R. and **Ross, J. G.** (1966) Origin of hemoglobin in the cecal contents of *Fasciola hepatica*. *Experimental Parasitology*, **19**, 151–154.

Torgerson, P. R., Pilkington, J., Gulland, F. M. D. and **Gemmell, M. A.** (1995) Further evidence for the long distance dispersal of taeniid eggs. *International Journal for Parasitology*, **25**, 265–267.

Trimble, J. J. III, Bailey, H. H. and **Nelson, E. N.** (1971) *Aspidogaster conchicola* (Trematoda: Aspidobothrea): histochemical localization of acid and alkaline phosphatases. *Experimental Parasitology*, **29**, 457–462.

Trimble, J. J. III, Bailey, H. H. and **Sheppard, A.** (1972) *Aspidogaster conchicola*: histochemical localization of carboxylic ester hydrolases. *Experimental Parasitology*, **32**, 181–190.

Tsang, V. C. W. and **Damian, R. T.** (1971) Demonstration and mode of action of an inhibitor for activated Hagman Factor (Factor XIIa) of the intrinsic blood coagulation pathway from *Schistosoma mansoni*. *Blood*, **49**, 619–633.

Tyson, E. (1683) *Lumbricus latus*, or a Discourse read before the Royal Society of the Joynted Worm, wherein a great many mistakes of former Writers concerning it, are remarked; its Natural History from more exact Observations is attempted; and the whole urged, as a Difficulty against the Doctrine of Univocal Generation. *Philosophical Transactions of the Royal Society*, **146**, 113–144.

Ubelaker, J. E. (1980) Structure and ultrastructure of the larvae and metacestodes of *Hymenolepis diminuta*, in *Biology of the Tapeworm* Hymenolepis diminuta, (ed. H. P. Arai), Academic Press, London, pp. 59–156.

Ubelaker, J. E. (1983) The morphology, development and evolution of tapeworm larvae, in *Biology of the Eucestoda*, vol. 1, (eds C. Arme and P. W. Pappas), Academic Press, London, pp. 235–296.

Ude, J. (1908) Beiträge zur Anatomie und Histologie der Süsswassertricladen. *Zeitschrift für wissenschaftliche Zoologie*, **89**, 308–370.

Uglem, G. L. (1980) Sugar transport by larval and adult *Proterometra macrostoma* (Digenea) in relation to environmental factors. *Journal of Parasitology*, **66**, 748–758.

Uglem, G. L. (1987) Environmental sodium regulates cutaneous sugar transport in a digenean fluke. *Parasitology*, **94**, 1–6.

Uglem, G. L. and **Aliff, J. V.** (1984) *Proterometra edneyi* n. sp. (Digenea: Azygiidae): behavior and distribution of acetylcholinesterase in cercariae. *Transactions of the American Microscopical Society*, **103**, 383–391.

Ulmer, M. J. (1951a) *Postharmostomum helicis* (Leidy, 1847) Robinson 1949, (Trematoda), its life history and a revision of the subfamily Brachylaeminae. Part I. *Transactions of the American Microscopical Society*, **70**, 189–238.

Ulmer, M. J. (1951b) *Postharmostomum helicis* (Leidy, 1847) Robinson 1949, (Trematoda), its life history and a revision of the subfamily Brachylaeminae. Part II. *Transactions of the American Microscopical Society*, **70**, 319–347.

Ulmer, M. J. (1957) Notes on the development of *Cotylurus flabelliformis* tetracotyles in the second intermediate host (Trematoda: Strigeidae). *Transactions of the American Microscopical Society*, **76**, 321–327.

Vanoverschelde, R. and **Vaes, F.** (1980) Studies on the life-cycle of *Himasthla militaris* (Trematoda: Echinostomatidae). *Parasitology*, **81**, 609–617.

Vaucher, C. (1971) Les cestodes parasites des Soricidae d'Europe. Etude anatomique, révision taxonomique et biologie. *Revue Suisse de Zoologie*, **78**, 1–113.

Viljoen, N. F. (1937) Cysticercosis in swine and bovines, with special reference to South African conditions. *Onderstepoort Journal of Veterinary Science and Animal Industry*, **9**, 337–570.

Vincent, J. F. V. and **Hillerton, J. E.** (1979) The tanning of insect cuticle – a critical review and a revised mechanism. *Journal of Insect Physiology*, **25**, 653–658.

Waite, J. H. (1990) The phylogeny and chemical diversity of quinone-tanned glues and varnishes. *Comparative Biochemistry and Physiology*, **97B**, 19–29.

Wakelin, D. (1984) *Immunity to Parasites*, Edward Arnold, London.

Walker, J. C. (1979) *Austrobilharzia terrigalensis*: a schistosome dominant in interspecific interactions in the mollusc host. *International Journal for Parasitology*, **9**, 137–140.

Walski, M., Cielecka, D., Chomicz, L. and **Grytner-Ziecina, B.** (1995) Transmission of horseradish peroxidase across the tegument of some hymenolepidid species (Cyclophyllidea). *Acta Parasitologica*, **40**, 206–210.

Wanson, W. W. and **Larson, O. R.** (1972) Studies on helminths of North Dakota. V. Life history of *Phyllodistomum nocomis* Fischthal, 1942 (Trematoda: Gorgoderidae). *Journal of Parasitology*, **58**, 1106–1109.

Ward, S. M., McKerr, G. and **Allen, J. M.** (1986) Structure and ultrastructure of muscle systems within *Grillotia erinaceus* metacestodes (Cestoda: Trypanorhyncha). *Parasitology*, **93**, 587–597.

Wardle, R. A. (1935) Fish-tapeworm. *Biological Board of Canada Bulletin*, **45**, 1–25.

Wardle, R. A. and **Green, N. K.** (1941) The rate of growth of the tapeworm *Diphyllobothrium latum* (L.). *Canadian Journal of Research*, **19**, 245–251.

Wardle, R. A. and **McLeod, J. A.** (1952) *The Zoology of Tapeworms*, University of Minnesota Press, Minneapolis, MN.

Wardle, W. J. (1988) A bucephalid larva, *Cercaria pleuromerae* n. sp. (Trematoda: Digenea), parasitizing a deepwater bivalve from the Gulf of Mexico. *Journal of Parasitology*, **74**, 692–694.

Warren, K. S. (1978) The pathology, pathobiology and pathogenesis of schistosomiasis. *Nature*, **273**, 609–612.

Watson, N. A. and **Rohde, K.** (1993) Ultrastructural evidence for an adelphotaxon (sister group) to the Neodermata (Platyhelminthes). *International Journal for Parasitology*, **23**, 285–289.

Watson, N. A. and **Rohde, K.** (1994) Two new sensory receptors in *Gyrodactylus* sp. (Platyhelminthes, Monogenea, Monopisthocotylea). *Parasitology Research*, **80**, 442–445.

Watson, N. A., Williams, J. B. and **Rohde, K.** (1992) Ultrastructure and development of the eyes of larval *Kronborgia isopodicola* (Platyhelminthes, Fecampiidae). *Acta Zoologica*, **73**, 95–102.

Webb, T. J. and **Hurd, D.** (1996) *Hymenolepis diminuta*: metacestode-induced reduction in the synthesis of the yolk protein, vitellogenin, in the fat body of *Tenebrio molitor*. *Parasitology*, **112**, 429–436.

Webbe, G. and **Jordan, P.** (1993) Control, in *Human Schistosomiasis*, (eds P. Jordan, G. Webbe and R. F. Sturrock), CAB International, Wallingford, UK, pp. 405–451.

Webber, R. A., Rau, M. E. and **Lewis, D. J.** (1987a) The effects of *Plagiorchis noblei* (Trematoda: Plagiorchiidae) metacercariae on the behaviour of *Aedes aegypti* larvae. *Canadian Journal of Zoology*, **65**, 1340–1342.

Webber, R. A., Rau, M. E. and **Lewis, D. J.** (1987b) The effects of *Plagiorchis noblei* (Trematoda: Plagiorchiidae) metacercariae on the susceptibility of *Aedes aegypti* larvae to predation by guppies (*Poecilia reticulata*) and meadow voles (*Microtus pennsylvanicus*). *Canadian Journal of Zoology*, **65**, 2346–2348.

Webster, J. D. (1949) Fragmentary studies on the life history of the cestode *Mesocestoides latus*. *Journal of Parasitology*, **35**, 83–89.

Webster, L. A. (1971) The flow of fluids in the protonephridial canals of *Hymenolepis diminuta*. *Comparative Biochemistry and Physiology*, **39A**, 785–793.

Weeks S. C. (1996) A re-evaluation of the Red Queen model for the maintenance of sex in a clonal–sexual fish complex (Poeciliopsis : Poeciliopsis). *Canadian Journal of Fisheries and Aquatic Sciences*, **53**, 1157–1164.

Weinstock, J. V. (1990) Schistosomiasis, in *Gastrointestinal Infections in the Tropics*, (ed. V. K. Rustgi), S. Karger, Basel, pp. 40–45.

Wells, T. A. G. (1964) *The Rat. A Practical Guide*, Heinemann, London.

West, A. F. (1961) Studies on the biology of *Philophthalmus gralli* Mathis and Leger, 1910 (Trematoda: Digenea). *American Midland Naturalist*, **66**, 363–383.

Weston, D. S. and **Kemp, W. M.** (1993) *Schistosoma mansoni* – comparison of cloned tropomyosin antigens shared between adult parasites and *Biomphalaria glabrata*. *Experimental Parasitology*, **76**, 358–370.

Wheeler, A. (1978) *Key to the Fishes of Northern Europe*, Frederick Warne, London.

Whitfield, P. J. (1982) Parasitic helminths, in *Modern Parasitology*, (ed. F. E. G. Cox), Blackwell, Oxford, pp. 34–83.

Whitfield, P. J., Anderson, R. M. and **Moloney, N. A.** (1975) The attachment of cercariae of an ectoparasitic digenean, *Transversotrema patialensis*, to the fish host: behavioural and ultrastructural aspects. *Parasitology*, **70**, 311–329.

Whitfield, P. J. and **Evans, N. A.** (1983) Parthenogenesis and asexual multiplication among parasitic platyhelminths. *Parasitology*, **86**, 121–160.

Whittington, I. D. (1987a) Hatching in two monogenean parasites from the common dogfish (*Scyliorhinus canicula*): the polyopisthocotylean gill parasite, *Hexabothrium appendiculatum* and the microbothriid skin parasite, *Leptocotyle minor*. *Journal of the Marine Biological Association of the United Kingdom*, **67**, 729–756.

Whittington, I. D. (1987b) Studies on the behaviour of the oncomiracidia of the monogenean parasites *Hexabothrium appendiculatum* and *Leptocotyle minor* from the common dogfish, *Scyliorhinus canicula*. *Journal of the Marine Biological Association of the United Kingdom*, **67**, 773–784.

Whittington, I. D. and **Horton, M. A.** (1996) A revision of *Neobenedenia* Yamaguti, 1963 (Monogenea: Capsalidae) including a redescription of *N. melleni* (MacCallum, 1927) Yamaguti, 1963. *Journal of Natural History*, **30**, 1113–1156.

Whittington, I. D. and **Kearn, G. C.** (1988) Rapid hatching of mechanically-disturbed eggs of the monogenean gill parasite *Diclidophora luscae*, with observations on sedimentation of egg bundles. *International Journal for Parasitology*, **18**, 847–852.

Whittington, I. D. and **Kearn, G. C.** (1989) Rapid hatching induced by light intensity reduction in the polyopisthocotylean monogenean *Plectanocotyle gurnardi* from the gills of gurnards (Triglidae), with observations on the anatomy of the oncomiracidium. *Journal of the Marine Biological Association of the United Kingdom*, **69**, 609–624.

Whittington, I. D. and **Kearn, G. C.** (1991) The adhesive attitudes of some gill-parasitic capsalid monogeneans. *Journal of Helminthology*, **65**, 280–285.

Whittington, I. D. and **Kearn, G. C.** (1993) A new species of skin-parasitic benedeniine monogenean with a preference for the pelvic fins of its host, *Lutjanus carponotatus* (Perciformes: Lutjanidae) from the Great Barrier Reef. *Journal of Natural History*, **27**, 1–14.

Whittington, I. D. and **Pichelin, S.** (1991) Attachment of eggs by *Concinnocotyla australensis* (Monogenea: Polystomatidae) to the tooth plates of the Australian lungfish, *Neoceratodus forsteri* (Dipnoi). *International Journal for Parasitology*, **21**, 341–346.

WHO Expert Committee on the Control of Schistosomiasis (1993) Public health impact of schistosomiasis: disease and mortality. *Bulletin of the World Health Organization*, **71**, 657–662.

Whyte, S. K., Secombes, C. J. and **Chappell, L. H.** (1991) Studies on the infectivity of *Diplostomum spathaceum* in rainbow trout (*Oncorhynchus mykiss*). *Journal of Helminthology*, **65**, 169–178.

Wilkins, H. A., Goll, P. H., Marshall, T. F. de C. and **Moore, P. J.** (1984) Dynamics of *Schistosoma haematobium* infection in a Gambian community. III. Acquisition and loss of infection. *Transactions of the Royal Society of Tropical Medicine and Hygiene*, **78**, 227–232.

Williams, A. and **Lethbridge, R. C.** (1990) The subcutaneous attachment of the monogenean *Heterobothrium elongatum* (Diclidophoridae) in the gills of *Torquigener pleurogramma* (Pisces: Tetraodontidae). *International Journal for Parasitology*, **20**, 769–777.

Williams, C. O. (1942) Observations on the life history and taxonomic relationships of the trematode *Aspidogaster conchicola*. *Journal of Parasitology*, **28**, 467–475.

Williams, H. H. (1959) The anatomy of *Köllikeria filicollis* (Rudolphi, 1819), Cobbold, 1860 (Trematoda: Digenea) showing that the sexes are not entirely separate as hitherto believed. *Parasitology*, **49**, 39–53.

Williams, H. H. (1960) The intestine in members of the genus *Raja* and host-specificity in the Tetraphyllidea. *Nature*, **188**, 514–516.

Williams, H. H. (1961) Observations on *Echeneibothrium maculatum* (Cestoda: Tetraphyllidea). *Journal of the Marine Biological Association of the United Kingdom*, **41**, 631–652.

Williams, H. H. (1964) Some new and little known cestodes from Australian elasmobranchs with a brief discussion on their possible use in problems of host taxonomy. *Parasitology*, **54**, 737–748.

Williams, H. H. (1966) The ecology, functional morphology and taxonomy of *Echeneibothrium* Beneden, 1849 (Cestoda: Tetraphyllidea), a revision of the genus and comments on *Discobothrium* Beneden, 1870, *Pseudanthobothrium* Baer, 1956, and *Phormobothrium* Alexander, 1963. *Parasitology*, **56**, 227–285.

Williams, H. H. (1968a) The taxonomy, ecology and host-specificity of some Phyllobothriidae (Cestoda: Tetraphyllidea), a critical revision of *Phyllobothrium* Beneden, 1849 and comments on some allied genera. *Philosophical Transactions of the Royal Society of London B, Biological Sciences*, **253**, 231–307.

Williams, H. H. (1968b) *Acanthobothrium quadripartitum* sp. nov. (Cestoda: Tetraphyllidea) from *Raja naevus* in the North Sea and English Channel. *Parasitology*, **58**, 105–110.

Williams, H. H. (1968c) *Phyllobothrium piriei* sp. nov. (Cestoda: Tetraphyllidea) from *Raja naevus* with a comment on its habitat and mode of attachment. *Parasitology*, **58**, 929–937.

Williams, H. H. (1969) The genus *Acanthobothrium*, Beneden 1849. *Nytt Magasin for Zoologi*, **17**, 1–56.

Williams, H. H., Colin, J. A. and **Halvorsen, O.** (1987) Biology of gyrocotylideans with emphasis on reproduction, population ecology and phylogeny. *Parasitology*, **95**, 173–207.

Williams, H. and **Jones, A.** (1994) *Parasitic Worms of Fish*, Taylor & Francis, London.

Williams, H. H. and **McVicar, A.** (1968) Sperm transfer in Tetraphyllidea (Platyhelminthes: Cestoda). *Nytt Magasin for Zoologi*, **16**, 61–71.

Williams, I. C., Ellis, C. and **Spaull, V. W.** (1973) The structure and mode of action of the posterior adhesive organ of *Pseudobenedenia nototheniae* Johnston, 1931 (Monogenea: Capsaloidea). *Parasitology*, **66**, 473–485.

Williams, J. B. (1978) Studies on the epidermis of *Temnocephala* III. Scanning electron microscope study of the epidermal surface of *Temnocephala dendyi*. *Australian Journal of Zoology*, **25**, 187–191.

Williams, J. B. (1980) Morphology of a species of *Temnocephala* (Platyhelminthes) ectocommensal on the isopod *Phreatoicopsis terricola*. *Journal of Natural History*, **14**, 183–199.

Williams, J. B. (1990) Ultrastructural studies on *Kronborgia* (Platyhelminthes: Fecampiidae): observations on the genital system of the female *K. isopodicola*. *International Journal for Parasitology*, **20**, 957–963.

Williams, J. B. (1993) Epidermal ultrastructure of the adult parasitic phase female and encapsulated larva of *Kronborgia isopodicola* (Platyhelminthes, Fecampiidae): phylogenetic implications. *International Journal for Parasitology*, **23**, 1027–1037.

Williams, M. O. (1966) Studies on the morphology and life-cycle of *Diplostomum* (*Diplostomum*) *gasterostei* (Strigeida: Trematoda). *Parasitology*, **56**, 693–706.

Willmer, P. (1990) *Invertebrate Relationships. Patterns in Animal Evolution*, Cambridge University Press, Cambridge.

Wilson, R. A. (1967a) The structure and permeability of the shell and vitelline membrane of the egg of *Fasciola hepatica*. *Parasitology*, **57**, 47–58.

Wilson, R. A. (1967b) The protonephridial system in the miracidium of the liver fluke, *Fasciola hepatica* L. *Comparative Biochemistry and Physiology*, **20**, 337–342.

Wilson, R. A. (1968a) The hatching mechanism of the egg of *Fasciola hepatica* L. *Parasitology*, **58**, 79–89.

Wilson, R. A. (1968b) An investigation into the mucus produced by *Lymnaea truncatula*, the snail host of *Fasciola hepatica*. *Comparative Biochemistry and Physiology*, **24**, 629–633.

Wilson, R. A. (1969a) Fine structure of the tegument of the miracidium of *Fasciola hepatica* L. *Journal of Parasitology*, **55**, 124–133.

Wilson, R. A. (1969b) Fine structure and organization of the musculature in the miracidium of *Fasciola hepatica*. *Journal of Parasitology*, **55**, 1153–1161.

Wilson, R. A. (1970) Fine structure of the nervous system and specialized nerve endings in the miracidium of *Fasciola hepatica*. *Parasitology*, **60**, 399–410.

Wilson, R. A. (1971) Gland cells and secretions in the miracidium of *Fasciola hepatica*. *Parasitology*, **63**, 225–231.

Wilson, R. A. (1987) Cercariae to liver worms: development and migration in the mammalian host, in *The Biology of Schistosomes*, (eds D. Rollinson and A. J. G. Simpson), Academic Press, London, pp. 115–146.

Wilson, R. A. (1990) Leaky livers, portal shunting and immunity to schistosomes. *Parasitology Today*, **6**, 354–358.

Wilson, R. A. and **Coulson, P. S.** (1986) *Schistosoma mansoni*: dynamics of migration through the vascular system of the mouse. *Parasitology*, **92**, 83–100.

Wilson, R. A. and **Coulson, P. S.** (1989) Lung-phase immunity to schistosomes: a new perspective on an old problem. *Parasitology Today*, **5**, 274–278.

Wilson, R. A. and **Denison, J.** (1970a) Studies on the activity of the miracidium of the common liver fluke, *Fasciola hepatica*. *Comparative Biochemistry and Physiology*, **32**, 301–313.

Wilson, R. A. and **Denison, J.** (1970b) Short-chain fatty acids as stimulants of turning activity by the miracidium of *Fasciola hepatica*. *Comparative Biochemistry and Physiology*, **32**, 511-517.

Wilson, R. A. and **Denison, J.** (1980) The parasitic castration and gigantism of *Lymnaea truncatula* infected with the larval stages of *Fasciola hepatica*. *Zeitschrift für Parasitenkunde*, **61**, 109–119.

Wilson, R. A., Draskau, T., Miller, P. and **Lawson, J. R.** (1978) *Schistosoma mansoni*: the activity and development of the schistosomulum during migration from the skin to the hepatic portal system. *Parasitology*, **77**, 57–73.

Wilson, R. A., Pullin, R. and **Denison, J.** (1971) An investigation of the mechanism of infection by digenetic trematodes: the penetration of the miracidium of *Fasciola hepatica* into its snail host *Lymnaea truncatula*. *Parasitology*, **63**, 491–506.

Wilson, R. A. and **Webster, L. A.** (1974) Protonephridia. *Biological Reviews*, **49**, 127–160.

Wilson, V. C. L. C. and **Schiller, E. L.** (1969) The neuroanatomy of *Hymenolepis diminuta* and *H. nana*. *Journal of Parasitology*, **55**, 261–270.

Wisniewski, L. W. (1932) *Cyathocephalus truncatus* Pallas. II. Allgemeine Morphologie. *Bulletin international de l'Académie des sciences et des lettres de Cracovie*, Série B, **2**, 311–327.

Wootton, D. M. (1957) The life history of *Cryptocotyle concavum* (Creplin, 1825) Fischoeder, 1903 (Trematoda: Heterophyidae). *Journal of Parasitology*, **43**, 271–279.

Wright, C. A. (1971) *Flukes and Snails*, Allen & Unwin, London.

Wright, C. A. and **Bennett, M. S.** (1964) The life cycle of *Notocotylus attenuatus*. *Parasitology*, **54**, 14P.

Wright, R. R. and **Macallum, A. B.** (1887) *Sphyranura osleri*: a contribution to American helminthology. *Journal of Morphology*, **1**, 1–48.

Xu, Y. and **Dresden, M. H.** (1986) Leucine aminopeptidase and hatching of *Schistosoma mansoni* eggs. *Journal of Parasitology*, **72**, 507–511.

Xu, Y. and **Dresden, M. H.** (1990) The hatching of schistosome eggs. *Experimental Parasitology*, **70**, 236–240.

Xylander, W. E. R. (1984) A presumptive ciliary photoreceptor in larval *Gyrocotyle urna* Grube and Wagener (Cestoda). *Zoomorphology*, **104**, 21–25.

Xylander, W. (1986) Zur Biologie und Ultrastruktur der Gyrocotylida und Amphilinida sowie ihre Stellung im phylogenetischen System der Plathelminthes. Doctoral thesis, University of Göttingen, Germany.

Yamaguti, S. (1938) Studies on the helminth fauna of Japan. Part 24. Trematodes of fishes, V. *Japanese Journal of Zoology*, **8**, 15–74.

Yamaguti, S. (1958) *Systema Helminthum, vol. I, The Digenetic Trematodes of Vertebrates, Part 1*, Interscience, New York.

Yamaguti, S. (1963) *Systema Helminthum, vol. IV, Monogenea and Aspidocotylea*, Interscience, New York.

Yamaguti, S. (1968) *Monogenetic Trematodes of Hawaiian Fishes*, University of Hawaii Press, Honolulu, HI.

Yamaguti, S. (1975) *A Synoptical Review of Life Histories of Digenetic Trematodes of Vertebrates*, Keigaku Publishers, Tokyo.

Yamasaki, H., Kominami, E. and **Aoki, T.** (1992) Immunocytochemical localization of a cysteine protease in adult worms of the liver fluke *Fasciola* sp. *Parasitology Research*, **78**, 574–580.

Yan, G., Stevens, L. and **Schall, J. J.** (1994) Behavioral changes in *Tribolium* beetles infected with a tapeworm: variation in effects between beetle species and among genetic strains. *American Naturalist*, **143**, 830–847.

Yokogawa, M. (1965) *Paragonimus* and paragonimiasis. *Advances in Parasitology*, **3**, 99–158.

Zdarska, Z., Giboda, M. and **Zenka, J.** (1992) The ultrastructure of the *Schistosoma mansoni* egg: the line of hatching. *Helminthologia*, **29**, 63–65.

Zuk, M. and **McKean, K. A.** (1996) Sex differences in parasitic infections: patterns and processes. *International Journal for Parasitology*, **26**, 1009–1024.

Index

Page references in **bold** refer to figures, those in *italics* refer to tables